Ionization of Solids
by Heavy Particles

NATO ASI Series
Advanced Science Institutes Series

A series presenting the results of activities sponsored by the NATO Science Committee, which aims at the dissemination of advanced scientific and technological knowledge, with a view to strengthening links between scientific communities.

The series is published by an international board of publishers in conjunction with the NATO Scientific Affairs Division

A	**Life Sciences**	Plenum Publishing Corporation
B	**Physics**	New York and London
C	**Mathematical and Physical Sciences**	Kluwer Academic Publishers
D	**Behavioral and Social Sciences**	Dordrecht, Boston, and London
E	**Applied Sciences**	
F	**Computer and Systems Sciences**	Springer-Verlag
G	**Ecological Sciences**	Berlin, Heidelberg, New York, London,
H	**Cell Biology**	Paris, Tokyo, Hong Kong, and Barcelona
I	**Global Environmental Change**	

Recent Volumes in this Series

Series B: Physics

Ionization of Solids by Heavy Particles

Edited by

Raúl A. Baragiola

University of Virginia
Charlottesville, Virginia

Springer Science+Business Media, LLC

Proceedings of a NATO Advanced Research Workshop on
Ionization of Solids by Heavy Particles,
held June 1–5, 1992,
in Taormina, Italy

NATO-PCO-DATA BASE

The electronic index to the NATO ASI Series provides full bibliographical references (with keywords and/or abstracts) to more than 30,000 contributions from international scientists published in all sections of the NATO ASI Series. Access to the NATO-PCO-DATA BASE is possible in two ways:

—via online FILE 128 (NATO-PCO-DATA BASE) hosted by ESRIN, Via Galileo Galilei, I-00044 Frascati, Italy

—via CD-ROM "NATO-PCO-DATA BASE" with user-friendly retrieval software in English, French, and German (©WTV GmbH and DATAWARE Technologies, Inc. 1989)

The CD-ROM can be ordered through any member of the Board of Publishers or through NATO-PCO, Overijse, Belgium.

```
        Library of Congress Cataloging-in-Publication Data
     _____

Ionization of solids by heavy particles / edited by Raúl A. Baragiola.
       p.   cm. -- (NATO ASI series. Series B, Physics ; v. 306)
     "Published in cooperation with NATO Scientific Affairs Division."
     "Proceedings of a NATO Advanced Research Workshop on Ionization of
 Solids by Heavy Particles, held June 1-5, 1992, in Taormina, Italy."
     Includes bibliographical references and index.
     ISBN 0-306-44489-5
     1. Condensed matter--Effect of radiation on--Congresses.
 2. Ionization--Congresses.  3. Heavy particles (Nuclear physics)-
 -Congresses.  4. Electrons--Emission--Congresses.  5. Auger effect-
 -Congresses.   I. Baragiola, Raúl A.  II. North Atlantic Treaty
 Organization.  Scientific Affairs Division.  III. NATO Advanced
 Research Workshop on Ionization of Solids by Heavy Particles (1992
 Taormina, Italy)  IV. Series.
 QC173.4.C65I67  1993
 530.4'16--dc20                                          93-8373
     _____  CIP
```

ISBN 978-1-4613-6229-6   ISBN 978-1-4615-2840-1 (eBook)
DOI 10.1007/978-1-4615-2840-1

© Springer Science+Business Media New York 1993
Originally published by Plenum Press, New York 1993

PREFACE

This book collects the papers presented at the NATO Advanced Research Workshop on "Ionization of Solids by Heavy Particles", held in Giardini-Naxos (Taormina), Italy, on June 1 - 5, 1992.

The meeting was the first to gather scientists to discuss the physics of electron emission and other ionization effects occurring during the interaction of heavy particles with condensed matter. The central problem in the field is how to use observations of electron emission and final radiation damage to understand what happens inside the solid, like excitation mechanisms, the propagation of the electronic excitation along different pathways, and surface effects.

The ARW began with a brief survey of the field, stressing the unknowns. It was pointed out that ionization theories can only address the very particular case of weak perturbations. For this problem, this meant high speed, low-charged projectiles (a perturbation treatment of interactions with slow, highly charged ions was later presented). Only semi-empirical models exist for velocities lower than the Fermi velocity in the solid, which can be used to predict kinetic electron emission yields. These models, however, do not address the basic questions about the mechanisms for electron excitation, transport and escape through the surface layer.

Theories of electron emission have not reached yet the point where they can be used to justify more predictive, semi-empirical models. The popular Monte Carlo descriptions of electron motion in solids still need to address conflicts with Heisenberg's principle. Structures in electron energy distribution, attributed to the decay of bulk and surface plasmons, do not follow theoretical expectations. Free-electron gas theories, which have been used successfully to predict electronic stopping do not pass the closer test of predicting individual electronic excitations. Data on extrapolated thresholds in the electron yields contradict these models and remain unexplained. New developments presented at the meeting, using atomic approaches may be able to solve this problem and show that electron excitation requires the presence of the atomic cores.

Other questions identified as part of the conclusions include: decay pathways of "hollow" atoms in solids, the description of electronic states at the surface in the presence of an outside charge, the description of elastic scattering surfaces of ionic crystals, the identification of excitation mechanisms at eV energies, the role of electron loss from the projectile in total electron yields, the origin of Z^3 deviations from the Born approximation, the atomic description in inelastic events at low velocities, the description of valence wave functions in solids far from equilibrium, the origin of non-Poisson statistics in electron emission, the origin of unusually high probability of no electron emission for slow ions traversing thin foils, the probability of multiple electron excitation in Auger neutralization at surfaces, the potential of ion tracks in solids and its time evolution, the role of structural effects in electron emission, and the explanation of anomalous ionization effects, electric fields and breakdown inside irradiated insulators.

This workshop was possible thanks to financial support from the NATO Scientific Affairs Division, the Italian Consiglio Nazionale delle Richerche, the University of Calabria and the University of Catania. I express my special gratitude to Tonino Oliva and Giovanni Marletta for obtaining financial support from Italian sources, for the smooth operation of the workshop and for showing that *savoir vivre* can coexist with successful scientific meetings. I would like to thank the other members of the International committee: Fernando Flores, Wolfgang O. Hofer and Tonino Oliva for their help and advice during the organization of the ARW. I am grateful to Pam Lockley and David Grosjean, who were very helpful during the preparation of this book. Finally, I am particularly thankful to the participants, the truly essential actors, who made this meeting a success through their cooperation, enthusiasm, and, above all, their piercing critique and critical inquiries.

Raúl A. Baragiola
Director of the NATO ARW on
Ionization of Solids by Heavy Particles

CONTENTS

THEORY OF IONIZATION AND ELECTRON EMISSION

AUGER PROCESSES AT METALLIC SURFACES

KINETIC AUGER PROCESSES AND SHELL EFFECTS

KINETIC ELECTRON EMISSION FROM THIN FOILS

SURFACE EFFECTS IN KINETIC ELECTRON EMISSION

SPIN POLARIZED ELECTRON EMISSION

ELECTRON EMISSION AND CHARGING OF INSULATORS

IONIZATION EFFECTS IN SEMICONDUCTORS AND INSULATORS

THEORY OF ELECTRON EJECTION FROM MATTER BY HIGHLY CHARGED ION IMPACT

J. H. Macek

Theoretical Physics, University of Tennessee, Knoxville, TN and
Oak Ridge National Laboratory, Oak Ridge, TN

INTRODUCTION

When highly charged ions collide with matter containing relatively loosely bound electrons, electrons typically become attached to the highly charged ion. In other words, capture is the most likely reaction. While this is generally the case, direct ejection of electrons is also an important process. In this report I will describe some recent calculations of the the electron energy distributions produced by highly charged ion impact. I will first discuss calculations pertinant to relatively high velocities. The basic idea is that electrons move in the field of the highly charged ion; in first approximation, and that the effect of the smaller potential which initially binds the electrons is treated as a perturbation.[1] This picture appears to be appropriate for fast electrons produced by collisions at high velocity. For lower energy collisions this picture is still applicable, but does not lead to a simple description of ionization. Here a theory due Solov'ev,[2] Ovchinnikov[3], and co-workers[4] is better adapted to describing the physical mechanisms. In particular, this work shows that at low velocities it may be necessary to explicitly consider that electrons move in the combined potential of the target and projectile. This latter picture may also apply to ion-surface interactions.

ION-ATOM COLLSIONS IN THE STRONG-POTENTIAL BORN APPROXIMATION

Briggs[1] pointed out that when a projectile of high charge Z_P interacts with an electron relatively loosely bound to a target of small nuclear charge Z_T, the ratio Z_T/Z_P is a natural expansion parameter. This insight forms the basis for a successful theory of ionization by highly charged ion impact for incident ion velocities v such that $Z_T/v << 1$ even though we may have $Z_P/v >> 1$. We consider that the electron moves in the field of the strong potential in first approximation, thus the theory is known as the strong-potential Born approximation[5] (SPB). A more complete version is known as the distorted-wave strong potential Born approximation[6] (DSPB). Approximate versions of these theories are the impulse approximation (IA), which has been used extensively for computations of electron capture,[7] but has also been used for ionization by Jakubassa-Amundsen.[8] Releted theories

Ionization of Solids by Heavy Particles, Edited by
R.A. Baragiola, Plenum Press, New York, 1993

are the continuum distorted wave approximation with an eikonal inital state[9] (CDW-EIS), and the CDW-CDW theory.[10] The latter theory is nearly identical to a peaked version of the IA theory for highly charged ions. The CDW-CDW amplitude is singular for ionization, thus it will not be considered further in this review. The CDW-EIS is also similar to the IA, but deletes the singular terms of the CDW approximation. Because some terms have been deleted, this theory cannot be interpreted in terms of electron motion in the strong potential and will not be further discussed here. It should be noted, however, that CDW-EIS electon distributions for highly charged ions are similar to the DSPB electron distributions reported here. Atomic units are used throughout unless otherwise indicated.

Consider a highly charged ion Z_P interacting with a one-electron ion with nuclear charge Z_T and suppose that $Z_P >> Z_T$. As is usual in heavy ion collisions we suppose that the nucleus-nucleus interaction has been removed by a phase transformation. Then the ion-atom interaction at large distances is given by

$$U(t) = < \varphi_i | \frac{Z_P}{|\mathbf{r} - \mathbf{R}(t)|} | \varphi_i >, \tag{1}$$

where φ_i is the initial electron eigenstate. The intial wave function is then

$$\Phi_i(t) = \varphi_i(t) e^{-i \int_{-\infty}^{t} U(t') dt'}, \tag{2}$$

and the exact wave function is given by

$$\Psi = \Phi_i(t) + \int_{-\infty}^{\infty} G(t, t')[V_P(t') - U(t')]\Phi_i(t') dt', \tag{3}$$

where $G(t, t')$ is the Green function for the operator

$$[-\frac{1}{2}\nabla^2 + V_P + V_T] - i\frac{\partial}{\partial t}$$

where

$$V_P = -Z_P/|\mathbf{r} - \mathbf{R}(t)|, \tag{4}$$

and

$$V_T = -Z_T/r. \tag{5}.$$

Here $\mathbf{r}$ is the coordinate of the electron relative to the target nucleus, and $\mathbf{R}$ is the coordinate vector of the projectile relative to the target nucleus.

The key approximation introduced by Briggs was to neglect the "small" potential V_T in G so that we have $G \approx G_P$ where G_P is the Green function for the operator

$$[-\frac{1}{2}\nabla^2 + V_P] - i\frac{\partial}{\partial t}$$

and the wave function for the system is now approximately

$$\Psi_i^{DSPB} = \Phi_i(t) + \int_{-\infty}^{\infty} G_P(t, t')[V_P(t') - U(t')]\Phi_i(t') dt'. \tag{6}$$

Except for the intial static electron distribution represented by φ_i, the electron motion is entirely described by the strong potential V_P. Since the electron dynamics relates entirely to the strong potential via $U(t)$ and $G_P(t, t')$, this wave function is known as the distorted-wave strong potential Born (DSPB) wave function. It has been used as an initial state to describe a variety of physical processes including capture and ionization. It is less useful for excitation or ionization to the low energy electron continuum since these processes can occur at large impact parameter where the "strong" potential actually only slightly perturbs the target eigenstate. In this case the second term on the right hand side of Eq.(6) should be small and the first Born wave function is recovered so that Eq.(6) is valid, although unecessarily complicated, for these latter processes. For ionization we employ a final state $\psi_{\mathbf{k}}^{-}(\mathbf{r} - \mathbf{R}(t))$ of an electron in the field of the projectile nucleus. This limits the region of validity of the theory to that portion of the electron velocity distribution such that $k < |\mathbf{k} + \mathbf{v}| Z_P/Z_T$ in order to insure that the electron does indeed move in the field of the projectile in the final state. For $|\mathbf{k} + \mathbf{v}|$ sufficiently small the inequality does not hold and it is necesary to employ target final states, as in excitation. The continuum-distorted wave (CDW) final state wave function interpolates between target and projectile eigenstates and has been used by Miraglia and Macek[11] to compute some electron distributions. Because the computations are computer intensive, no extensive calculations over a large energy and angular range are presently available.

The ionization amplitude in the DSPB approximation is given by

$$A^{DSPB} = \int_{-\infty}^{\infty} <\psi_{\mathbf{k}}^{-}(t)|V_T|\Psi_i^{DSPB}> . \tag{7}$$

The corresponding T-matrix element in the time independent representation has been evaluated approximately to give

$$T^{DSPB} = \int \tilde{V}(\mathbf{J} + \mathbf{s}) < \psi_{\mathbf{k}}^{-}|e^{-i(\mathbf{J}+\mathbf{s})\cdot\mathbf{r}}|\psi_{\mathbf{s}-\mathbf{v}}^{+} > \tilde{\varphi}_i(\mathbf{s})d^3s \tag{8}$$

where the tilde denotes the Fourier transform, $\psi_{\mathbf{k}}^{\pm}$ are time independent Coulomb wavefunctions of the electron in the field of the strong potential V_P, and $\mathbf{J}$ is the momentum of the recoiling target nucleus in the final state. This expression has a simple physical interpretation, namely $\check{\varphi}(\mathbf{s})$ represents the momentum distribution of the electron in the initial state. The quasi-free electron of momentum $\mathbf{s}$ has a momentum $\mathbf{s} - \mathbf{v}$ in the field of the projectile with velocity $\mathbf{v}$. This motion is described by the intial Coulomb continuum function $\psi_{\mathbf{s}-\mathbf{v}}^{+}$. The weak potential represented by $\tilde{V}(\mathbf{J} + \mathbf{s})e^{-i(\mathbf{J}+\mathbf{s})\cdot\mathbf{r}}$ causes transitions to the final projectile eigenstate $\psi_{\mathbf{k}}^{-}$, where $\mathbf{k}$ is the electron momentum in the projectile frame. We thus see that ionization is represented in terms of free-free transitions of the electron in the strong field of the projectile nucleus due to the presence of the weaker potential V_T.

The doubly differential cross section is given by

$$\frac{d\sigma}{dE_e d\Omega_e} = \frac{(2\pi)^4 k}{v^2} \int |T^{DSPB}|^2 d^2 J_{\perp} \tag{9}$$

where $J_{\perp}$ is the component of $\mathbf{J}$ perpendicular to $\mathbf{v}$. The component of $\mathbf{J}$ parallel to $\mathbf{v}$ is given by

$$J_{\parallel} = \frac{2|\varepsilon_i| + k^2 - v^2}{2v} \tag{10}$$

where $|\varepsilon_i|$ ionization potential of the target.

The energy distribution of electrons ejected in the forward direction have been computed for H^+ and F^{9+} at 1.5 MeV/amu colliding with He and H-atom targets. Fig.(1)

compares Eq.(9) with the data of Lee and co-workers[12] for F^{9+}. The agreement over the entire electron energy range is good, in particular, the prominant features of the energy distribution are accurately represented. The CEC cusp at electron energy $E_e = \frac{1}{2}v^2$ and the binary encounter peak at $E_e \approx 2v^2$, for example are accurately represented. Even the low energy part of the distribution is well described by theory.

The origins of the prominant features in the theory are easily identified. The final state wave function $\psi_{\mathbf{k}}$ has the normalization constant

$$|N|^2 = \frac{2\pi Z_P}{k} \frac{1}{(1 - exp\frac{-2Z_P\pi}{k})} \qquad (11)$$

which becomes infinite at $k = 0$, i. e. when the velocity of the electron in the laboratory frame matches the velocity of the projectile. This gives rise to the continuum electron capture cusp.

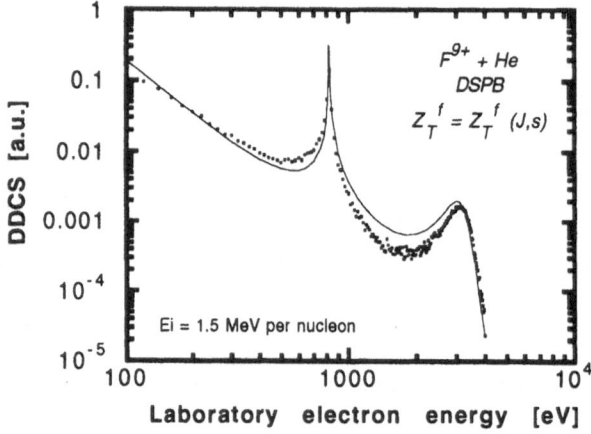

Figure 1. Electron energy distribution for electrons ejected at 0^0 by 1.5 MeV/amu F^{9+} ion impact on He. The solid curve is the DSPB theory and the dots are the experimental data from Ref.(12).

The binary encounter peak near $v_e = 2v$ is described by all theories, including the first Born theory as shown in Fig.(2). In contrast to the DSPB theory the first Born theory greatly underestimates the cross section in the region away from the binary encounter peak. The DSPB theory differs from the first Born theory in another, more subtle, way. This is seen when we examine the *amplitude* rather than the cross section in the binary encounter region. The binary encounter peak occurs when no momentum is transferred to the target nucleus so that $\mathbf{J} \approx \mathbf{0}$. In this region one can show that, to a good approximation, the DSPB T-matrix is given by

$$T^{DSPB} \approx T^{Ruth}(\mathbf{k}, -\mathbf{J} - \mathbf{v})\tilde{\phi}_i(-\mathbf{J}) \qquad (12)$$

where T^{Ruth} is the *amplitude* for Rutherford scattering of the target electron from the projectile nucleus. This is just the expression that is expected on intuitive grounds, namely, the quasi-free electron with momentum distribution $\tilde{\varphi}_i(-\mathbf{J})$ has an initial momentum $-\mathbf{J}-\mathbf{v}$ in the projectile frame. It scatters elastically from the projectile and aquires a final momentum $\mathbf{k}$, also in the projectile frame. The elastic scattering is represented by the Rutherford amplitude. All other theories obtain similar expressions for the *magnitude* of the T-matrix element, but only the DSPB recovers both the magnitude and *phase* of the Rutherford amplitude. That the DSPB recovers Eq.(12) when $J \approx 0$ strongly indicates that it is the correct description of ionization of atoms by highly charged ions, provided the electron velocity is sufficiently large. In this connection note that the DSPB theory seems to work down to $120eV$ for F^{9+}.

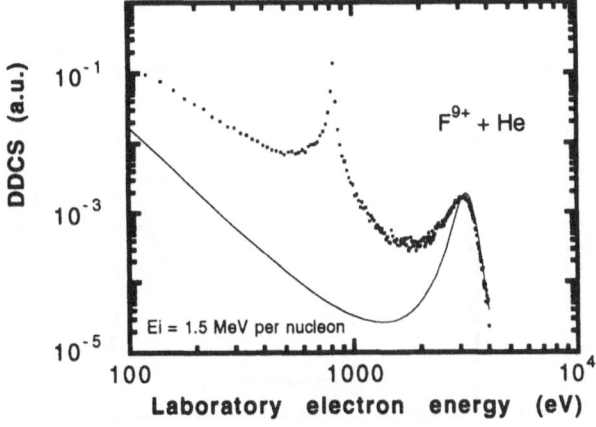

Figure 2. Electron energy distribution for electrons ejected at 0^0 by 1.5 MeV/amu F^{9+} ion impact on He. The solid curve is the conventional first Born approximation with target eigenstates and the dots are the experimental data of Ref.(12).

Although the theory was formulated for highly charged ions it has been applied to ionization by protons. Fig.(3) compares the data of Ref.(12) with the DSPB theory. Here the agreement is surprisingly good, although there is some indication of disagreement at the lower electron veolcities. This is expected since the final state wave function should not be appropriate for electron energies below 400 eV. That the theory agrees well down to somewhat lower energies is encouraging, and shows that for fast electrons, projectile final eigenstates are more appropriate than target eigenstates. Further comparisons are needed in order to delineate the region of applicability of the DSPB theory, however, it appears that the electron dynamics for collisions involving highly charged projectiles is accurately described by considering that the electron moves in the field of the highly charged ion.

LOW ENERGY ION-ATOM COLLISIONS

In principle, the DSPB theory is applicable at all energies since it requires only that $Z_T/Z_P << 1$, provided one employs appropriate final eigenstates. This latter requirement has not been fully resolved since, at low impact energy, the duration of the collision is sufficiently long that the electron moves in the field of both the target and projectile ions in the final state. In addition, the DSPB amplitude is not readily evaluated for slow collisions, thus it does not identify ionization mechanisms. Here fundamental work of Solovev[2] and computations of Peiksma and Ovchinnikov[4] have uncovered two important mechanisms that operate at low energy.

The top-of-barrier mechanism

The potential energy surface for an electron in the field of two positive charges has a saddle point where the net force on the electron vanishes. If an electron gets to this region

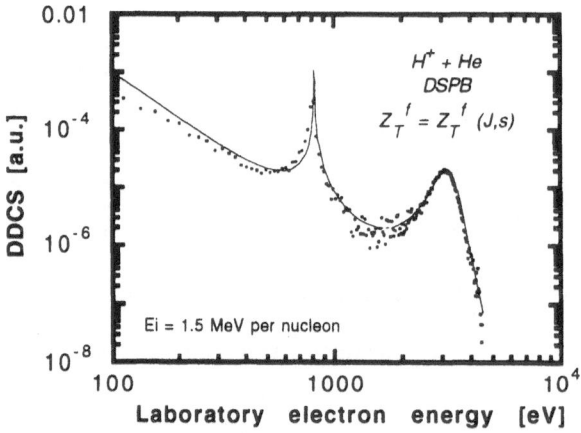

Figure 3. Energy distribution for electrons ejected at 0^0 by 1.5 MeV/amu H^+ impact on He. The solid curve is the DSPB approximation and the dots are the data of Ref.(12).

and has zero velocity relative to the saddle point, it stays there classically and appears as a free electron of low energy in the laboratory frame. This provides an ionization mechanism similar to the Wannier[13] theory for ionization by electron impact as recognized by many workers.[14,15] In the quantum theory, an electron near the saddle point is represented by a transient molecular eigenstate of the electron in field of both target and projectile. Ionization is then represented in terms of transitions between these transient molecular eigenstates. Solov'ev and co-workers have shown how to approximately compute these transitions employing the theory outlined in the textbook by Landau and Liftshitz.[16] Computations by Peiskma and Ovchinnikov for $He^{2+} + H$ show that this mechansim contributes substantially to ionization for incident ion velocities between 1 and 2 atomic units but that for velocities below 1 au another mechanism, namely a united atom electron promotion, dominates.

United atom promotion

The second mechanism operating at low velocities is also described in terms of transitions between molecular eigenstates, but now refers to to the region where target and projectile nuclei are close together, i. e. in the united atom region. Again the transitions can be computed using the theory of Ref.(16). The theory has been interpreted in terms of the promotion of electrons due to the rapid change of the angular momentum barrier experienced by electrons in molecular states near the united atom limit. The mechanism

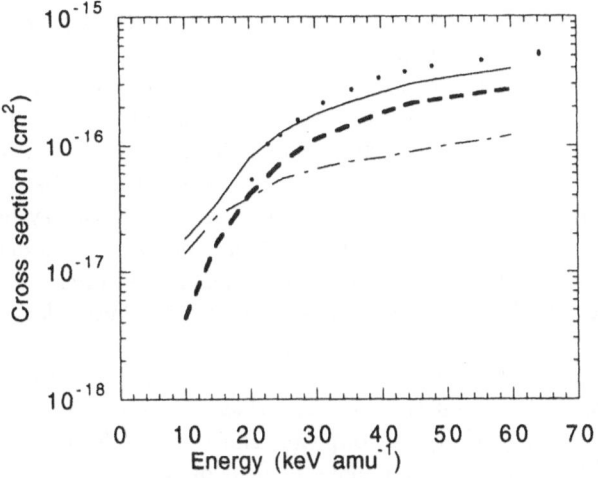

Figure 4. Ionization cross section vs. ion impact energy for $He^{2+} + H$ collisions. The dashed curve gives the T-series contribution, the dot-dashed curve is the S-series contribution, and the solid curve is the sum. The theory is from Ref.(4) and the dots represent the experimental data of Ref.(17).

in this case is referred to as S-series promotion. Computations have been reported for $He^+ + H$. Fig.(4) shows the ionization cross section as a function of projectile velocity for incident ion energies ranging from 10 to 60 keV/amu. Theory is compared with the data of Shaw and coworkers.[17] Notice that the two mechanisms account fairly well for the total ionization cross section, and that below 25 keV/amu corresponding to $v = 1au$ the S-series promotion is most imnportant, but that above this range the T-series becomes increasingly significant.

United atom eigenstates for highly charged ions incident on atomic hydrogen are approximately represented by eigenstates of the highly charged ion alone since $Z_P + 1 \approx Z_P$ for $Z_P >> 1$ thus the S-series promotion is consistant with the remark that electrons move in the field of the strong potential. This has not been demonstrated quantitatively, however, owing to the difficulty in evaluating the matrix element in Eq.(7) for low velocities. Even so, we can anticipate that the basic framework of section I still holds. Alternatively the T-series promotion does not represent electron motion in the strong potential only, rather both potentials play equivalent roles. Thus, to the extent that the T-series motion is important, the DSPB theory of section I does not apply. No calculations of the T-series mechanism have been reported for $Z_P >> Z_T$ thus it is not known if this mechanism limits the basic SPB framework.

ION-METAL SURFACE COLLISIONS

The sucess of the Z_T/Z_P expansion for ion-atom interactions is limited to physical situations where the "active" electrons are thought to move in the strong potential. The top-of-barrier mechansim for slow collisions in one electron systems provides one example which cannot be described in terms of electron motion in the strong potential. Here, the saddle point region was critical for the ionization process, although not important for the overall description of the electron motion. Ion-metal surface collisions provide another example where the SPB ideas do not apply. Here the infinite source of loosely bound electrons provide a means of screening the high charge, thus the electrons that become free move in the potential of a screened charge. Then the effective projectile charge is small, possibly of the order of one atomic unit, and the DSPB theory become inappropriate. This is almost certainly the case at low velocities, but here a theory based on adiabatic molecular states may apply. This short review concludes with a model which may have some validity for ion-suface collisions, but which is put forward here mainly to emphasize that the strong-potential Born approach requires more than just a large initial charge for the projectile ion.

When a highly charged ion approachs a metal surface electrons concentrate on the surface closest to the ion so that the potential that they produce outside of the surface is represented by an image charge of opposite sign to the charge of the ion. This already indicates that electrons do not move in field of the projectile alone, even in first approximation. At some distance from the surface electrons may move from the surface region to the ion where they initially populate states of high principle quantum number, according to the usual description of resonant electron transfer. Most likely, complete neutralization is not achieved and electrons outside the surface including those in the high n states of the projectile move in the field of the screened ionic charge and its image. This suggests a molecular model with an electron moving in the field of a positive charge and its image.[18] The image charge need not equal the projectile charge owing to screening by the electron itself. In any event an approximate description uses a model of a negative antiproton nucleus, a positve He^{++} projectile nucleus and a single electron. For simplicity we consider a positve H^+ ion in place of He^{++} since much information is available for that system. Then the "molecular" curves have no bound states at distances less than $\approx 0.7au$, athough this point is not essential for the model. It is essential that at distances approximate double this value the molecular states change rapidly so that there is some transfer of electrons to other states including continuum states. In fact the radius where the rapid change takes place depends upon the angular momentum ℓ of the initial state of the electron in a high n state of the positive ion and increases very rapidly with ℓ. In any event, at some distance direct transfer of electron into the continuum occurs. The computation of this transition probability follows exactly as in the transient molecular state model for ion-atom collisions The energy distribution of the electrons turns out to be nearly independent of the charge state, the initial angular momentum ℓ and the initial principal quantum number n. Thus we can illustrate the energy distribution by considering antiprotons colliding with a hydrogen atom in the ground state. Fig.(5) shows the electron energy distribution computed for three different impact energies. The main noteworthy feature of these distributions is the exponential dependence upon electron energy for high electron energy. It remains to be seen whether this model can be applied to ion-surface interactions. The key point here is that it shows how the presence of a large number of screening electrons must modify the assumption that for highly charged ions it is only necessary to consider electron motion in the strong potential.

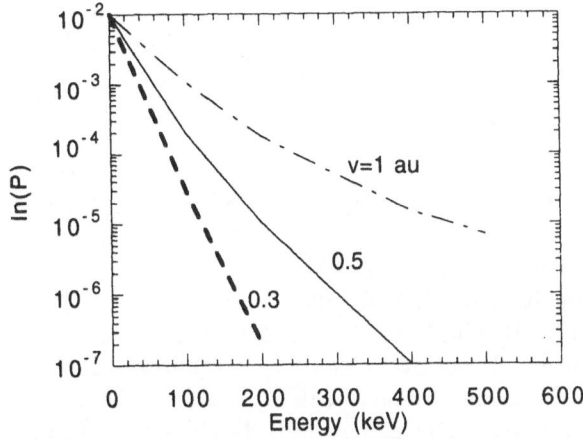

Figure 5. Plot of the log of ionization probability vs electron energy for the ionization of hydrogen atoms by antiproton impact at three different velocities. The log scale is relative.

SUMMARY

The interaction of fast, highly charged ions with few electron atoms has been successfully modelled using a theory that treats dynamical electron motion in the strong potential of the ion to all orders, but treats motion in the weak potential only in first order. This theory derives fundamentally from expansions of the exact Green's function in powers of Z_T/Z_P. The theory may apply at low velocities where $Z_T/v \gg 1$, however the possible presence of a top-of-barrier mechanism, even for highly charged ions, indicates that the theory is incomplete for low impact energies. For ion-metal surface interactions, the collective motion of a large number of electrons may screen the high charge of the projectile so that a completely new picture is needed. A picture based upon electron motion in the screened ionic charge and an image charge is one such computable model which provides predictions for the electron energy distribution.

ACKNOWLEDGMENT

Support for this research by the National Science Foundation under grant number PHYS-89-18713 is gratefully acknowledged. Discussions with Dr. S. Y. Ovchinnikov were most helpful in preparing this lecture.

REFERENCES

1. J. S. Briggs, J. Phys. B **10**, 3075 (1977).
2. E. A. Solov'ev, Sov. Phys.-JETP **54**, 893 (1981).
3. S. Y. Ovchinnikov and E. A. Solov'ev, Commonets on Atomic and Molecular Physics **XXII**, 69 (1988).
4. M. Pieksma and S. Y. Ovchinnikov, J. Phys. B. **24**, 2699 (1991).
5. J. H. Macek and S. Alston, Phys. Rev. A **26**, 2670 (1986) and J. H. Macek, J. Phys. B **10**, L71 (1985).
6. K. Taulbjerg, R. O. Barrachina and J. H. Macek, Phys. Rev. A **41**, 207 (1990).
7. D. H. Jakubassa-Amundsen and P. A. Amundsen, Z. Phys. **A297**, 203 (1980).
8. D. H. Jakubassa-Amundsen, J. Phys. B **16**, 1767 (1983).
9. P. D. Fainstein, V. H. Ponce and R. D. Rivarola, J. Phys. B **24**, 3091 (1991).
10. Dz Belkic, J. Phys. B **11**, 3529 (1978).
11. J. E. Miraglia and J. H. Macek, Phys. Rev. A **43**, 5919 (1991).
12. D. H. Lee, P. Richard, T. J. Zouros, J. M. Sanders, J. L. Shinpaugh and J. Hidmi, Phys. Rev. A **42**, 4816 (1991).
13. G. H. Wannier, Phys. Rev. **90**, 817 (1953).
14. R. E. Olson, Phys. Rev. A **33**, 4397 (1986).
15. T. G. Winter and C. D. Lin, Phys. Rev. A **29**, 3071 (1984).
16. L. D. Landau and E. M. Lifshitz, *Quantum Mechanics, Non-Relativistic Theory 3rd ed.* (Addison-Wesley Publishing Company, INC., Reading, MA 1977), p196.
17. M. B. Shah, D. S. Elliott, P. McCallion and H. B. Gilbody, J. Phys. B **21**, 2455 (1988).
18. S. Y. Ovchinnikov, Private Communication.

DYNAMIC INTERACTION OF IONS WITH
CONDENSED MATTER USING A LCAO APPROACH

F. Flores, J.J. Dorado, F.J. García-Vidal, J. Ortega and R. Monreal

Departamento de Física de la Materia Condensada C-XII.
Facultad de Ciencias. Universidad Autónoma de Madrid.
E-28049 Madrid. Spain.

ABSTRACT

A free-parameter LCAO method is used to calculate: (a) the stopping power for He moving in alkali metals in the low velocity limit; (b) the dynamical charge transfer processes between He$^+$ and He* with Al-metal. Our approach allows us to calculate the stopping power dependence of He on the impact parameter for channeling conditions. The dynamical charge transfer processes are discussed as a function of the resonance between the He-2s level and the metal.

I. INTRODUCTION

The purpose of this communication is to analyze the interaction of <u>slow</u> He-atoms with condensed matter in the bulk of metals and near their surfaces. Most of the work done in the field of particles interacting with solids has been addressed to understanding the high velocity limit, whereby ions move with a velocity much larger than the mean velocity of the valence electrons of the solid [1]. The low and intermediate limits have been recently started to be analyzed by introducing the different electron loss and capture processes associated with the interaction of the projectile with the target [2]. The difficulty of the low velocity limit is due to the strong electronic effects controlling the ion-solid interaction: this is a limit where the ion-solid chemistry is going to play the most important role. Motivated by these considerations, we have tried to developed a free-parameter LCAO method [3-4] that allows to calculate the different ion-solid dynamic interactions; this approach tries to obtain these interactions using a LCAO framework that is closely related to the atom-solid chemisorption problem.

In this paper we will be concerned with two different problems:(i) first, we shall analyze the stopping power for He moving in a metal for very low velocities [5]. (ii)

Ionization of Solids by Heavy Particles, Edited by
R.A. Baragiola, Plenum Press, New York, 1993

On the other hand, we shall discuss the different dynamical processes appearing for a He-atom approaching a metal surface [6-10].

As regards the first problem we shall mention the work done [5] by Echenique, Nieman and Ritchie (ENR); these authors have calculated very accurately the stopping power of He moving in metals using a LDA approach. This calculation has been done, however, only for a uniform electron gas. Our approach presents the advantage of allowing us to analyze also a metal with its full structure. Regarding the second problem, we shall mention the work of Nordlander and colaborators [6]; using a jellium model and a LDA approach, they have discussed the dynamical processes associated with He approaching a metal surface. Our approach introduces the whole metal structure and allows us to analyze in full detail the electronic structure of the moving atom interacting with the metal.

In section II we shall discuss the stopping power of He, while in section III , the dynamical interaction of He with metal surfaces will be analyzed.

II. STOPPING POWER FOR He MOVING IN METALS

In this section, we shall present our LCAO approach by considering the simple case of He moving slowly inside a metal. For the sake of simplicity, it is worth starting our discussion by considering the static interaction of He with an alkali metal atom. The metal band is simulated by a s-orbital occupied by one electron that is interacting with the doubly occupied 1s-orbital of He.

The discussion of ref 3 shows that the metal and the He-levels interact with each other due to the overlap, $S = < \psi_M \mid \psi_{He} >$, between both wavefunctions. This overlap introduces some increase in the kinetic energy of the system and a hopping, T, that tends to transfer an electron between both levels. Both effects modify the energy levels of the two wavefunctions; in a calculation, up to second order in the overlap, it was shown that the metal and the He levels are shifted by the following amounts:

$$\delta E_M = S^2(E_M - E_{He}) \tag{1}$$
$$\delta E_{He} = 0, \tag{2}$$

these values yielding the following repulsive energy between both atoms:

$$V_{repulsive}^{one-electron} = S^2(E_M - E_{He}). \tag{3}$$

Many-body contributions have also been discussed in ref [3]. Then, equation (3) is modified to the following form:

$$V_{repulsive} = S^2(E_M - E_{He} - \frac{3}{8}\beta + J_o). \tag{4}$$

In this equation $(-\frac{3}{8}\beta S^2)$, includes the bare coulomb and exchange interactions between the metal atom and the He-electrons, while $J_o S^2$ is the correction to the interaction due to the metal-He overlap.

Consider,now, the case of an alkali cristal with different orbitals, ψ_i, at each site. When a He-atom is moving inside the metal, electrons in the conduction band are promoted through the Fermi level to the empty state. In this case, we are interested in calculating how the different hopping terms, $T_{ii'}$, are modified by the presence of the external atom. This problem is similar to the previous one, where a metal level, E_M, is shifted by its interaction with He (this corresponds to taking $i = i'$ in the $T_{ii'}$ interaction). An analysis similar to the one presented in ref 3 shows that the effect of He on $T_{ii'}$ is to introduce the following correction:

$$\delta T_{ii'} = S_{i\ He}S_{He\ i'}(E_M - E_{He} - \frac{3}{8}\beta + J_o), \qquad (5)$$

an equation that can be obtained from (4) by replacing S^2 by $S_{i\ He}S_{He\ i'}$.

Equation (5) is the basis of our analysis for obtaining the stopping power for He. In our approach, we consider $\delta T_{ii'}$ and δE_{ii} as the perturbative hamiltonian that presents a time-dependence through the He-position that appears in $S_{i\ He}$.

The stopping power at a given time, t, is obtained by introducing the total hamiltonian

$$\hat{H} = \hat{H}_o + \delta\hat{H} \qquad (6)$$

where H_o is the cristal contribution and $\delta\hat{H}$ is the He-perturbation defined by $T_{ii'}$ and $\delta E_{ii'}$. Then, we introduce the hamiltonian eigenstates, $\mid n >$, such that

$$\hat{H} \mid n >= E_n \mid n > \qquad (7)$$

and obtain the stopping power as:

$$S = \frac{1}{v}\frac{dE}{dx} = -\frac{2}{v^2}Re\sum_n \int_{-\infty}^t dt' \frac{e^{-iw_{no}(t-t')}}{\hbar w_{no}} < o \mid \frac{d\hat{H}(t)}{dt} \mid n >$$

$$< n \mid \frac{d\hat{H}(t')}{dt'} \mid o > . \qquad (8)$$

In our actual case $\delta\hat{H}$ can be neglected for defining the eigenstates $\mid n >$ and the eigenvalues E_n, and start from the unperturbated hamiltonian, $\hat{H}_o$. Thus:

$$S = 2\pi\sum_{\vec{G}} e^{i\vec{G}\cdot\vec{R}} \int \frac{d^3k}{(2\pi)^3} \int \frac{d^3q}{(2\pi)^3}\theta(k_F - k)\theta(k' - k_F)\frac{\vec{q}\cdot\vec{v}}{v^2}$$

$$< \vec{k}' \mid \delta\hat{H}(\vec{q}) \mid \vec{k} >< \vec{k} \mid \delta\hat{H}(\vec{q}+\vec{G}) \mid \vec{k}' > \delta(w_{kk'} + \vec{q}\cdot\vec{v}) \qquad (9)$$

where $\hat{H}(\vec{q})$ is the fourier-transform of the perturbative hamiltonian, $\vec{G}$ a reciprocal crystal lattice vector, $\vec{R}$ the He-position, $\vec{k}' = \vec{k} - \vec{q}$, $\vec{k}$ and $\vec{k}'$ being the momenta of the crystal eigenfunctions, and $w_{kk'}$ their energy difference. From equation (9) we can write S in the general form:

$$S(\vec{R}) = S_o + \sum_{\vec{G}} S(\vec{G})e^{i\vec{G}\cdot\vec{R}} \qquad (10)$$

where S_o represents the mean stopping power for atoms moving at random in the crystal. For channeling directions, only the $\vec{G}$-vectors perpendicular to these directions should be considered in equation (10).

Let us first comment on our results for S_o. We have calculated the case of He moving inside Li, Na, K and Rb crystals. Table I shows our figures and the ones calculated in LDA. We see that the agreement between both calculations is very good except for slight differences for Na.

Table 1. Stopping power in a. u. for He in different alkali metals.

	Our results	LDA
Li	0.110	0.100
Na	0.068	0.053
K	0.023	0.023
Rb	0.014	0.016

We have also explored the case of a channeled He-atom moving along the (100)-direction in Na. We have obtained for the first $\vec{G}$-vector ($\vec{G} = \frac{2\pi}{a}(0,1,1)$) the following value:

$$S(0,1,1) = 0.016 a.u. \tag{11}$$

This is the most important contribution to S, as the next terms, $S(\vec{G})$, decrease with $\vec{G}$ aproximately as $\frac{1}{G^4}$. The results for S_o and $S(\vec{G})$ show that the stopping power for the moving atom in Na presents an important dependence on the impact parameter. In general, a channeled ion would have an oscillatory motion , with its impact parameter changing along its trajectory. This oscillation would tend to reduce the effect of channeling on the stopping power; we expect, however, that the important changes found above for the channeled ions could be seen experimentally.

III. DYNAMICAL INTERACTION OF He WITH METAL SURFACES

The interaction of simple atoms with metal surfaces has been discussed in ref. 4. The basic approach is the introduction of a tight-binding hamiltonian:

$$
\begin{aligned}
\hat{H} &= \sum_{i,\sigma} E_i^\sigma \hat{n}_{i\sigma} + \sum_{\sigma,(i,j)} T_{ij}^\sigma (\hat{c}_{i\sigma}^\dagger \hat{c}_{j\sigma} + \hat{c}_{j\sigma}^\dagger \hat{c}_{i\sigma}) + \\
&+ \sum_i U_i^{(0)} \hat{n}_{i\uparrow} \hat{n}_{i\downarrow} + \frac{1}{2} \sum_{i,j\neq i,\sigma} (J_{ij}^{(0)} \hat{n}_{i\sigma} \hat{n}_{j\sigma} + \\
&+ \bar{J}_{ij}^{(0)} \hat{n}_{i\sigma} \hat{n}_{j\sigma}) + \sum_{j\neq i} \frac{Z_1 Z_2}{d_{ij}}
\end{aligned}
\tag{12}
$$

where the different levels, E_i, hopping interactions, T_{ij}, and coulomb interactions, $U^{(0)}$, $J_{ij}^{(0)}$ and $\bar{J}_{ij}^{(0)}$, can be obtained from the atomic orbitals, ψ_i, associated with each i-site. For instance, E_i is given by:

$$
\begin{aligned}
E_i &= E_i^{(0)} + \delta E_i \\
\delta E_i &= -\sum_{j\neq i}' S_{ij} T_{ij} + \frac{1}{4} \sum_{j\neq i} S_{ij}^2 (E_i^{(0)} - E_j^{(0)})
\end{aligned}
\tag{13}
$$

where $S_{ij} = < \psi_i \mid \psi_j >$, and $E_i^{(0)}$ is the level of the ψ_i-atomic wavefunction. On the other hand, T_{ij}, can be related to the Bardeen tunneling current, T_{ij}^B, given by:

$$T_{ij}^B = -\frac{h}{2m} \int_{\sigma_{ij}} (\psi_i \bar{\nabla} \psi_j - \psi_j \bar{\nabla} \psi_i) d\bar{S} \tag{14}$$

Many-body contributions appearing in eqn. (12) are treated introducing a Slater-like potential [4].

Here, we report on our calculation for the interaction of He$^+$ or He* with an Al-surface, in the limit of very <u>slow</u> velocities. In this problem, the He-1s level is assumed to be singly occupied. Our interest is focused on the calculation of the interaction of the He-2s level with the metal. This interaction defines how the electronic charge is transferred between the approaching ion and the metal surface. It is worth commenting that He$^+$ interacting with the metal is similar to the case of Li: the He$^+$-nucleus is almost identical to the Li-one.

Fig. 1 shows the selfconsistent He*-2s level as a function of the distance between the last metal layer and the ion. In this calculation, we have solved the spin-polarized equations that appear by assuming an empty He-2s spin down level. In this way, the selfconsistent level of fig.1 is related to the first electron (assumed to have spin up) captured by the He$^+$ when the ion approaches the surface or, equivalently, to the last electron lost by the He*-2s level.

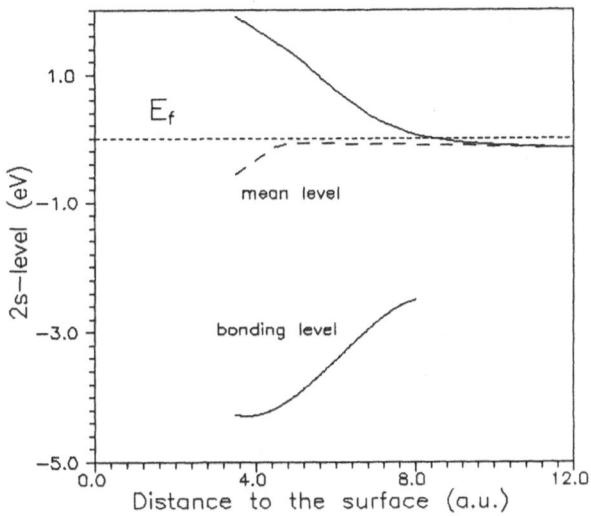

Figure 1. *He-2s level as a function of the He-metal distance. The "bonding" level only appears for $d \leq 8a.u.$*

Fig. 2 shows the mean occupancy of the He-2s spin up level as a function of the distance between the ion and the metal surface.

Figs. 1 and 2 are closely related: at long distances, the 2s-level is located below the Fermi level; then, the level occupancy goes to 1. At closer distances, the 2s-level starts moving up in energy, eventually crossing the Fermi energy at $d = 8a.u.$; at the same time, the level occupancy, Q, decreases and $Q \approx 0.5e^-$ for $d \approx 8a.u.$. When the ion gets closer to the surface, the simple picture of an electronic level resonating with the conduction band starts to change. This is shown in fig. 3 where the 2s-local density of states is shown for $d = 3.5a.u.$ and $d = 9a.u.$. For $d = 3.5a.u.$ a new resonance around 4 eV below the Fermi energy has evolved, while a strong resonance above the Fermi energy still remains. This effect is shown in fig.1 in the "bonding" resonance that appears for $d \leq 8a.u.$. (the dashed level of fig.1 is the mean 2s-level). In fig. 2, we find a minimun in the 2s-occupancy that is due to the new resonance that allows for some electron charge to be transferred from the metal to the ion. This implies that for He penetrating the solid, we can expect a 2s-resonance to evolve in the occupied part of the conduction band; this charge is going to neutralize partially the ion.

Previous calculations have been performed in the static limit. He$^+$ neutralization can be studied by combining the different capture and loss probabilities in the rate

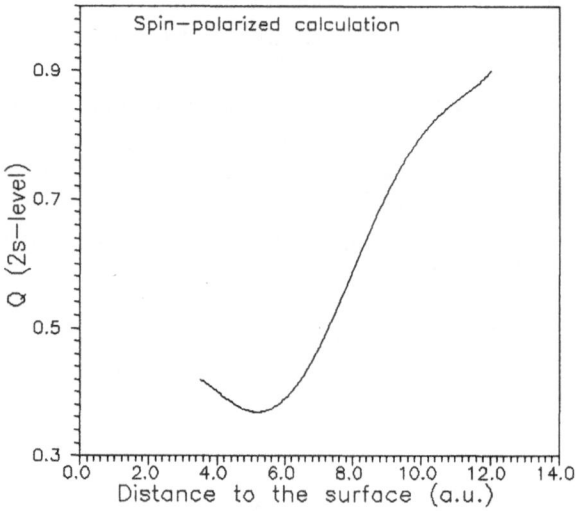

Figure 2. *Mean occupancy for He-2s level as a function of the He-metal distance.*

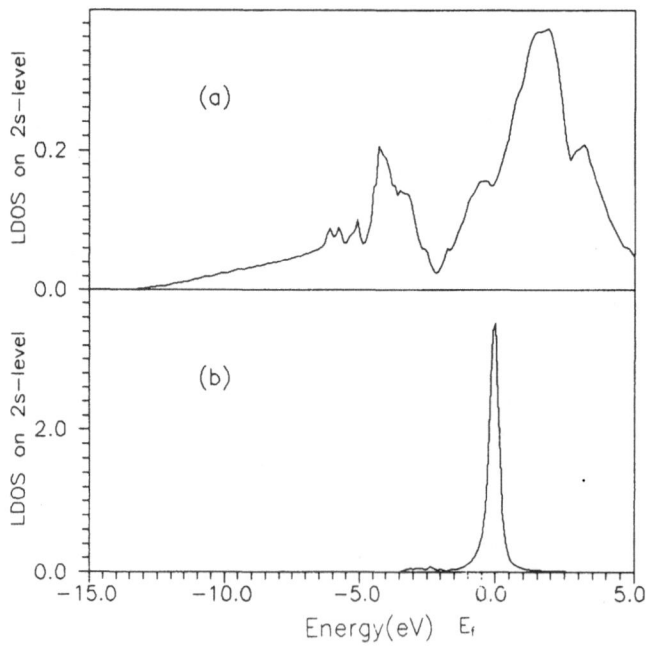

Figure 3. *Local density of states on the He-2s level for: (a) d = 3.5a.u.; (b) d = 9a.u.*

equations for the different charge fractions [11]. This can be done in the following way. Consider the He-charge states: $n(He^+)$, $n(He^{*\uparrow})$ and $n(He^{*\downarrow})$, corresponding to the ion and the excited He*-states with the 2s-electron having spin up or down, respectively. Introducing the lifetimes for loss and capture, $\frac{1}{\tau_L}$ and $\frac{1}{\tau_C}$, values that can be deduced from the previous calculations, we can write:

$$
\begin{aligned}
\frac{dn^+}{dt} &= \frac{1}{\tau_L}n^{\uparrow} + \frac{1}{\tau_L}n^{\downarrow} - \frac{2}{\tau_C}n^+ \\
\frac{dn^{\uparrow}}{dt} &= \frac{1}{\tau_C}n^+ - \frac{1}{\tau_L}n^{\uparrow} \\
\frac{dn^{\downarrow}}{dt} &= \frac{1}{\tau_C}n^+ - \frac{1}{\tau_L}n^{\downarrow}
\end{aligned}
\tag{15}
$$

These linear equations can be solved if we assume $\frac{1}{\tau_L}$ and $\frac{1}{\tau_C}$ to depend on z like $exp(-\gamma z)$ (a reasonable approximation for $d \geq 6a.u.$), and introduce the boundary conditions $n^+ = 1$, $n^{\uparrow} = n^{\downarrow} = 0$ for $t \to -\infty$. Tipically, we find from our calculations $\gamma \approx 0.72a.u.$, and obtain that the dependence of n^+ on z is the following:

$$
n(He^+) = \frac{1-q_0}{1+q_0} + \frac{2-q_0}{1+q_0} exp\left(-(1+q_0)\frac{1}{v_0\tau(z)\gamma}\right)
\tag{16}
$$

where q_0, that has been taken constant, is the 2s-level occupancy around the most probable neutralization distance. This yields the following values for the most probable neutralization distance:

He$^+$-velocities	most probable neutralization distance
$v_{thermal}$	7.6 Å
10^{-3} a.u.	5.5 Å
0.1 a.u.	5.2 Å

A similar analysis can be done for a He*-singlet initial state; in this case, He* ends up with two different spin polarized states, the process occurring at distances similar to the ones obtained for the He$^+$ initial state.

IV. CONCLUSIONS

In conclusion, we have shown how to use a free-parameter LCAO approach to calculate the interaction of slow He-states with condensed matter. In particular, the stopping power of neutral He moving inside alkali metals have been analyzed and the results have been found to be in good agreement with other independent calculations. Moreover, we have shown how to use the method to calculate the stopping power of channeled atoms as a function of the impact parameter.

We have also discussed He$^+$ and He* interacting with an Al-metal surface. We have shown how the 2s-level evolves as a function of the ion-metal surface and used these results to calculate the most probable neutralization distance of the ion. We have found that for He$^+$ interacting with Al, a bonding resonance appears in the metal conduction band, this effect modifying the simple picture of a 2s-level crossing the Fermi level at a given distance.

ACKNOWLEDGMENTS

This work has been partially funded by the Spanish CICYT under contract PB-89-165. J.J.D. thanks the "Universidad Autónoma de Madrid" for its finantial support. F.F. acknowledges support by Iberdrola S.A.

REFERENCES

[1] P. M. Echenique, F. Flores and R. H. Ritchie, "Solid State Physics series", 44, 229 (1990)

[2] A. Arnau, M. Peñalba, P. M. Echenique, F. Flores and R. H. Ritchie, *Phys. Rev. Lett.* 65, 1024 (1990); F. Flores in "Interaction of Charged Particles with Solids and Surfaces", Eds. Gras-Martí et al. (Plenum Press) (1991)

[3] E. C. Golberg, A. Martín-Rodero, R. Monreal and F.Flores, *Phys. Rev. B 39*, 5684 (1989)

[4] F. J. García-Vidal, A. Martín-Rodero, F. Flores, J. Ortega and R. Pérez, *Phys. Rev. B 44*, 11412 (1991)

[5] P. M. Echenique, R. M. Nieminen and R. H. Ritchie, *Solid State Commun. 37*, 779 (1981)

[6] F. B. Dunning, P. Nordlander and G. K. Walters, *Phys. Rev. B 44*, 3246 (1991)

[7] P. A. Zeijlmans, V. Emmichoven, P.A.A.F. Wouters and A. Niehaus, *Surf. Sci. 195*,115 (1988)

[8] B. Woratschek, W. Sesselman, J. Kuppers, G. Ertl and H. Haberland, *Phys. Rev. Lett.* 55, 611 (1985)

[9] R. Hemmen and H. Conrad, *Phys. Rev. Lett. 67*, 1314 (1991)

[10] H. Brenten, H. Muller and V. Kempter, *Z. Phys. D 22*, 563 (1992)

[11] R. Zimmy, H. Nienhaus, H. Winter, *Rad. Effects and Defects in Solids 109*, 9 (1989)

PROMOTION OF ELECTRONIC LEVELS IN BOMBARDED SOLIDS

Z. Šroubek[1], G. Falcone[2]

[1]Czechoslovak Academy of Sciences
 Prague 8, Chaberská 57, CSFR
[2]Department of Physics, University of Calabria
 870 36 Cosenza, Italy

INTRODUCTION

The formation of electron holes in deep electron levels in solids bombarded by atomic particles has been demonstrated by observation of Auger electrons induced by ion bombardment of metal surfaces [1,2,3]. The mechanism of the hole formation is basically the Fano-Lichten electronic level promotion process which is for atom-atom collisions well understood. However, quantitative description of this process in solids, i.e. the estimation of probabilities of deep hole excitations in metals and of one-electron excitations around the Fermi level has not been carried out in detail. In this paper we will outline the method which we have used for analysing electronic excitation in solids by level promotion processes with specific application on deep hole formation in Al metal.

THEORETICAL DESCRIPTION

We assume that the nuclear motion in the studied system can be described by a classical trajectory $R(t)$. Thus the Hamiltonian describing the electronic motion is not only a function of electronic coordinates r but also of the nuclear coordinates $R(t)$. Through the time-dependence of $R(t)$ the Hamiltonian $H(r, R(t))$ of the electronic

Ionization of Solids by Heavy Particles, Edited by
R.A. Baragiola, Plenum Press, New York, 1993

system becomes time-dependent. We now define two sets of normalized electronic basic functions ψ_k and ψ_l which should be, in principle, complete. The functions ψ_k belong to the continuum of states of the valence band of the solid whereas ψ_l are the localized functions, in our case the wave functions of deep levels. The corresponding self-energies are assumed to be:

$$\langle \psi_k(r,t) | H(r,R(t)) | \psi_k(r,t) \rangle = \epsilon_k \tag{1}$$

$$\langle \psi_l(r,t) | H(r,R(t)) | \psi_l(r,t) \rangle = \epsilon_l(t) . \tag{2}$$

Moreover, it is assumed that $\psi_l(t)$ are the eigenfunctions of $H(r,R(t))$ at any t, i.e.

$$H(r,R(t))\psi_l(r,t) = \epsilon_l(t)\psi_l(r,t) \tag{3}$$

and that at any time t the functions $\psi_k(r,t)$ are made orthogonal to $\psi_l(r,t)$, i.e.

$$\langle \psi_l(r,t) | \psi_k(r,t) \rangle = 0 . \tag{4}$$

We can define a wave function $\chi(r,t)$ describing the electronic motion and satisfying the time-dependent Schrödinger equation

$$H(r,R(t))\chi(r,t) = i\frac{\partial \chi(r,t)}{\partial t} \tag{5}$$

as a linear combination of ψ_l and ψ_k,

$$\chi(r,t) = \sum_k c_k(t)\psi_k(r,t)\exp(-i\epsilon_k t) + \sum_l c_l(t)\psi_l(t) . \tag{6}$$

Our objective is to estimate $c_k(t)$ and $c_l(t)$. The absolute values squared, $|c_k(t)|^2$ and $|c_l(t)|^2$, determine the probability of occupation of individual states.

To simplify the notation we assume only one localized state is important in the excitation (usually the highest antibonding state formed during the binary atomic collision) and we drop the summation over l. If (6) is substituted in (5) and the conditions (1)-(4) are considered, the following set of equations is obtained for $c_l(t)$ and $c_k(t)$:

$$i\frac{\partial c_l}{\partial t} = \sum_k -ic_k\langle\psi_l|\frac{\partial\psi_k}{\partial t}\rangle \exp(-i\epsilon_k t) + c_l\epsilon_l(t) \tag{7}$$

$$i\frac{\partial c_k}{\partial t} = -ic_l\langle\psi_k|\frac{\partial\psi_l}{\partial t}\rangle \exp(i\epsilon_k t) . \tag{8}$$

In (7) and (8) we have neglected the matrix elements between k-states assuming that the states in the continuum are coupled only to the localized states. This approximation is not necessary but simplifies the problem. It should be noted that when the basic sets of electronic wave functions are chosen in such a way that the functions are orthogonal without the constrain (3), the leading interaction term is $\langle\psi_k|H|\psi_l\rangle$. When $\langle\psi_k|H|\psi_l\rangle$ replaces $-i\langle\psi_k|\partial\psi_l/\partial t\rangle$, the set of equation (7) and (8) corresponds to the dynamical Anderson Hamiltonian [4]. The Anderson Hamiltonian is known to have the solution in a closed form, when the "slowness approximation" is applied [5]. Though the slowness approximation is probably applicable also in our case we have solved (7) and (8) directly using Runge-Kutta method. The continuum is replaced by 150 energy levels separated by 0.15 eV. The direct solution of 151 complex equations is time consuming but avoids uncertainties with slowness approximation and allows further generalization.

The Hamiltonian $H(r,R(t))$ has been chosen as follows: We adopt the picture of a free electron model for the states in the conduction band of the solid (metal), thus ψ_k are the free-electron wave functions. The two colliding atoms in the solid are represented by properly screened atomic potentials. Specifically we have used for ion cores the independent-particle-model (IPM) potential of Green, Sellin and Zachor [6] screened by free electrons. The formation of adiabatic quasimolecular levels for various interatomic distances has been described with variable-screening model of Eichler and Wille [7]. The effective single-electron potential is in this model obtained by smoothly interpolating the screening parameters of IPM potentials between the united-atom limits. The model allows to calculate with reasonable precision the wave functions and energies of deep levels, in our case ψ_1 and ϵ_1 of the highest antibonding level in (7) and (8).

The IPM potential (in a.u.) of an ion in vacuum is assumed to have the same form as in [6]

$$V_i(r) = ((N-1)\Gamma - Z)/r, \tag{9}$$

where

$$\Gamma = 1 - \Omega(r) \tag{10}$$

with

$$\Omega(r) = ((\eta/\xi)(\exp(\xi r) - 1) + 1)^{-1}, \tag{11}$$

where N is the total number of electrons in the atom (or ion) plus one electron which is calculated and Z is the nuclear charge. The parameters η and ξ are found for given Z and N in refs [6]. To the potential (9) one has to add the potential of $(Z-N+1)$ conduction electrons. When the conduction electrons are homogenously spread within a sphere of the radius R_a this additional potential is equal to [8]

$$\frac{3(Z-N+1)}{2R_a}(1 - \frac{1}{3}(\frac{r}{R_a})^2) \qquad for \ \ r < R_a$$

$$\tag{12}$$

$$\frac{Z-N+1}{r} \qquad for \ \ r > R_a$$

The total potential is then

$$V(r) = \frac{Z}{r}(\frac{(N-1)\Gamma - Z}{Z} + \frac{3r(Z-N+1)}{2R_aZ}(1 - \frac{1}{3}(\frac{r}{R_a})^2)) \qquad for \ \ r < R_a$$

$$\tag{13}$$

$$V(r) = 0 \qquad for \ \ r > R_a$$

which can be also written as

$$V(r) = -\frac{Z}{r}\phi(r, \xi, \eta, R_a, Z, N) \tag{14}$$

where $\phi(r, \xi, \eta, R_a, Z, N)$ is the parametrical screening function.

The radius R_a can be calculated from the condition that the electron density should be equal to the conduction electron density in the metal. To improve the potential we use however R_a as an additional fitting parameter such that (13) satisfies the Friedel sum rule. It should be noted that the fitted potential (13) gives transport cross-sections σ_l in close agreement with σ_l calculated with the best density-functional potentials.

With the aim of calculating molecular correlation diagrams and wave functions we consider as optimal the variable screening model [7]. This model is based on the idea that the mutual screening of the atoms making up the quasimolecule may well be approximated by spherical, but variable (as a function of the internuclear distance D) screening functions. The electronic states of the separated and the united atom may be generated to a good approximation by phenomenological single-electron potentials $V(r)$ = $(-Z/r) \, \Phi \, (r,\alpha)$ depending on sets of parameters collectively described by α^{sa} and α^{ua}, respectively. If we adopt the interpolation scheme [7]

$$\alpha_i^{eff} = \frac{\alpha^{sa}\lambda^2 + \alpha^{ua}\rho_i^2}{\lambda^2 + \rho_i^2} \quad , \tag{15}$$

where $\lambda^2 = 3$ and $\rho_i = 2r_i/D$, we can expect that

$$V^{eff}(r_1, r_2, D) = -\frac{Z}{r_1}\phi(r_1, \alpha_1^{eff}) - \frac{Z}{r_2}\phi(r_2, \alpha_2^{eff}) \tag{16}$$

is a good approximation to the molecular single-electron potential. r_1 and r_2 are distances of the electron from the nuclei.

APPLICATION ON Al

For the two colliding aluminium atoms which become a Fe atom when united the parameters are the following [6]

$$\begin{aligned}
\eta^{sa} &= 3.544 \\
\xi^{sa} &= 2.496 \\
R_a^{sa} &= 3.4 \\
\eta^{ua} &= 4.36 \\
\xi^{ua} &= 1.733 \\
R_a^{ua} &= 2.2
\end{aligned} \tag{17}$$

For a given interatomic separation D we can calculate the energy levels $\epsilon_l(D)$ and the wave functions $\psi_l(D)$ by solving the Schrödinger equation

$$(-\frac{1}{2}\nabla^2 + V^{eff}(r_1, r_2, D))\psi_l = \epsilon_l \psi_l \; . \tag{18}$$

In order to solve (18) for relevant energy levels, i.e. levels originating from 2s,2p Al deep levels we expand the wave function ψ_l in terms of four basis function $\psi_{2p\sigma}(r_1)$, $\psi_{2p\sigma}(r_2)$, $\psi_{2s}(r_1)$, $\psi_{2s}(r_2)$ and use simple Slater orbitals (in a.u.)

$$\psi_{2p\sigma} = 34.79\, z\, \exp(-5.2r)$$

$$\psi_{2s} = 6.4 \exp(-6.8r) - 3\, 2r\, \exp(-5.2r) \tag{19}$$

The calculated energy of the highest antibonding orbital as a function of the interatomic distance D for two colliding Al atoms in the Al metal is shown in Fig.1.

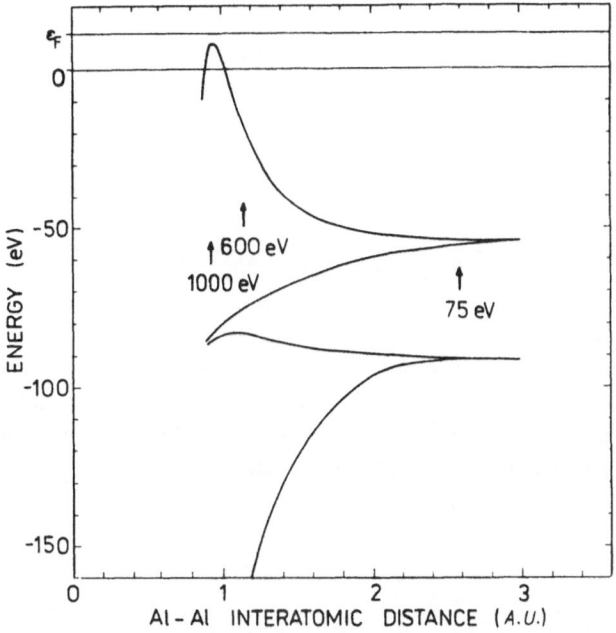

Fig.1 The energies of 2p-2s levels of an Al-Al molecule embedded in Al metal as a function of the interatomic distance D. The distance is given in a.u., the energies in electronvolts. Also marked are the bottom of the conduction band of Al ($\epsilon = 0$) and Fermi energy ($\epsilon = \epsilon_F$). The energies have been calculated with the use of the variable-screening model [7]. Some center-of-mass (c.o.m.) kinetic energies of the colliding Al atoms are indicated by arrows at the corresponding distances of closest approach. For c.o.m. energy of 75eV the 2p interaction is already so strong that a 2p hole would be shared in less than 10^{-15} sec.

The corresponding wave function ψ_1 changes rapidly its character with decreasing D, from almost pure 2p antibonding orbital to an orbital with a strong admixture of 2s orbitals at small D. The dependence of ϵ_1 and ψ_1 on D can be converted into the time-dependence by calculating the classical path trajectory R(t) and by obtaining D as a function of time. We have calculated the trajectories for various center-of-mass kinetic energies using the Tietz inter-atomic potential [9]. In this paper calculations only for head-on collisions are reported.

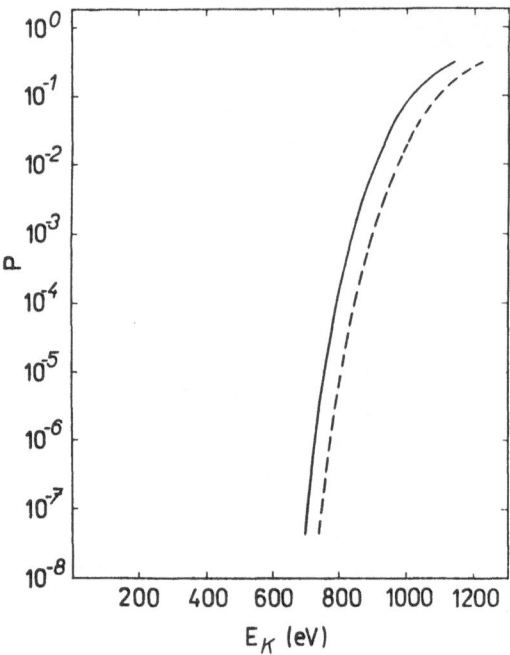

Fig.2 The probability of the hole formation in the 2p level of Al as a function of center-of-mass kinetic energy E_k of two atoms colliding inside Al metal. Only head-on collision are considered. The metal is represented by a simple, free-electron like system of levels. The full line is calculated with the Tietz potential, the dashed line with the unscaled Moliére potential.

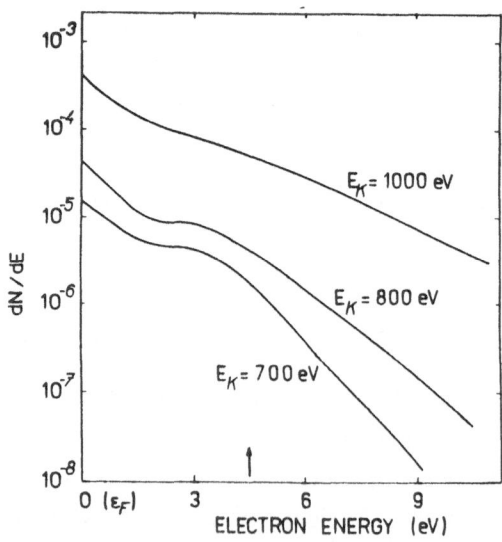

Fig.3 The energy spectrum of one-electron excitations above the Fermi level for various Al-Al colliding energies (in units of the number of excited electrons per electronvolt). The spectra were calculated using the same model as was used for the determination of the 2p-hole excitation probability in Fig.2. The vacuum level is marked by an arrow.

24

Using the knowledge of $\epsilon_i(t)$ and of the time derivative of $\psi_i(t)$ we have solved the set (7) and (8) in order to get two quantities.

Firstly we have calculated $|c_i(+\infty)|^2$ for the initial conditions $c_k(-\infty) = 1$, where k are the wave vectors of unoccupied states ($|k| > |k_F|$) above the Fermi level. The resulting sum of $|c_i(+\infty)|^2$ gives probability that during the collision a hole will be created in the deep 2p level. This probability as a function of the center-of-mass kinetic energy of colliding atoms is shown in Fig.2. The use of the Moliere potential instead of the Tietz potential shifts the graph slightly to higher energies.

Secondly, we have calculated $|c_k(+\infty)|^2$ (where $|k| > |k_F|$) for initial conditions $c_{k'}(-\infty) = 1$ (where $|k'| < |k_F|$). The sum of resulting $|c_k(+\infty)|^2$ gives the probability of electron excitation above the Fermi level. This is the true one-electron "kinetic excitation" and is shown for various impact energies in Fig.3. The results are certainly underestimated by neglecting in (7) and (8) all matrix experiments between k-states. However, the one-electron energy spectrum is not expected to be changed significantly by including these terms [10] except, of course, of a large increase of excitation just above the Fermi level, typical for electron-particle binary collisions.

CONCLUSION

We have calculated the probability of formation of a 2p-hole in head-on collisions of two Al atoms in Al metal. The probability is plotted as a function of the kinetic energy of the center of mass of colliding atoms. Also calculated is the energy spectrum of one-electron excitations above the Fermi level. The character of the one-electron spectra is similar to that obtained in kinetic electron emission models when deep levels are accounted for by orthogonalization processes [10]. It is obvious that high energy electrons can be effectively emitted only as a results of Auger neutralization of deep holes. Similar observation seems to be also valid for valence band excitations.

REFERENCES

1. J.J.Vrakking and A.Kroes, Surf. Sci. **84**, 153 (1979).
2. R.A.Baragiola, E.V.Alonso and H.Raiti, Phys. Rev. A25, 1969 (1982).
3. T.D.Andreadis, J.Fine and J.A.D.Matthew, Nucl. Instr. and Meth. **209/210**, 495 (1983).
4. Z.Sroubek, Surface Sci. **44**, 47 (1974); A.Blandin, A.Nourtier and D.W.Hone, J. Phys. **37**, 369 (1976).
5. W.Bloss and D.Hone, Surface Sci. **72**, 277 (1978); G.Blaise and A.Nourtier, Surface Sci. **90**, 495 (1979); R.Brako and D.M.News, J.Phys. C**14**, 3065 (1981).
6. A.E.S.Green, D.L.Sellin and A.S.Zachor, Phys. Rev. **184**, 1 (1969); R.H.Garvey, C.H.Jackman and A.E.S.Green, Phys. Rev. A12, 1144 (1975).
7. J.Eichler and U.Wille, Phys. Rev. A11, 1973 (1975).
8. W.Heine and D.Waire, Solid Stat. Physics vol.24, page 260, Academic Press, New York and London, 1970.
9. T.Tietz, Nuovo Cimento 1, 955 (1955).
10. G.Falcone and Z.Sroubek, Nucl. Instrum. and Meth. B58, 313 (1991).

THEORY OF ION-INDUCED KINETIC ELECTRON EMISSION FROM SOLIDS

Max Rösler

Project: Particle-induced electron emission from solids
Supported by KAI e.V., Berlin
Karl-Pokern-Str. 12, O-1162 Berlin, Germany

INTRODUCTION

The emission of electrons as a result of the bombardment of the solid surface by ions plays a fundamental role in the physics of particle beam-solid interaction. Both, atomic physics and solid state physics have stimulated the development in this field.

Independent from the great variety of solids which can be used as target materials with respect to their chemical composition and their structure there is a lot of possibilities and special conditions to observe the electron emission from the solid surface. At first we can use different ions: H^+, He^+,..., molecular ions, multicharged ions, charged clusters. Secondly, we can use different impact energies E_o. Depending on the impact energy we can decide between the so-called potential emission and the kinetic emission. In the first case (at low impact energy) the different projectile properties, especially their potential energy, are responsible for the electron emission. In the second case energy and mass of the projectile are of importance. The electron emission is governed by the transfer of kinetic energy from the projectile to the target electrons. Thirdly, it is very interesting to vary the angle of incidence. Depending on the impact energy different physical processes are responsible for the electron emission if we go from perpendicular to gracing incidence. Especially, in this way it is possible to investigate the transition from bulk to surface related properties as well as the interrelation between atomic like and solid state specific processes.

The present contribution is devoted only to the theoretical description of the ion-induced kinetic electron emission. The experimental situation in the case of kinetic emission is comprehensively discussed recently by Hofer (1990) and Hasselkamp (1991). The current state of the theory is represented in several reviews (Schou, 1988; Devooght et al., 1991; Rösler & Brauer, 1991). Concerning the potential emission and, especially, the problem of separation of potential and kinetic emission we refer to the excellent review paper by Varga and Winter (1991).

It is beyond the scope of this contribution to formulate a general theory of ion-induced electron emission which includes all types of solids and experimental conditions. Different restrictions are taken into consideration. The time variable as well as the spin is completely ignored. With respect to the spin variable it should be mentioned, however, that the inclusion of this additional degree of freedom is of special interest in the investigation of magnetic materials.

Collisions between the impinging ion and target atoms can lead to a sizeable fraction of energetic recoil atoms which generate an additional contribution to the number of emitted electrons. This contribution is important for heavy ions at low impact energies as shown by Sigmund and Tougaard (1981). For protons, however, the contribution of recoil atoms can be neglected for impact energies above 10 keV considered here.

A microscopic theory of ion-induced electron emission for nearly-free-electron (NFE) metals based on the transport equation formalism is represented. Only in the case of these metals we can start from first principles in order to calculate the measurable quantities. Proton impact will be discussed in detail, because in this case complications due to projectile electrons are missing.

In the following we will restrict us to perpendicular incidence of the ion beam. This leads to substantial mathematical simplifications for the solution of the transport equation. Furthermore, we formulate a theory for polycrystalline metals. In this case the mathematical complications related to the crystal structure can be avoided to a large extend. It should be noted that most of the experiments were carried out on polycrystalline targets. Because the range of protons $R(E_0)$ is large compared with the maximum escape depth L of excited electrons it seems to be justified to take into account bulk processes only. Nevertheless, processes which are directly related to the existence of the surface can be seen in some cases.

In spite of these different restrictions we expect that most of the statements about the relative importance of various excitation and scattering mechanisms obtained here are also valid under other conditions including different types of target materials and ions as well as different angles of incidence.

BASIC QUANTITIES

The maximum information about the emission process can be obtained by measuring the energy and angle dependent current density $j(E,\vec{\Omega})$. This is the number of electrons with energy E emitted per second from 1 cm^2 of the surface in the direction $\vec{\Omega}$. This quantity is related to the unit current of ions impinging on the surface with energy E_0. The other important experimental quantities are the energy distribution of emerging electrons

$$j(E) = \int j(E,\vec{\Omega})d\Omega \qquad (1)$$

the angular distribution

$$j(\vec{\Omega}) = \int j(E,\vec{\Omega})dE \qquad (2)$$

and the electron yield

$$\gamma = \int \int j(E,\vec{\Omega})d\Omega dE = \int j(E)dE \qquad (3)$$

With respect to the special conditions mentioned in the introduction we will show some experimental results for NFE metals. In fig. 1 the impact energy dependence of the electron yield is shown for proton impact on polycrystalline aluminum. The results obtained by different groups are collected. The behavior of yield vs. ion energy is similar for other ion target combinations. However, only for light ions the maximum of the yield curve has been obtained so far.

It is interesting to note that the impact energy dependence of the electron yield shows strong resemblance with the behavior of the stopping power $(-dE_0/dx)$ vs. ion energy. The maxima of both quantities are located at the same energy: $E_0^{max} \approx 55$ keV for Al (Brauer & Rösler, 1985).

Fig. 2 shows the energy distribution of ejected electrons from Magnesium for different projectiles at $E_0 = 200$ keV (H^+, Ar^+) and $E_0 = 800$ keV (Xe^+). These unpublished results for Mg obtained by Hippler (1988) show the typical behavior for NFE metals.

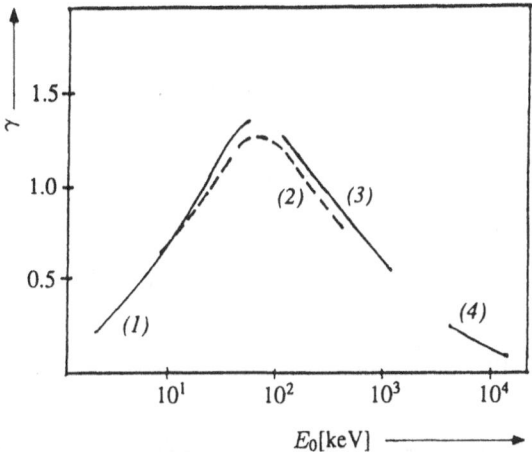

Figure 1. Electron yield as a function of the ion energy for proton impact on Al. Experimental values: (1) Baragiola et al. (1979), (2) Svensson & Holmen (1981), (3) Hasselkamp et al. (1981), (4) Koyama et al. (1981)

Besides the so-called cascade maximum sligthly below 3 eV a weak shoulder slightly below 8 eV was found. This shoulder is clearly seen in the differentiated spectrum in the lower part of fig. 2. This structure can be related to the excitation of conduction electrons by decay of plasmons via interband transitions. These plasmons are generated by the impinging ion or by excited electrons. It should be noted that the plasmon shoulder is also obtained for heavier ions at the same position. The measurements of Hippler (1988) for Mg show no surface plasmon in contrast to the measurements for Al (Hasselkamp & Scharmann, 1982). The energetic position of the shoulder is given approximately by the plasmon energy $\hbar\omega_p$ and the work function Φ: $\hbar\omega_p-\Phi$. For Mg with $\hbar\omega_p = 10.8$ eV and $\Phi = 3.6$ eV we have $\hbar\omega_p-\Phi = 7.2$ eV which is nearly at the position given by the experiment. The same is true for Al. With $\hbar\omega_p = 15$ eV and $\Phi = 4.3$ eV we have $\hbar\omega_p-\Phi = 10.7$ eV which is something below the value given by Hasselkamp and Scharmann (1982).

MATHEMATICAL DESCRIPTION OF THE ION-INDUCED KINETIC ELECTRON EMISSION

All the measurable quantities mentioned before can be obtained in a simple way if we know the density of inner excited electrons at the surface $N(E,\Omega) = N(x=0;E,\vec{\Omega})$.

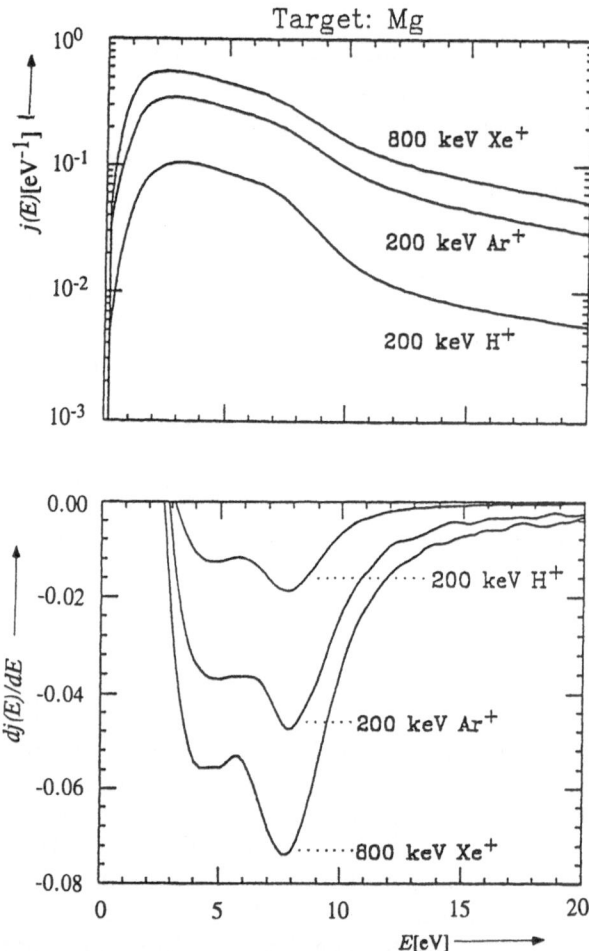

Figure 2. Energy distribution of ejected electrons and the corresponding derivative spectra from Mg for different projectile ions and impact energies (Hippler, 1988)

General Considerations

The density of inner excited electrons $N(x;E,\vec{\Omega})$ will be determined by solving Boltzmann's transport equation taking into account suitable boundary conditions at the surface. In order to formulate these boundary conditions we describe the escape process of excited electrons using the standard model of a planar surface barrier and free electrons inside the metal.

Surface $x=0$

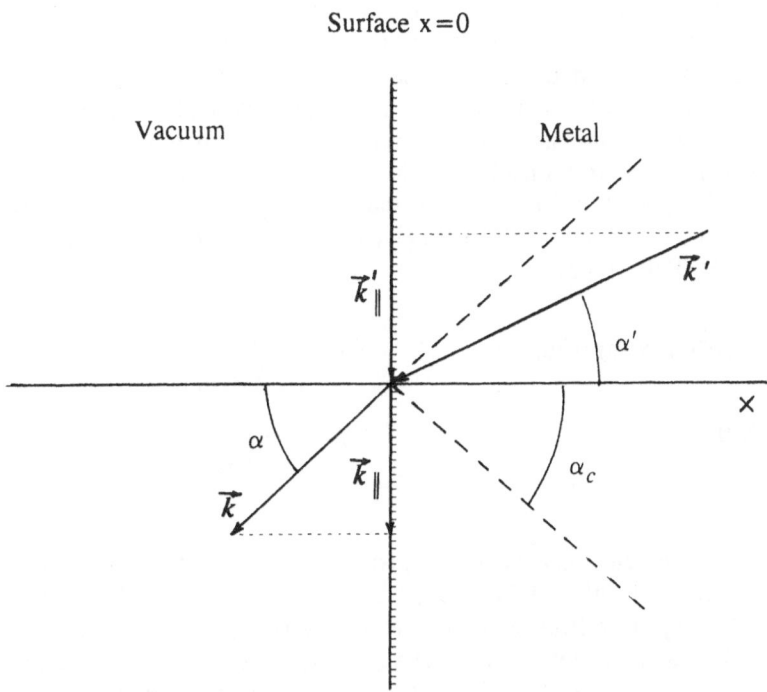

Figure 3. Momentum diagram for the escape process. $\vec{k}'$ and $\vec{k}$ are the momenta of the electron at both sides of the surface. α_c defines the so-called escape cone

In metals the barrier high W is determined by the Fermi energy $E_F=\hbar^2 k_F^2/2\,m$ and the work function Φ, i.e. $W=E_F+\Phi$. k_F is the Fermi momentum. From the conservation laws for energy and parallel momentum (see fig. 3) connecting inner and outer variables we obtain the escape conditions

$$E'>W$$
$$\cos\alpha'>\cos\alpha_c=\sqrt{\frac{W}{E'}} \tag{4}$$

For $\cos\alpha'>\cos\alpha_c$ the electrons are specularly reflected at the surface and cannot leave the target.

In the case of electron impact the Boltzmann equation was given for the first time by Puff (1963). This equation can be used also in the case of ion impact. In a planar geometry we may write

$$v(E)\cos\alpha \frac{\partial N(x;E,\vec{\Omega})}{\partial x} = S(x,E_0;E,\vec{\Omega}) - \frac{v(E)}{l(E)}N(x;E,\vec{\Omega})$$
$$+ \int\int dE'd\Omega' W^o(E,\vec{\Omega};E',\vec{\Omega}')N(x;E',\vec{\Omega}')$$

(5)

The excitation function $S(x,E_0;E,\vec{\Omega})$ expresses the number of electrons in the state $\vec{k}(E,\vec{\Omega})$ at the depth x created by the impinging ion. The second term on the right hand side denotes the number of electrons removed from the state $\vec{k}$ by elastic and inelastic collisions. The third term denotes the number of electrons entering the state $\vec{k}$ by collisions. This number is determined by the transition function $W^o(E,\vec{\Omega};E',\vec{\Omega}')$. $v(E)$ and $l(E)$ are the velocity and the total mean free path (mfp) of electrons, respectively. Transition functions and mfp are solely determined by the properties of the target. Density and excitation function are normalized to the unit primary current.

Except for the condition $N(\infty;E,\vec{\Omega})=0$ the boundary conditions at the surface which must be fulfilled by the solution of (5) are given for the ensemble of electrons with $E' > W$ according to (4) (Puff, 1963) by

$$N(0,E',\vec{\Omega}')\equiv N(0,E',\alpha',\phi)=N(0,E',\pi-\alpha',\phi) \quad for \quad \sqrt{\frac{W}{E'}}>\cos\alpha'>-\sqrt{\frac{W}{E'}}$$

$$N(0,E',\vec{\Omega}')=0 \quad for \quad -\sqrt{\frac{W}{E'}}>\cos\alpha'>-1$$

(6)

No further discussion which includes realistic excitation functions, scattering functions, and mfp was given by Puff (1963). This problem was discussed in detail in a number of papers by the group from Brussels (Dubus et al., 1986; Devooght et al., 1987; Dubus et al., 1987; Dubus et al., 1990; Devooght et al., 1991; Devooght et al., 1992). Especially, in the last paper of Devooght et al. (1992) the solution of the Boltzmann equation for homogeneous excitation in an infinite medium was compared with a Monte-Carlo calculation for a semi-infinite target. It is shown that the electron yield is overestimated by about 25% within the so-called infinite slowing down (ISD) model. This effect is due to the surface. Nevertheless, we will use this model in the further considerations because it is difficult to calculate the surface corrections using our scattering cross sections calculated from first principles.

Infinite Slowing Down Model

For impact energies about 20 keV the range of the protons $R(E_0)$ is large compared to the maximum escape depth L of excited electrons. In this case the assumption of a homogeneous excitation is justified. Then, instead of (5) we will start with the equation

$$\frac{v(E)}{l(E)}N(E,\vec{\Omega}) = S(E_0;E,\vec{\Omega}) + \int\int dE'd\Omega' W^o(E,\vec{\Omega};E',\vec{\Omega}')N(E',\vec{\Omega}')$$

(7)

The part of the distribution of inner excited electrons which are able to leave the target can be obtained by the solution of (7) taking into account the escape conditions (4). Then, the relevant part of the flux density of inner excited electrons can be written as

$$j_i(E',\vec{\Omega}')=v(E')\cos\alpha'\Theta(E'-W)\Theta(\cos\alpha'-\cos\alpha_c)N(E',\vec{\Omega}') \tag{8}$$

Within this formulation the three step model (excitation, transport, escape) of the emission phenomenon will be obvious because the escape process is not enclosed in complicate boundary conditions for solving Boltzmann's equation as mentioned in the discussion of (5).

If we denote by $j(E,\vec{\Omega})$ the flux density measurable in free space then we obtain from particle conservation at the surface

$$j(E,\vec{\Omega})dEd\Omega=j_i(E',\vec{\Omega}')dE'd\Omega' \tag{9}$$

using $E'=E+W$ and

$$\cos\alpha'=\sqrt{\frac{E\cos^2\alpha+W}{E+W}} \tag{10}$$

the final expression

$$j(E,\vec{\Omega})=v(E')\left(1-\frac{W}{E'}\right)\sqrt{\frac{E'\cos^2\alpha'-W}{E'-W}}\Theta(E'-W)\Theta(\cos\alpha'-\cos\alpha_c)N(E',\vec{\Omega}') \tag{11}$$

The angular dependence of the problem is treated by expansion into Legendre polynomials. With

$$N(E,\vec{\Omega})=\sum_{l=0}^{\infty}(-1)^lN_l(E)P_l(\cos\Theta) \tag{12}$$

$$S(E_0;E,\cos\Theta)=\sum_{l=0}^{\infty}S_l(E_0,E)P_l(\cos\Theta) \tag{13}$$

and

$$W^o(E,E',\cos\vartheta)=\sum_{l=0}^{\infty}W_l^o(E,E')P_l(\cos\vartheta) \tag{14}$$

we obtain from (7) a set of independent integral equations

$$\frac{v(E)}{l(E)}N_l(E)=(-1)^lS_l(E_0,E)+\int_E^{E_{max}}dE'W_l^o(E,E')N_l(E'); \qquad l=0,1,2,... \tag{15}$$

Depending on the behavior of the excitation function the energy integration is extended up to a suitable upper limit E_{max}.

Inserting (12) into (11) the energy distribution of emerging electrons and the electron yield γ are given by

$$j(E) = 2\pi\nu(E')\Theta(E'-W)\sum_{l=0}^{\infty} A_l(E')N_l(E') \tag{16}$$

and

$$\gamma = 2\pi\sum_{l=0}^{\infty}\int_{W}^{E_{\max}} dE'\nu(E')A_l(E')N_l(E'), \tag{17}$$

respectively, where

$$A_l(E') = \int_{\sqrt{\frac{W}{E'}}}^{1} dx\,x P_l(x) \tag{18}$$

In the case of a spatially uniform excitation of electrons if the ion trajectories are straight lines and their energy loss within the escape depth can be neglected the angle of incidence can be incorporated in the model in a simple way (Devooght et al., 1991).

SCATTERING PROCESSES WITHIN THE SOLID

The excitation of solid state electrons by the impinging ion as well as the transport of inner excited electrons are directly related via the screened Coulomb interaction to the complex wave number and frequency dependent dielectric function

$$\varepsilon(q,\omega) = \varepsilon_1(q,\omega) + i\varepsilon_2(q,\omega) \tag{19}$$

of the solid. In general, we have to taken into account the lattice structure of the solid. This means, we have instead of (19) a dielectric matrix $\epsilon(\vec{q}+\vec{K},\vec{q}+\vec{K}',\omega)$, where $\vec{K}$ and $\vec{K}'$ are reciprocal lattice vectors. In order to obtain the screened interaction potential in a periodic lattice a matrix inversion is necessary.

As we will see in the following it is possible to describe the different processes which are responsible for the emission within the free-electron-gas model or within a model of a real solid which allows a simple model potential calculation of the band structure.

Here we will consider only the mean free paths and the transition functions which are important for the transport of excited electrons starting from a general expression for the transition probability between Bloch states for two interacting screened point charges

$$W_{\vec{k}_1\vec{k}_2-\vec{k}_1'\vec{k}_2'} = \frac{32\pi^3 e^4}{h^4}\frac{1}{\Omega}\sum_{\vec{q}}\frac{|<\vec{k}_1'|e^{i\vec{q}\vec{r}}|\vec{k}_1>|^2|<\vec{k}_2'|e^{-i\vec{q}\vec{r}}|\vec{k}_2>|^2}{q^4|\epsilon(q,E_{\vec{k}_1}-E_{\vec{k}_1'})|^2}\delta(E_{\vec{k}_1'}+E_{\vec{k}_2'}-E_{\vec{k}_1}-E_{\vec{k}_2}) \tag{20}$$

Ω is the normalization volume. The description of different excitation processes of target electrons by the incident ion is given in the next chapter. Also in this case we start from the transition probability (20).

Free-Electron-Gas Model

As a first step in the description of the system of conduction electrons in metals the free-electron-gas model is well established. In this case the sole parameter which characterizes the metal r_s (radius of a sphere, in units of Bohr radius, containing on the average one electron) is determined by the electron density. For real metals r_s varies between 1.8 (Be) and 5.5 (Cs). In fig. 4 the elementary excitation spectrum is shown in the random phase approximation (RPA). The corresponding dielectric function (proposed

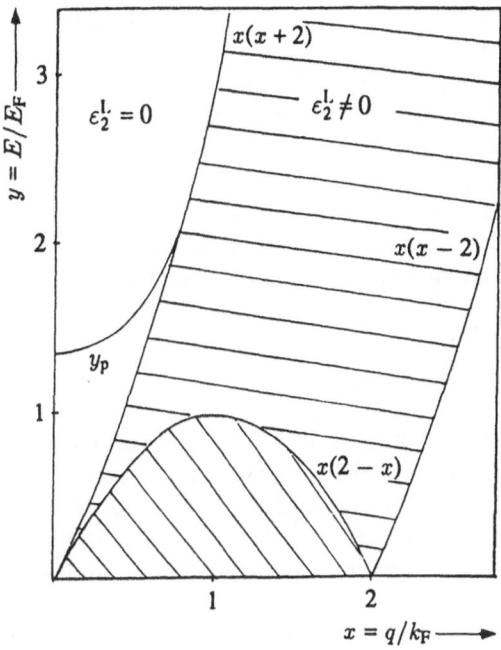

Figure 4. Elemantary excitation spectrum of the homogeneous free-electron gas (jellium) in random phase approximation.

by Lindhard (1954)) is given by

$$\epsilon_1(x,y) = 1 + \frac{2\alpha r_s}{\pi x^3}[1 + R(x,y) + R(x,-y)] \tag{21}$$

$$\epsilon_2(x,y) = \begin{cases} \dfrac{\alpha r_s}{x^3} y & \text{for } y \le x(2-x) \\[2ex] \dfrac{\alpha r_s}{x^3}\left[1 - \left(\dfrac{x^2-y}{2x}\right)^2\right] & \text{for } x|2-x| \le y \le x(2+x) \\[2ex] 0 & \text{otherwise} \end{cases} \tag{22}$$

where

$$R(x,y) = \frac{1}{2x}\left[1 - \left(\frac{x^2+y}{2x}\right)^2\right]\ln\left|\frac{x^2+2x+y}{x^2-2x+y}\right| \tag{23}$$

and $\alpha = (4/9\pi)^{1/3}$. Wave number and energy are measured in units of Fermi momentum k_F and Fermi energy E_F: $x = q/k_F$, $y = \hbar\omega/E_F$.

As shown in fig.4 there is a region of individual excitations which is bounded by two parabolas obtained from energy and momentum conservation. Besides these excitations there is for wave numbers below the so-called cut-off wave number q_c a plasmon mode $y_p(x) = \hbar\omega_p(q)/E_F$. For the finite value of the plasmon energy at zero wave number we obtain for Al with $r_s = 2.07$: $\hbar\omega_p(0) = 15.7$ eV which is something above the experimental value: $\hbar\omega_p(0) = 15.0$ eV. This difference may be attributed mainly to the contribution of interband processes as well as core polarization effects (Sturm, 1982).

In order to describe different metallic quantities with higher accuracy we have as a first possibility to take into account the real lattice structure. A first attempt in this direction is discussed in the next section. The other possibility is to describe the interacting system of free electrons with higher accuracy including exchange and correlation effects. Exchange and correlation lead to strong modifications of the RPA result at large momentum transfer. Including these corrections the dielectric function may be written as

$$\varepsilon(q,\omega) = 1 + \frac{\varepsilon^{RPA}(q,\omega) - 1}{1 - G(q,\omega)(\varepsilon^{RPA}(q,\omega) - 1)} \tag{24}$$

Different approximations for the so-called local field correction $G(q,\omega)$ are discussed in the literature. Using for instance the fit formula given by Vashishta & Singwi (1972) a significant reduction of the number of emitted electrons (about 20%) is obtained (Devooght et al., 1991). Nevertheless, we will use in the following the RPA dielectric function (21, 22) in order to describe the different scattering properties which are related to the system of conduction electrons.

Nearly-Free-Electron Metals. Plasmon Damping

In the RPA there is no damping of bulk plasmons for wave numbers below q_c. However, a finite plasmon line width is observed in real metals. There is no satisfactory explanation of this finite plasmon damping within the free-electron-gas model. As shown by Paasch (1969) and Sturm (1976, 1977, 1982) interband transitions primarily determine the plasmon damping in simple metals. It is obvious that these interband processes which are responsible for the plasmon damping lead on the other hand to an excitation of conduction electrons.

In NFE metals the electronic structure is well described within a model potential scheme. Then, according to Sturm (1982), the diagonal element of the inverse dielectric matrix may be expressed by an effective dielectric function

$$\varepsilon^{-1}(\vec{q},\vec{q},\omega) = \frac{1}{\varepsilon_{eff}(\vec{q},\omega)} \tag{25}$$

$\varepsilon_{eff}(\vec{q},\omega)$ is given by the diagonal element of the dielectric matrix $\varepsilon(\vec{q},\vec{q},\omega)$ modified by local field corrections. These local field corrections can be neglected in calculating the plasmon damping as shown by Sturm (1977). The plasmon line width is defined in the usual way using the energy loss function

$$\frac{1}{2}Im\frac{1}{\epsilon_{eff}(\vec{q},\omega_p(\vec{q}))}=Im\frac{1}{\epsilon_{eff}(\vec{q},\omega_p(\vec{q}))\pm\dfrac{\Gamma(\vec{q})}{2h}}\tag{26}$$

Then we obtain for the plasmon damping

$$\Gamma(\vec{q})=2h\frac{Im\,\epsilon_{eff}(\vec{q},\omega_p(\vec{q}))}{\left|\dfrac{\partial\epsilon_{eff}(\vec{q},\omega)}{\partial\omega}\right|_{\omega=\omega_p(\vec{q})}}\tag{27}$$

The interband contribution to the imaginary part of $\epsilon_{eff}(\vec{q},\omega)$, which is quadratic in the weak model potential may be expressed in terms of Bloch integrals

$$Im\,\epsilon_{eff}(\vec{q},\omega)=\frac{4\pi^2e^2}{q^2}\frac{1}{\Omega}\sum_{\vec{k},\vec{K}}\theta[E_{\vec{k}+\vec{q}+\vec{K}}-E_F]\theta[E_F-E_{\vec{k}}]\,|B^{\vec{K}}(\vec{k},\vec{k}+\vec{q}+\vec{K})|^2\delta(E_{\vec{k}+\vec{q}+\vec{K}}-E_{\vec{k}}-h\omega)\tag{28}$$

where the Bloch integral is defined as

$$B^{\vec{K}}(\vec{k},\vec{k}')=\frac{1}{\Omega_0}\int_{\Omega_0}u_{\vec{k}}^*(\vec{r})u_{\vec{k}'}'(\vec{r})e^{i\vec{K}\vec{r}}d^3\vec{r}\tag{29}$$

Ω_0 is the volume of the unit cell, $u_{\vec{k}}(\vec{r})$ is the periodic part of the Bloch function, and $\vec{K}$ is a vector of the reciprocal lattice. In calculating the plasmon damping the Bloch energies in (28) are replaced by simple free electron expressions. Using perturbation theory for nearly degenerate states (two-band model) we obtain for the square of the Bloch integral

$$|B^{\vec{K}}(\vec{k},\vec{k}')|^2=|V_{\vec{K}}|^2\frac{(D_{\vec{k}}+D_{\vec{k}'})^2}{(D_{\vec{k}}^2+|V_{\vec{K}}|^2)(D_{\vec{k}'}^2+|V_{\vec{K}}|^2)}\tag{30}$$

with $D_{\vec{k}}=E_{\vec{k}}-\hat{E}_{\vec{k}}$. $V_{\vec{K}}$ are the Fourier coefficients of the local model potential. $\hat{E}_{\vec{k}}$ means the well-known square-root expression for the Bloch energy (Ashcroft & Mermin, 1976). In the denominator of (27) ϵ_{eff} can be replaced by the RPA dielectric function. In order to obtain a formula for polycrystalline materials the expression (28) should be averaged over all directions of $\vec{K}$ $(d\Omega_{\vec{K}}=d\cos\vartheta d\varphi,\ \vartheta=\angle(\vec{K},\vec{q}+\vec{K}))$. Then we arrive at the final formula for the plasmon damping (Rösler & Brauer, 1981)

$$\Gamma(q)=\frac{h}{\pi a_B q^3\left|\dfrac{\partial\epsilon_1(q,\omega)}{\partial\omega}\right|_{\omega_p(q)}}\sum_{[\vec{K}]}n_{[\vec{K}]}\frac{1}{|\vec{K}|}\int_0^{k_F}dkk\times$$

$$\times\int_{|k-q|}^{k+q}dp\Theta[p-\left||\vec{K}|-\sqrt{k^2+\frac{2m}{h}\omega_p(q)}\right|]\int_0^{2\pi}d\phi\,|B^{\vec{K}}(\vec{k},\vec{k}+\vec{q}+\vec{K})|^2\tag{31}$$

where $p=|\vec{k}+\vec{q}|$. $n_{[\vec{K}]}$ is the number of reciprocal lattice vectors of equal length and a_B is the Bohr radius. Contributions from nearest, next nearest, and third nearest neighbours must be included in the calculation of the plasmon damping. But only the Fourier coefficients $|V_{K_1}|$ and $|V_{K_2}|$ are defined by de Haas-van Alphen data. Most of the relations given here must be used in the calculation of the excitation of conduction electrons by plasmon decay.

Transition Functions and Mean Free Paths

Mean free paths (mfp) and transition functions are the basic quantities for the description of transport of inner excited electrons. All these quantities can be calculated starting from the transition probability (20). The total transition function which appears in the transport equation (7) can be written as

$$W^\sigma(\vec{k},\vec{k}') = W^{\sigma(inel)}(\vec{k},\vec{k}') + W^{el}(\vec{k},\vec{k}') \tag{32}$$

An atomic picture will be used in order to describe the elastic scattering. We have for the elastic transition function in suitable variables

$$W^{el}(E,E',\cos\vartheta) = N_{at}\nu(E)\frac{d\sigma(E,\vartheta)}{d\Omega}\delta(E-E') \tag{33}$$

where $\vartheta = \measuredangle(\vec{k},\vec{k}')$. N_{at} is the density of randomly distributed atoms. The differential scattering cross section for elastic scattering can be obtained from a partial wave analysis

$$\frac{d\sigma(E,\vartheta)}{d\Omega} = \frac{1}{k^2}\left|\sum_{l=0}^{\infty}(2l+1)\sin\delta_l e^{i\delta_l}P_l(\cos\vartheta)\right|^2 \tag{34}$$

P_l is the lth Legendre polynomial and δ_l is the phase shift related to the lth partial wave. The phase shifts for Al used here are obtained from an improved version of Pendry's (1974) program based on a muffin-tin scheme including exchange and correlation. The density of atoms is taken as $N_{at}=6.07\cdot10^{22}$ cm^{-3}.

By summation of (20) over the initial states $\vec{k}_2$ ($<k_F$) (including spin) and different final states we obtain the transition probability

$$W_{\vec{k}_1-\vec{k}_1'} = \sum_{\substack{\vec{k}_2(<k_F) \\ \vec{k}_2'(>k_F)}} W_{\vec{k}_1\vec{k}_2-\vec{k}_1'\vec{k}_2'} \tag{35}$$

and the excitation probability

$$W^s_{\vec{k}_1-\vec{k}_2'} = \sum_{\substack{\vec{k}_2(<k_F) \\ \vec{k}_1'(>k_F)}} W_{\vec{k}_1\vec{k}_2-\vec{k}_1'\vec{k}_2'} \tag{36}$$

Then, the total inelastic transition function is given by both contributions

$$W^{\sigma(inel)}(\vec{k}',\vec{k}) = W_{\vec{k}-\vec{k}'} + W^s_{\vec{k}\vec{k}'} \tag{37}$$

In the description of transport of excited electrons we will only taken into account the single particle processes and plasmon processes which are related to the system of conduction electrons. Explicit expressions for the different transition and excitation probabilities (W_e, W_e^s, W_p, W_p^s) are given by Rösler & Brauer (1991).

For the total mfp of an electron with energy E we may write

$$\frac{1}{l(E)} = \frac{1}{l_{inel}(E)} + \frac{1}{l_{el}(E)} \tag{38}$$

Using the total elastic cross section

$$\sigma_{el}(E) = \int d\Omega \frac{d\sigma(E,\vartheta)}{d\Omega} \tag{39}$$

the elastic mfp can be written as

$$\frac{1}{l_{el}(E)} = N_{at}\sigma_{el}(E) \tag{40}$$

Within the partial wave analysis we have the well-known representation for σ_{el}:

$$\sigma_{el}(E) = \frac{4\pi}{k^2} \sum_{l=0}^{\infty} (2l+1)\sin^2\delta_l \tag{41}$$

The elastic mfp is obtained using the phase shifts of Pendry mentioned before. There are only small differences between these results and the corresponding ones using in other calculations (Ganachaud & Cailler, 1979; Devooght et al., 1991) starting in the case of Al from the atomic potential of Smrčka (1970).

The inelastic mfp is defined in the usual way by the transition probability (35)

$$\frac{1}{l_{inel}(E)} = \frac{1}{v(E)} \sum_{\vec{k}'} W_{\vec{k}-\vec{k}'} \tag{42}$$

Different scattering mechanisms contribute to l_{inel}

$$\frac{1}{l_{inel}(E)} = \frac{1}{l_e(E)} + \frac{1}{l_p(E)} + \frac{1}{l_c(E)} \tag{43}$$

The first two contributions which are related to the system of conduction electrons can be obtained from

$$W_{\vec{k}-\vec{k}'} = -\frac{8\pi e^2}{h} \frac{1}{\Omega} \sum_{\vec{q}} \frac{1}{q^2} |<\vec{k}'|e^{i\vec{q}\vec{r}}|\vec{k}>|^2 Im \frac{1}{\epsilon(q,E_{\vec{k}}-E_{\vec{k}'}+i0^+)} \tag{44}$$

using the RPA dielectric function (21 to 23). Explicit formulas are given by Quinn (1962).

In fig. 5 we have plotted the different contributions to l_{inel} together with the total inelastic mfp. Also the core contribution l_c is shown using the values given by Ashley et al. (1979). In a first attempt we can neglect core excitations in the transport process because $l_c(E) \gg l_e, l_p$. in the 100 eV range. For the same reason the contribution of core processes to the transition function $W^{\sigma(inel)}$ was not included.

It is shown in fig. 6 that both elastic and inelastic mfp are of the same order of magnitude in the relevant energy range. Only at very low energies l_{el} is dominant.

EXCITATION PROCESSES

The interaction of the incident ion with the system of target electrons leads to different possibilities of generating excited electrons. In this way the parameters of the incident beam enter the description of ion induced electron emission. Four excitation mechanisms

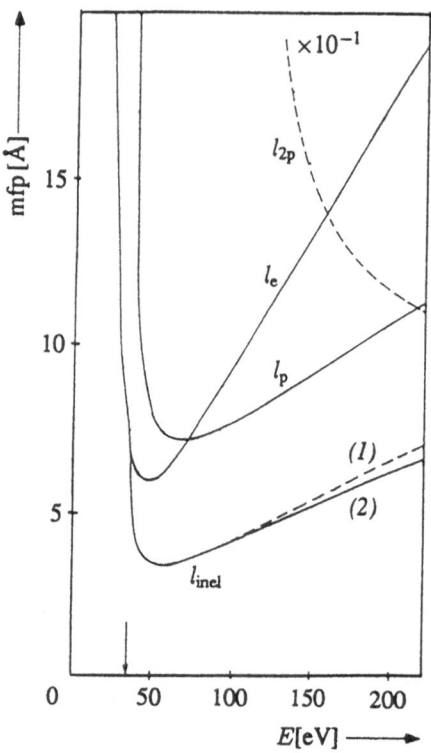

Figure 5. Energy dependence of the inelastic mfp of electrons in Al. Comparison of different contributions. l_{inel} is the total inelastic mfp including (curve 2) and neglecting (curve 1) the core contribution. l_c is almost completely given by the 2p-contribution l_{2p}. The arrow indicates the plasmon threshold

will be discussed:

- excitation of conduction electrons via screened ion-electron interaction (e)
- excitation of conduction electrons by decay of bulk plasmons generated by the incident ion (p)
- excitation of core electrons (c)
- excitation by Auger processes (a) which are related to the excitation of inner shell electrons

40

Besides the excitation processes S_e and S_p which are related to the system of conduction electrons the other excitation processes S_c and S_a are related to the excitation of core electrons. This will be shown schematically in fig. 7.

The total excitation function may be expressed as

$$S(E_0;\vec{k}) = \sum_i S_i(E_0;\vec{k}); \qquad i=e,p,c,a \tag{45}$$

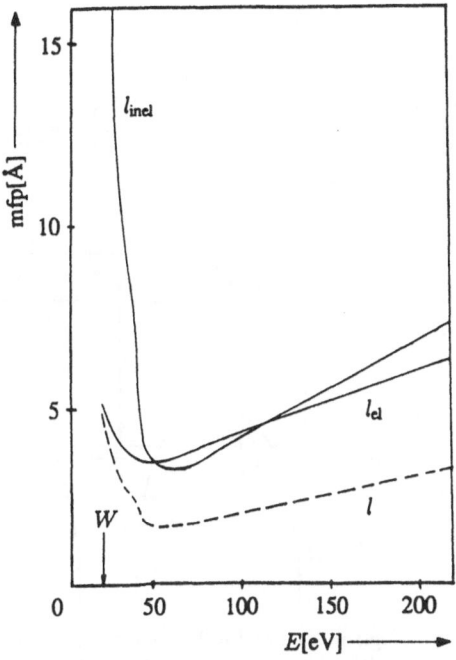

Figure 6. Energy dependence of the total mfp of electrons $l(E)$ in Al (dashed curve). Comparison of elastic and inelastic mfp. The arrow indicates the vacuum level

The number of electrons entering the state $\vec{k}$ by the impinging beam is given by the excitation probability (36)

$$S(E_0;\vec{k}) = \frac{1}{v_0} W^s_{\vec{k}_0,\vec{k}} = \frac{1}{v_0} \sum_{\vec{k}'_0,\vec{k}'(<k_F)} W_{\vec{k}_0\vec{k}'-\vec{k}'_0\vec{k}} \tag{46}$$

where $\vec{k}_0$ and v_0 denote the wave number and the velocity of the incoming ion.

With the simplifying assumption that the impinging ion before $(\vec{k}_0)$ and after $(\vec{k}_0')$ the scattering process is in a plane wave state

$$<\vec{k}_0'|e^{i\vec{q}\vec{r}}|\vec{k}_0>=\delta_{\vec{k}_0',\vec{k}_0+\vec{q}} \; ; \qquad E_{\vec{k}}^i=\frac{h^2k^2}{2M} \; ; \qquad E_0=E_{\vec{k}_0}^i \qquad (47)$$

we obtain the expression

$$S(E_0;\vec{k})=\frac{64\pi^3e^4}{hv_0}\frac{1}{\Omega^2}\sum_{\vec{q},\vec{k}'(<k_F)}\frac{|<\vec{k}|e^{-i\vec{q}\vec{r}}|\vec{k}'>|^2}{q^4|\epsilon(q,E_0-E_{\vec{k}_0+\vec{q}})|^2}\delta(E_0+E_{\vec{k}'}-E_{\vec{k}_0+\vec{q}}-E_{\vec{k}}) \qquad (48)$$

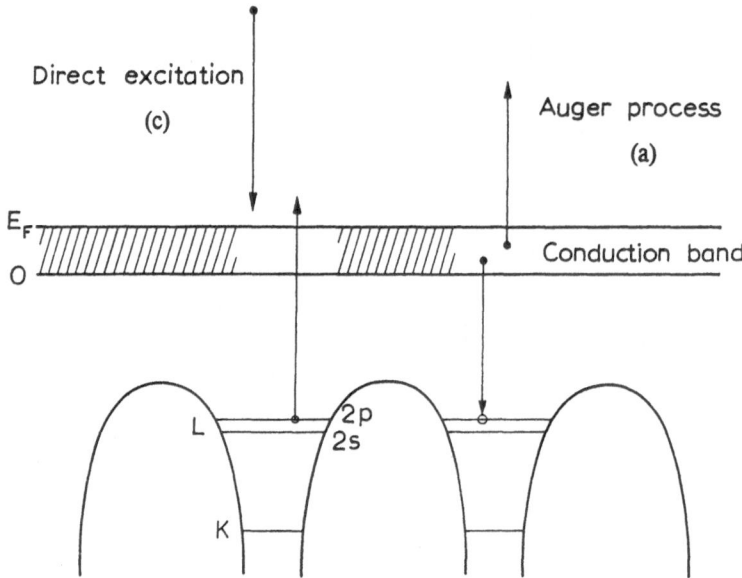

Figure 7. Schematic energy level diagram of a metal including direct excitation of core electrons as well as the related Auger processes with participation of conduction electrons

In order to evaluate the different excitation functions various approximations are necessary concerning the electronic transition matrix element, the dielectric function, and the electron energies.

Excitation of Single Conduction Electrons

In this case the free-electron-gas model is sufficient in a first attempt to describe the excitation rate. Then, the transition matrix element is calculated with plane waves and the energies are given by simple parabolic expressions. The screening is described using the RPA dielectric function (21-23). The following expression results

$$S_e(E_0;\vec{k})=\frac{64\pi^3e^4M}{hk_0}\frac{1}{\Omega}\sum_{\vec{k}'(<k_F)}\frac{1}{|\vec{k}'-\vec{k}|^4|\epsilon(|\vec{k}'-\vec{k}|,E_{\vec{k}'}-E_{\vec{k}})|^2}\delta(E_{\vec{k}_0+\vec{k}'-\vec{k}}^i+E_{\vec{k}}-E_{\vec{k}_0}^i-E_{\vec{k}'}) \qquad (49)$$

It can be shown (Rösler & Brauer, 1984, 1991) the excitation is restricted to angles $\Theta = \measuredangle(\vec{k}, \vec{k}_0)$ which fulfill the relation

$$\cos\Theta_1 \leq \cos\Theta \leq Min[\cos\Theta_2, 1] \tag{50}$$

with

$$\cos\Theta_{1,2} = \frac{1}{2kk_0}\left[(k^2 - k_F^2)\left(\frac{M}{m} + 1\right) \mp 2k_F\sqrt{k_0^2 - \frac{M}{m}(k^2 - k_F^2)}\right] \tag{51}$$

In fig. 8 the angular dependence of the excitation from the conduction band is shown at different impact energies. In contrast to the case of electron impact there is an excitation in the direction of the primary beam owing to the large mass difference of colliding particles. At higher impact energies, however, the excitation takes place preferably perpendicular to thr direction of the primary beam at all relevant excitation energies.

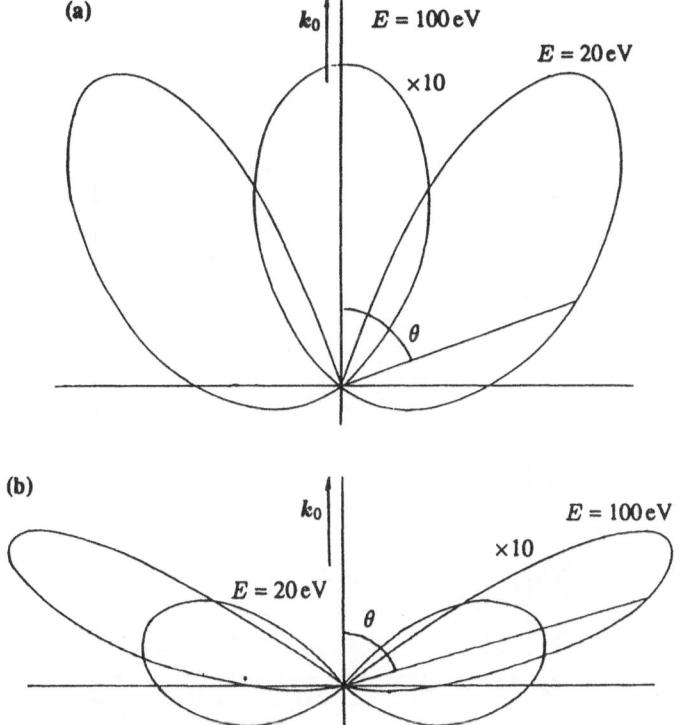

Figure 8. Angular dependence of excitation by dynamically screened proton-electron interaction for $E = 20$ eV and 100 eV at $E_0 = 40$ keV (a) and 500 keV (b). $\vec{k}_0$ is the wave vector of the incident proton and Θ is the excitation angle

Excitation of single conduction electrons is possible if $\cos\Theta_1 < 1$. This leads to an upper boundary E_m for the excitation energy

$$E_m = \frac{h^2}{2m}\left[\frac{2k_0 + k_F(\frac{M}{m}-1)}{\frac{M}{m}+1}\right]^2 \cong 2m\left(v_0 + \frac{v_F}{2}\right)^2 \qquad (52)$$

where v_F is the Fermi velocity. The condition that E_m must be larger than the vacuum level results in the well-known formula for the threshold velocity of the impinging ion to excite conduction electrons

$$v_{th} = \frac{v_F}{2}\left[\sqrt{1 + \frac{\Phi}{E_F}} - 1\right] \qquad (53)$$

given for the first time by Baragiola et al. (1979).

In fig. 9 we have plotted the upper boundary E_m for the excitation of single conduction electrons by different ions as a function of the impact energy. With increasing ion mass the upper boundary decreases for a given E_0.

In fig. 10 an energy momentum diagram is shown which contains besides the elementary excitations of the free electron gas (fig. 4) some curves for the maximum

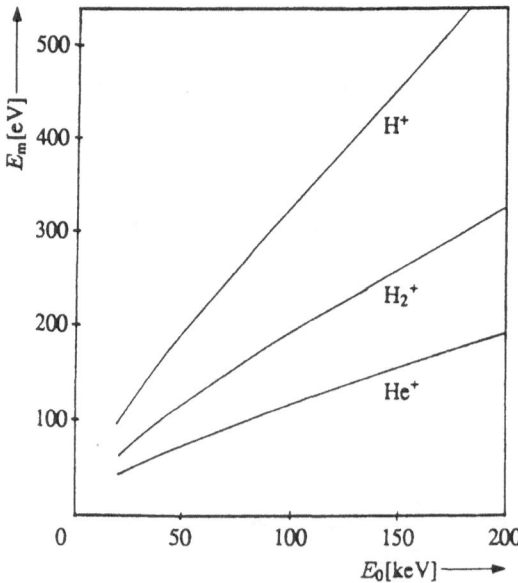

Figure 9. Upper boundary for the excitation energy of single conduction electrons as a function of impact energy for different ions

energy transfer of the ion (H^+ in our case)

$$\Delta_q(E_0) = \frac{\hbar^2 q}{2M}(2k_0 - q) \qquad (54)$$

corresponding to different impact energies. We obtain from the crossing point of the maximum energy transfer curve with the high momentum border of the pair excitation spectrum the upper boundary of the pair excitation energy E_m^{pair} (see fig. 9). If we add the Fermi energy we comes out exactly with the same expression (52) as before.

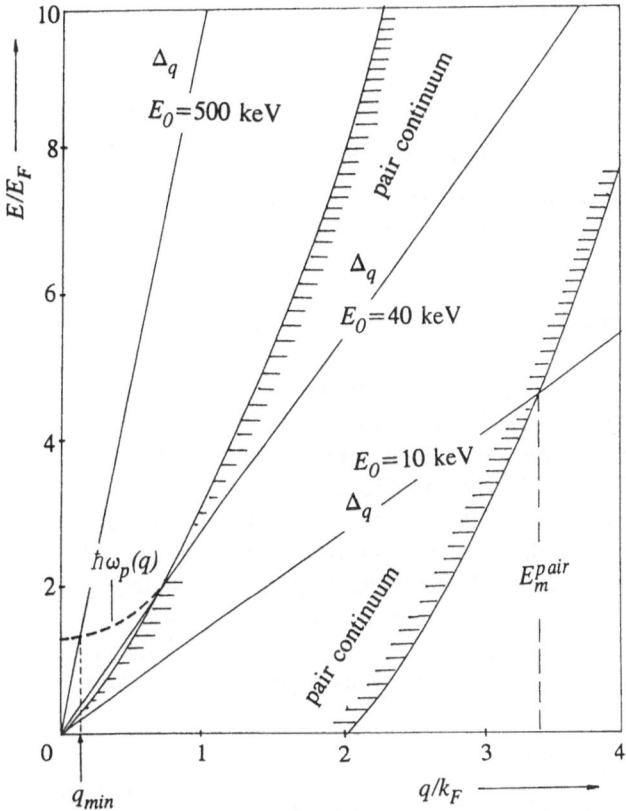

Figure 10. Elementary excitation spectrum of the system of conduction electrons in metals in relation to the maximum energy transfer (54) for different impact energies. E_m^{pair} is the maximum pair excitation energy (depending on the impact energy). q_{min} is the minimum wave number related to decay of plasmons via interband transitions in the case of plasmon excitation by the incident ion (see next chapter)

Finally, in fig. 11 the energy dependent excitation function which is defined by the angular integral of the excitation function is shown at different impact energies. We obtain in all cases a pronounced peak around 35 eV (above the bottom of conduction band) due to the resonant behavior of $1/|\epsilon(q,\omega)|^2$.

In order to calculate the number of excited electrons with sufficient accuracy it is important to take into account dynamic screening. Neglecting screening (i.e. $\epsilon = 1$) we obtain an overestimation of excited electrons at low energies. In the case of Thomas-Fermi screening the excitation rate is underestimated at all excitation energies.

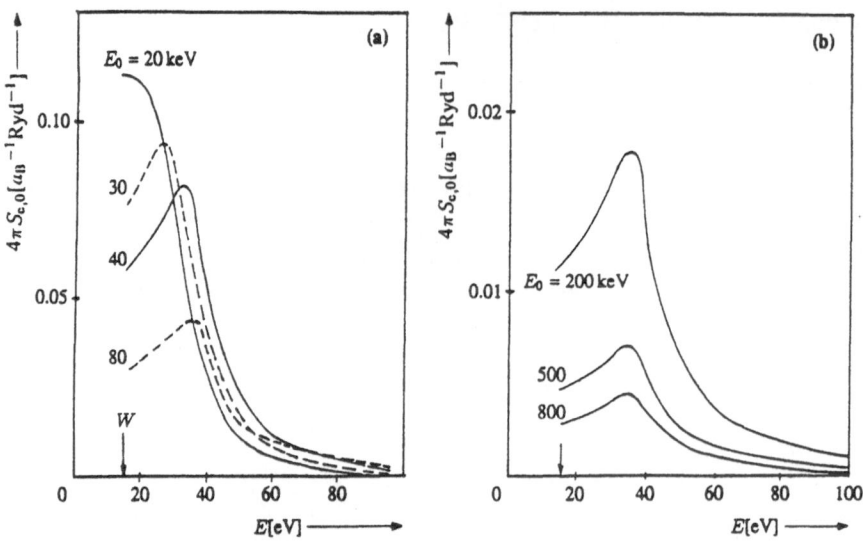

Figure 11. Energy dependent excitation function for dynamically screened ion-electron interaction. Proton impact on Al at low (a) and medium (b) impact energies. The arrow indicates the vacuum level

Excitation by Decay of Plasmons

In the following we consider the excitation of conduction electrons by decay of plasmons generated by the incident ion. The first microscopical description of this excitation process is given by Chung and Everhart (1977). Here we will follow the representation given by Rösler and Brauer (1981, 1984, 1991) which differs in some details from the evaluation of the excitation rate obtained by Chung and Everhart (1977).

As mentioned in the chapter before interband transitions of conduction electrons are the most important processes which determine the plasmon damping. In order to calculate their contribution to the total excitation rate we can start from (48). In analogy to the evaluation of the interband contribution to the imaginary part of the dielectric function the transition matrix element which describes the excitation process may be expressed by Bloch integrals (29). We have

$$S_p(E_0, \vec{k}) = \frac{64\pi^3 e^4 M}{h^2 k_0} \frac{1}{\Omega} \sum_{\vec{K}, \vec{q}(<q_c)} \frac{|B^{\vec{K}}(\vec{k}, \vec{k}+\vec{q}+\vec{K})|^2}{q^4 |\epsilon(q, E_0 - E^i_{\vec{k}_0+\vec{q}})|^2} \Theta(E_F - [E^i_{\vec{k}_0+\vec{q}} - E_0 + E_{\vec{k}}]) \\ \times \delta(E^i_{\vec{k}_0+\vec{q}} + E_{\vec{k}} - E_0 - E_{\vec{k}+\vec{q}+\vec{K}})$$

(55)

Using the representation (30) of the square of the Bloch integral the different steps of the further evaluation of (55) can be performed as described by Rösler and Brauer (1984, 1991). Especially, we take advantage the resonance structur of $1/|\epsilon(q,\omega)|^2$ in the frequency region about $\omega_p(q)$

$$\frac{1}{|\epsilon(q,\Delta)|^2} = \frac{h^2}{\left(\left|\frac{\partial \epsilon_1(q,\omega)}{\partial \omega}\right|_{\omega_p(q)}\right)^2} \frac{1}{(\Delta - h\omega_p(q))^2 + \left(\frac{\Gamma(q)}{2}\right)^2}$$

(56)

46

in order to carry out the integral over the energy transfer Δ. The relevant energy range is restricted on a region of width $\Gamma(q)$ about the plasmon energy. The momentum integration is restricted to the region from q_{min} to q_c. q_{min} is the maximum of the crossing points of the plasmon line with the maximum energy transfer or the lower boundary of the Δ-integration (which follows from the unit step function in (55): $\Theta(\Delta-[E_{\bar{k}}-E_F])$).

For the further discussion of the excitation by plasmon decay fig. 10 is helpful. The excitation of conduction electrons by plasmon decay via interband transitions can only take place if the Δ-curve intersects the region of the damped plasmon for $q < q_c$. This leads to a minimum ion impact energy E_0^{min} for plasmon excitation

$$E_0^{min} = E_F \frac{M}{m}\left(1+\frac{q_c}{2k_F}\right)^2 \tag{57}$$

We obtain for different projectile ions: $E_0^{min} \approx 40$ keV for H^+, $E_0^{min} \approx 80$ keV for H_2^+, and $E_0^{min} \approx 1600$ keV for Ar^+. It follows from (57) that for impact energies below this value the ion is not able to excite plasmons. Nevertheless, we observe also in such cases plasmon shoulders in the energy distribution of ejected electrons for Ar^+-impact at $E_0=200$ keV and Xe^+-impact at $E_0=800$ keV (see fig. 2). In these cases the plasmon shoulder should be attributed to the decay of plasmons which are generated by energetic excited electrons.

Fig. 12 shows the angular dependence of the excitation by plasmon decay. At low impact energy the excitation takes place preferably in the forward direction. With increasing impact energy also excitation in the backward direction occurs.

The energy dependent excitation function is shown in fig. 13 for proton impact on Al. We obtain remarkable differences in the behavior of the energy distribution at low and high impact energies. These differences are related to the relative positions of the limits of the energy transfer Δ ($E-E_F \le \Delta \le \Delta_q(E_0)$) compared with the plasmon line as shown in fig. 10. The position of the plasmon shoulder is determined by the strong decrease of the

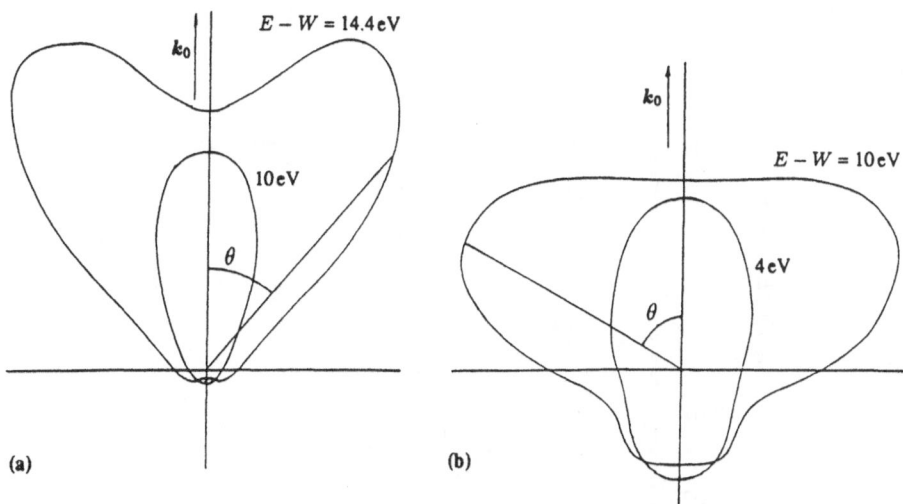

Figure 12. Angular dependence of the excitation by plasmon decay for different excitation energies. Proton impact on AL at $E_0=60$ keV (a) and 500 keV (b). $\vec{k}_0$ is the wave vector of the incident proton and Θ is the excitation angle

excitation rate above $\hbar\omega_p(q_{min})+E_F$ where q_{min} (shown in fig. 10) is given by the crossing point of the energy transfer limits with the plasmon line. By reason of plasmon dispersion we obtain with lowering E_0 a shift of q_{min} and, therefore, of $\hbar\omega_p(q_{min})$ to higher values. This leads to a small shift of the plasmon shoulder to somewhat higher energies with decreasing impact energy. This tendency can be seen in the experimental results of Hippler (1988) as shown in fig. 14. With increasing impact energy we obtain $\hbar\omega_p(q_{min}) \Rightarrow \hbar\omega_p(0)$ which means that the position of the plasmon shoulder is independent from E_0. In every case the upper limit of excitation energy is given by $\hbar\omega_p(q_c)+E_F \approx 36$ eV.

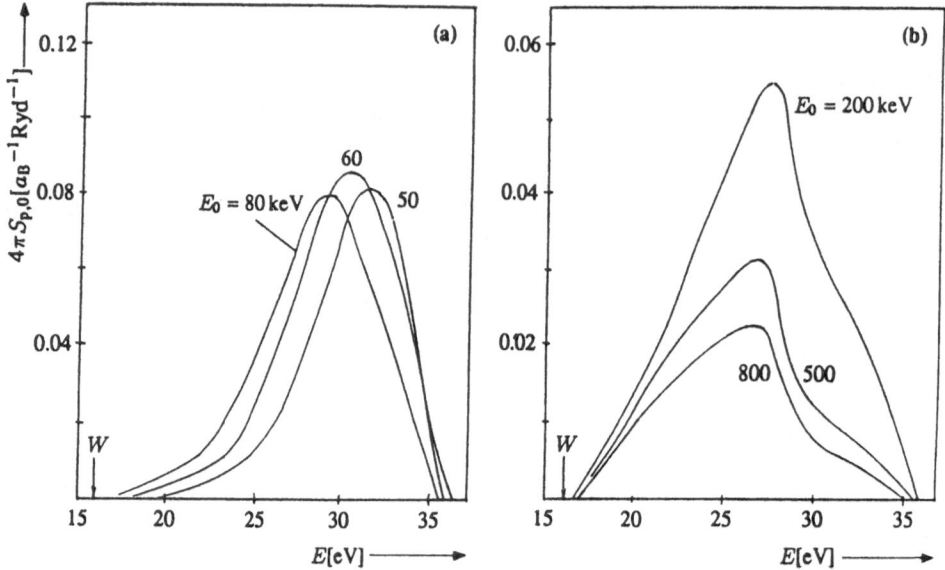

Figure 13. Energy dependent excitation function by plasmon decay for proton impact on Al at low (a) and medium (b) impact energy. The arrow indicates the vacuum level

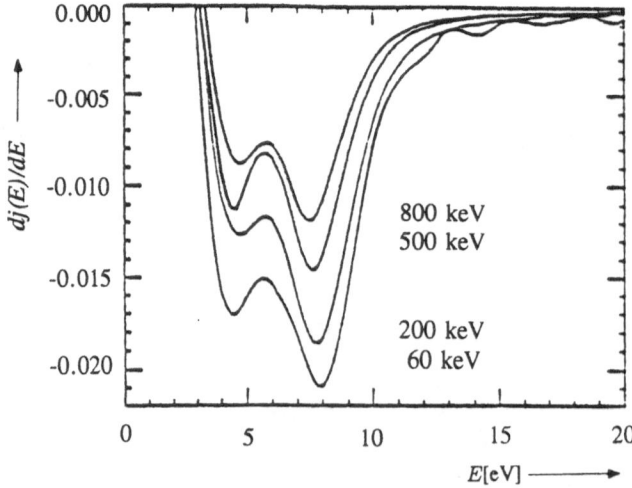

Figure 14. Derivative spectra of the energy distributions of ejected electrons for proton impact on Mg at different projectilec energies (Hippler, 1988)

48

Excitation of Core Electrons

Also in this case we can start from (48) for the evaluation of the excitation rate neglecting screening by reason of the large frequency argument in the dielectric function. We will employ a orthogonalized plane wave (OPW)-formalism as formulated in the theory of secondary electron emission for the first time by Fischbeck (1966) and Arendt (1969). The same formalism was applied by Rösler and Brauer (1984) to calculate the corresponding excitation rate in the case of ion impact on metals. In the transition matrix element we describe the core states by Bloch sums

$$\psi_{\vec{k}\nu} = G^{-\frac{3}{2}} \sum_{\vec{R}} e^{i\vec{k}\vec{R}} \Phi_{\nu}(\vec{r} - \vec{R}) \tag{58}$$

ν is the band index and G^3 is defined by $G^3 = \Omega/\Omega_0$. $\Phi_{\nu}(\vec{r} - \vec{R})$ is the atomic wave function centered at the lattice point $\vec{R}$. The excited states are described by OPW's

$$\psi_{\vec{k}}(\vec{r}) = \frac{1}{\sqrt{\Omega}} e^{i\vec{k}\vec{r}} - \sum_{\vec{k}',\nu',\vec{K}} b_{\vec{k}\nu'} \psi_{\vec{k}'\nu}(\vec{r}) \delta_{\vec{k},\vec{k}'+\vec{K}} \tag{59}$$

With

$$b_{\vec{k}\nu} = \frac{1}{\sqrt{\Omega_0}} \int_{\Omega_0} e^{i\vec{k}\vec{r}} \phi_{\nu}^{*}(\vec{r}) d^3\vec{r} \tag{60}$$

the different states entering the transition matrix element are orthogonal. We approximate the Bloch energies for the core states and the excited states by the corresponding atomic levels $E_{\vec{k}\nu} \approx E_{\nu} = E_{nl}$ (n, l are atomic quantum numbers) and simple parabolic expressions, respectively. In the evaluation of the transition matrix element overlap integrals of atomic functions centered at different lattice points can be neglected. In this way we obtain for the excitation function in suitable variables

$$S_c(E_0; E, \cos\theta) = \frac{k}{\pi^3 e^2 a_B^3 k_0^3} \left(\frac{M}{m}\right)^2 \sum_{\nu} \int_{k_0 - k_0'}^{k_0 + k_0'} \frac{dq}{q^3} \int_0^{2\pi} d\varphi \, |B_{\nu}(\vec{k}, \vec{q})|^2 \tag{61}$$

where the transition matrix element is defined by

$$B_{\nu}(\vec{k}, \vec{q}) = \int_{\Omega_0} e^{i\vec{q}\vec{r}} \phi_{\nu}(\vec{r}) \left[\frac{e^{-i\vec{k}\vec{r}}}{\sqrt{\Omega_0}} - \sum_{\nu'} b_{\vec{k}\nu}^{*} \Phi_{\nu}^{*}(\vec{r}) \right] d^3\vec{r} \tag{62}$$

and $\vec{q} = \vec{k}_0 - \vec{k}_0'$. k_0' is determined by the energy conservation law and φ is the polar angle of $\vec{k}_0'$ with respect to $\vec{k}_0$. Actual calculations was carried out for Al using Herman-Skillman atomic wave functions.

Fig. 15 shows the angular dependence of the excitation from core states. At low excitation energies (≤ 50 eV) the excitation is nearly isotropic. With increasing energy the excitation takes place preferably in the forward direction.

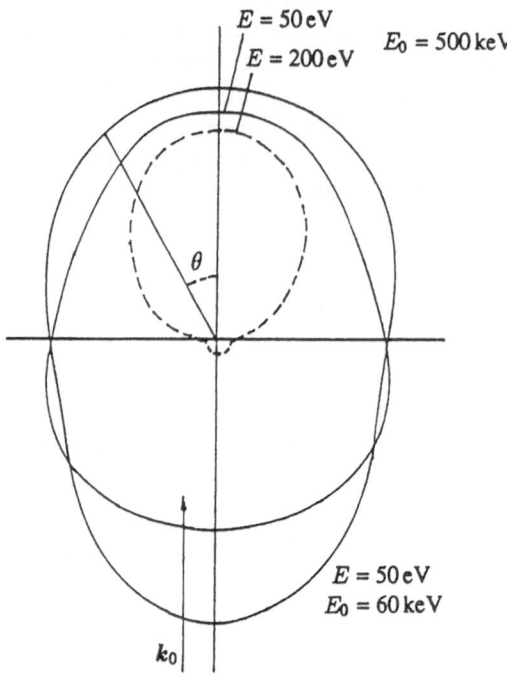

Figure 15. Angular dependence of inner shell excitations at different excitation energies. Proton impact on Al at E_0=60 keV and 500 keV. $\vec{k}_0$ is the wave vector of the incident proton and θ is the excitation angle

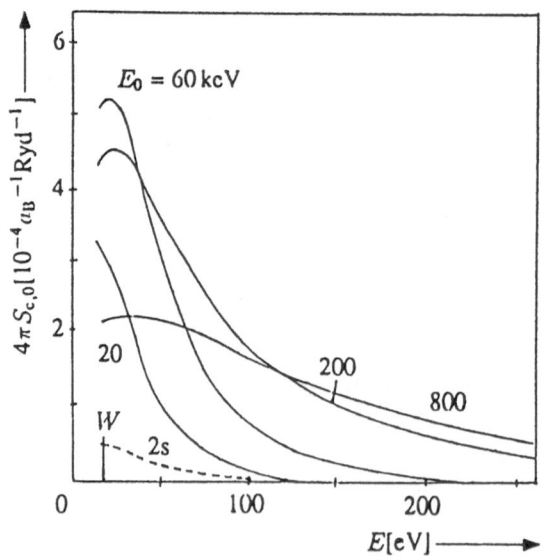

Figure 16. Energy dependent excitation function for inner shell excitations including the 2p- and the 2s-contributions. Proton impact on Al at different impact energies. For E_0=60 keV the 2s-contribution is shown (dashed line). The arrow indicates the vacuum level

The energy dependent excitation function is shown in fig. 16 for different proton impact energies. K-shell contributions can be neglected. The excitation rate is almost completely determined by the 2p-contribution. In order to confirm this the 2s-contribution for $E_0 = 60$ keV is also shown in this figure.

Excitation of Conduction Electrons via Auger Processes

The bombardment of metals with energetic ions produces via excitation of core electrons inner shell vacancies. These vacancies (e.g. in the L-shell of Al) are filled in a very short time by electrons from the conduction band. At the same time another conduction electron would be excited. This Auger process is shown schematically in fig. 7. In order to obtain statements about the role of Auger excitation processes with respect to the energy distribution of emerging electrons at low energies and the electron yield we start with a three-parameter model (Rösler, 1991; Rösler and Brauer, 1991)

$$S_a^v(E_0;E) = A_v(E_0) \frac{1}{\pi} \frac{\Gamma_v}{(E - E_v^a)^2 + \Gamma_v^2/4} \tag{63}$$

or we use a simple monoenergetic excitation

$$S_a^v(E_0;E) = A_v(E_0) \delta(E - E_v^a) \tag{64}$$

Width and position of the excitation are determined by the experimentally observed peak of directly emitted Auger electrons. The strength of the excitation $A_v(E_0)$ is determined by the assumption that the number of inner shell vacancies produced by the incoming ion is equal to the number of electrons excited via the Auger processes. This may be expressed as

$$\frac{1}{4\pi} \int S_c^v(E_0;\vec{k}) d^3\vec{k} = \int S_a^v(E_0;E) dE = A_v(E_0) \tag{65}$$

The integral on the left hand side is simply related to the contribution of core states v to the reciprocal mfp. In this way we obtain for the strength of excitation

$$A_v(E_0) = \frac{1}{4\pi l_v(E_0)} \tag{66}$$

For the positions of the L-Auger lines we have used $E_{2p}^a = 82.9$ eV (measured from bottom of conduction band) and $E_{2s}^a = 127.9$ eV, respectively. The width of the excitation function (63) is of minor importance. We obtain practically the same results for the number of inner excited electrons at low energies using (63) with $\Gamma_{2p} = 10$ eV or the δ-like excitation (64).

Comparison of Different Excitation Mechanisms

Because all the discussed excitation processes occur simultaneously it is interesting to compare the various excitation rates with respect to their energy and angular dependence. As shown by fig. 8, 12, and 15 the excitation processes S_e, S_p, and S_c are more or less anisotropic in the relevant energy range. Therefore, it will be interesting what happens as a result of the interplay of excitation and scattering processes with respect to the emission characteristics at low energies. This question will be answered if we discuss the results obtained by the solution of the transport equation.

With respect to the energy dependence the various excitation functions show a remarkable different behavior at all impact energies. In fig. 17 the energy dependent excitation functions are summarized at $E_0=500$ keV and 10 MeV.

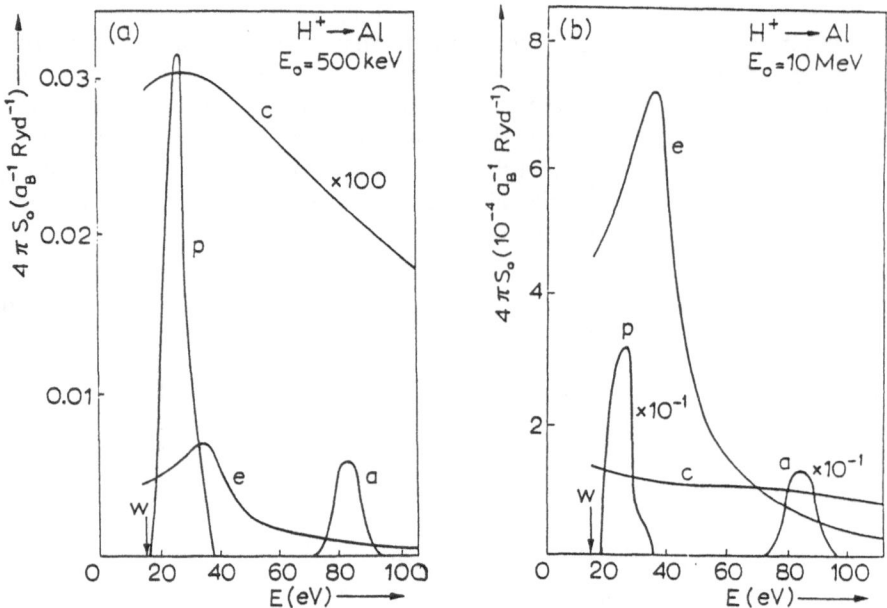

Figure 17. Energy dependence of different excitation processes for proton impact on Al at $E_0=500$ keV (a) and 10 MeV (b). Excitation mechanisms: excitation of single conduction electrons (e), excitation by plasmon decay (p), excitation of inner shell electrons (c), and excitation via Auger processes (a). The arrow indicates the vacuum level

Excitation by decay of plasmons and the excitation via Auger processes are restricted to relatively small energy intervals. The width of the excitation function by plasmon decay is determined in essence by the upper limit of excitation energies $\hbar\omega_p(q_c)+E_F$ (≈ 36 eV for Al). As mentioned before the width of the excitation via Auger processes is of minor importance with respect to the emission properties at low energies. By way of contrast for the excitation of single conduction and single core electrons we obtain broad energy distributions which are extended to relatively high excitation energies. We will see later that in the case of inner shell excitations the small number of electrons with high excitation energies lead to an important contribution to the number of escaping low energy electrons.

SOLUTION OF THE TRANSPORT EQUATION

If we know the excitation functions as well as the scattering functions and mean free paths we are able to solve the transport equation (7) in the ISD model. The treatment of the angular dependence of the emission problem by expansion into Legendre polynomials leads to the set of independent Volterra integral equations of second kind (15). These equations can be solved by standard numerical methods. The special features of the different excitation and scattering functions in relation to the solution procedure are discussed by Rösler and Brauer (1988, 1991).

At first we will discuss the results of a model calculation using a monoenergetic isotropic excitation (Rösler and Brauer, 1992). The model of excitation via Auger processes given by (64) is of this type. It is interesting in this case to compare the full solution of (15) (for $l=0$) with the different steps of an iterative solution of (15). This is shown in fig. 18. The iterative procedure can be related to real physical processes. Every step in the iteration procedure ($n=0,1,2,...$) corresponds to the inclusion of the corresponding number of scattering events for the excited electron before it leaves the target.

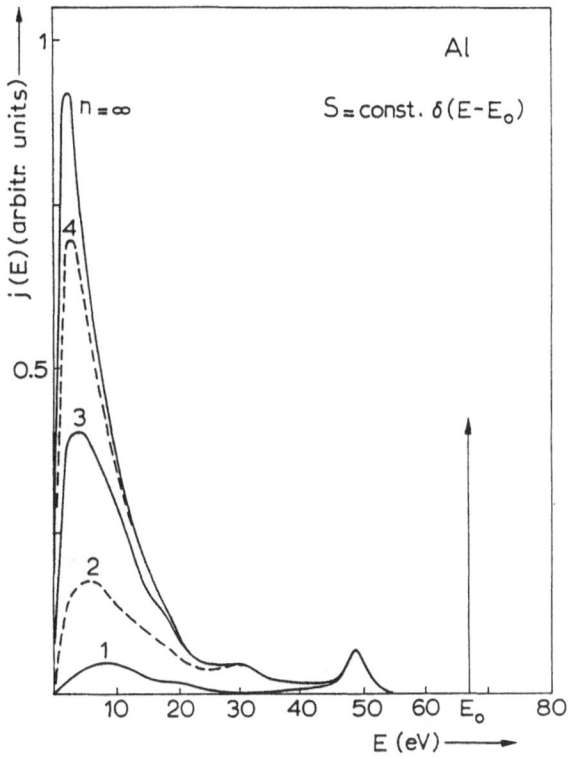

Figure 18. Accumulation of the energy distribution of ejected electrons in the case of an isotropic monoenergetic excitation function. The number n denotes the energy distribution including n-fold scattered electrons.

From fig. 18 it is obvious that despite the different energy dependence of the various excitation functions as shown in fig. 17 we can expect almost the same low energy distribution of emerging electrons. Furthermore, this figure demonstrates the important role of the cascade process for the accumulation of the energy distribution of emitted electrons.

In order to approach the full solution of (15) the number of iterative steps and therefore the number of scattering events for the excited electrons rises with increasing energy of the primary excited electrons. In contrast to the excitation by plasmon decay, which is restricted to low excitation energies, cascade processes are of special importance in the case of inner shell excitations because this excitation rate exceeds all the other contributions at high excitation energies.

A short comment to the convergence of the expansion of (7) with respect to Legendre polynomials is given in the last chapter. This problem is closely related to the influence of the angular dependence of the excitation rates on the emission properties.

RESULTS FOR ALUMINUM

With the solution of the transport equation the different measurable quantities can be obtained from the energy and angular dependent outer current density (11). Using the expansion (12) the energy distribution of emerging electrons as well as the electron yield are given by (16) and (17), respectively.

In fig. 19 the total energy distribution of escaping electrons is shown at low and intermediate impact energies. All the excitation mechanisms discussed before are included. Here we are interested only to the low energy range (below 20 eV). Special features which are related to the directly emitted electrons via Auger processes are not discussed here. With respect to the total number of escaping electrons these processes are of minor importance.

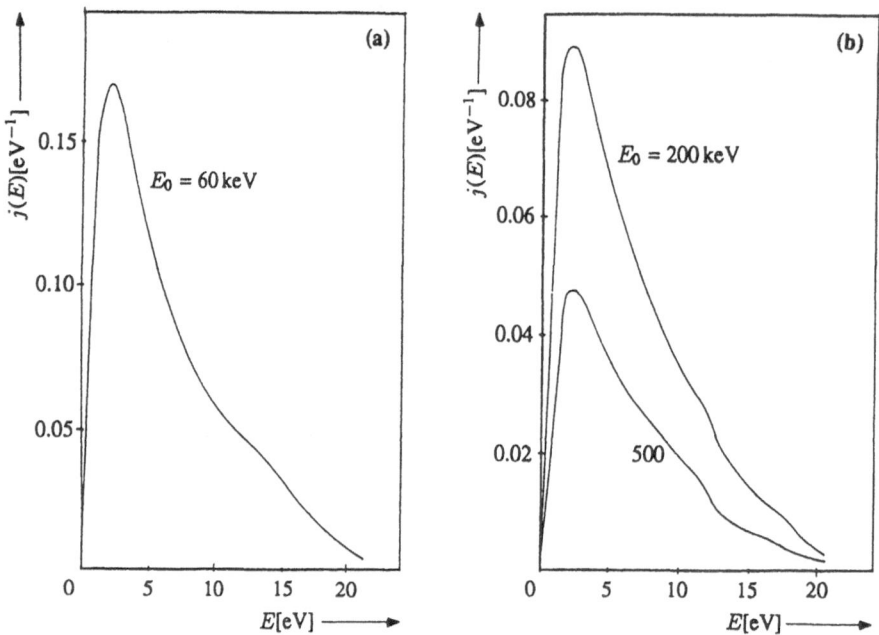

Figure 19. Energy distribution of emitted electrons at low (a) (E_0=60 keV) and intermediate (b) impact energies

In accordance with the experiments of Hasselkamp and Scharmann (1982) we obtain for al a maximum of the energy distribution of ejected electrons at ≈ 2 eV, the plasmon shoulder at 11 to 12 eV, and for the half-width of the distribution ≈ 8.5 eV. The predicted shift of the plasmon shoulder with decreasing impact energy to higher values (as can be seen in fig. 19a) is not confirmed by the experiments up to now. Only in the measurements of proton impact on Mg by Hippler (1988) a small shift of the plasmon shoulder with decreasing impact energy was found as shown in fig. 14.

In fig. 20 we have plotted the impact energy dependence of the different contributions (e,p,c,a) to the electron yield γ. At low impact energies (including the region of the stopping power maximum) the emission properties are governed only by the contributions from the conduction band (e,p) (Rösler and Brauer, 1992). At intermediate and high impact energies all the excitation mechanisms must be taken into account. For very high impact energies the excitation processes with participation of core states (c,a) dominate the electron emission.

The assumption seems to be justified that in the case of ion-induced kinetic electron emission the emission properties of all metals are determined at low impact energies by the excitations from the conduction band and at very high impact energies by the excitation via core states (Rösler and Brauer, 1992).

In fig. 21 we compare our results with the experimental ones shown in fig. 1. We obtain agreement with the measured yields at low impact energies. Also the position of the yield maximum at $E_0^{max} \approx 55$ keV is in accordance with the measurements of Svensson and

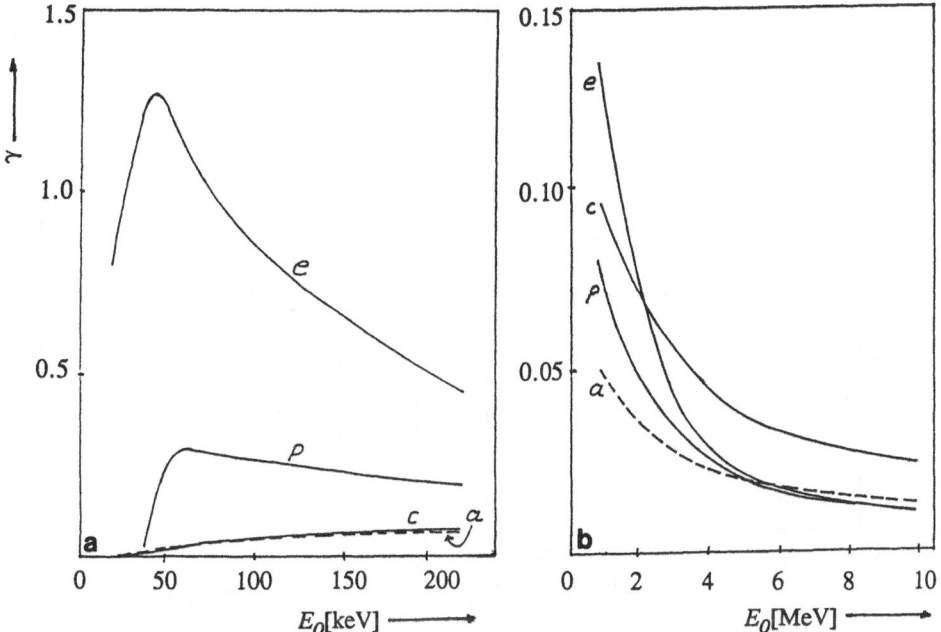

Figure 20. Impact energy dependence of the electron yield γ at low (a) and high (b) impact energies. Contributions from different excitation processes: excitation of single conduction electrons (e), excitation by plasmon decay (p), excitation of inner shell electrons (c), and excitation of conduction electrons via Auger processes (a)

Holmen (1981). However, it should be noted that also in the case of proton impact the charge state of the pojectile must be taken into account at low proton velocities (Echenique et al., 1988). This leads to a reduction of the excitation rate at low impact energies compared with the model of a point charge used in our treatment.

At high impact energies the theoretical results are clearly below the experimental values. One possible reason for this discrepancy is our choice of E_{max} in equation (15). We have used a value slightly above 400 eV. However, it seems to be probably that in the case of excitation of core electrons we must taken into account the electrons with excitation energies above this limit.

Besides the inelastic electron-electron scattering which is essential for the accumulation of electrons at low energies (cascade maximum) we will emphasize the important role of the elastic scattering processes. The elastic scattering is mainly responsible for the nearly isotropic angular distribution of internal electrons. This will be demonstrated by fig. 22. For the case of the strongly anisotropic excitation of single conduction electrons the angular distribution of the density of inner electrons $N(E, \vec{\Omega})$ is shown.

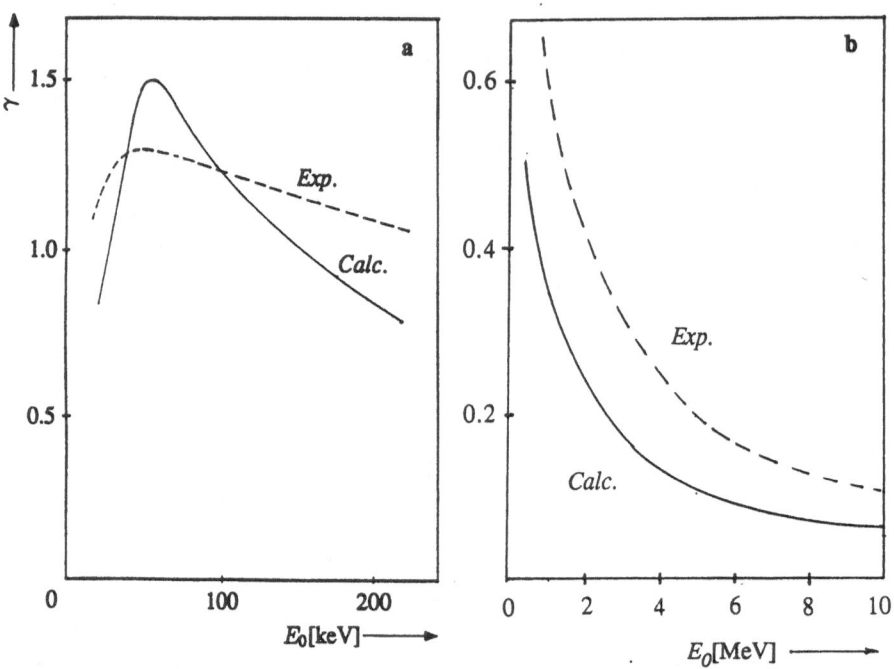

Figure 21. Impact energy dependence of the electron yield γ. Comparison of our calculations with experimental results (see fig. 1) at low (a) and high (b) impact energies

Neglecting elastic scattering we obtain an angular distribution of inner electrons which shows similarities with the anisotropic behavior of the corresponding excitation function. Taking into account elastic scattering we obtain a nearly isotropic distribution. However, the current density of escaping electrons $j(E, \vec{\Omega})$ is determined by that part of the internal distribution which is restricted to the escape cone. Within the escape cone we obtain a drastic enhancement of the number of inner excited electrons by reason of elastic scattering. This enhancement effect can be clearly seen in fig. 23. In this figure the impact energy dependence of the yield is shown at low impact energies. For all excitation mechanisms (e,p,c) we obtain a considerable enhancement of the yield if we take into account elastic scattering.

Finally , we will give a short comment to the angular dependence of the emission problem caused by the anisotropy of the excitation processes. When taking into account the elastic scattering it is sufficient to restrict the expansion of equation (7) with respect

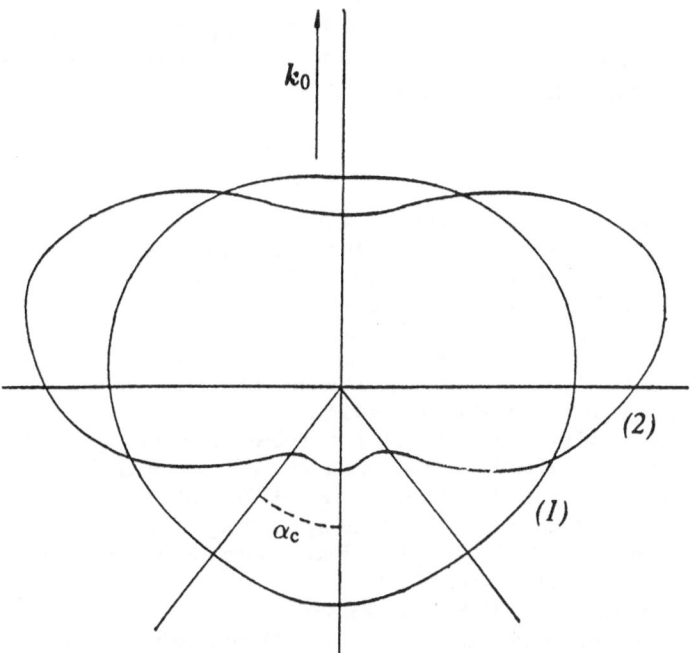

Figure 22. Angular distribution of the density $N(E,\overline{\Omega})$ of inner excited electrons for the case of excitation of single conduction electrons including (1) and neglecting (2) elastic scattering. $E = W + 10$ eV. $\overline{k}_0$ is the wave vector of the proton and α_c defines the escape cone

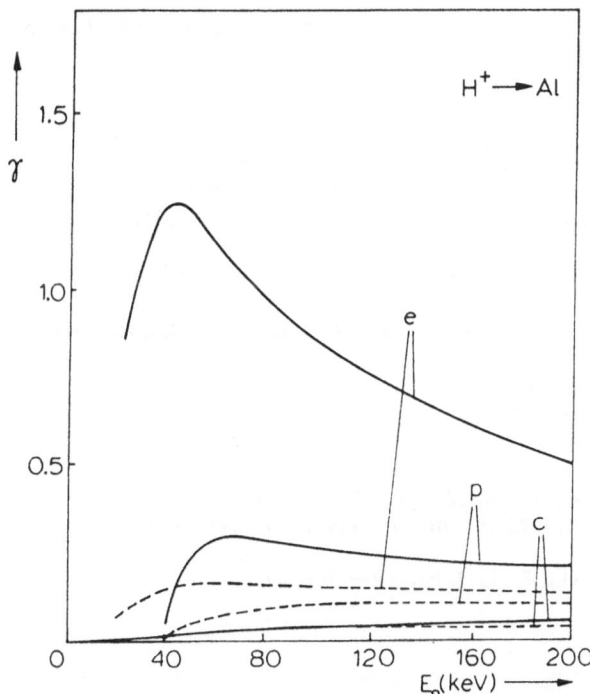

Figure 23. Impact energy dependence of the electron yield γ for different excitation mechanisms (e,p,c). Effect of elastic scattering at low impact energies. dashed lines without, solid lines with elastic scattering

to Legendre polynomials to the contributions from $l \leq 2$. Therefore, the strong anisotropy of the excitation process (e,c) is smeared out to a large extend by the combined action of inelastic and elastic scattering processes.

REFERENCES

Arendt,P., 1969, Phys. Status Solidi 31, 713

Ashcroft,N.W.,and Mermin,N.D., 1976, in *Solid State Physics* (Holt, Rinehart,and Winston, New York)

Baragiola,R.A.,Alonso,E.V.,and Oliva Florio,A., 1979, Phys.Rev. B 19, 121

Brauer,W.,and Rösler,M., 1985, Phys. Status Solidi (b) 131, 177

Chung,M.S.,and Everhart,T.E., 1977, Phys. Rev. B 15, 4699

Devooght,J.,Dubus,A.,and Dehaes,J.C., 1987, Phys. Rev. B 36, 5093

Devooght,J.,Dehaes,J.C.,Dubus,A.,Cailler,M.,and Ganachaud,J.P., 1991, in *Particle induced electron emission I,* Springer Tracts in Modern Physics 122, 67

Devooght,J.,Dubus,A.,and Dehaes,J.C., 1992, to be published in Nucl. Instrum.&Methods

Dubus,A.,Devooght,J.,and Dehaes,J.C., 1986, Nucl.Instrum.&Methods B 13, 623

Dubus,A.,Devooght,J.,and Dehaes,J.C., 1987, Phys. Rev. B 36, 5110

Dubus,A.,Devooght,J.,and Dehaes,J.C., 1990, Scanning Microscopy 4, 1

Echenique,P.M.,Flores,F.,and Ritchie,R.H., 1988, Nucl. Instrum.&Methods B 33,91

Fischbeck,H.J., 1966, Phys. Status Solidi 15, 387

Ganachaud,J.P.,and Cailler,M., 1979, Surf. Sci. 83, 498

Hasselkamp,D.,Lang,K.G.,Scharmann,A.,andStiller,N., 1981, Nucl.Instrum.&Methods 180, 349

Hasselkamp,D.,and Scharmann,A., 1982, Surf. Sci. 119, L388

Hasselkamp,D., 1991, in *Particle induced electron emission II,* Springer Tracts in Modern Physics 123, 1

Hippler,S., 1988, Thesis, Giessen, FRG

Hofer,W., 1990, in *Fundamental Beam Interactions with Solids for Microscopy, Microanalysis and Microtopography,* Scanning Microscopy Suppl. 4, 265

Koyama,A.,Shikata,T.,and Sakairi,H., 1981, Jpn. J. Appl. Phys. 20, 65

Lindhard,J., 1954, K. Dan. Vidensk. Selsk. Mat.-Fys.Medd. 18, 1

Paasch,G., 1969, Phys. Status Solidi 38, K123

Pendry,J.B., 1974, in *Low Energy Electron Diffraction,* (Academic Press, London and New York)

Puff,H., 1964, Phys. Status Solidi 4, 125

Quinn,J.J., 1962, Phys. Rev. 126, 1453

Rösler,M., 1991, Nucl.Instrum.&Methods B 58, 309

Rösler,M.,and Brauer,W., 1981, Phys. Status Solidi (b) 104, 161 and 575

Rösler,M.,and Brauer,W., 1984, Phys. Status Solidi (b) 126, 629

Rösler,M.,and Brauer,W., 1988, Phys. Status Solidi (b) 148, 213

Rösler,M.,and Brauer,W., 1991, in *Particle induced electron emission I,* Springer Tracts in Modern Physics 122, 1

Rösler,M.,and Brauer,W., 1992, Nucl. Instrum.&Methods B 67, 641

Schou,J., 1988, Scanning Microscopy 2, 607

Sigmund,P.,and Tougaard,S., 1981, in *Inelastic Particle-Surface Collisions,* Springer Series in Chemical Physics 17, 2

Sturm,K., 1976, Z. Phys. B 25, 247

Sturm,K., 1977, Z. Phys. B 28, 1

Sturm,K., 1982, Adv. Phys. 31, 1

Smrčka,L., 1970, Czech. J. Phys. B 20, 291

Svensson,B.,and Holmen,G., 1981, J. Appl. Phys. 52, 6928

Varga,P.,and Winter,H., 1992, in *Particle induced electron emission II,* Springer Tracts in Modern Physics 123, 149

Vashishta,P.,and Singwi,K.S., 1972, Phys. Rev. B 6, 875

PARTICLE-INDUCED ELECTRON EMISSION:

OPEN QUESTIONS, PITFALLS, AND A FEW ATTEMPTS AT ANSWERS

Peter Sigmund

Physics Department
Odense University
DK-5230 Odense M, Denmark

INTRODUCTION

This contribution addresses charged-particle-induced kinetic electron emission from solids in the classical geometry: An ion, electron, or positron beam hits a thick target, and the flux of electrons emitted from the bombarded surface is observed. Much available experimental information about this topic has been collected recently in two well-written reviews (Hofer, 1990; Hasselkamp, 1992). In addition, these papers also provide an overview over various attempts to arrive at a theoretical understanding. More theoretically-oriented surveys are likewise available (Sigmund & Tougaard, 1981; Schou, 1988).

Electron emission is a ubiquitous phenomenon in particle-solid interaction and therefore warrants understanding. Quantitative predictions are required in view of numerous applications. Yet the phenomenon is complex, and so is the theoretical analysis.

In principle, a considerable amount of input is needed for a comprehensive theory. This includes atomic data such as ionization cross sections for target atoms or molecules exposed to primary radiation. Even when the source of primary radiation is not an electron beam, cross sections for ionization by electrons are needed because of secondary ionizations. Also transport properties of fast and slow electrons enter on the input side, such as cross sections for elastic and inelastic scattering. Assumptions enter about the interplay between single-particle and collective excitations in the material. The significance of coherent scattering and the role of target impurities need to be considered. Surface properties are known to be important, in particular the energy barrier for emerging electrons and its dependence on surface morphology and chemical composition.

Ionization of Solids by Heavy Particles, Edited by
R.A. Baragiola, Plenum Press, New York, 1993

With all this input, what can one get on the output side? Well, kinetic electron emission is predominantly a multiple scattering phenomenon, i.e., deconvolution is notoriously difficult or impossible. On the positive side, one's hope is the existence of reasonably universal scaling laws. Indeed, electron yields show a dependence on particle energy not unlike that of the stopping power. The dependence of the yield on angle of incidence Θ is like $\cos^{-1}\Theta$ or a bit steeper up to fairly large angles. Energy spectra of emitted electrons show a considerable degree of universality when plotted in suitable variables, and angular distributions are cosine-like except for single-crystalline targets. The possibility of obtaining experimental data on emission statistics generated expectations to improve the understanding of fundamental processes, but so far, such data have been most powerful as a means of determining reliable values of *average* emission yields.

It seems that measurements of "true" secondary-electron currents (i.e., electrons emitted with energies less than some upper limit which is often taken to be 50 eV) from thick targets do not provide detailed information on parameters that are related to what is called the input side above: Ionization cross sections are best studied on gas targets or, if necessary, on thin films. Pertinent inelastic cross sections for electrons moving in condensed matter may be determined from photoelectron spectroscopy (Tougaard, 1988). Cross sections for elastic scattering of low-energy electrons in solids are hard to get at by any means, but measurements on secondary electrons offer particularly little promise in that respect. Numerous techniques are available to measure surface properties, and most of them offer more specific information than what can be deduced from true secondary-electron currents.

Conversely, the rather universal behavior on the output side must imply that comparatively few parameters determine the gross features of emitted electron currents. It appears worthwhile to trace these key quantities and to establish pertinent scaling laws. This has been the guideline in the theoretical work that I have conducted or supervised over limited periods (Schou, 1980; Sigmund & Tougaard, 1981). Evidently, a small number of representative materials need to be studied in detail to make sure that nothing essential is missing. This could perhaps be aluminium, germanium, and an insulator such as solid hydrogen or some biological material. There is hardly a solid enough basis yet for computing secondary-electron currents element after element and compound after compound.

The present paper is intended to be a followup to our 1981 summary. The main reason for the long delay is the fact that our work had implications on electron spectroscopy for surface analysis which we thought were more immediately important at the time (Tougaard & Sigmund, 1982, Tougaard, 1988). There have been significant theoretical developments in the meantime, such as model calculations on free-electron metals by Rösler & Brauer (1991) and several Monte Carlo studies that were summarized by Devoogt et al. (1991). At the same time, much of the material presented in our 1981 paper appears still relevant and by and large correct to the best of my knowledge. Therefore, only a few selected aspects will be taken up here. The reader who is interested first to get a wider view is referred to our earlier paper or to one of the other summaries mentioned above.

Electron emission is commonly described in terms of three main stages: Primary excitation, transport, and escape. This paper addresses the first two aspects and omits the third item. In addition, attention will be given to emission statistics in view of recent experimental activities. The paper concludes with more speculative considerations on recently observed tails of very-high-energy electrons emitted during low-energy heavy-ion bombardment.

PRIMARY EXCITATION

A quantity of central interest in radiation physics is W, the energy to create an ion pair, defined as

$$W = \frac{E}{\nu}, \tag{1}$$

where E is the energy available for electronic excitation and ionization and ν the number of ion pairs (positive ion + electron). The importance of this quantity is reflected in the fact that a whole monograph has been devoted to it as a single item (ICRU, 1979).

Most striking may be the experience that the magnitude of W is almost independent of the type of ionizing radiation. Deviations exceeding a few per cent occur mainly near the threshold for ionization, and for heavy-ion bombardment under conditions where elastic scattering becomes an important channel of energy deposition (ICRU, 1979).

Another interesting feature is the observation that W varies more smoothly across the periodic table than does the ionization potential. One might have expected a strong correlation between the two quantities. The main reason for this behavior was pointed out many years ago (Fano, 1946): Atoms with a high ionization potential show oscillator-strength spectra that favor ionization relative to excitation (table 1). In other words, the more energy is needed to ionize an atom, the less energy is wasted into excitation. A more recent study (Inokuti et al., 1974) indicates that the ratio W/I for elements is reasonably constant within a column of the periodic table but varies from group to group, with values ≈ 1.7 for noble gases and about twice as much for some metal vapors.

These findings are relevant for secondary electrons. Disregard for a moment the fact that we deal with condensed matter instead of a gas. A prime quantity in electron emission is the mean number of electrons liberated by the incoming radiation within a certain depth

Table 1. Ionization potential I and oscillator strength of all excitations for three elements (Fano, 1946).

	H	He	Li
Ionization potential (eV)	13.5	24.5	5.4
Total oscillator strength of all excitations	0.57	0.21	0.88

from the surface, commonly called the escape depth. According to eq. (1), that number should be given by

$$\nu = \frac{\Delta E}{W} ,\qquad (2)$$

where ΔE is the mean electronic energy deposited within the escape depth by the beam, independent of the details of the deposition process. Deviations from this simple relationship should be expected mainly near the threshold for ionization.

Deposited Energy and Stopping Power

Eq. (2) is a fairly precise statement, but it is not operational: We need to know how to determine ΔE. This is the domain of transport theory (Schou, 1980). What follows below can be deduced from standard transport equations, but I prefer to apply qualitative arguments.

The deposited energy per depth is related to the stopping power dE/dR, but it is not identical with it. Several differences, obvious and more subtle ones, need to be noted.

Consider first the idealized model of an infinite medium and a projectile in uniform motion (Fig. 1). In view of the translational invariance of this system, the energy deposited in a layer of thickness Δx near some reference plane $x = 0$ must be given by

$$\Delta E = \frac{|dE/dR|\ \Delta x}{\cos\Theta} ,\qquad (3)$$

where Θ is the angle between the beam and the x-axis. The assumption of translational invariance underlies the treatment by Rösler & Brauer (1991). The simplification is essential in the performance of that calculation: It causes all spatial variables to drop out

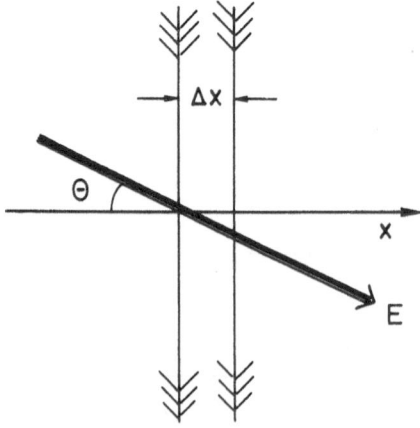

Figure 1. Uniformly moving projectile in an infinite medium.

from the Boltzmann transport equation, thereby reducing the number of variables to a manageable number.

More or less pronounced deviations from eq. (3) may be found. Firstly, angular scattering of the projectile increases the density of energy deposition. A correction may be determined by means of standard multiple scattering theory. Secondly, in case of substantial slowing-down of the projectile during its passage through the escape depth, the density of deposited energy may increase or decrease, depending on whether the stopping power increases or decreases with decreasing projectile speed.

Transients show up when a target surface is included. Conceptually, it might be easiest to stick to the infinite medium but to assume energy dissipation to set in only after the projectile passes the plane $x = 0$. Since the ionization cascade has a spatial extension (fig. 2a), some energy is deposited in the region where $x < 0$. That energy is missing in the region $x > 0$ (fig. 2b). This reduction of the density of deposited energy in comparison with the stopping power is expressed by a factor β (Schou, 1980) which is somewhat smaller than 0.5 if only inward-moving projectiles are considered. Reflected projectiles give rise to an increase and may, in fact, cause β to increase above $\beta = 1$. All this is very similar to what is known from sputtering where the corresponding factor was called α (Sigmund, 1969). The depth dependence of the deposited energy sketched in fig. 2b also implies that for oblique incidence, i.e., $\Theta > 0$, the energy deposited near the surface increases faster than

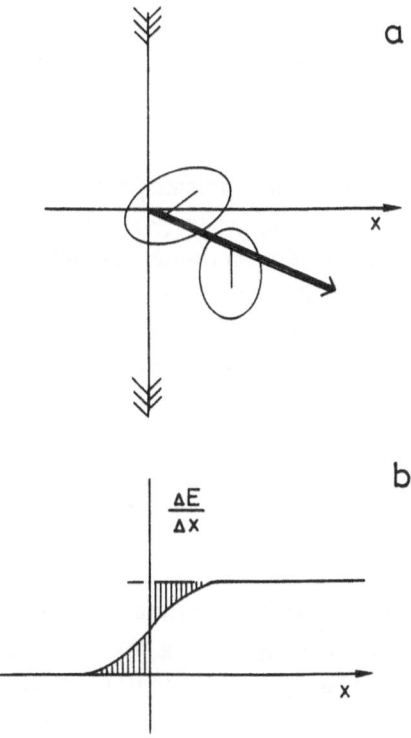

Figure 2. Projectile motion starting in the plane $x = 0$.
a) Trajectories of a heavy projectile and two energetic electrons with their associated ionization cascades (schematically).
b) Distribution in depth of deposited energy (schematically).

the $\cos^{-1}\Theta$ relation indicated in eq. (3). This effect, again well known from sputtering, can be approximated by a $\cos^{-f}\Theta$ dependence for not too large angles, with f being somewhat greater than 1 (Sigmund, 1969). It is well documented experimentally for electron emission (Hofer, 1990; Hasselkamp, 1992). These authors list four different items as possible (and very plausible) reasons, but the one outlined above was not amongst them.

An effort has been made by Rösler and Brauer (1991) to estimate the contribution of plasmon decays to the primary ionization spectrum. While the existence of the effect appears to be supported by experimental evidence, the available theoretical treatments are limited to momentum space and do not consider the dimensions of a plasmon in real space. In a treatment assuming translational invariance from the beginning, this simplification is of no importance. In the more realistic situation sketched in fig. 2b, it appears essential to incorporate a feasible model for the plasmon in real space which allows an estimate of the depth of origin of energetic electrons excited by plasmon decay. Such an estimate appears important since a plasmon is a collective excitation and, as such, must have a spatial extension. I have no answer ready, and I am not sure whether anybody else has. But it is a problem that needs clarification (Ritchie et al., 1989,1990).

Clearly, transients occur when the charge state of incident ions deviates from equilibrium, in particular so for highly-charged ions.

Finally, ionization cascades may be truncated when intersecting a (real) surface. The effect is a complication in analytic cascade theory: The equations can still be handled but things tend to get messy. The numerical significance of this effect is more economically estimated by comparison between otherwise identical Monte Carlo simulations for an infinite and a semi-infinite medium.

Shell Effects

Now that the distinction between deposited energy and stopping power has been spelled out, we may concentrate on an aspect where the common features of the two quantities should be predominant. For theoretical modelling, it is useful to know the relative contributions of various shells to the electron yield. Ignoring transport and escape, just consider the number ν of electrons liberated within the escape depth. Approximating ΔE in eq. (2) by the stopping power, and splitting the latter into contributions from various shells you may write

$$\nu = \frac{1}{W} N \Delta x \sum_j \int d\sigma_j(T) \, T \qquad (4)$$

as a first approximation. Here, $d\sigma_j(T)$ is the differential cross section for energy transfer (T,dT) and $\int d\sigma_j(T) \, T$ the corresponding stopping cross section, both for the j'th shell, and N the number density of target atoms. The integral extends over all possible energy transfers to electrons of the shell.

Table 2 shows percentage contributions to the total stopping power of aluminium vapor for protons. The data were calculated by Oddershede & Sabin (1984) on the basis of the kinetic theory of electronic stopping (Sigmund, 1982) which, for the present purpose, should be a reasonably accurate approximation to the Bethe theory including shell corrections. It is seen that at high projectile speed, the magnitude of these contributions reflects the number of electrons in the various shells. At lower velocities, contributions from outer shells increase on a relative scale at the cost of the inner shells.

More fundamentally, the following expression for ν would seem more appropriate,

$$\nu = N \ \Delta x \ \sum_j \int_{U_j}^{T_{max}} d\sigma_j(T) \ \nu_e(T-U_j) \ , \tag{5}$$

where U_j is the ionization energy of the j'th shell and $\nu_e(E)$ the mean number of electrons liberated by an electron of energy E. By definition, ν_e includes the electron itself, i.e., $\nu_e = 1$ for $E > 0$ up to the lowest ionization potential.

If Rutherford's cross section were inserted for the $d\sigma_j$ in eq. (5), the contributions from inner shells would be noticeably reduced in comparison with eq. (4), and the percentages in table 2 would shift toward higher contributions from outer electrons. This is

Table 2. Relative contribution of various shells to the stopping cross section of a bare proton in aluminium vapor. Data extracted from Oddershede & Sabin (1984).

ν/ν_0	1s	2s	2p	3s	3p
1	0.0004	0.0081	0.0223	0.5412	0.4281
10	0.0262	0.1142	0.4177	0.2656	0.1762
50	0.089	0.121	0.457	0.201	0.129

presumably the main reason for the widely accepted view that conduction electrons heavily dominate the flux of secondary electrons. However, Rutherford's law is inadequate to determine total as well as differential ionization cross sections. In view of Fano's observation mentioned above, the oscillator strength distribution of inner shells favors transitions into the continuum, and this tendency counteracts the larger binding energy. Thus, a significant part of the oscillator strength making up the integrals entering eq. (4) lies above U_j and hence also contributes to the "ionization stopping power" that governs eq. (5). As a result, table 2 is likely to be more representative for ionization phenomena than what one might have suspected. Evidently, the details may vary dependent on the material, but table 2 clearly indicates that core electrons cannot be disregarded and could easily dominate the liberation process even in materials with a low bandgap.

The case of slow-heavy-ion bombardment is special and will be discussed below.

Remarks about Stopping Power

The close relation between the electron yield and the stopping power of the bombarding particle has been recognized long ago (Bethe, 1941). It has frequently served as input and occasionally emerged as output in theoretical descriptions of electron emission. Attempts have been made to document such a relationship empirically by comparing the measured dependence of electron yields on projectile energy with that of the stopping power. Quite apart from the fact that it is the deposited energy and not the stopping power that counts, such comparisons should never be done without a critical look at the origin of the stopping power table utilized for the purpose.

With few exceptions, those stopping-power data that are needed in electron emission refer to energies where genuine stopping measurements are difficult or impossible and where stopping theory is either absent or questionable. Extensive comparisons have been done for proton bombardment (Hasselkamp, 1992). Measured stopping powers for protons in solid targets are quite accurate in the energy range down to several hundred keV. A few reliable data on selected targets cover lower energies down to somewhat below 100 keV (Bauer, 1987; Mertens, 1987). Experimental data for even lower energies refer to gas targets. Such data are not necessarily representative for solids.

Fig. 3 shows an illustrative example of what may happen. The tabulated stopping power for protons on helium (Andersen & Ziegler, 1977) is proportional to the projectile speed in the low-energy regime up to about 30 keV. Measurements by Golser & Semrad (1991) revealed a clear threshold-type of behavior. The latter is consistent with a prediction based on the Bethe theory (Mikkelsen et al., 1992), the former follows the well-established formula by Lindhard & Scharff (1961). The difference is very pronounced. The matter gets more complicated when helium is replaced by hydrogen (Golser & Semrad,

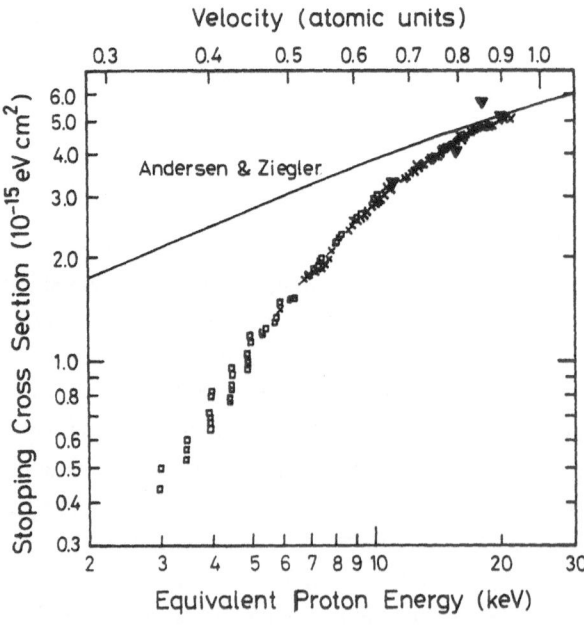

Figure 3. Measured stopping cross sections for protons and deuterons in helium (Golser & Semrad, 1991), compared to tabulation (Andersen & Ziegler, 1977).

1992). I do not wish to go into details but just to issue a warning against uncritical use of stopping-power tables in this context.

For ions other than protons, most available data for electron yields refer to fairly low projectile velocities where tabulated electronic-stopping data for solid targets are not based on measurements but on scaled theoretical predictions. The scaling is done on the basis of empirical data that mostly refer to gases. Empirical electronic stopping powers in solids for low energies may in principle be extracted from range measurements, but the separation of electronic from nuclear stopping is difficult in case of dominance of the latter and gives rise to noticeable errors.

In this connection, it appears relevant to mention the Barkas effect, i.e., the difference in stopping power between a particle and its antiparticle at equal speed. This effect occurs at velocities that are low enough to allow for noticeable polarization of the target during collision. Target polarization gives rise to enhanced stopping for positive particles and smaller stopping for negative particles in comparison with the Bethe theory. The effect is quite pronounced for protons/antiprotons (fig. 4) near the stopping maximum, and a reasonable theoretical understanding is now available. There was some uncertainty for a while as to whether the Barkas effect was mainly due to soft, distant interactions (Basbas, 1984). If so, it was not obvious whether that part of the stopping power would influence the yield of emitted electrons. This issue has been resolved, it is not (cf. Mikkelsen & Sigmund, 1990 and references quoted there). Hence, a noticeable difference should be observed between antiproton- and proton-induced electron emission.

Differences in electron emission between electron and positron bombardment have been known for a long time: This difference has been ascribed exclusively to the different

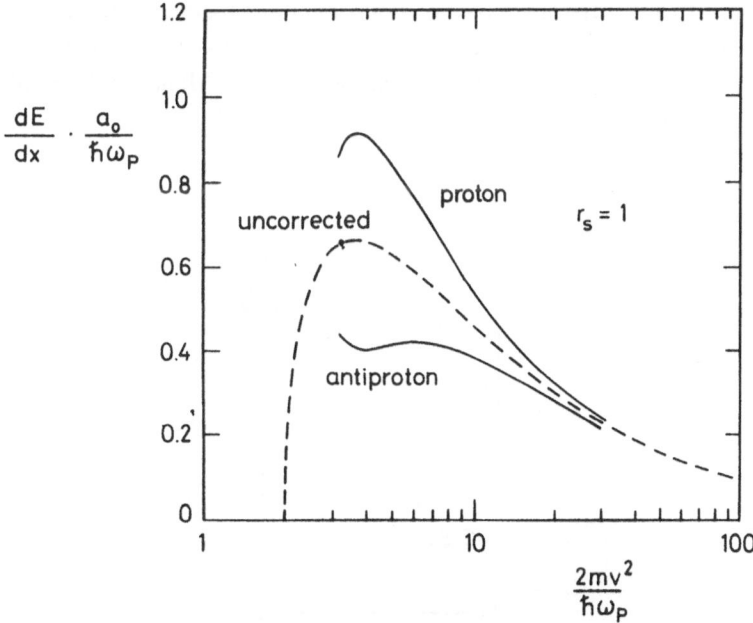

Figure 4. Calculated stopping power for protons and antiprotons in uniform electron gas. From Esbensen & Sigmund (1990). The behavior at the very lowest energies is an artefact caused by the truncation of the Born series.

role of exchange in electron-electron and positron-electron collisions. That effect that does not occur for proton/antiproton bombardment. On the other hand, the Barkas effect *must be* relevant also for electron/positron bombardment. The relative significance of the two effects in electron/positron bombardment is not known.

ELECTRON DEGRADATION AND TRANSPORT

This section addresses the question of what happens to an electron that has been liberated, in particular what fraction of the liberated electrons can come close enough to the target surface to get emitted.

Specifically, I shall treat the following problem (fig. 5). Consider an electron at a depth x from a plane target surface, moving in a random direction with some kinetic energy E. What is the probability $P(E,x)$ for this electron to reach the surface and get emitted? And what is its distribution in energy ε and angle θ when it arrives at the target surface?

Incidentally, only the *number* of liberated electrons has been discussed so far while nothing has been said about their distribution in energy, depth of origin, and angle. The latter questions belong to the subject of the theory of ionization cascades, discussed in detail by Schou (1980) and Sigmund & Tougaard (1981). In brief, secondary ionizations need to be considered, i.e., ionization by liberated electrons as well as higher-order ionizations. As a result, the energy spectrum of all liberated electrons deviates from the primary excitation spectrum, and the depth of origin is more or less affected by processes of the type indicated in fig. 2a. The energy spectrum at the point of liberation should be close to E^{-2} (Sigmund & Tougaard, 1981), and the distribution in depth of origin should resemble the one indicated in fig. 2b under bombardment conditions well above threshold. This spectrum could easily be determined by a Monte Carlo simulation. While splitting up the problem into several stages like primary excitation stage and transport is not a necessary step in such simulations, it might be useful in assessing the validity of analytical estimates and, therefore, be a way to condense the results of Monte Carlo simulations into pertinent scaling rules.

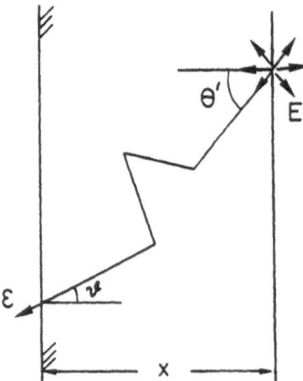

Figure 5. Definition of variables for escape probability $P(E,x)$ for monochromatic isotropic source at depth x from surface.

A frequently used answer to the problem posed in the beginning of this section is a survival probability

$$P(E) \simeq e^{-x/\Lambda\cos\Theta'} \tag{6}$$

where Λ is some escape depth, used as an adjustable parameter, and Θ' the angle between the direction of motion at depth x and the outward surface normal. An escape probability so defined is a key quantity in the application of Auger and photoelectron spectroscopy to surface analysis (Tougaard, 1988). In that case, the electron spectrum is analyzed in the vicinity of a sharp primary energy E corresponding to some Auger or photoelectron peak. Eq. (6) delivers the probability for emission without energy loss if Λ is understood as the mean free path for inelastic scattering. It ignores angular deflection by elastic scattering and is therefore valid only over a fairly limited depth range (Tougaard & Sigmund, 1982). That, however, is often sufficient to allow deductions on surface composition and near-surface depth profiles.

Qualitative Picture

Secondary electrons form a continuum spectrum and typically lose a sizable fraction of their initial energy on the way from the point of liberation to the surface. Electrons that are allowed to slow down over a substantial fraction of their maximum range experience angular deflection, and their trajectories may be characterized by a diffusion picture. Angular deflections are caused by elastic screened-Coulomb interactions with the atoms of the medium that are largely independent of the excitation and ionization processes that are responsible for the slowing-down.

A feasible description of an electron trajectory looks as follows: Ignore initially all energy loss and consider the effect of multiple angular deflections on the electron density distribution in space and time. The time variable can be converted into an energy variable if the stopping is treated within the continuous-slowing-down approximation, i.e., if fluctuations in energy loss can be ignored. Such fluctuations are nonvanishing but less central to the problem in question than angular deflections.

A calculation along these lines was outlined previously (Sigmund & Tougaard, 1981) but was applied mostly to photoelectron spectroscopy (Tougaard & Sigmund, 1982) where its range of validity is limited. It will be used here for the purpose that it was originally designed for, i.e., to determine the escape probability P(E,x) under conditions of strong elastic scattering. That assumption appears feasible for all solids, but it is particularly relevant for *insulators*, where the majority of emitted electrons has energies in the subexcitation range. This is the regime where electronic stopping is zero and electron ranges become excessively large (Inokuti, 1990).

Elastic Scattering in Infinite and Semi-Infinite Medium

Consider initially a number of electrons located in a plane x = 0 in an infinite medium. Give each of them a velocity v (or energy E) at time t = 0 in a random direction. The spatial distribution will broaden as a function of time. Initially, the width will increase linearly with time, but eventually, after some multiple angular deflections, a $\sqrt{t}$ behavior must be expected.

This can be deduced from the linear Boltzmann transport equation. If energy loss is neglected, one finds (Sigmund & Tougaard, 1981)

$$<x^2> = \frac{2v}{3N\sigma_1} \left[t + \frac{1}{vN\sigma_1} (e^{-vN\sigma_1 t} - 1) \right] , \qquad (7)$$

where N is again the density of scattering centers (atoms), σ_1 the transport cross section for elastic scattering,

$$\sigma_1 = \int d\sigma(\phi) [1 - \cos\phi] , \qquad (8)$$

and $d\sigma(\phi)$ the differential cross section for a scattering angle $(\phi, d\phi)$ in the laboratory system.

Eq. (7) behaves like

$$<x^2> = \begin{cases} \frac{1}{3} (vt)^2 & \text{small} \\ & \text{for t} \\ 2Dt & \text{large} \end{cases} \qquad (9)$$

as expected, with the diffusion coefficient D defined by

$$D = \frac{1}{3} \lambda_1 v , \qquad (10)$$

and the transport mean free path

$$\lambda_1 = 1/N\sigma_1 . \qquad (11)$$

Knowing the diffusion coefficient we may now consider a plane isotropic source of electrons at depth x in a *semi-infinite* medium. The electron density in such a medium, with an absorbing surface at x = 0, can be constructed by the image method, and the current through the surface reads

$$J = - \frac{x}{(4\pi Dt^3)^{1/2}} e^{-x^2/4Dt} \qquad (12)$$

Within the diffusion picture, this current must be distributed cosine-like in the angle of emergence, i.e., a factor $d^2\Omega \cos\theta/\pi$ could be added if desired, with $d^2\Omega = \cos\theta \, d\theta \, d\chi$ and χ an azimuth.

Unlimited integration of eq. (12) over t yields the escape probability per electron,

$$P' = \int_0^\infty dt \; \frac{x}{\sqrt{4\pi Dt^3}} \; e^{-x^2/4Dt} = 1 \; , \qquad (13)$$

i.e., all electrons will escape sooner or later. This unphysical result originates from the neglect of energy loss.

Escape Depth

In order to repair for this shortcoming, note first that according to eq. (10), we have

$$D dt = \frac{1}{3} \lambda_1 \; dR \; , \qquad (14)$$

where $dR = v dt$ is a path-length segment. When energy-loss straggling is neglected, one may represent the electron energy ε versus travelled path length R and initial energy E by some function

$$\varepsilon = \varepsilon(E, R) \; , \qquad (15)$$

which may be determined from the stopping power. Since σ_1 or λ_1 depend on energy, an additional path-length dependence may enter. Disregarding the latter for a moment, one may integrate eq. (13) subject to the boundary condition that $\varepsilon > U$, where U is the surface barrier. This yields

$$P(E, x) = erfc(x/\Lambda) \qquad (16)$$

with

$$\Lambda = 2 \; (\lambda_1 R_0/3)^{1/2} \; . \qquad (17)$$

Here, R_0 is the path length for slowing down from E to U. For analytical convenience, an angular-independent surface barrier was assumed in eq. (16). This is important only for E approaching U.

In cases where the dependence of λ_1 on energy cannot be neglected, (17) remains valid after the substitution

$$\lambda_1 R_0 \rightarrow \int_0^{R_0} dR \; \lambda_1(R) \; . \qquad (18)$$

There are several striking differences between eqs. (16,17) and eq. (6). Firstly, eq. (6) implies a strong correlation between the initial direction of motion and the angle of emergence. Any such correlation is wiped out in the diffusion limit of eq. (9). Secondly, the exponential has been replaced by a complementary error function. The difference is not very pronounced, but the latter is more triangular-shaped than an exponential (fig. 6).

Finally, and most important, eq. (17) provides an explicit estimate of an energy-dependent escape depth $\Lambda(E)$. It is essentially the geometric mean between the transport mean free path λ_1 for elastic scattering and the path length R_0 travelled from E to U. That pathlength is governed by the pertinent stopping power. It will typically be electronic, but for insulators at energies below the lowest ionization threshold, i.e., for subthreshold electrons (Inokuti, 1990), the pathlength is governed by nuclear collisions.

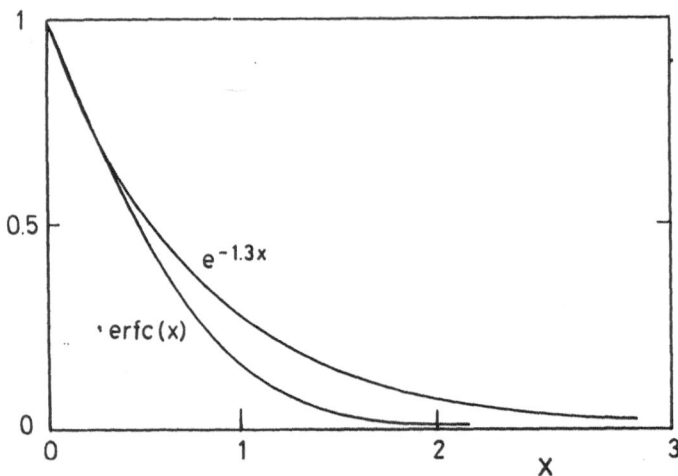

Figure 6. Comparison of exponential with error-function-like escape probability.

The present result assumes explicitly that the electron can undergo a series of angular deflections without losing a substantial fraction of its energy. This situation is very different from sputtering where angular deflection of a target recoil is accompanied by substantial energy loss. Therefore, the escape depth in sputtering is governed primarily by the energy loss (Sigmund, 1969; Falcone & Sigmund, 1981).

Differential Escape Depth

One may also inquire about a doubly differential escape probability $P(E,x,\varepsilon,\theta)\,d\varepsilon\,d^2\Omega$, where ε is the energy at the internal surface. That quantity follows similarly from eq. (12) and reads

$$P(E,x,\varepsilon,\theta) = \frac{\cos\theta}{\pi} \frac{1}{|d\varepsilon/dR|} \left[\frac{3x^2}{4\pi\lambda_1 R'^3} \right]^{1/2} e^{-3x^2/4\lambda_1 R'}, \quad (19)$$

where $d\varepsilon/dR$ is the stopping power at energy ε, and

$$R' = R(E,\varepsilon) = R(E) - R(\varepsilon) \quad (20)$$

is the mean path length travelled from energy E to energy ε. Here, R(E) is the conventional range, i.e., the path length travelled from energy E to rest.

In view of (9), eq. (20) breaks down for ε approaching E, i.e., for R' $< 2\lambda_1$. In that case, the conventional continuous-slowing-down picture (Schou, 1980) is more adequate.

The above estimate is probably less accurate than what could be provided by a Monte Carlo simulation, but presumably comparable in accuracy with the age diffusion model utilized by Devoogt et al. (1987) and Dubus et al. (1987). At any rate, it shows what appears to be essential. Several recent simulation codes involve vast amounts of elastic-scattering data determined by partial-wave analysis of electron scattering on more or less questionable scattering potentials. Eqs. (17) and (19) confirm the dominating role of the transport cross section and the electron range. The transport cross section is an integral quantity that is rather insensitive to small-angle scattering events which are most frequent and therefore consume most of the processing time in a Monte Carlo simulation. In fact, the transport cross section is a very robust quantity, just like the stopping power: You can do a lot to the cross sections within physically acceptable limits without affecting the transport cross section. The same is true for the range which is an integral over the inverse stopping power.

Eqs. (16,19) have practical implications on the magnitude of electron yields and the shape of emitted spectra in comparison with Schou's (1980) theory which will have to be worked out quantitatively. It would also be desirable to make a systematic comparison between these analytic estimates and an explicit Monte Carlo computation of the transport problem.

STATISTICS OF ELECTRON EMISSION

It has been possible for some time to experimentally determine statistical distributions of the number of electrons emitted per bombarding particle. The experimental situation has been summarized recently by Hofer (1990). Poisson distributions have commonly served as the standard of reference to be compared to experimental distributions. More or less pronounced deviations have been frequently found, most often in the direction of *broader* spectra.

While the statistics of penetrating particles hinges on Poisson's law, this does by no means imply that the yield of emitted electrons is Poisson-distributed. In order to appreciate this, consider first the total number of electrons liberated in a *thick* target by an impinging charged particle. Under conditions of dominating electronic stopping, the mean value is given by eq. (2), with ΔE replaced by the total energy E, and the fluctuation is given by

$$< (v - <v>)^2 > \simeq <v> F , \qquad (21)$$

with the Fano factor F which is commonly of the order of 0.2. When this is so, the distribution of the number of liberated electrons is sub-poissonian. An example in the opposite category is found for heavy-ion bombardment under conditions of sizable nuclear stopping (Lindhard & Nielsen, 1962), where fluctuations may be very large and obey quite different scaling relations.

As discussed above, electron emission is governed by the energy ΔE deposited within the escape depth. The fluctuation of that quantity is determined primarily by energy-loss straggling and, perhaps, by multiple angular scattering of the projectile. Denote the distribution of energy loss ΔE_p of the projectile in a layer Δx by $f(\Delta E_p, \Delta x)\, d(E_p)$. Ignoring the Fano factor you may convert ΔE_p into a number of liberated electrons in the layer Δx, i.e., $\nu = \Delta E_p/W$, where W is the appropriate conversion factor for the target under consideration. Then, the probability distribution for ν reads

$$f(\nu W, \Delta x)\, d(\nu W)\ . \tag{22}$$

Consider a very-high-energy beam for a moment, such as 100 MeV protons impinging on aluminium. With a stopping power of ~ 0.1 eV/Å, the *mean* energy deposited over a depth of 50 Å is about 5 eV while the *maximum* energy that can be transferred to a target electron is about 200 keV. Such a spectrum is well described by a Landau distribution which, in this case, is a very skew distribution with a narrow peak somewhere below 5 eV and a long Rutherford-like tail that goes like ΔE_p^{-2} up to ~ 200 keV, corresponding to several thousand liberated electrons. Only a fraction of those electrons will be emitted, but neither transport nor the surface barrier will be able to drastically affect the ν^{-2} spectrum. Ejection of an electron from an aluminium surface requires several

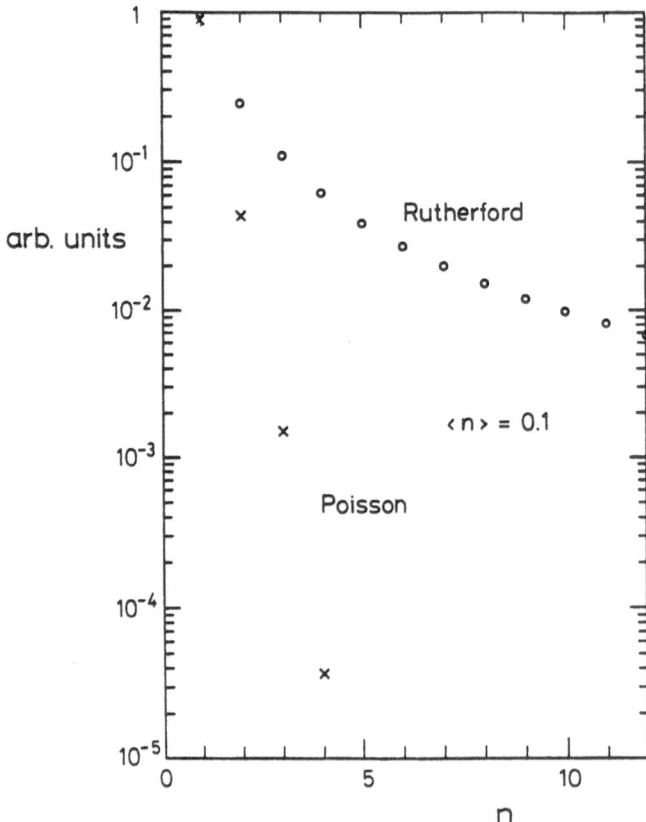

Figure 7. Statistics of electron emission. Comparison of Poisson distribution with a distribution governed by a Landau-like straggling distribution. An emission yield of 0.1 has been adopted which is *less extreme* a case than the one discussed in the text.

eV. With the empirical conversion factor between stopping power and electron yield (Hasselkamp, 1992), an average electron yield $\gamma \sim 0.01$ is expected. The vast majority of events will not lead to electron emission at all, and an event where one electron is emitted will already be well described by the Rutherford tail which should extend up to many hundred electrons. The upper part may not be visible in the noise, but it must be there (fig. 7).

It may be instructive to perform such a measurement at a suitable accelerator in order to once and for all bury the Poisson distribution as a reference standard in this context: You need not ask why a measured distribution *deviates* from Poisson; you need to ask for the reason why a given distribution *coincides* with Poisson, if it does.

HIGH-ENERGY TAILS IN ELECTRON SPECTRA UNDER LOW-ENERGY ION BOMBARDMENT

I like to conclude this contribution by discussing a specific item in a more speculative manner.

Electron emission under low-energy heavy-ion bombardment is known to differ substantially from the case of electron or high-velocity light-ion bombardment. Past and current thinking about the topic has been summarized by Hofer (1990), and I have little to add to the general picture. Most interesting from the point of view of fundamental processes is the observation that no mechanisms are known that would excite outer-shell and conduction electrons directly up to energies that would allow them to escape through the surface. Core electrons, on the other hand, may be promoted up to high levels and give rise to emission of electrons. Much information about primary spectra has been gathered from single-collision experiments, although not all aspects are well understood. The relation between electron yields and electronic stopping power, on the other hand, is less obvious for heavy-ion bombardment because conventional stopping theories are based on the assumption that the most weakly bound electrons are responsible for stopping.

A new aspect has come up recently when Baragiola et al. (1992) reported secondary-electron spectra for metal targets bombarded by 1 - 6 keV argon and helium ions. Fig. 8 shows spectra for Ar^+-Mg for several incident energies. These spectra show electrons with energies up to half the incident-ion energy. Those high-energy electrons form an exponential tail at an extremely low intensity level. The slope of these tails decreases with increasing incident-ion energy.

Baragiola and coauthors expressed the expectation that these tails originate in the primary interaction between the incident ion and a target atom and refer to observations on ion-atom collisions (Clapis & Kessel, 1990). However, even the fastest electrons seen in ion-gas collisions receive only a small percentage of the incident-ion energy. Baragiola et al. argue that long tails are present also in gases but cannot be detected there because of lacking intensity. This problem should be less serious in solids because of higher density.

The claim that such high-energetic electrons are generated in a single ion-atom collision may not contradict standard conservation laws, but a process that would promote electrons into such giant excitation levels is not readily at hand, not even at the minute intensity level indicated in fig. 7.

A hint at an alternative explanation may be taken from the exponential shape of all observed tails. Assume that electrons may be promoted in primary interactions up to 50 -

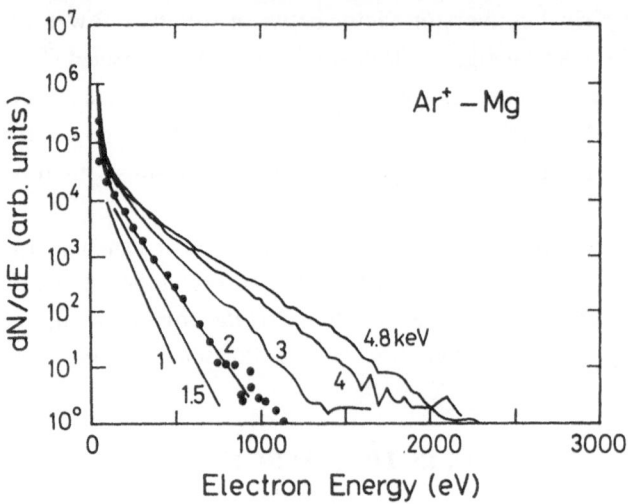

Figure 8. Energy spectra emitted from magnesium under bombardment with argon ions (Baragiola et al., 1992).

100 eV, and that they may subsequently be *accelerated* in repeated collisions with the incident ion by means of a shuttle mechanism of the type that has been discussed in connection with cluster-induced fusion.

It is essential to note that electrons liberated by promotion move typically much faster than the incident ion. A 50 eV electron has a velocity $\sim 2v_0$ (v_0 being the Bohr velocity), while a 4 keV argon ion has a velocity $\sim 0.07\ v_0$. Thus, an electron liberated shortly after the entrance of an ion into the target can get reflected from a target atom without much energy loss, hit the incident ion head-on and thus receive a velocity increment up to twice the ion velocity. This process would have to be repeated about 50-100 times in order to accelerate the electron above 1 keV. This has low probability, but it does not appear impossible, and such a sequence of events leads clearly to an exponential spectrum (Hautala et al., 1991).

Several features appear consistent with the above model.

1. The maximum velocity increment experienced by an electron in a single shuttle increases with increasing ion energy. The pertinent scattering cross section for elastic scattering is essentially unaffected since it is determined by the electron velocity. Therefore, the slope of the exponential part of the spectrum should decrease with increasing ion energy. This is seen in fig. 8.

2. In experiments with gold bombarded with helium and argon ions *at the same energies*, the tails for helium bombardment are consistently flatter than the ones observed with argon (fig. 9). A 4 keV helium ion has about three times the velocity of a 4 keV argon ion, and hence produces a correspondingly flatter tail.

3. Bombardment of four different targets from beryllium to gold with 2.5 keV argon ions shows similar tails at similar levels (Baragiola et al., 1992). This feature would appear very hard to explain in terms of a single primary interaction. It follows naturally from the shuttle mechanism since the target atoms only play the passive role of providing a wall for backscattering of electrons.

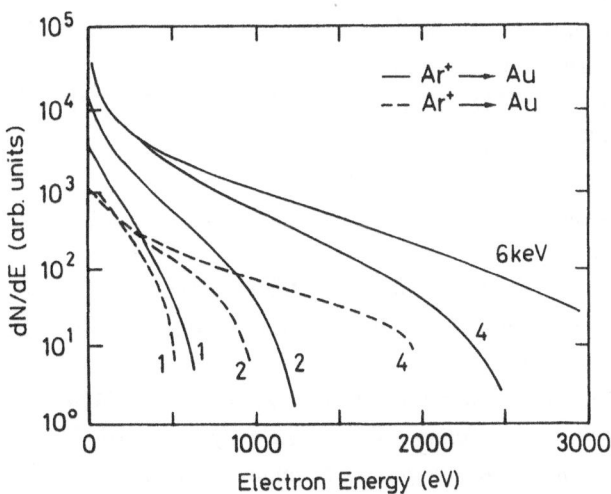

Figure 9. Energy spectra emitted from gold
under bombardment with argon and helium ions (Baragiola et al., 1992).

4.   It is essential for the shuttle process to become operational that excited electrons have
a starting velocity far above the ion velocity. This implies that tails will not be found under
conditions of pure Coulomb excitation, i.e., under bombardment with fast protons.

While this explanation may look less probable than an experimental artefact, there
should be enough interesting physics to have a closer look. An estimate of the slope should
be within reach, and some information on absolute intensities of the observed tails should
be provided by the experimentalists.

Acknowledgements

Thanks are due to Mitio Inokuti for an enlightening discussion, to Jørgen Schou for
going carefully through the manuscript, and to Raul Baragiola for his kind invitation to
participate in and contribute to a very stimulating workshop. Financial support was
received from the Danish Natural Science Research Council (SNF).

REFERENCES

Andersen, H.H., and Ziegler, J.F., 1977, *Hydrogen Stopping Powers and Ranges in All
        Elements*, Pergamon Press, New York
Baragiola, R.A., Alonso, E.V., Oliva, A., Bonnano, A., and Xu, F., 1992, Phys. Rev. A **45**,
        5286
Basbas, G., 1984, Nucl. Instr. Methods B **4**, 227
Bauer, P., 1987, Nucl. Instr. Methods B **27**, 301
Bethe, H.A., 1941, Phys. Rev. **59**, 940
Clapis, P., and Kessel, Q.C., 1990, Phys. Rev. A **41**, 4766
Devoogt, J., Dubus, A., and Dehaes, J.-C., 1987, Phys. Rev. **36**, 5093

Devoogt, J., Dehaes, J.-C., Dubus, A., and M.Cailler, 1991, in *Particle Induced Electron Emission I*, Springer Tracts in Modern Physics **122**, 67

Dubus, A., Devoogt, J., and Dehaes, J.C., 1987, Phys. Rev. **36**, 5110

Esbensen, H., and Sigmund, P., 1990, Ann. Physics **201**, 152

Falcone, G., and Sigmund, P., 1981, Appl. Phys. **25**, 307

Fano, U., 1946, Phys. Rev. **70**, 44

Golser, R., and Semrad, D., 1991, Phys. Rev. Lett. **66**, 1831

Golser, R., and Semrad, D., 1992, Nucl. Instr. Methods B **69**, 18

Hasselkamp, D., 1992, in *Particle Induced Electron Emission II*, Springer Tracts in Modern Physics **123**, 1

Hautala, M., Pan, Z., and Sigmund, P., 1991, Phys. Rev. A **44**, 7428

Hofer, W.O., 1990, in *Fundamental Beam Interactions with Solids for Microscopy, Microanalysis and Microtopography*, Scanning Microscopy Suppl. **4**, 265

ICRU, 1979, *Average Energy to Create an Ion Pair*, International Commission on Radiation Units and Measurements, Report no. **31**

Inokuti, M., Dehmer, J.L., and Saxon, R.P., 1974, in *Radiological and Environmental Division Research Report*, R.E.Rowland, editor, Argonne National Laboratory, p. 16

Inokuti, M., 1990, in *Molecular Processes in Space*, T.Watanabe et al., editors, Plenum, New York, p. 65

Lindhard, J. and Nielsen, V., 1962, Phys. Lett. **2**, 209

Lindhard, J., and Scharff, M., 1961, Phys. Rev. **124**, 128

Mertens, P., 1987, Nucl. Instr. Methods B **27**, 315

Mikkelsen, H., Meibom, A., and Sigmund, P., 1992, submitted to Phys. Rev. A

Mikkelsen, H.H., and Sigmund, P., 1990, Phys. Rev. A **40**, 101

Oddershede, J., and Sabin, J.R., 1984, Atomic Data and Nucl. Data Tables **3**, 275

Ritchie, R.H., Hamm, R.N., Turner, J.E., Wright, H.A., Ashley, J.C., and Basbas, G.J., 1989, Nucl. Tracks Radiat. Meas. **16**, 141

Ritchie, R.H., Howie, A., Echenique, P.M., Basbas, G.J., Ferrell, T.L., and Ashley, J.C., 1990, Scann. Microsc. Suppl. **4**, 45

Rösler, M., and Brauer, W., 1991, in *Particle Induced Electron Emission I*, Springer Tracts in Modern Physics **122**, 1

Rothard, H., Groeneveld, K.O., and Kemmler, J., 1992, in *Particle Induced Electron Emission II*, Springer Tracts in Modern Physics **123**, 97

Schou, J., 1980, Phys. Rev. B **22**, 2141

Schou, 1988, Scanning Microscopy **2**, 607

Sigmund, P., 1969, Phys. Rev. **184**, 383

Sigmund, P., 1982, Phys. Rev. A **26**, 2497

Sigmund, P., 1991, in *Interaction of Charged Particles with Solids and Surfaces*, A.Gras-Marti et al., editors, NATO ASI Series (Plenum Press), vol. B **271**, 73

Sigmund, P., and Tougaard, S., 1981, in *Inelastic Particle-Surface Collisions*, E.Taglauer and W.Heiland, editors, Springer Series in Chemical Physics **17**, 2

Tougaard, S., 1988, Surf. Interf. Anal. **11**, 453

Tougaard, S., and Sigmund, P., 1982, Phys. Rev. B **25**, 4452

Varga, P., and Winter, H., 1992, in *Particle Induced Electron Emission II*, Springer Tracts in Modern Physics **123**, 149

AUGER PROCESSES AT SURFACES

Arend Niehaus

Debye Institute, Utrecht University
Buys Ballotlaboratorium, PB 80.000
3508 TA Utrecht, The Netherlands

INTRODUCTION

Any electronic system that contains two or more electrons and is prepared in a state lying in a one-electron continuum of positive energy, will decay with a characteristic lifetime into this continuum by ejecting an electron. In principle, many electrons may participate in such a spontaneous auto ionization process. In case of the lowest order process, which involves only two "active" electrons that exchange their energy, one may say that the ejected electron gets its positive energy because the other electron gets more strongly bound, or "fills a hole". The term "Auger-process" for such a two electron process was originally used for the case of the decay of an isolated atom, prepared by some ionization process with a "hole" in an inner shell. However, it has become customary to use the term "Auger-process" quite generally for any autoionization process involving two active electrons. Auger-processes in this general sense are the subject of this contribution.

When an ion approaches a metal surface, the combined system automatically is prepared in an excited state, because a metal electron usually can be more strongly bound if it recombines with the ion. If the energy that can be gained by this recombination is higher than the work function of the metal, the system can decay by several types of Auger-processes, either directly from the state prepared, or from states modified by resonant one electron exchange processes. Depending on the velocity of the ion-surface approach, this leads to a characteristic time evolution of the ion-surface system, determined by the probabilities for the various electronic processes. In cases of doubly- or multiply charged ions this time evolution becomes rather complex. If the ion-surface velocity is sufficiently low compared to the characteristic velocities of the electrons involved, the evolution of the system as a function of time or ion-surface distance should be describable in terms of well defined electronic potential energy curves and transition probabilities between these curves. Under such conditions the electrons ejected in Auger-type processes during the interaction directly reflect the time evolution of the system. In fact, for any given ionization process the corresponding electron energy spectrum is determined essentially by three distance dependent quantities, (i) the population of the initial state of the transition, (ii) the transition probability to the final state, and (iii) the energy difference between initial- and final state potential curves. An appropriate analysis of measured total electron spectra, therefore, should allow one to determine these quantities, and in this way to reconstruct the complete time evolution of the system.

Ionization of Solids by Heavy Particles, Edited by
R.A. Baragiola, Plenum Press, New York, 1993

In the present contribution we demonstrate that, to a certain extent, we have been able to carry out such a reconstruction of the time evolution of ion-surface systems. We discuss the analysis of electron spectra from Auger-type processes for the systems $Ar^{++}/Pb(111)$, and He^{++}, $He^{+}/W(110)$. First, the experimental conditions are briefly outlined, and examples of the electron spectra to be analyzed are presented. The following, main part of the contribution is then devoted to the description of the theoretical model that links the electron spectra to the theoretical quantities used. Finally, the application of the model to the concrete cases mentioned is discussed.

EXPERIMENTAL

The experimental electron spectra that will be discussed have been obtained at very low perpendicular velocities ($v_\perp$) of the projectile ion with respect to the surface. This is achieved by using low incidence angles (ψ) of a well collimated ion beam with the surface, at beam energies (E_{beam}) of the order 0.1 to 1 keV. For ideally flat surfaces in this way the projectile trajectories are virtually independent of the impact position, with a turning point well above the first layer of surface atoms. Typical perpendicular energies ($E_\perp$) for the electron spectra discussed are of the order:

$$E_\perp = (M/2)v_\perp^2 = E_{beam} \times \sin^2\psi \sim 1 \text{ to } 10 \text{ eV}. \tag{1}$$

For an analysis of electron spectra due to Auger-type ionization processes the use of a low perpendicular velocity ($v_\perp$) has several advantages. First of all, since all trajectories occurring in the experiment are virtually identical, structures in electron spectra resulting for particular trajectories are not obscured by the necessary integration over the contributions from all trajectories occurring. Other advantages are, that the general broadening caused by the time - energy uncertainty relation at higher vertical velocities is avoided, and that contributions from kinetic electron emission to the spectra is negligible.

Electron spectra that will be discussed have been measured by two different groups, by our group in Utrecht (Eken et al.,1992), and by the group in Clausthal Zellerfeld (Brenten et al.,1992a,1992b). In the following, we will call the two experiments "experiment 1" and "experiment 2", respectively. The principle of the experimental procedure applied is essentially the same for the two experiments. Details of the experimental setups used are well documented in earlier publications (Eeken et al., 1988, Schall et al.,1989, Brenten et al.,1991). We therefore only give a brief outline here. A well collimated mono energetic and mass selected ion beam is directed to a single crystal surface at grazing angles of incidence. Electrons are detected perpen-dicular to the ion beam with high energy resolution. The single crystal surfaces used are prepared by standard techniques and their quality controlled intermittently during the experiments. The background pressure of the vessel containing the setup is approximately one to two times 10^{-10} torr. In experiment 2, partial coverage of the surface by alkali atoms can be achieved "in situ" in a controlled way. In this way the work function of the W(110) surface can be lowered down to approximately 1.4 eV. In order to compensate for the retarding contact potential arising between surface and spectrometer at low work function values, the spectrometer is biased in such a way that electrons leaving the surface of a work function of 1.4 eV arrive at the spectrometer with approximately 2 eV. If the bias voltage is fixed, the variation of the work function upon adsorption of alkali atoms shows up as a corresponding variation of the "zero energy edge" of the electron spectra. The electron spectra from experiment 2 are corrected for an energy dependent transmission. In experiment 1, measurements are carried out only for the clean Pb(111) surface. No bias between crystal and spectrometer is applied. The detection efficiency for electrons below approximately 4 eV decreases with decreasing energy due to spurious fields, leading to a fall off towards zero energy for all electron spectra. Above 4 eV the electron spectra are believed to be only negligibly influenced by transmission effects.

ELECTRON SPECTRA AND THEIR QUALITATIVE INTERPRETATION

A series of electron spectra obtained in experiment 1 for $Ar^{++}/Pb(111)$ at constant

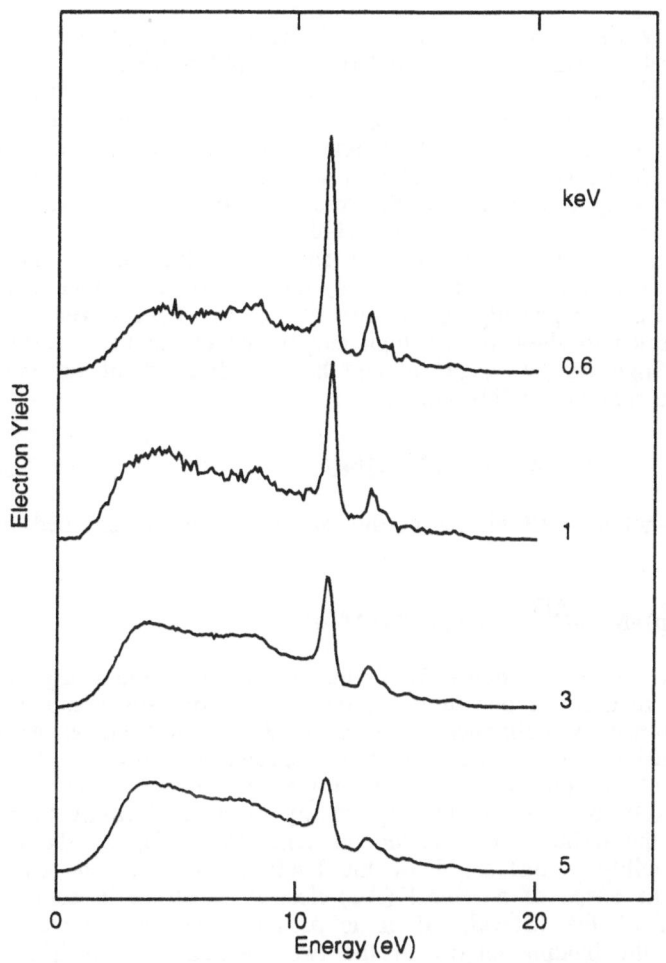

Fig.1. Measured electron spectra for Ar^{++}/Pb(111), at four different ion energies, and at an elevation angle of 2°.

incidence angle of $\psi=2°$ and beam energies varying between 0.6 and 5 kev is shown in fig.1 (Eken et al.,1992). Different kinds of features are observed, peaks, "edges", and broad distributions. For a qualitative interpretation of the spectra it is important that these different kinds of features can be related to the three different kinds of Auger processes which can be distinguished:

(i) Both active electrons "come from" the atomic particle. In this case one has an autoionization of the perturbed atomic particle, and expects a peak-like spectrum. We call this process simply autoionization (AU). Because of the energy position of the main peak in the spectra of fig.1, we ascribe this peak to the (AU)-process

$$Ar^{**}(3p^4)^3P(4s^2)+M \xrightarrow{\text{AU}} Ar^+(3p^5)+M+e^- \qquad (2)$$

The other clearly visible peak around 13.3 eV we ascribe to (AU) of $Ar^{**}(3p^4)^1D(4s^2)$, and the tiny peak around 17 eV to (AU) of $Ar^{**}(3p^4)^1S(4s^2)$.

(ii) One active electron "comes from" the metal, and the other "comes from" the atomic particle. In accordance with literature, we call this process an Auger-deexcitation (AD). Since the metal electron can be bound at an energy anywhere between the Fermi-level and the bottom of the conduction band, while the atomic electron has a rather well defined binding energy before the transition, the spectrum corresponding to this process is expected to have a certain width, reflecting the width of the conduction band, and a certain shape, reflecting in some way the density of electronic states within the conduction band at the surface, the so called "surface density of electronic states (SDOS)". Especially, if the density at the Fermi-level is high, (AD)-processes are expected to show up via high energy "edges" in the spectrum. There is one clearly visible high energy "edge" around 8.5 eV. Because of its energy position we ascribe it to the following (AD)-process:

$$Ar^{+*}(3s3p^6)+M \xrightarrow{\text{AD}} Ar^+(3s^23p^5)+M^++e^-. \qquad (3)$$

Another edge, not so well visible in the spectra shown, is ascribed to the (AD)-process:

$$Ar^{+*}(3p^4)^3P(4p)+M \xrightarrow{\text{AD}} Ar^+(3p^5)+M^++e^-. \qquad (4)$$

(iii) Both active electrons "come from" the metal. Since the charge state of the atomic particle is reduced by one in this process, the term Auger capture (AC) has been introduced (Zeijlmans van Emmichoven et al.,1988). The well known Auger neutralization of singly charged ions (Hagstrum, 1954) is a special case of (AC). The electron energy distribution resulting from (AC) should in general be broad and less structured than the one for (AD), because the binding energy of both electrons before the transition can vary over the width of the conduction band. Depending on the way in which the transition probability is influenced by the binding energy of the two electrons, it is expected that the shape of a typical (AC)-electron spectrum in some way reflects the autoconvolution of the (SDOS). If it is assumed that the transition probability is independent of the binding energy of the two electrons, and if it is further assumed that the (SDOS) is unperturbed by the projectile, the (SDOS) can be extracted from measured (AC)-spectra by some deconvolution procedure. This has been the basis of attempts to determine the (SDOS) from (AC)-spectra for singly charged ions (Sesselmann et al., 1987). In the present case of a doubly charged projectile ion, one expects contributions from (AC) connected with the "filling" of the 3p-hole in Ar^{++}, and Ar^+, respectively. A broad distribution extending from ca.15eV to zero energy, which is thought to underlie the structures of the observed spectra, is ascribed to the (AC)-process

$$Ar^{++}+M \xrightarrow{\text{AC}} Ar^++M^{++}+e^-, \qquad (5)$$

and the "hump" showing up below 5eV electron energy is ascribed to the (AC)-process

$$Ar^++M \xrightarrow{\text{AC}} Ar+M^{++}+e^- \qquad (6)$$

Qualitative interpretations of electron spectra from singly- and doubly charged ion-surface collisions in terms of the Auger-type processes (AU,AD,AC) have been carried out earlier (Varga, 1989). We present such a qualitative interpretation here for the Ar^{++}/Pb case, because it is the basis for the quantitative model we have developed in recent years (Zeijmans van Emmichoven,et al.,1988, Zeijlmans van Emmichoven and Niehaus,1990, Eeken et al.,1992), and which we want to discuss here.

In order to to make the symbols indicating the different Auger-type processes more specific, we add to the symbol a bracket indicating the number of "holes", and the number of electrons in excited states for the atomic particle before the transition.

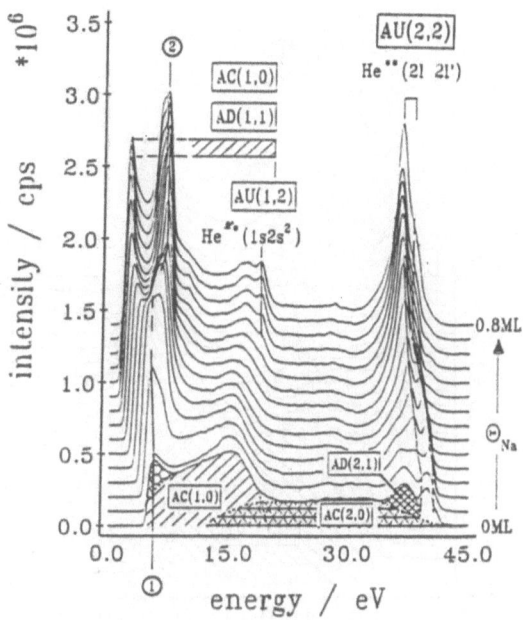

Fig.2. *Electron spectra for He^{++}/W(110) at an ion energy of 100 eV, and at an elevation angle with the surface of 5°. Different spectra are obtained for different Na-coverage. The shift of the low energy edge of the spectra is a measure of the change of the work function. The symbols indicate the processes that are responsible for the features. The meaning of the symbols is explained in the text.*

The symbol for process (5), for instance, in this way becomes AC(2,0), in contrast to the symbol for process (6), which becomes AC(1,0). In case of (AD)-processes, in addition to a process as (3) or (4), which arises after capture of one excited electron into the doubly charged ion and is indicated by AD(2,1), a process AD(2,2) following the capture of two electrons into excited states should be possible. In the case of the Ar** state invoked in the AU(2,2) process (2), this means that there arises competition with AD(2,2).

A series of electron spectra for the systems He^{++}/W(110), and He^{+}/W(110), at varying (Na)-coverage obtained in experiment 2 are reproduced in fig.2, and fig.3, respectively (Brenten,et al.,1992a,1992b,1992c).

A decrease of the work function of the surface upon Na-coverage from the nominal value of 5.2 eV for the clean surface, to a value of ca 1.4 eV is reflected quantitatively by the shift of the low energy edge of the spectra. The qualitative interpretation of the spectra in terms of Auger-type processes, as carried out by the authors, is indicated. The symbols explained above have been used. Drastic changes of the electron spectra with varying Na-coverage are observed for both systems. The interpretation given implies a general behavior of the the changes observed: At low coverage and high work function the dominating Auger-type processes

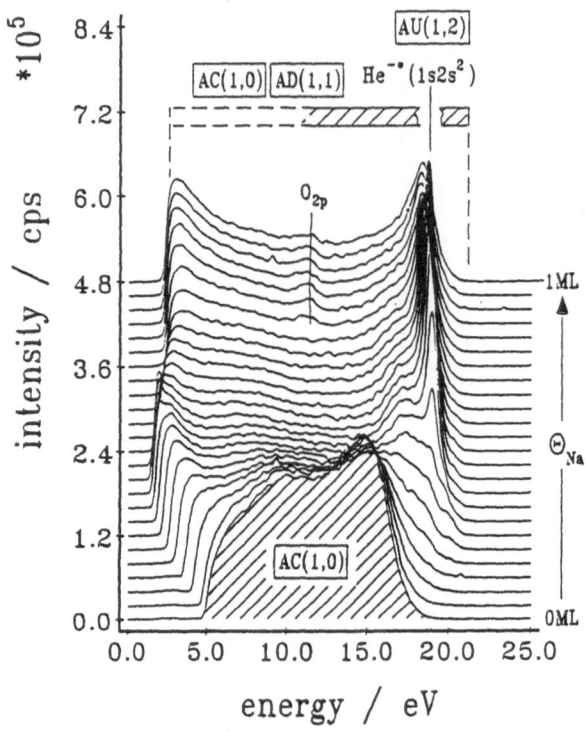

Fig.3. *Electron spectra for He$^+$/W(110) at an ion energy of 50 eV, and an elevation angle with respect to the surface of 5^0. Different spectra are obtained for different Na-coverage of the surface as indicated.*

are AC-processes that do not involve a captured electron. The process labeled by "1" is ascribed to AC(2,0) with He^{+*}(n=2) as the final state. At higher coverages and lower work function the processes involving captured excited electrons dominate. It is interesting to note, that the capture of two electrons into a singly charged ion, leading to the formation of the negative, doubly excited ion He^{-**}(n=2, n'=2), which subsequently decays by AU(1,2), is suggested. The process labeled "2" is ascribed to an AD(2,1)-process in which the captured electron in He^{+*} is deexcited from (n=3) to (n=2). At the higher work functions, this process cannot lead to positive ejected electron energies.

DESCRIPTION OF THE MODEL

General Considerations

The model is developed for the case of low collision velocities, and especially of low perpendicular velocities. Under these conditions it may be assumed that the metal electronic system responds to a good approximation adiabatically to the slow changes of the projectile surface distance. and that therefore the combined electronic system may be described in terms of Born-Oppenheimer type potential curves for the relevant electronic states, and in terms of transitions between these curves. In principle these potential curves are dependent on the lateral impact position of the projectile. However, at sufficiently low perpendicular velocity the processes to be described occur at large distances, where the trajectory is determined by the distance alone. In the model, therefore, one dimensional potential curves $V_i(z)$, with (z) the projectile-surface distance, are used. The ionizing transitions between these potential curves are further assumed to be "vertical", in the sense that the projectile kinetic energy is conserved and consequently the ejected electron energy (ε_{ik}) is given by

$$\varepsilon_{ik}(z)=V_i(z)-V_k(z). \tag{7}$$

In order to make a semiclassical description of the problem possible, it is assumed that a scattering potential can be defined which determines the projectile trajectory (z(t)) independently of the electronic processes that occur.

At the level of approximations indicated above, it would in principle be possible to carry out semiclassical calculations of electron spectra, similarly as has been done for the case of spontaneous ionization during atom-atom collisions (Gerber and Niehaus,1976, Niehaus,1990). In such calculations the phases of the different transition amplitudes contributing to a certain electron intensity at energy (ε) are taken into account. Since in the experimental electron spectra for ion-surface collisions no indications of interference structures have been observed, the more simple classical formulation has been chosen. In this classical formulation the time- or distance evolution of the electronic system is described by population probabilities ($N_i(z)$), and transition probabilities ($G_{ik}(z)$). Here we label by the indices (i) and (k) the states of the atomic particle. The situation is complicated by the fact that, for given atomic states, the metal electrons involved in the various transitions can have different binding energies. These binding energies determine the transition energy directly and can be taken into account in relation (7). In order to account for the influence of the binding energies on the transition probabilities, we have to distinguish between the different types of transitions. In the case of resonance transitions, where resonance at a certain distance arises only for a well defined binding energy $E_b(z)$, the problem is easily solved by including the dependence on $E_b(z)$ into the z-dependence of $G_{ik}(z)$. However, in the case of AD and AC, where electrons with different binding energies are involved in transitions (i→k) at a certain distance, the the functions $G_{ik}(z)$ to be used in the rate equation for the populations are integrals over differential transition probabilities. Calling these differential probabilities $g_{ik}(E_b,z)$, and $g_{ik}(E_b^1,E_b^2,z)$ for AD, and AC, respectively, one has the following relations for the contributions to the electron spectrum arising from transitions at a certain distance:

$$p(z,E_b)_{ik}= \left| \frac{N_i(z)\ g_{ik}(E_b,z)}{v(z)\ d\{\varepsilon_{ik}(z,E_b)\}/dz} \right|_{\varepsilon=\varepsilon_{ik}(z,E_b)} \quad , \text{ for (AD)} \tag{8}$$

$$p(z,E_b^1,E_b^2)_{ik}= \left| \frac{N_i(z)\ g_{ik}(E_b^1,E_b^2,z)}{v(z)\ d\{\varepsilon_{ik}(z,E_b^1,E_b^2)\}/dz} \right|_{\varepsilon=\varepsilon_{ik}(z,E_b^1,E_b^2)} \quad , \text{ for AC)} \tag{9}$$

The transition probabilities to be used in the rate equations are for (AD) and (AC), respectively:

$$G_{ik}(z) = \int_{\phi}^{\phi+E_f} g_{ik}(E_b,z)dE_b \; , \text{ and } \; G_{ik}(z) = \int_{\phi}^{\phi+E_f} g_{ik}(E_b^1,E_b^2,z)dE_b^1dE_b^2 \; , \qquad (10)$$

where the integration extends from the work function (ϕ) to $(\phi+E_f)$, with (E_f) the Fermi energy. For resonance transitions (RT) the analogous expression is:

$$G_{ik}(z) = \int_{\phi}^{\phi+E_f} g_{ik}(E_b,z)\delta(E_b-E_{ba}(z))dE_b = g_{ik}(E_{ba}(z),z). \qquad (11)$$

Once the differential transition probabilities (g_{ik}) are known, the transition probabilities (G_{ik}) connecting the various atomic states are obtained using relations (10, 11), and the set of rate equations can be solved for the population probabilities $N_i(z)$ of these states. Once the $N_i(z)$ are known, the electron spectrum can be constructed using relations (8,9). To apply these relations, not only the functions g_{ik} have to be known, but also the potential curves $V_i(z)$ from which the electron energies ε_{ik} are obtained.

In our model the approximate classical description outlined above is used in the following way:

(i) The functions g_{ik} are represented as a product of a constant and a function, as

$$g_{ik} = A_{AD}*h_{ik}(AD) \; , \; g_{ik} = A_{AC}*h_{ik}(AC), \; g_{ik} = B*h_{ik} \; (RT), \; g_{ik} = C_{ik} \; (AU). \qquad (12)$$

The functions h_{ik} are estimated, while the two constants for all Auger-type transitions, and the one constant for all resonance transitions, are free parameters to be adapted by comparing calculated spectra with experimental ones. The autoionization probabilities are assumed to be constant. In the case of the He-projectile the constants (C_{ik}) for the relevant (AU) processes are taken to be the known values for the transitions of the isolated atom.

(ii) The potential curves $(V_i(z))$ for all states are estimated.

(iii) The average projectile-surface potential necessary to calculate the velocity function $v(z)$ is estimated.

Details concerning the estimates mentioned are given below (atomic units will be used). We would like to emphasize that these estimates are rather rough, but in our opinion appropriate for use in the present, very approximate, formulation of the complex problem.

Resonant One-Electron Processes

One of the crucial quantities entering the estimates of transition probabilities is the density of electronic states of the metal electrons at the surface, the so called (SDOS). We represent these functions in terms of a product of a normalized distribution function $(P(E_b))$, and the number (n_{el}) of electrons per metal atom contained in the conduction band:

$$\rho(E_b) = n_{el}* P(E_b). \qquad (13)$$

It is assumed that the (SDOS) is not influenced by the projectile, and therefore the functions $P(E_b)$ are taken from literature. This is of course a problem, because here we need the distribution function for the first layer of surface atoms, while the functions available in the literature are obtained using the methods of (UPS) or (XPS) which means that they are an average over several of the first layers.

Our estimate of the transition probability function for (RT) is based on the assumption that the transitions are proportional to the square of the overlap integral of the one electron wave function of the active electron in initial and final state of the transition. Since we want to approximate this overlap integral only for large projectile-surface distances, we further assume that its distance dependence may be described in terms of the exponential tails of the wave functions. As usual we approximate the exponent of the tail as $\beta(z)=[2E_b(z)]^{1/2}$. Since in case of a resonance transition the binding energies are the same in initial and final state, one obtains for the exponent describing the distance dependence of the square of the overlap integral:

$$\gamma(z,E_b)_{RT}= \left(\frac{1}{2\beta(z)} + \frac{1}{2\beta(z)}\right)^{-1} = [2E_b(z)]^{1/2} \qquad (14)$$

The dependence of the binding energy on (z) is given by the condition $E_b(z)=E_{ba}(z)$. The binding energy of the electron on the projectile depends on initial an final state of the projectile. The way in which we estimate the variation with (z) of the binding energy $E_b^{ik}(z)$, will be outlined further below. In order to be able to use one constant (B) for transitions to projectile states of different "size", we have to take into account that the surface area that is effective for tunneling transitions is proportional to $(r_k)^2$, with (r_k) the "radius" of the electron orbital in question. We approximate (r_k) using "Slaters rules" (Slater,1930). The final expression we obtain for the probability of resonance capture (RC) is:

$$G_{ik}(z)_{RC}= B\ n_{el}\ P(E_{ba}^{ik}(z))\ (r_k/r_0)^2\ exp(-z\ \gamma^{ik}(z)_{RT}), \qquad (15)$$

with (r_0) the radius of the orbital to which (B) is normalized. The situation is schematically indicated in fig.4.in terms of potential curves. Due to the different

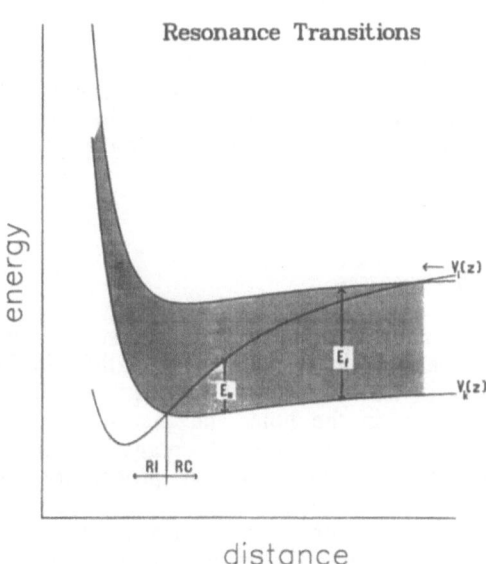

Fig.4. *Potential curve diagram for resonance transitions. A resonance capture transition (i→k) at a certain distance leads to excitation of the metal by energy E∗, which can range from zero up to the Fermi-energy E_f. E∗ is assumed to be dissipated to the metal. Below a certain distance only resonance ionization is possible.*

projectile-surface interaction in states (i) and (k), the potential curve $V_i(z)$ crosses the band of curves $V_k(z)$. A certain energy position in the band indicates the extra binding energy (E∗) below the Fermi-level of the "hole" created when the electron is captured by the projectile. If the system is initially in state $V_i(z)$, an electron can be resonance captured (RC) as long as $V_i(z)$ lies within the band. It is now assumed, that the capture process is irreversible, because the "hole" will be available only for a time much shorter than the time corresponding to the transition rate. The "time of availability of a hole", relevant in the present context, is on the one hand given by the electronic decay, and on the other hand by the mobility of the hole. This leads to times of the order of a few atomic units (Zeijlmans van Emmichoven and Niehaus,1990), which are indeed short compared to the times corresponding to tunneling rates at distances of a few atomic units. The excitation energy of the metal, which results when a metal electron of a binding energy larger than the work function is captured, may therefore be assumed to be dissipated. In the figure this is indicated by the vertical arrow to the ground state. When the system is prepared in the state $V_k(z)$- either from the beginning, or by (RC)- an electron may be lost to empty states above the Fermi-level - resonance ionized (RI)- as soon as the distance becomes smaller than the crossing distance indicated in the figure. For the probability of (RI) the (z)-dependence is in the present approximation the same as the one given by expression (15), but the constants have to be modified to account for the different direction of the tunneling process. In the original version of the model (Zeilmans van Emmichoven et al.,1988, Eeken et al.,1992), where the proportionality to $P(E_b)$ had not yet been taken into account, the (RI)-probability was taken to be proportional to the (RC)-probability, with the value of the factor of proportionality as a free parameter. It was found, that a value of the order of unity gave best agreement. The more refined estimate implied in expression (15), asks for a different approach. In the present version of the model, which will be applied to the analysis of He-spectra, we estimate the necessary modification of (15). This is done by accounting (i) for the different electron density relevant in (RI), and (i) for the fact that the binding energy of the atomic electron has a well defined value at given (z). The expression for the (RI)-probability we obtain in this way is:

$$G_{ik}(z)_{RI} = B \ (r_k/r_0)^2 \ (r_s/r_k)^3 \ \exp(-z \ \gamma^{ik}(z)_{RT}) \ . \tag{16}$$

(r_s) is the radius of the spherical volume available for one electron of the conduction band. The value of this quantity is taken from literature (Ashcroft and Mermin, 1981). Successive (RT) processes, and (RT) processes in different electronic states of the projectile are all treated in the same way, using expressions (15,16), and the extra conditions outlined above.

Auger-Type Two-Electron Processes

Like in the one-electron resonance transitions, also in the two-electron Auger-type transitions (AD) and (AC) a hole in the projectile particle is filled by a metal electron. The main difference is that, the binding energy of the electron in the metal differs from its binding energy in the hole, and that a second electron has to participate in the transition in order to account for energy conservation. As a crude estimate, the distance dependence of the transition probability for these two electron processes is usually assumed to be proportional to the square of the overlap of the wave functions of the "down electron"- i.e. of the electron that fills the hole- before and after the transition. This estimate is supported by theoretical calculations and by semiempirical analyses of electron spectra for both, interatomic Auger-type processes, and Auger-type processes during atom-surface interaction. We also use this estimate in our model. At the level of approximation chosen for (RT)-transitions, we therefore obtain the following exponent of the exponential decrease of an Auger-type transition (i→k):

$$\gamma^{ik}(z,E_b)AUG= \left(\frac{1}{2[2E_b]^{1/2}} + \frac{1}{2[2E_{ba}^{ik}(z)]^{1/2}} \right)^{-1} \qquad (17)$$

Regarding the dependence of the transition probability on the "size" of the hole filled by the metal electron in the final state (k), we follow the argumentation outlined above for the (RC)-transitions. In case of an (AD) process we thus obtain the following expression for the differential transition probability:

$$g_{ik}(z,E_b)AD= A_{AD} \, (r_k/r_0)^2 \, n_{el} \, P(E_b) \, exp(-z \, \gamma^{ik}(z,E_b)AUG) \qquad (18)$$

Assuming that, in case of an (AC)-process caused by the transition of a "down" electron of binding energy E_b^1, the doubly differential transition probability is proportional to the density at binding energy E_b^2 of the second active metal electron, the expression analogous to (18) in case of (AC) becomes:

$$g_{ik}(z,E_b^1,E_b^2)AC= A_{AC} \, (r_k/r_0)^2 \, n_{el}^2 \, P(E_b^1) \, P(E_b^2) \, exp(-z \, \gamma^{ik}(z,E_b)AUG) \qquad (19)$$

The constants in the expressions (15),(18), and (19) are defined in such a way, that their relative values for a certain ion-surface system, and the variation of these values for different systems, can be interpreted in a physically meaningful way.Due to the shifts of the electronic levels with distance, there arise certain boundary conditions regarding the possibility of the Auger-type transitions. These boundary conditions can be taken from the potential curve diagrams shown in fig.5 for (AD), and in fig.6 for (AC). The potential curves shown are defined as the total electronic energy

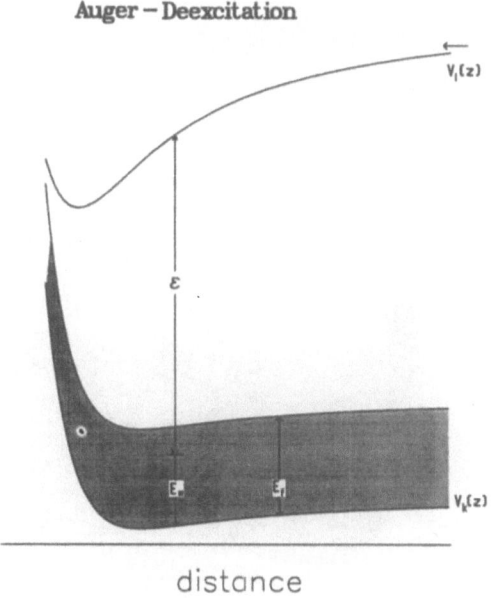

Fig.5. Potential curve diagram for Auger- deexcitation processes (AD). For a vertical transition (i→k), the metal can be excited by an energy E*, depending on the binding energy of the "down-electron". The possible excitation ranges from zero up to the Fermi-energy E_f.

for all bound electrons in the initial (i) and final (f) state. In the final state of an ionizing transition there is of course one bound electron less, and instead one electron of positive energy (ε). The difference $V_i(z)-V_k(z)$ indicated by the arrow is thus the energy (ε) of the ionized electron. The excitations arising in the metal when the active electrons that are lost by the metal had been bound below the Fermi-level, is indicated by the band of possible final state potential curves. For (AD) this band has a width equal to the Fermi-energy (E_f), and for (AC) a width of $2E_f$. For the construction of the electron spectra of course only transitions leading to positive energies ($\varepsilon_{ik}(z)$) have to be considered. The corresponding boundary conditions are evident from the diagrams. No ionization occurs for distances smaller than the distance at which $V_i(z)$ crosses the bottom of the band. On the other hand, in the case of metals with empty states above the Fermi-level, transitions may occur to negative energies of the ejected electron. These are transitions for which the ejected electron has an energy below the vacuum level and above the Fermi-level. For the calculation of the transition rates such transitions with negative electron energies, up to the limit ($-\varepsilon_{ik}(z)=\phi$), are taken into account.

Estimate of Potential Curves

To estimate the potential curves $V_i(z)$, we consider the charge-image charge interaction, and the image charge- induced dipole interaction:

$$V_i(z)=V_i^{im}(z)+V_i^{pol}(z) \tag{20}$$

At large distances, the first term is well approximated by the standard expression $(-q^2/(4z))$ for a point charge (q) at distance (z) from the image plane. As usual, the

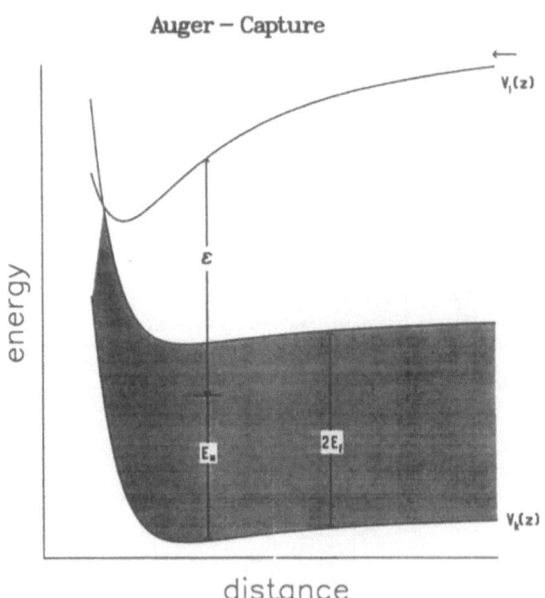

Auger – Capture

Fig.6. Potential curve diagram for Auger-capture transitions (AC). For a vertical transition (i→k) the metal can be excited by an energy E∗, depending on the binding energies of the two active metal electrons. The excitation energy ranges from zero up to two times the Fermi-energy E_f.

image plane is taken to lie a certain distance (d) outside the plane of the first surface layer. A correction which accounts for the screening of the charge by the metal electrons at small (z) is introduced by the factor $(1-\exp(-z/\lambda))$, with (λ) the typical screening length. In accordance with literature, we assume $\lambda \sim 1$ (Jennings et al.,1988). The factor leads to a finite value of the image charge attraction for $(z \to 0)$. Since in the presently treated systems the atomic particles sometimes have captured electrons in orbitals of large size, we include in addition a correction of the charge (q) by introducing an effective charge $(Z_i^{eff}(z))$:

$$Z_i^{eff}(z) = q + \sum n_r \exp(-(z+d)/l_r) \tag{21}$$

By the correction term we account for the fact that at distances of the order of the size (l_r) of an electron orbital, one positive charge becomes partly visible at the surface. In (21) n_r is the number of electron in some excited state labeled by (r), and l_r is the size-parameter of the orbital (r), which is approximated using "Slaters rules" (Slater,1930). The charge-image charge interaction term thus becomes:

$$V_i^{im}(z) = -\frac{1-\exp(-z/\lambda)}{4z} (Z_i^{eff}(z))^2 \tag{22}$$

Expressions (21,22) are also straight forwardly applied to estimate the potential curve for $He^{**}-(1s2s^2)$.
If the dipole polarizability (α_i) of the projectile particle in state (i) is large, the corresponding polarization interaction is taken into account. This interaction leads to an induced dipole of the projectile, and this dipole induces in turn the corresponding image dipole. The interaction energy of these two dipoles with the effective image charge and charge, respectively, we approximate as (Eeken et al.,1992):

$$V_i^{pol}(z) = -\frac{(1-\exp(-z/\lambda))^4}{32z^4} (1-\exp(-z/l_r)) \, \alpha_i \, (Z_i^{eff}(z))^2 \tag{23}$$

By the factor $(1-\exp(-z/l_r))$ we account for the reduction of the polarizability with decreasing distance when the distance is of the order of the size of the polarized orbital.
In case of the systems with He-projectiles, we have to consider the projectile-surface interaction for nearly degenerate hydrogenic states $He^{+*}(n=2)$. In this case we replace the polarization interaction term (23) by an estimate of the Stark-shifts for the three states - a Σ-state, a Π-state, and the Σ^*-state. The estimated potentials we then obtain are:

$$V_\Sigma^{im}(z) = V_i^{im}(z) - S(z), \quad V_\Pi(z) = V_i^{im}(z), \quad V_{\Sigma^*}(z) = V_i^{im}(z) + S(z), \tag{24}$$

with $S(z) = \dfrac{(1-\exp(-z/\lambda))^2}{z^2} (3/8) (1-\exp(-z/l_r))$

The last factor accounts for the vanishing of the Stark-shift for $(z \to 0)$.

Transition Energies

At the level of approximation adopted so far, the energy of the ejected electron should be equal to the vertical difference between the relevant potential curves, as given by relation (7). However, our earlier analyses of the electron spectra for the system $He^{++}-Cu(110)$ had shown that (Zeijlmans van Emmichoven et al.,1988), in case of the AC-process involving the transition $He^{++} \to He^+$, the spectrum could not be satisfactorily explained if the ejected electron energy was assumed to be given by relation

(7). It was concluded that some extra energy losses arise because the local distribution of the metal electrons cannot adapt during the short time the ionization takes. A parameter was introduced into the model to describe this extra "diabatic" energy loss. This parameter was used to interpolate between the adiabatic behavior, leading to electron energies $\varepsilon_{ik}^{a}(z)$, calculated using relation (7), and a "diabatic" behavior, leading to electron energies $\varepsilon_{ik}^{d}(z)$. This diabatic behavior was modeled by assuming (i) that the holes created in the conduction band in Auger-type transitions would not decay and stay localized during the ionization time, and (ii) that the spatial distribution of the electrons corresponding to the image charge in the initial state would not change during the ionization time. The difference between the adiabatic- and the diabatic energy of the ejected electron then becomes:

$$\Delta_{ik}(z)=\varepsilon_{ik}^{a}(z)-\varepsilon_{ik}^{d}(z)= \frac{(1-\exp(-z/\lambda))}{z}\left(Q \, Z_k^{eff}(z)+ \frac{(Z_i^{eff}(z)-Z_f^{eff}(z))^2}{4} \right) \qquad (25)$$

The first term in the brackets accounts for the Coulomb interaction of the charge of the projectile with the positive holes of total charge Q created in the process. The value of Q is 0, 1, and 2, for AU, AD, and AC, respectively. The second term accounts for the image charge effect explained above. In case of doubly charged primary ions, by far the largest effect due to a partly diabatic behavior is predicted by (25) for (AC) leading from the doubly charged ion to the singly charged ion, because then Q=2, and $Z_f^{eff}(z)\sim1$. This is in agreement with the observations. In a later publication on the analysis of spectra for A^{++}/Pb(111), the free interpolation parameter was replaced by an estimated parameter. This estimated parameter is also used for the the He^{++}-spectra discussed here. What is estimated, is the ratio of the "ionization time" (t) to the "response time" of the metal (τ). In terms of this ratio the electron energies are corrected in the following way:

$$\varepsilon_{ik}(z)=\varepsilon_{ik}^{a}(z)-\Delta_{ik}(z) \, \exp(-t/\tau) \qquad (26)$$

The response time of the metal is estimated as the reciprocal surface plasmon frequency, and the ionization time is estimated as the time the ejected electron needs to cover a distance equal to the screening distance (r_s) of the metal. The ratio of these times can then be expressed in terms of (r_s) and $\varepsilon_{ik}^{a}(z)$:

$$t/\tau \sim [3/(4 \, r_s \, \varepsilon_{ik}^{a}(z))]^{1/2} \qquad (27)$$

For the screening distance (r_s) literature values are available (Ashcroft and Mermin, 1981).

Estimate of the Projectile Trajectory

At the low vertical velocities used in the present experiments, most of the Auger-type processes occur at rather large projectile-surface distances, where the interaction potential is mainly determined by the charge-image charge interaction. For the repulsive interaction of the projectile with the target atoms, therefore, a rather crude approximation is sufficient. We use a planar potential derived from a Moliere-type pair potential (Gemmel, 1974). The general expression, which can be adapted to different projectile-surface systems is:

$$V_{pt}(z)=2\pi n Z_p Z_t a \; 1.17 \, \exp(-0.3(z+d)/a) \quad \text{with} \; a=0.8853 \, (\sqrt{Z_p}+\sqrt{Z_t})^{-2/3} \qquad (28)$$

with Z_p and Z_t the atomic numbers for projectile and target atom, (n) the areal density of the surface atoms, and (a) the screening length used in the Moliere potential. The attractive part of the interaction potential is determined mainly by the charge of the projectile which leads to the charge-image charge attraction. Since the charge of the projectile is not a well defined quantity at a given distance, because due to the various electronic transitions the projectile will in general be in different charge states at the same time, there is no unique attractive interaction potential at finite (z). Only asymptotically, the attractive potential is well defined and corresponds to the charge prepared in the experiment. We approach the problem in the following way. At a given vertical velocity and initial charge state of the projectile, the set of rate equations for the population of projectile states is solved along the trajectory, starting at large (z), and using the asymptotic potential to calculate the vertical velocity v(z). When the population of the projectile drops below 0.5, the charge state used in the expression for the charge-image charge attraction is diminished by one. If the projectile had initially been doubly charged, the charge state is changed from one to zero at a distance where the population of the neutral ground state of the projectile becomes larger than 0.5. In this way a certain effective charge state is introduced which is not only dependent on (z), but also on the asymptotic vertical velocity, and on the system.

Remarks on the Calculation of the Electron Spectra

The spectra, constructed in the described way from the quantities defined in the previous sections, do not contain any angular distributions of the components of the various processes. On the other hand, the measured spectra are obtained angle resolved at an angle perpendicular to the projectile beam, i.e., to a good approximation along the surface normal. Deviations between calculated and measured spectra therefore can, to a certain extent, be due to anisotropic angular distributions. For the He^{++}/Cu(110) system some angular distributions have been measured in experiment 1 (Niehaus,1989). Except for a somewhat more pronounced angular distribution for the (AU) peak ascribed to $He^{**}(2p^2)^1D$, the main measured anisotropy observed could be explained by an angle dependent transmission through a surface potential barrier. The presence of such a barrier is expected, because the electrons are emitted, with some intrinsic angular distribution, from positions slightly above the first surface layer. The height of the barrier may be process dependent, but should be of the order of a fraction of the work function for all processes. Assuming that the intrinsic angular distribution for (AU) of $He^{**}(2s^2)^1S$ is isotropic, it was concluded from an analysis of the measured angular distribution, that the barrier for the corresponding electrons should be approximately 3 eV. Not knowing any more details, we introduce a barrier (b) of half the work function for all processes into the model. For observation of electrons in the direction of the normal, this results in the following transformation of the calculated spectrum:

$$P_{lab}(\varepsilon)=P_{calc}(\varepsilon)\ \varepsilon/(\varepsilon+b) \tag{29}$$

If, due to a contact potential, or to a bias applied between crystal and spectrometer, a potential difference (c) arises between the vacuum levels of the surface and of the spectrometer, a similar transformation has to be applied in order to account for an energy dependant influence of such a difference. The calculated spectrum as measured in the spectrometer then becomes:

$$P_{sp}(\varepsilon'=\varepsilon+c)=P_{lab}(\varepsilon)\ (\varepsilon+c)/\varepsilon=P_{calc}(\varepsilon)\ (\varepsilon+c)/(\varepsilon+b) \tag{30}$$

In measurements of experiment 2, a bias is chosen which leads to a positive value of (c) for all work functions. The low energy edge observed in these spectra corresponds to electrons emitted at $\varepsilon=0$ and observed at energy $\varepsilon'=c$.

COMPARISON OF CALCULATED AND MEASURED SPECTRA

Results for the System Ar^{++}/Pb(111)

Before the model can be applied, one has to choose the atomic states that are thought to contribute to the electron spectra. In the case of Ar^{++}-projectiles a small problem arises, because the beam contains, besides ground state $Ar^{++}(..3p^4)^3P$, also fractions of the metastables $Ar^{++}(..3p^4)^1D,^1S$. Within the model, there is no difference regarding the transition probabilities for these different projectiles, only the transition energies are influenced. This means that the spectrum calculated for the mixed beam can be obtained by adding the different components. We therefore discuss the reaction scheme only for the ^3P-ground state component.

The choice of projectile states that are thought to contribute is in the first place based on considerations regarding the potential curves. Most of the higher excited states with one or two captured electrons will not be populated at smaller distances because the critical distance below which only resonance ionization is possible, is large. In the present case only $Ar^{+*}(..3p^4nl)$ and $Ar^{**}(..3p^4nl,n'l')$ with n=4 have to be considered. In addition, as has been indicated in a previous section, a qualitative interpretation of measured spectra can lead to a further selection. Details for the $Ar^{++}/Pb(111)$ system can be found elsewhere (Eeken et al,1992). The reaction scheme which has been used for calculations is shown in fig.7.

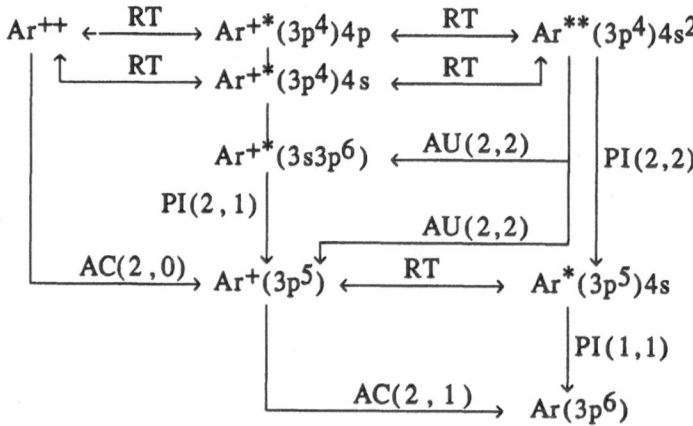

Fig.7. Reaction scheme used for $Ar^{++}/Pb(111)$

The free parameters of the model were adapted by trial and error. For the comparison with experimental spectra it was assumed, that the ground state- and the two metastable Ar^{++} ions are contained in the beam in a proportion given by their statistical weights, namely, 9:5:1 for $^3P:^1D:^1S$. A series of spectra for different energies, calculated using one chosen set of parameters, is compared in fig.8 to a series of measured spectra. The main features of the spectra, as well as their variation with energy, are rather well reproduced. In fig.9 it is shown how, according to the model, the total spectrum is composed of the components due to the various processes. The (AD) processes are here called (PI) (Penning Ionization).At the low vertical energy of 1.2 eV, the components of all possible processes are of similar importance. At higher vertical energies, all processes which occur after capture of one or two electrons become less probable, and the two AC-processes become dominant. In accordance with the measurement, only one clearly visible high energy edge due to a (PI)-process shows up in the calculations, namely, the one for (PI) caused by $Ar^{+*}(..3s3p^6)$. The edges of the other (PI)-components are less well visible, because they are smeared out due to the distance dependence of the transition energy, and in addition, because contributions

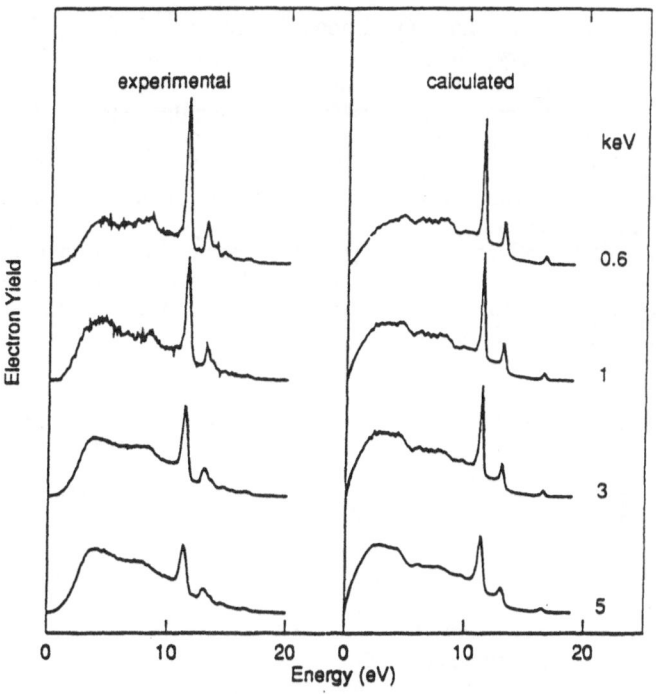

Fig.8. Comparison of measured and calculated electron spectra for Ar++/Pb(111), at four different ion energies, and at an elevation angle with the surface of 2°.

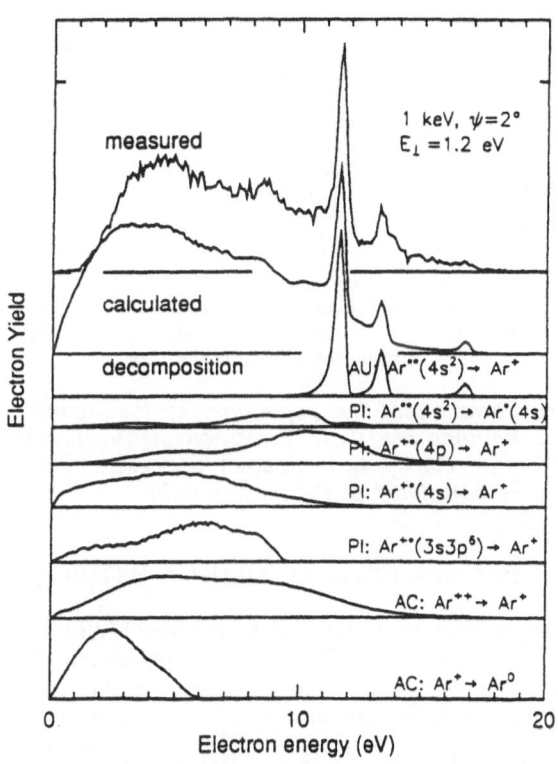

Fig.9. Decomposition of the Ar++/Pb(111) spectrum into the contributions from the various Auger- type processes, for an ion-energy of 1 keV and an elevation angle of 2°. For comparison also the measured spectrum is shown. The strong deviation of the measured spectrum from the calculated spectrum at electron energies below ca. 4 eV is due to a decrease of the transmission of the spectrometer.

from the different core states are superimposed. For $Ar^{+*}(3s3p^6)$, the distance dependence of the transition energy is much less pronounced because of the much smaller polarizability of the core excited state, and because there arises of course no superposition of components from different projectile states. The variation of the population of the various atomic states during the approach at a vertical velocity corresponding to 1.2 eV is shown in fig.10.. The neutralization to the ground state is seen to be completed at the tur- ning point of the trajectory, which lies at 0.6 (a.u.) in front of the image plane, and at 3.1 (a.u.) in front of the first layer of surface atoms. One notices that all processes occur in a rather narrow region of distances. As the vertical velocity is increased, this transition region shifts towards smaller distances. Further details on the results for the $Ar^{++}/Pb(111)$ are reported elsewhere (Eeken et al.,1992).

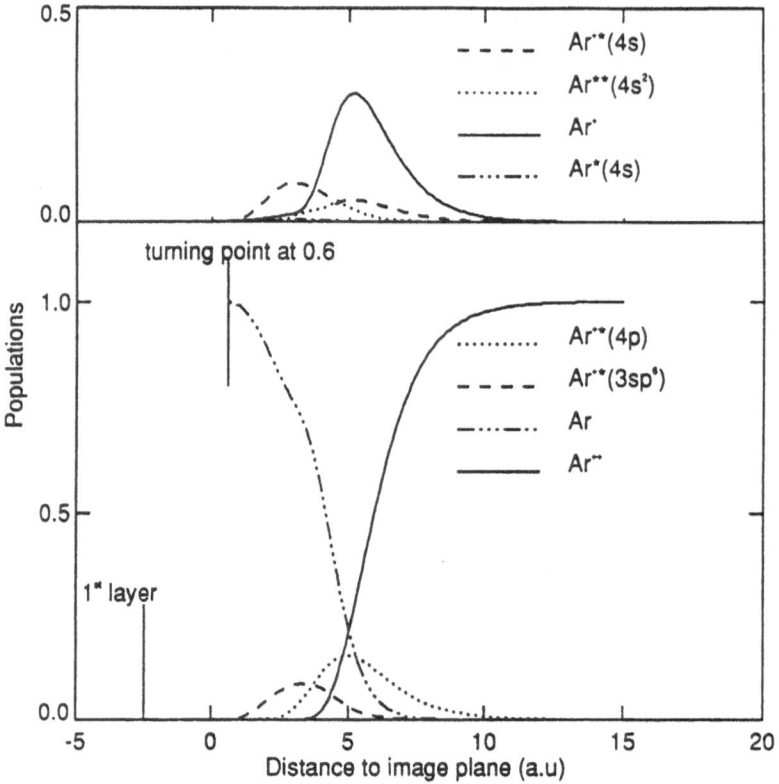

Fig.10. The calculated variation of the population of the atomic states as a function of the distance. The conditions are those corresponding to 1 keV ion energy and an elevation angle of 2^0. For clarity two windows are used. The neutralization to the ground state is seen to be completed at the turning point.

Results for He^{++} and He$^+$ with CLean or Alkalated W(110)

Some results of the application of the model to the He^{++}/W(110) system are presented in (Brenten et al.,1992a), and in a forthcoming publication (Brenten et al.,1992c). The reaction scheme used was based on a qualitative interpretation of the features observed in the spectra, as indicated in one of the previous sections. The reaction scheme used in the present analysis is essentially the same, except for inclusion of the AC(1,0)-process He$^+$→ He*(n=2), and of an AD(1,2)-process He**→ He which leads to the emission of two electrons. The (AC)-process was included because it becomes energetically possible at the very low work functions realized experimentally, and the (AD)-process was included because there is no physical reason why it should not occur. There arises of course a difficulty because the two electrons which become free when the He-1s hole is filled by a metal electron will share the excess energy in some way, so that the resulting spectrum cannot be constructed in a straight forward way. We make the following guess: (i) we assume that an electron can have energies extending from $(-\phi)$ to the maximum energy (ε_0), available when the other electron has energy $(-\phi)$. (ii) we assume that the extreme energies are preferred, and introduce one parameter (D) to account for this preference.(iii) we describe the decrease of the intensity as a function of $(\varepsilon - \varepsilon_0)$ by a Lorentz-function. The spectral component including both electrons then becomes:

$$P(\varepsilon,\varepsilon_0)=N\ (1/((\varepsilon-\varepsilon_0)^2+D^2)+1/(\varepsilon+\phi)^2+D^2),\quad -\phi < \varepsilon < \varepsilon_0,\ \int P(\varepsilon,\varepsilon_0)\ d(\varepsilon)=2 \qquad (31)$$

We realize of course that the described procedure to account for the (AD)-decay of the negative ion is highly arbitrary. On the other hand, the negative ion is only populated at the lowest work functions. The constant is chosen to have a value $D=0.5$ (a.u.). The scheme we used for the calculations reported here is shown in fig.11. The

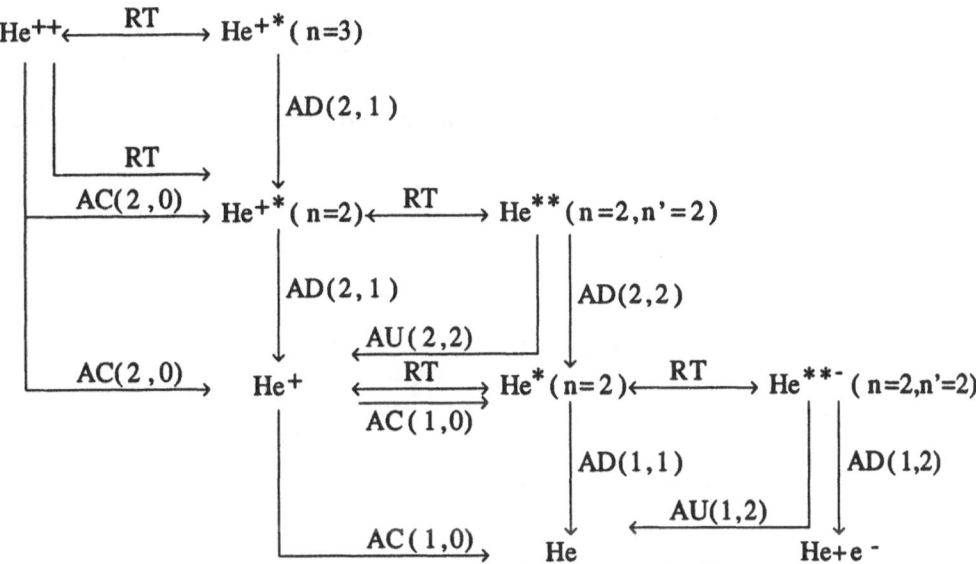

Fig.11. Reaction scheme used for the He/W(110) system

same scheme is of course used for He^{++}-, and He^{+}-projectiles. The three free parameters , A_{AC}, A_{AD}, and B, fixing the absolute transition probabilities for (AC)-, (AD)-, and (RT)-processes, are adapted by trial and error. Calculated spectra using the set $A_{AC}=0.02$, $A_{AD}=0.01$, and $B=0.1$, are shown in fig.12. and fig.13. for the case of He^{++}-, and He^{+}-projectiles. These spectra should be compared to the spectra of fig.2 and fig.3, respectively. The work function (ϕ) is varied in the same range as in the experiment, and also the vertical energy is the same in the calculations and in the experiment. With the fixed parameters, all quantities entering the description of the system are determined. Also, the variation of observable quantities upon a variation of the experimental conditions can be predicted. For instance, electron spectra for metastable He colliding at thermal energies with some metal surface can be calculated. Here we only give a few examples for conditions used in the presently discussed expe-

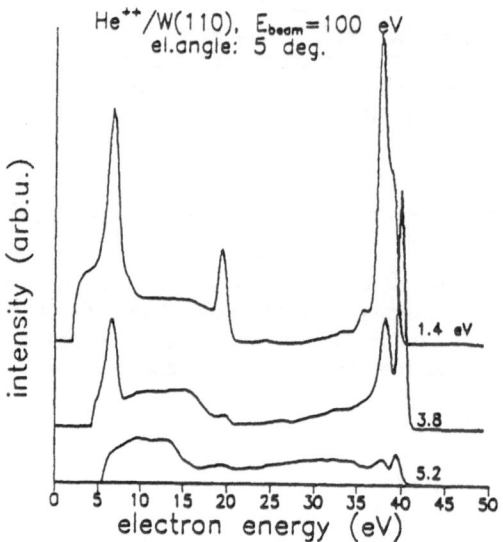

Fig.12. Calculated electron spectra to be compared to experimental spectra of fig.2.

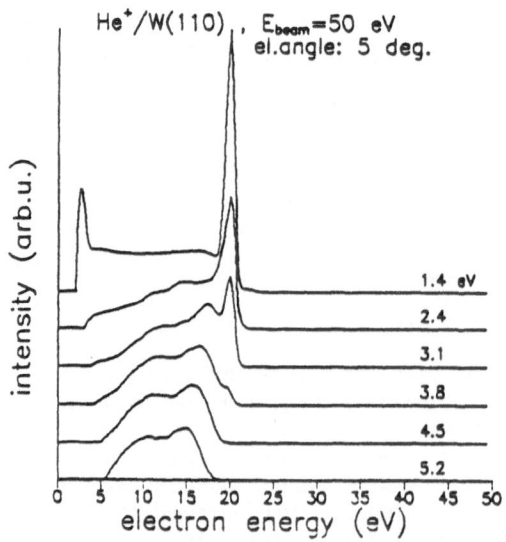

Fig.13. Calculated electron spectra to be compared to experimental spectra of fig.3.

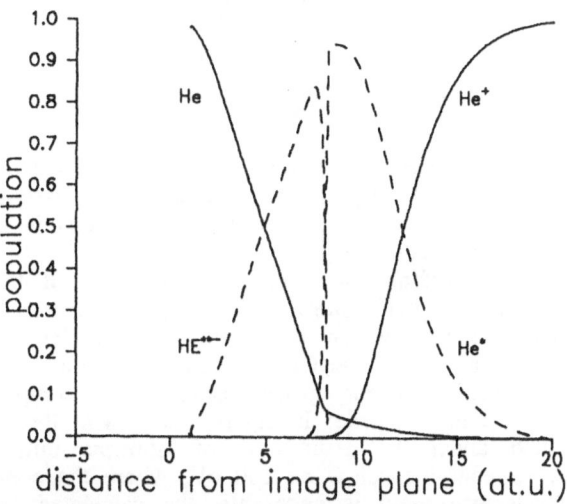

He$^+$/W(110), E_{beam}=50 eV,
elevation angle: 5 deg.
work function: 1.4 eV

Fig.14. *Calculated variation of the populations of projectile states in collisions of He$^+$ with a clean W(110)-surface of a work function of f=5.2 eV. Only Auger-capture transitions leading directly to He in the ground state occur.*

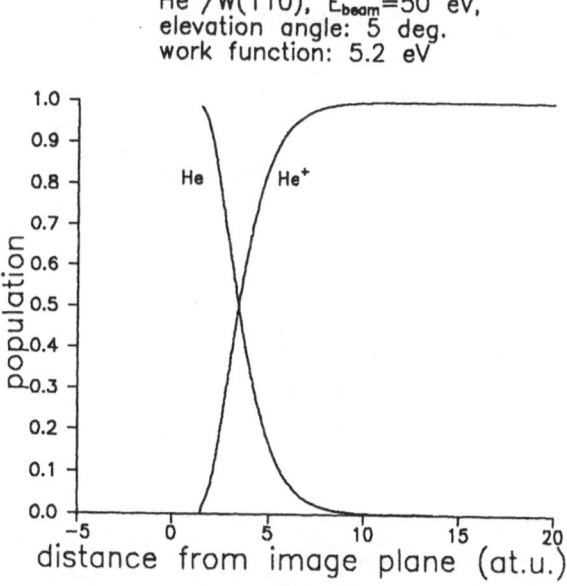

He$^+$/W(110), E_{beam}=50 eV,
elevation angle: 5 deg.
work function: 5.2 eV

Fig.15. *Calculated variation of the populations of projectile states in collisions of He$^+$ with a clean W(110)-surface of a work function of 1.4 eV. Most of the projectiles capture two electrons to form the negative ion He^{**-} (1s2s^2). The neutralization into the ground state then procedes via autodetachement (AU) and Auger-deexcitation (AD).*

99

riments. In fig.14., and fig.15. we show the population probabilities for the case of He$^+$-projectiles, for ϕ=5.2 eV, and ϕ=1.4 eV, respectively. The corresponding figures for He^{++}-projectiles are too complex to be easily represented. In fig.16 the average collision potential for He^{++} is shown for work-functions of f=1.4 eV, and =5.2 eV. Finally we give in fig.17. "reduced" transition probability functions for (RT),(AD), and (AC). Plotted are the following quantities:

$$A_{AC} \exp(-z \; \gamma^{ik}(\phi,E_k))_{AUG} \quad \text{for He}^{++} \longrightarrow \text{He}^+(n=1),$$

$$A_{AD} \exp(-z \; \gamma^{ik}(\phi,E_k))_{AUG} \quad \text{for He}^{+*}(n=2) \longrightarrow \text{He}^+(n=1), \tag{32}$$

$$B \exp(-z \; \gamma^{ik}(\phi,E_k))_{RT} \; (r(n=2)/r(n=1))^2 \quad \text{for He}^{++} \longrightarrow \text{He}^{+*}(n=2)$$

In order to allow a better comparison, the functions γ^{ik} are calculated assuming tha the active metal electron is bound by the work function (ϕ) before the transition in all three cases, and that the binding energy of the "hole" (E_k) is independent of (z). In this way the γ^{ik} become constants.

If assumptions are made regarding the angular distributions of the spectral components, also the total electron yield due to the Auger-type transitions can be extracted from the model calculations. In the described formulation this is, however, only possible for sufficiently low vertical velocities. If the trajectory enters the metal, and if the neutralization to the ground state is not completed at that point, additional assumptions have to be made. In fig.18 we present the calculated yield for He$^+$-projectiles, predicted under the assumption of isotropic initial angular distributions of the electrons. At energies below ca 100 eV, where the contribution of kinetic emission to the total electron yield is negligible, the calculated yield may be compared to experimentally determined values (Aumayr and Winter, 1993).

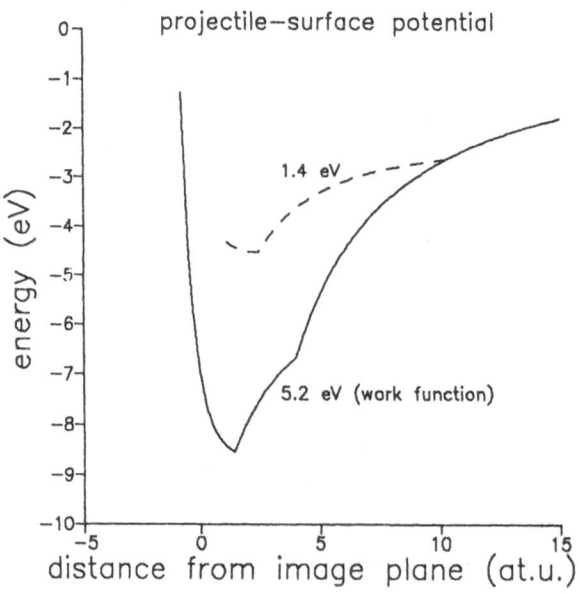

Fig.16. Average projectile-surface potential used for for He^{++}/W(110) in the calculations. The potential is dependent on the conditions of the collision. It corresponds to 100 eV ion energy at an elevation angle of incidence of 5°. It is shown for two work-function values of the metal, 1.4 eV, and 5.2 eV. The discontinuities correspond to the neutralization steps He^{++}$_{\rightarrow}$ He$^+$$_{\rightarrow}$ He.

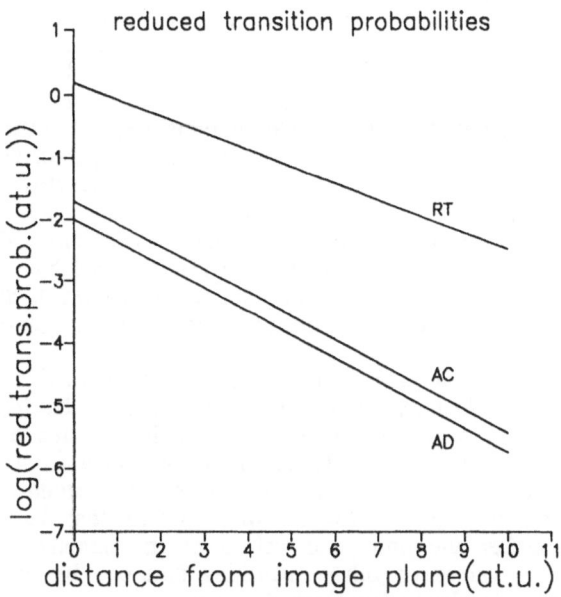

Fig.17. *The reduced transition probabilities for He^{++}/W(110) (see text).*

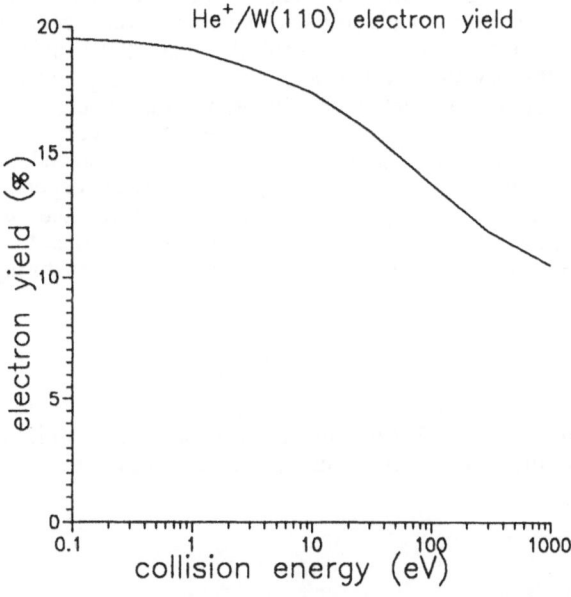

Fig.18. *Calculated electron yield for clean He$^+$/W(110) at perpendicular ion incidence. According to the calculations, the neutralization is completed before the projectile enters the first surface layer in the whole energy range of the figure*

SUMMARY AND CONCLUSIONS

It has been shown that the rather complex electron spectra due to Auger processes occurring during the slow approach of doubly charged ions to metal surfaces can be rather well reproduced by a theoretical model. This theoretical model uses transition probabilities for the electronic processes, and population probabilities for the electronic states of the projectile. The electron spectra are constructed assuming "vertical" transitions. Except for three free parameters, all quantities entering the model are estimated or taken from other sources. Two parameters fix the absolute values of the transition probability functions for the two-electron processes Auger-deexcitation and Auger-capture, and one parameter fixes the absolute transition probability functions for the one-electron resonance transitions. The model may be seen as a further development of a model originally introduced by Hagstrum(1978). On the other hand, it may also be seen as a consequent application of approximate methods developed to describe spontaneous ionization processes in atom- atom collisions (Niehaus,1990).

At the present stage, the main goal of analyses of experimental spectra by the model should be, to find out to what extent the "physics" put into the model is correct. One important criterion here is of course, whether the main features of measured electron spectra can be reproduced. But the conclusions that are drawn from the quality of some reproduction achieved in the calculations, should take into account the number of adapted parameters in relation to the amount of structure that had to be reproduced. Other criteria are, whether the numerical values of the adapted parameters are "realistic", and whether the values of other quantities that can be calculated by the model are in conflict with knowledge from other sources.

The present analysis of the spectra for $Ar^{++}/Pb(111)$, $He^{+}/W(110)$, and $He^{++}/W(110)$, has shown that all prominent features in the spectra were reproduced. This was possible by adapting a set of three parameters. The set of the same three parameters could be used for the two He-systems. In addition, in the case of He, the dramatic change of the spectra with the value of the work function, and in the case of Ar the characteristic change of the spectra with vertical collision velocity, were well reproduced. Further, the values of the absolute transition probabilities calculated using the adapted parameters, as well as the absolute value and the energy dependence resulting for the electron yield, are not in conflict with literature. Based on these results we conclude that the "physics" put into the model is essentially correct. On the other hand we do not claim, that the numerical values of the parameters adapted have now been accurately determined, and that, for instance, the distance dependence of the transition rates cannot, in reality, significantly deviate from the estimated functions. Nevertheless, we believe that the model constitutes a significant step forward in the understanding of the time evolution of ion-surface systems during slow ion-surface approach.

Acknowledgement

The author would like to thank V.Kempter, H.Brenten, and H.Müller, for communicating their experimental results prior to publication, and for fruitful discussions.

REFERENCES

Ashcroft,N.W.and Mermin,N.D.,1981:"Solid State Physics",CBS Publishing,Tokyo
Aumayr,F.,Winter,H.P.,1993,(this volume)
Brenten,H.,Müller,H.,Kempter,V.,1992a,Surf.Sci.22(in press)
Brenten,H.,Müller,H.,Kempter,V.,1992b,Z.Phys.D22:563
Brenten.H.,Müller,H.,Niehaus,A.,Kempter,V.,1992c,Surf.Sci.278:183
Brenten,H.Müller,H.Knorr,K.H.,Kruse,D.,Kempter,V.,1991,Surf.Sci.243:309
Eken,P.,Fluit,J.M.,Niehaus,A.,Urazgil'din,I., 1992,Surf.Sci.273:160

Gemmel,D.S.,1974,Rev.Mod.Phys.46:129

Gerber,G.and Niehaus,A.,1976,J.Phys.B9:123

Hagstrum,H.D.,1954,Phys.Rev.96:336

Hagstrum,H.D.,1978:"Electron and Ion Spectroscopy of Solids",Plenum,New York

Jennings,P.J.,Jones,R.O.,Weinert,M.,1988,Phys.Rev.B37:6113

Niehaus,A.,1989,in:Proc.12th Werner Brandt Int.Conf.,San Sebastian,Spain:361

Niehaus,A.,1990,Physics Reports 186(4):149

Sesselmann,W.,Woratschek,B.,Küppers,J.,Ertl,G.,Haberland,H.,1987,Phys.Rev.B35:8348

Schall,H.,Huber,W.,Hörmann,H.,Maus-Friedrichs,W.Kempter,V.,1989,Surf.Sci.210:163

Slater,J.C.,1930,Phys.Rev.36:57

Varga,P.,1989,Comm.At.Mol.Phys.XXIII 3:111

Zeijlmans van Emmichoven,P.A.,Wouters,P.A.A.F.,Niehaus,A.,1988,Surf.Sci.195:115

Zeiljmans van Emmichoven,P.A., and Niehaus,A.,1990,Comm.At.Mol.Fhys.XXIV:65

ELECTRONS FROM INTRA- AND INTERATOMIC AUGER PROCESSES IN LOW-ENERGY COLLISIONS OF SINGLY AND DOUBLY CHARGED INERT GAS IONS WITH W(110) SURFACES PARTIALLY COVERED BY ALKALI ATOMS AND NACL MOLECULES

H. Brenten, H. Müller, and V. Kempter

Physikalisches Institut der
Technischen Universität Clausthal
Leibnizstraße 4
W-3392 Clausthal-Zellerfeld, Germany

ABSTRACT

Electron energy spectra from slow He^+, He^{++}, and Ar^+ ions colliding under grazing incidence with W(110) surfaces are reported. The surface density of states and the surface work function are varied by the exposure of the W(110) surface to alkali atoms and NaCl molecules. For clean W(110) the sequence of electronic transitions during the collision is similar as reported previously for other clean metals: two electron Auger Capture processes dominate both for He^+, Ar^+ and He^{++} collisions. For sufficiently large coverages by alkali atoms resonant capture of one or two surface electrons by the projectile ions leads to the formation of core excited states with two electrons occupying outer-shell orbitals. These states decay by Auger Deexcitation (Penning ionization) and intra-atomic Auger processes (autoionization and autodetachment), respectively.

The influence of the kinetic energy onto the sequence of electronic transitions is studied by comparing 50 and 1000 eV Ar^+ collisions with W(110) partially covered by K atoms. It is shown that Ar^+ 3p vacancies are created during the collision provided the projectile energy surmounts about 300 eV. Contributions from both potential and kinetic emission can than be seen in the spectra of emitted electrons.

For W(110) fully covered by NaCl vacancies are formed in the NaCl valence band by electronic transitions. The Coulomb interaction of the charged projectile and these vacancies leads to electron spectra which in the case of He^{++} collisions strongly differ from those obtained for collisions with metallic surfaces.

INTRODUCTION

Bombardment of solids can result in the emission of electrons initiated by the potential energy (potential emission) or kinetic energy (kinetic emission) of the projectile. For both grazing incidence and low kinetic energy of the projectile the emission results

Ionization of Solids by Heavy Particles, Edited by
R.A. Baragiola, Plenum Press, New York, 1993

primarily from potential emission; electronic transitions between projectile and surface lead to electron emission before the projectile can make close contact with the surface. Electronic transitions involved under these circumstances can engage several electrons. Resonant electron transfer (RT), as for instance resonant neutralisation, often acts as a precursor for processes in which one or more electrons are ejected. Interatomic Auger processes involve at least one surface electron. In Auger deexcitation (AD) a surface electron fills a vacancy at the projectile ion thereby transferring the available excess energy to an excited projectile electron which is ejected. In Auger Capture (AC) this excess energy is rather transferred to a second surface electron which as a consequence may be emitted.

Intra–atomic Auger processes, as AUtoionization and AUtodetachment (AU), will take place after formation of multiply (at least doubly) excited projectile states via the resonant transfer of electrons from the surface to the projectile.

During the impact of inert gas ions with surfaces all the above mentioned processes can be studied simultaneously. This has already been done successfully in a number of cases, mainly involving clean metal surfaces, either via the spectroscopy of the emitted electrons,[1-4] or via the statistics for the ion-induced electron emission.[5-7] Electron spectra for collisions of singly and doubly charged inert gas ions with metal surfaces partially covered by alkali atoms were reported in.[8-11]

We report here the electron energy spectra for the impact of slow He^+ (50eV) and He^{++} (100eV) ions at W(110) partially covered by Cs atoms (notation: $Cs(\Theta)/W(110)$). Through the adsorption of alkali atoms the work function of the surface and the surface density of states can be varied considerably. We demonstrate that under these conditions the resonant capture of two electrons by the projectile from the surface becomes of major importance. Emission due to Auger Deexcitation (Penning Ionization) and intra-atomic Auger processe (autoionization and autodetachment) dominates the electron spectra under thes conditions.

By comparing 50 and 1000eV Ar^+ collisions on W(110) partially covered by K atoms (notation: $K(\Theta)/W(110)$) we demonstrate that Ar^+ 3p vacancies are produced during the collision provided the Ar^+ projectile energy surmounts about 300eV. Contributions from both potential and kinetic emission can be observed in the spectra of the emitted electrons.

For the case of an insulator surface (W(110) covered by NaCl, notation: NaCl/W(110)) we show for slow He^{++} (100eV) collisions that the collision-induced vacancies produced in the valence band strongly influence both the probabilities and transition energies for the various Auger processes. This is caused by the strong Coulomb interaction between the vacancies in the surface layer and the charged projectile.

APPARATUS AND MEASUREMENTS

The apparatus was documented previously.[12,13] Inert gas ions (He^+, He^{++} and Ar^+) are formed in a gas discharge source capable to produce He^+, Ar^+ (He^{++}) ion current densities of the order of 10^{-7} (10^{-9}) A/mm^2 at the W(110) single crystal at 1keV beam energy. The ion beam is mass/charge selected by a Wien filter. It impinges upon a W(110) crystal held at room temperature during the measurements. The incidence angle of the beam can be varied while the electrostatic analyser recording the energy distribution of the ejected electrons is positioned at a fixed emission angle of 90 degs with respect to the beam axis. The spectrometer (type EA10/100 of Leybold Inc.) records the energy spectra at constant pass energy ($\Delta E = 0.2$ eV). The spectra of the slow electrons were taken in the following way: the difference in the work functions (WF) of the crystal (typically partially alkalated) and the analyser was overcompensated by biasing the electrostatic analyser in such a way that electrons leaving the crystal with zero energy at WF = 1.5 eV (0.6 monolayer (ML) coverage) arrive at the electrostatic analyser with approximately 2 eV. This procedure affects the collection efficiency below about 10 eV electron energy, but it was checked that this procedure did not produce artefacts in the low energy part of the spectra.[11] Therefore the spectra reported in this paper, typically at an angle of incidence of 5 degs with respect to the W(110) surface, do not represent truly angle–resolved electron spectra, but are to some extent angle–integrated at energies below 10 eV. A (1/E) correction has been applied to the spectra in order to compensate the energy dependence of the electrostatic analyser's transmission.

Alkali atoms (Cs, K) are offered to the surface by means of SAES Getters Inc. dispenser sources. The partial alkali coverage (in units of completed monolayers (ML) at room temperature) is determined by WF measurements. The relation between WF and the alkali coverage is established by combining WF measurements with AES results.[14,15]

NaCl is offered to the surface by evaporating molecules from chips of NaCl single crystals at about 700K.[16,17] From ELS, UPS, and metastable impact electron spectroscopy (MIES) studies[16,17] we conclude that a dense overlayer of NaCl molecules (corresponding to 1 A.U. exposure in **Figs. 6 and 7**) can be formed in this way. The results are compatible with the assumption that two-dimensional islands grow laterally with increasing exposure to NaCl; the molecular axes in the first double layer are parallel to the surface normal and the orientation of neighbor molecules alternates.[17]

The procedure for cleaning the W(110) crystal prior to the preparation of the adlayer is described elsewhere.[12] The cleanliness is checked by AES and WF measurements. In addition ISS with the inert gas ion beams is employed to search for C and O impurities.

The experimental results display energy spectra of the ejected electrons for grazing incidence under 5 degs with respect to the surface in order to avoid penetration effects of the primary ions. The change of the position of the low–energy cut–off of the spectra reflects the change of WF upon exposure to alkali atoms and NaCL molecules starting from the value of the clean W(110) surface (5.3 eV).

DISCUSSION

The discussion of the data obtained for partially alkalated W(110) will be based on the reaction scheme presented in **Fig. 9**.[2,8,10] It anticipates that a number of well defined intermediate charge states of the projectile are populated via the various electronic transitions as indicated by the acronyms given with the arrows interconnecting the states between which the transition occurs. The numbers given in the brackets denote the number of inner shell vacancies and valence electrons, respectively, in the projectile prior to the indicated electronic transition. The same notation is also used in the Figures

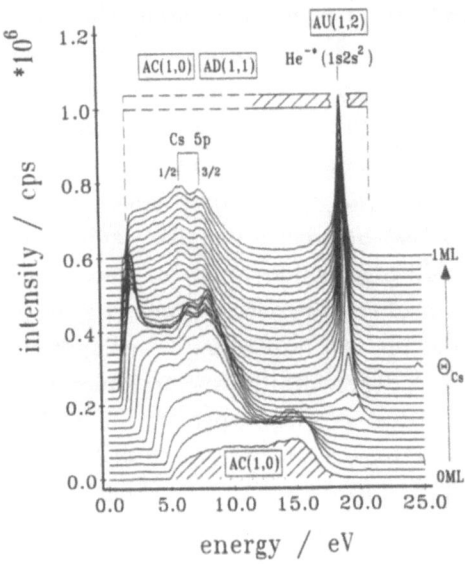

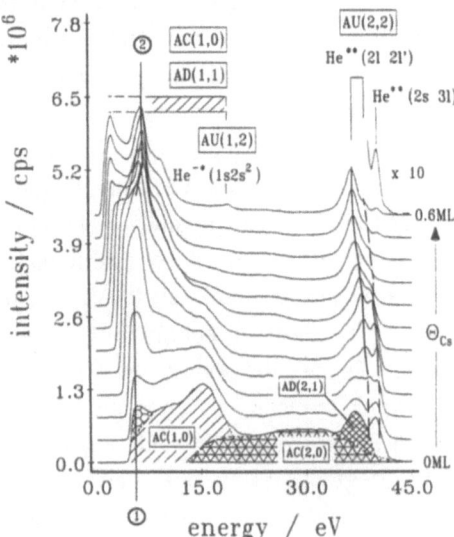

Figure 1. Electron energy spectra after 50eV He$^+$ impact energy for collision with Cs(Θ)/W(110) as a function of the Cs coverage Θ. The angle of incidence with respect to the surface is $\Phi = 5$ degs.

Figure 2. Electron energy spectra after 100eV He^{++} impact with Cs(Θ)/W(110) as a function of the Cs coverage Θ. The angle of incidence with respect to the surface is $\Phi = 5$ degs.

with experimental data in order to identify the observed features.

Potential Emission In He$^+$, He^{++} And Ar$^+$ Collisions With Alkalated W(110)

The analysis of the results obtained for 50eV He$^+$ and Ar$^+$ collisions (see **Figs. 1 and 3**) can be summarized as follows:[8,11] for small coverages, in particular when scattering from clean W(110) (see bottom curve of **Figs. 1 and 3**), the projectile ion is essentially neutralized by AC involving W(5d) electrons. At sufficiently large coverages (at least below a WF of 2.5eV) the neutralization proceeds via intra– and interatomic Auger processes involving the alkali s–valence electron, and, in the case of the Cs adsorbate, also the Cs 5p electrons (see **Fig. 1**): at intermediate coverages (up to 0.7ML) the high energy part of the spectra is dominated by the intra–atomic Auger process: autodetachment of He^{-*} (1s 2s^2 ^{2}S) and Ar^{-*} (3p^5 4s^2) which are formed from He$^+$ and Ar$^+$ by the resonant capture of two electrons.[28,8] The narrow peak (0.5eV FWHM) due to the autodetachment of of He^{-*} resides on top of a background which originates at energies above about 12eV mainly from AD of He* (1s 2s) formed from He$^+$ via single resonant electron capture. The near-zero energy electrons formed around the WF minimum may originate from the AD of higher excited He* states (formed by resonant capture) when they decay into the He* (1s2s).[18]

From the analysis of the 100eV He^{++} collisions (**Fig. 2**) we come to the following conclusions:[8,10] for small alkali coverages, in particular for clean W(110), the He^{++} is mainly neutralized in two consecutive AC processes involving W(5d) electrons. There are however also events in which the first AC process produces He^{+*} (n=2) excited ions which suffer an AD process to the He^+ ground state. The He^+ is finally neutralized again by AC. The first step in the neutralization sequence results in very slow electrons (structure (1) in **Fig. 2**). It is remarkable that in the sequence of events three electrons are emitted during the collision process per impinging He^{++} projectile.[5-7,10] For large coverages resonant capture of one electron from the surface by the He^{++} projectile dominates, and leads to the formation of He^{+*} (n=3) ions. Structure (2) in **Fig. 2** is due to AD of He^{+*} (n=3) to the He^{+*} (n=2) states.[8,10] Resonant capture of a second electron by the He^{+*} (n=2) leads to the formation of He^{**} (2l 2l'). The autoionization (AU) of these doubly excited states of neutral He is reflected by the prominent structure near the high energy end of the spectra.[1-4,8,10] The smooth contribution to the low-energy part (below 20eV) of the spectra is mostly due to AD of He^* atoms formed as an intermediate state during the neutralization sequence (see **Fig. 9**).

In order to put our interpretation on a firmer basis we have simulated the energy spectra emitted in slow He^+ and He^{++} collisions with alkalated W(110) by using the model introduced in[2,4] where it was applied to He^{++} collisions with clean metal surfaces. Basically, it assumes that in the course of the collision process a number of well defined intermediate states (see **Fig. 9**) are occupied. They are characterized by their time (or distance) dependent occupation probabilities. Coupled rate equations, one for each involved intermediate state, take into account the various electron transfer processes which do cause transitions between these states. Details of the model and calculations

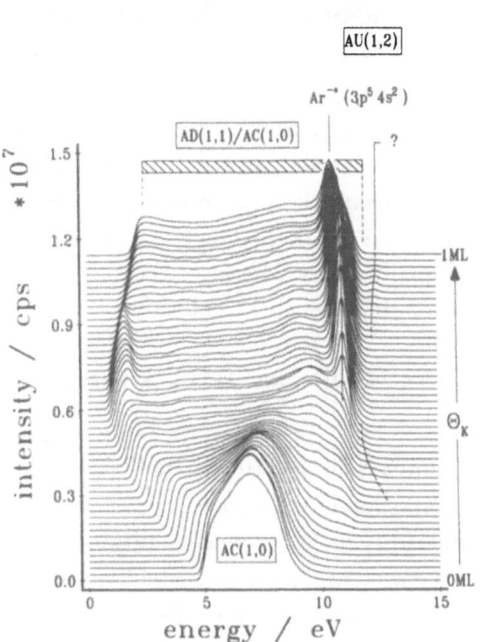

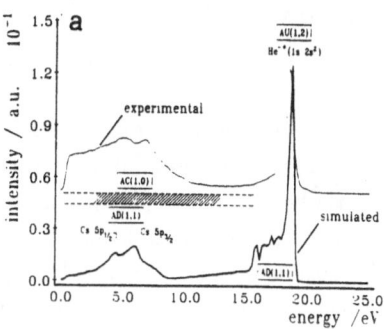

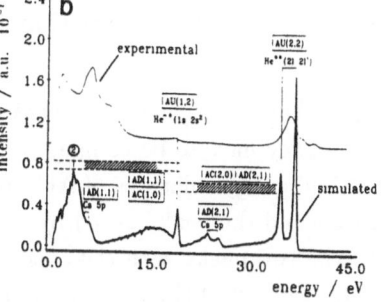

Figure 3. Electron energy spectra at 50 eV Ar^+ impact energy for collision with $K(\Theta)$/W(110) as a function of the K coverage Θ. The angle of incidence with respect to the surface is $\Phi = 5$ degs.

Figure 4. Simulated and measured electron energy spectra for (a) 50 eV He^+ impact (b) 100 eV He^{++} impact with cesiated W(110) (cesium coverage 0.6ML). The angle of incidence with respect to the surface is $\Phi = 5$ degs.

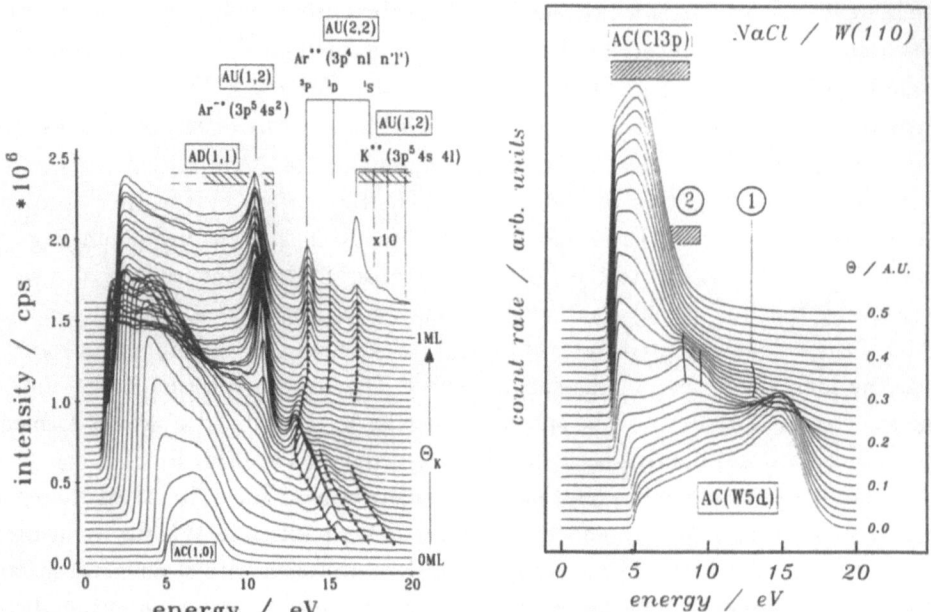

Figure 5. Electron energy spectra at 1 keV Ar$^+$ impact energy for collision with K(Θ)/W(110) as a function of the K coverage Θ. The angle of incidence with respect to the surface is $\Phi = 5$ degs.

Figure 6. Electron energy spectra at 50 eV He$^+$ impact with NaCl(Θ)/W(110) as a function of the NaCl exposure Θ. At 1 A.U. the substrate is completely covered with NaCl. The angle of incidence with respect to the surface is $\Phi = 5$ degs.

are discussed in the previous paper.[4] **Fig. 4(a) and (b)** show a comparison of the experimental and the simulated spectra for slow collisions He$^+$ and He^{++} ions colliding with partially cesiated W(110). The black dots designate the peak positions while the lines connect the peaks obtained as a function of the alkali coverage.[19] It is obvious that the observed structures can be reproduced qualitatively. Moreover the simulations confirm the reaction sequences proposed above for He$^+$ and He^{++} collisions.

Potential And Kinetic Emission In 1keV Ar$^+$ Collisions With Alkalated W(110)

Fig. 5 shows energy spectra for Ar$^+$ (1keV) impact on W(110) partially covered by K atoms. They have to be compared with the corresponding low energy results of **Fig. 3**. The high energy collisions of Ar$^+$ with K/W(110) (**Fig. 5**) produce additional peaks located beyond the peak caused by autodetachment of Ar^{-*}. They can be identified as follows: the two peaks with the excitation energies 12 and 13.2eV (with respect to the low energy cut–off of the spectra) agree well with those attributed to autoionization of Ar** (3p^4 nl nl') with the core states ^{3}P and ^{1}D after collisions of doubly charged Ar^{++} ions with clean Cu(110).[4] In the present situation these autoionizing states are formed via RT into the Ar 4s level after an additional core vacancy 3p^{-1} has been generated via electron promotion. Results at other projectile energies indicate a threshold energy of about 200eV for this process.[19] The third peak at 14.8eV is not due to autoionization of Ar** (3p^4 (^{1}S) 4s^2) which should appear at 15.8eV,[22] and should be much weaker if the

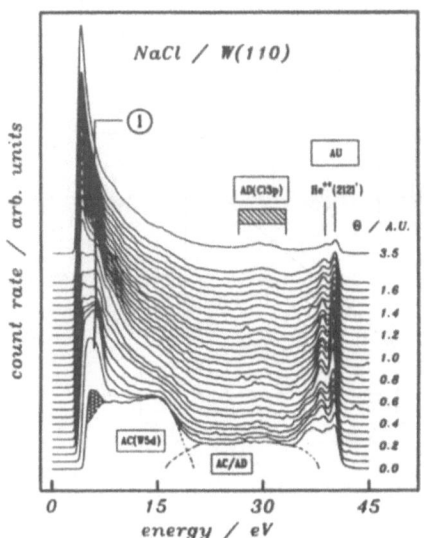

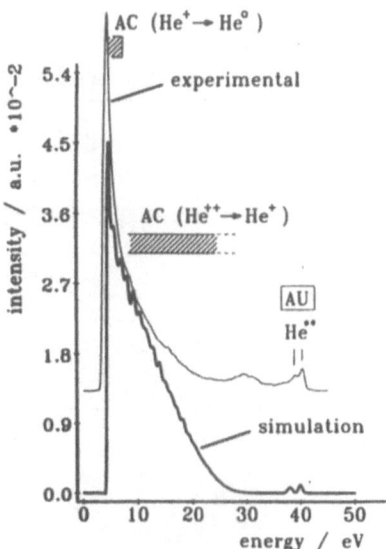

Figure 7. Electron energy spectra after 100eV He^{++} impact with NaCl(Θ)/W(110) as a function of the NaCl exposure Θ. Otherwise as in Fig. 5.

Figure 8. Simulated and measured electron energy spectra after 100eV He^{++} impact with W(110) fully covered (3.5 A.U.) by NaCl. The angle of incidence with respect to the surface is $\Phi = 5$ degs.

three core states would be populated statistically. By comparison with the results of[20] we attribute it to autoionization of K** ($3p^5 4s^2$). This is supported by the occurence of additional weak features in **Fig. 5** at even higher energies. As for the collisions of alkali ions (Li$^+$, K$^+$) with alkalated W(110) the production of inner shell vacancies, here in the 3p shell of Ar and K, can be explained by the promotion of electrons induced by the kinetic energy of the projectile:[20,21] during the collision event the projectile forms transient quasi-molecular states with that particular adatom with which it comes into closest contact. Under the influence of the kinetic energy of the projectile promotion of electrons from occupied into empty orbitals of the quasi-molecule may occur at intermolecular distances where these orbitals are nearly degenerate. This model has been extremely useful for describing electronically inelastic processes in binary ion-atom collisions,[23-25] and was already successfully applied to ion–surface collisions.[20,21,26] Therefore, besides the features due to potential emission (see **Fig. 3**) those due to kinetic emission (originating from electron promotion) can be seen above 200eV projectile energy. Part of the smooth low energy contribution (below 10eV) may also be due to kinetic emission.[20,21]

Potential Emission In He$^+$ And He^{++} Collisions With W(110) Exposed To NaCl

The analysis of the results for He$^+$ (50eV) impact on W(110) exposed to NaCl (**Fig. 6**) can be summarized as follows:[16,17]

for small coverages the sequence of electronic transitions was described in section 3.1. For W(110) fully covered by NaCl molecules the He$^+$ ion is neutralized in an AC process involving two Cl 3p electrons from the NaCl adlayer. Obviously, this process creates two

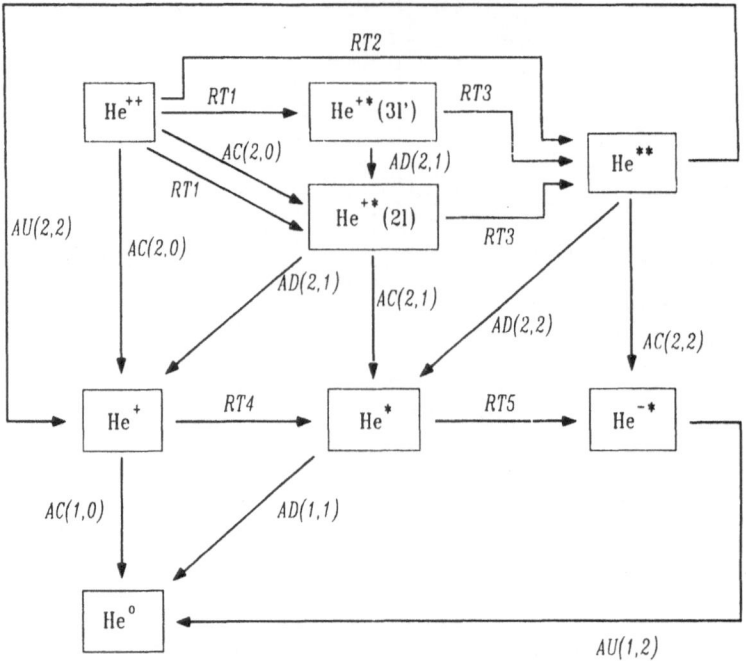

Figure 9. Scheme for the main processes, i.e. resonant transitions (RT), Auger Deexcitation (AD) and Capture (AC) and autoionization/autodetachment (AU), occuring in He^{++} impact with low work function surfaces. For the abbreviations see text.

vacancies per neutralization event. At intermediate coverages additional weak features (structures (1) and (2)) are seen due to the AD of He* formed by resonant capture and AC processes (involving both W(5d) and Cl(3p) electrons). This interpretation of the spectra of **Fig. 6** is again confirmed by simulations of the type described in section 3.1.[27]

A preliminary analysis of the results for He^{++} ions colliding with NaCl/W(110) is as follows: for small exposures to NaCl the sequence of electronic transitions was described in section 3.1. Compared to the results for clean W(110) the low-energy contribution (feature (1)) caused by AC (He^{++} → He^{+*} (n=2)) gains in importance due to the WF decrease by 1.2eV upon exposure to NaCl. For fully covered W(110) the spectra display a strong low-energy component and a comparatively small contribution of electrons with energies beyond 5eV. Simulations (**Fig. 8**) can reproduce the spectra in a qualitatively correct way only when assuming that electron emission is mostly due to Auger Capture of He^{++} to the He$^+$ ground state (Auger Capture to the He^{+*} (n=2) is inhibited as soon as the band gap of NaCl has developed), and due to autoionization after simultaneous double capture of He^{++} forming He** (2l 2l'). Thus only electrons from the very first step in the transition sequence by which He^{++} is neutralized and those from intra-atomic Auger processes appear at energies beyond 5eV. The explanation may be that the additional Coulomb repulsion between the collisionally induced vacancies in the Cl 3p valence band and the still charged projectile, in particular He^{+*} (n=2),

(1) shifts the transition energies to rather low energies, and

(2) prevents that the projectile stays close enough to the surface in order for further electronic transitions to occur.

OUTLOOK

We have presented the electron energy spectra induced by the impact of slow inert gas ions with partially alkalated surfaces. At low work functions the fate of the projectile is determined by the large probability for resonant capture into excited states leading in particular to strong features due to intra-atomic Auger processes (autoionization and autodetachment) in the electron spectra. A glance at **Fig. 10** indicates that only a comparatively small change of the work function is required to bring excited states of the inert gas atom into resonance with occupied states of the surface. We also realize from **Fig. 10** that the same is true for reactive ions as in particular H⁺ and N⁺. Again the

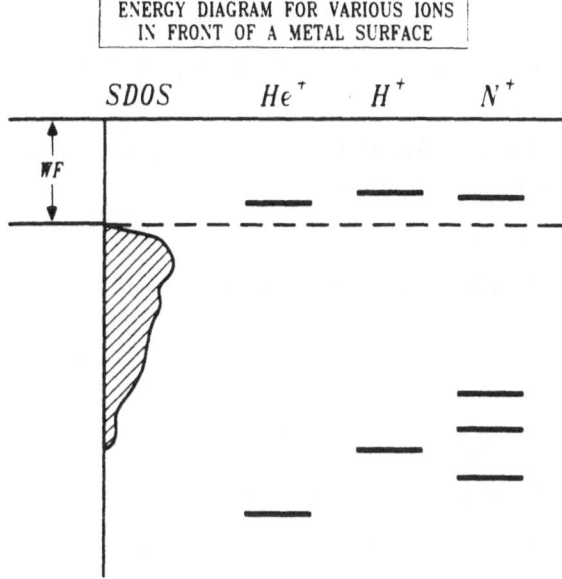

Figure 10. Energy level diagram relevant to the discussion of electronic transitions between a metal and He⁺, H⁺, N⁺ ions. The level shift and broadening in the vincinity of the surface are ignored.

formation of core projectile states subject to intra-atomic Auger processes is expected when decreasing the work function. However, these processes may populate several final states as indicated for N⁺ in **Fig. 10**. Therefore the electron spectra may look more complicated due to the availability of several channels for intra-atomic Auger decay. For molecular projectile ions, such as N_2^+, the formation of molecular core excited states (Feshbach type resonances) can be expected which are again subject to autodetachment. Additional channels leading to the dissociation of the projectile may stay in competition though.

The study of the electron emission after slow collisions with insulationg surfaces may turn into a promising tool to characterize long lived vacancies created at the surface during the collision. In addition processes initiated by these vacancies, as for example desorption and sputter processes, may be studied.

ACKNOWLEDGEMENTS

Support of this work came from the Deutsche Forschungsgemeinschaft and from the BMFT. We thank A. Niehaus for the permission to use his computer code for the simulation of the collision-induced electron energy spectra.

REFERENCES

1. H.D. Hagstrum, G.E. Becker, Phys. Rev. **B8** (1973) 107.

2. P.A. Zeijlmans van Emmichoven, P.A.A.F. Wouters, A. Niehaus, Surf. Sci. **195** (1988) 115.

3. S. Schippers, S. Oelschig, W. Heiland, L. Folkerts, R. Morgenstern, P. Eeken, I.F. Urazgil'din, A. Niehaus, Surf. Sci. **257** (1991) 289.

4. A. Niehaus, this volume.

5. H.P. Winter, in: XVII ICPEAC (1991–Brisbane/Australia) Progress Report.

6. F. Aumayr, this volume.

7. H.P. Winter, this volume.

8. H. Brenten, H. Müller, V. Kempter, Z. Phys. **D22** (1992) 563.

9. H.D. Hagstrum, P. Petrie, E.E. Chaban, Phys. Rev. **B38** (1988) 10264.

10. H. Brenten, H. Müller, V. Kempter, Surf. Sci. (in print).

11. H. Brenten, H. Müller, A. Niehaus, V. Kempter, Surf. Sci. (in print).

12. H. Schall, W. Huber, H. Hörmann, W. Maus–Friedrichs, V. Kempter, Surf. Sci. **210** (1989) 163.

13. H. Brenten, H. Müller, K.H. Knorr, D. Kruse, H. Schall, V. Kempter, Surf. Sci. **243** (1991) 309.

14. W. Maus–Friedrichs, H. Hörmann, V. Kempter, Surf. Sci. **224** (1989) 112.

15. H. Brenten, H. Müller, W. Maus–Friedrichs, S. Dieckhoff, V. Kempter, Surf. Sci. **262** (1992) 151.

16. S. Dieckhoff, W. Maus–Friedrichs, V. Kempter, Nucl. Instr. Meth. **B65** (1992) 488.

17. S. Dieckhoff, H. Müller, H. Brenten, W. Maus–Friedrichs, V. Kempter, Surf. Sci. (in print).

18. H. Brenten, H. Müller, V. Kempter, Nucl. Instr. Meth. **B58** (1991) 328.

19. H. Brenten, doctoral thesis, Clausthal (1992).

20. H. Brenten, H. Müller, V. Kempter, Z. Phys. **D21** (1991) 11.

21. H. Brenten, H. Müller, V. Kempter, Surf. Sci. **271** (1992) 103.

22. P. Eeken, J.M. Fluit, A. Niehaus, I. Urazgil'din, Surf. Sci. **273** (1992) 160.

23. U. Fano, W. Lichten, Phys. Rev. Lett. **14** (1965) 627.

24. M. Barat, W. Lichten, Phys. Rev. **A6** (1972) 211.

25. M. Barat, NATO ASI Series **B103** (1983) 389.

26. I. Terzić, Z. Rakočević, M.M. Tošić, Surf. Sci. **260** (1992) 200

27. A. Niehaus, privat communication.

28. R. Hemmen, H. Conrad, Phys. Rev. Lett. **67** (1991) 1314.

INNER SHELL TARGET IONIZATION BY THE IMPACT OF

N^{6+}, O^{7+} AND Ne^{9+} ON Pt(110)

S. Schippers[1,2], S. Hustedt[1] and W. Heiland[1],
R. Köhrbrück[2], J. Kemmler[3], D. Lecler[4] and N. Stolterfoht[2,4]

[1] Universität Osnabrück, W-4500 Osnabrück, Germany
[2] Hahn-Meitner-Institut, W-1000 Berlin 39, Germany
[3] Centre Interdisciplinaire de Recherche avec les Ions Lourds,
   F-14040 Caen Cedex, France
[4] Laboratoire de Spectroscopie Atomique, ISMRa Université Caen,
   F-14021 Caen, France

ABSTRACT

Highly charged ions from the test bench of the 14GHz Electron Cyclotron resonance (ECR) ion source at the Grand Accelerateur National des Ions Lourds (GANIL), Caen, were used to study the ion-induced electron emission (IEE) from a single crystal Pt(110) surface. Mass and charge selected beams of N^{6+}, O^{7+} and Ne^{9+} with currents of 150nA, 60nA and 15nA are obtained. The IEE spectra are measured with an electrostatic analyzer with an energy resolution of $\Delta E/E = 2.5\%$. The dependencies of the IEE yields are studied as function of the impact angle and of the electron observation angle, i. e. the analyzer is rotable around the target in the scattering plane. The primary energy of the ions is 10q-keV, where q is the charge of the ion. Projectile velocities are hence 0.4 v_0 (v_0 = Bohr velocity). The energy component perpendicular to the surface $E_\perp = E_0 \sin^2 \psi$ reaches values below 2keV, when lowering the grazing angle of incidence ψ below 10°. The IEE spectra show beside the "δ-electron" background line emission due to Auger processes in the projectile and in the target atoms. We identify KLL and LMM projectile electrons and NNV, NOO, NVV and OVV target Auger electrons. An interesting result is the finding that Ne^{9+} does not excite the Pt-NOO and -NNV Auger emission. The analysis of possible vacancy transfer mechanisms for a K-shell projectile vacancy to the target N-shell is based on an atomic structure calculation and on a diabatic potential curve crossing model. The target N-shell ionization is identified as a vacancy transfer of the Landau-Zener type.

INTRODUCTION

Multiply charged ions are a new tool in the study of interaction of particles with atoms, molecules and solids. The multiply charged ions bring high potential energies into a given collision. The potential energy of such an ion can be larger than its kinetic energy. Naturally highly charged ions occur in the corona of stars[1]. Man made they populate partly plasmas of the fusion experiments, e. g. tokomaks[2]. For the use in scattering experiments multiply charged ions are produced in electron cyclotron ion sources (ECRIS)[3] or electron beam ion sources (EBIS)[4]. First experiments with the simplest highly charged ion, i. e. He^{++} interacting with a solid surface were done with an electron impact ion source[5]. At kinetic energies below 79eV this ion fulfills the condition $E_{kin} < E_{pot}$, where $E_{pot} = E_{ions}$ the ionisation energy of the doubly ionised He. In the discussion of the electron spectra produced by the interaction with solid surfaces Hagstrum and Becker use the theoretical model proposed by Hagstrum[6] to understand the observed effects. The main events in the interaction are resonant capture and loss processes between the ion and the conduction band of a metal, Auger processes involving the ground state of e. g. He and again the conduction band, and Auger de-excitation processes within the ion. In case of He^{++} also electrons from the autoionisation of doubly excited electrons are found. These spectra were reproduced in an experiment equipped with an electrostatic analyzer[7]. Hagstrum used a retarding field analyzer. The rotable electrostatic analyzer affords double differential spectroscopy. By

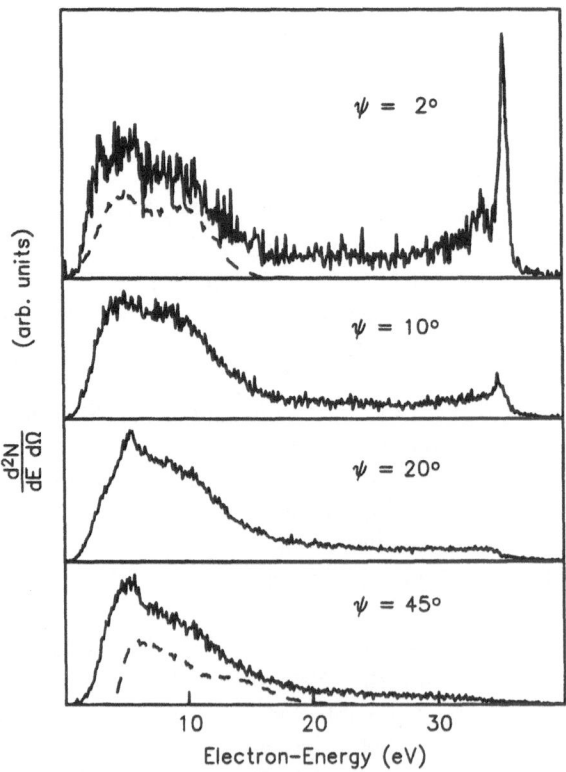

Figure 1. Electron energy spectra from the interaction of 1keV He^{++} with a Pb(111) surface at different glancing angles of incidence ψ. The peak at 35eV is due to autoionization of He^{**}, the broad peak at low energies is due to a variety of Auger capture processes. The dashed line at $\psi=2°$ and $\psi=45°$ are the electron spectra from 1keV He^+ [8].

studying the angular dependencies the time scale of the capture processes becomes a quantity accessible to the experimentalist. Fig. 1 shows angular resolved energy spectra of electrons released during the interaction of He^{++} with a Pb surface. When increasing the grazing angle of incidence the intensity of the autoionisation lines from the He^{++} at 33eV and 35eV decreases. The low energy part of the spectra is due to Auger capture processes into the doubly charged and singly charged He, i. e. the "reaction" with the surface is an Auger capture (AC 2,0) $He^{++} + e^- \rightarrow He^+$ followed by a second Auger capture (AC 1,0) $He^+ + e^- \rightarrow He^\circ$. Other processes involving excited states of He play a minor role. At grazing incidence the He^{++} has enough time to capture two electrons into excited states and relax via autoionisation. At larger impact angles the ion moves to fast into the Auger capture regime such that the excited states cannot be formed. From the experiments and the theoretical model[7] transition rates of $1.7 \cdot 10^{15}$ s^{-1} for AC(2,0) and $3.0 \cdot 10^{15}$ s^{-1} for AC(1,0) are evaluated[8]. Resonant capture rates to form He^{**} are of the order of 4 to $9 \cdot 10^{14}$ s^{-1}, the autoionisation transition rate is known from gas phase experiments.

A final aspect of the charge exchange between ions and surfaces is the quasi resonant charge exchange found first for the interaction between He^+ and Pb[9]. This effect is manifested by an oscillatory dependence of the (angular resolved) ion yield on the ion velocity. It is a case of a Stueckelberg oscillation theoretically explained by Tully for the case of ion solid interaction [10]. An interesting aspect of the theory is the "competition" between the Auger neutralisation and resonant exchange between the Pb 5d states and the He 1s state.

The first theory developed for highly charged ions beyond He^{++} was based on Auger neutralisation too[11]. In the model electrons are transferred into an outer shell of the ion, from there an Auger cascade is initiated until all levels are filled. From the experiments of the last years (for reviews see ref. 12-14) studying the total electron yield, the statistics of the electron emission, the electron spectra, X-ray emission and the charge state of the backscattered particles it became clear that the model is insufficient.

One of the major setbacks of the model is an underestimate of the rate of the resonant capture into the outer shells of the impinging ion. It is now recognized that the resonant capture for sufficiently slow ions may lead to a complete neutralisation of the ion. These particles are however in a quite unusual excited state, i. e. all electrons are almost in one outer shell with large main quantum number. From there autoionisation, Auger-de-excitation and inter-atomic Auger processes cause the electron emission and lead to the filling of the inner shells of the ions.

In this paper we will report experimental results of the electron emission of a clean Pt(110) surface under the impact of N^{6+}, O^{7+} and Ne^{9+} ions. We identify electron emission from the projectiles: KLL and LMM Auger emission. We find target Auger emission induced by a vacancy transfer from the K shell of the projectile into N-states of the Pt target atoms. These results give new insight into the neutralisation scheme of the highly charged ion after penetration into the solid.

EXPERIMENT

The experiments were done with a 14GHz Electron Cyclotron Resonance (ECR) ion soure at the Grand Accelerateur National des Ions Lourds (Ganil) at Caen. Ion currents of 150nA (Ne^{6+}), 60nA (O^{7+}) and 15nA (Ne^{9+}) were used. For the experiments we built a Ultra High Vacuum (UHV) system with two internal sections. The upper section is used for target cleaning and target analysis. The lower section is screened by μ-metal. The magnetic field measured in that part is below 10mG. Here a tandem parallel plate electron energy analyzer[15] is mounted on a rotable platform. Angular resolved energy

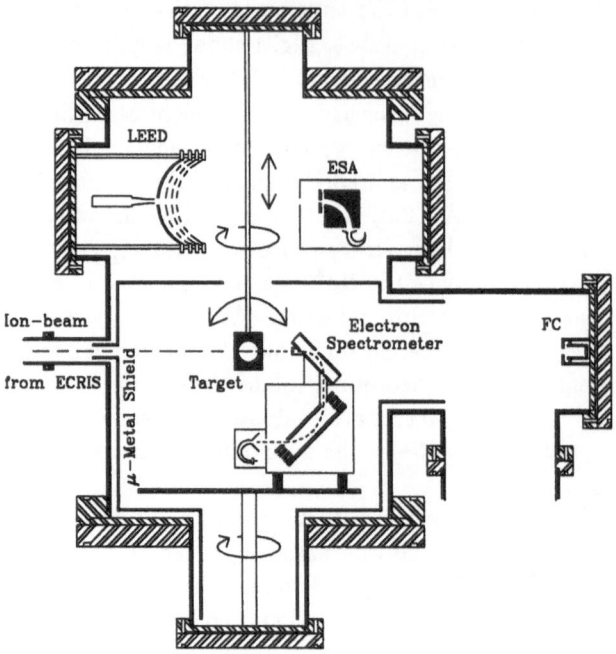

Figure 2. Experimental system for the angular resolved electron spectroscopy from the interaction of highly charged ions with surfaces. LEED = low energy electron diffraction used for target control (surface periodicity). ESA = electro static analyser used for the control of the target cleanliness by ion scattering spectrometry. ECRIS = electron cyclotron resonance ion source. FC = Faraday cup.

spectra can be measured in the range from 0° to 180°. The energy resolution of the analyzer is $\Delta E/E = 2.5\%$. Fig. 2 shows a section through the vacuum chamber. The target, Pt(110), is mounted on a two-axis goniometer. For all results reported here the plane of scattering defined by the incident beam and the analyzer was oriented parallel to a direction 18° of the $[\bar{1}10]$ surface direction. In such a direction channeling effects expected for low index directions are minimized.

RESULTS

Our results are angular resolved electron energy spectra. Fig. 3 shows three spectra due to the impact of N^{6+}, O^{7+} and Ne^{9+} on Pt(110). The grazing angle of incidence is $\psi = 10°$, the spectrometer is placed at $\Theta = 120°$ with respect to the beam direction, i. e. almost perpendicular to the target surface (20° off the surface normal). The particle energy is 10keV times the charge state. The velocity of the ions is 0.4 v_0 ($v_0 = 2.18 \times 10^8$cm s^{-1}), the perpendicular velocity component $v_\perp = v \sin \psi$ is hence below 0.1 v_0. The spectra have two major features, peaks around the respective projectile KLL Auger energies and a low energy peak. The dashed lines mark target Auger lines which appear for the N and O projectiles only. The low energy peak and the background extending up to the projectile Auger peaks is due to kinetic electron emission, secondary electrons produced by the KLL projectile electrons and Auger electrons from outer shells as in case of He$^+$ and He^{++} (Fig. 1).

In order to subtract the background we use an approximation by a 4th power polynominal. This is suggested when plotting the data on log-log scales, the background becomes a straight line. It is also possible to use e. g. a N^{4+} spectrum to generate a background function, because N^{4+} produces neither projectile nor target Auger electrons in the energy range of interest. Similar procedures have been used by the Oak Ridge Group [16,17]. The background subtraction produces the spectra shown in Fig. 4. The KLL Auger peaks of the projectiles appear at their proper energies of 340eV, 450eV and 740eV for O, N and Ne respectively. At lower energies (marked by a vertical line) we find the projectiles LMM Auger peaks and target Auger peaks at 40eV and 60eV. For nitrogen and oxygen the target Auger peaks marked already in Fig. 3 appear now clearly at 135eV and 220eV (Fig. 4a,b). For oxygen we also find a weak target Auger structure at 370eV.

A remarkable finding is that Ne^{9+} does not produce these target Auger lines. This is clear evidence that these lines are not produced in a "violent" collision, e. g. as

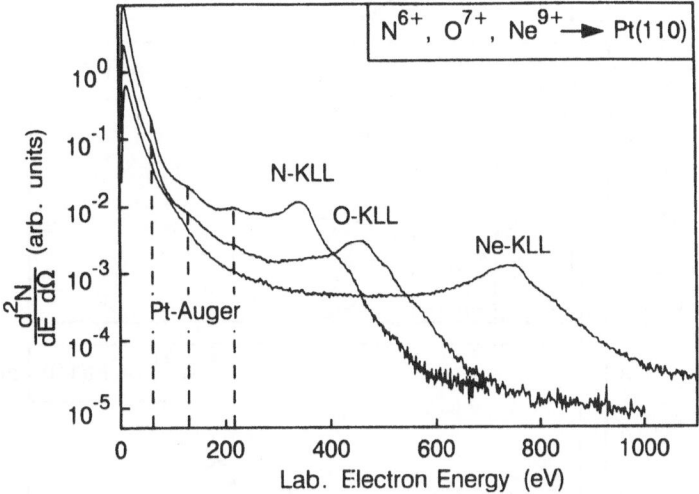

Figure 3. Electron energy spectra from hydrogen like ions, N^{6+}, O^{7+} and Ne^{9+}, interacting with Pt(110). The glancing angle is $\psi=10°$, the spectrometer is placed at $\Theta=120°$ with respect to the primary beam direction. The dashed lines mark target Auger electron emission.

observed in sputtering experiments[18]. Since all three projectiles produce the 40eV and 60eV Pt-Auger peaks these may indeed be due to a direct ionisation process [19].

Verification of the nature of the projectile and target Auger peaks is obtained from experiments with a variation of the impact and observation angles. The energetic position of the target lines is almost independent of these parameters. The projectile lines show shifts related to the fact that the electrons are emitted from a moving projectile. These measurements are discussed in detail elsewhere [20].

An interesting finding is the dependence of the target line intensity and the projectile line intensity on the grazing angle of incidence (Fig. 5), i. e. both types of lines show the same dependence of the impact angle. This indicates that both types of lines are related to the K-shell hole in the projectile. At grazing incidence both the projectile KLL and the target NOO and NNV Auger electrons are generated close enough to the surface, or "above" the surface that they can reach detector. At perpendicular incidence ($\psi = 90°$) the primary ions penetrate into the solid beyond the escape depth of the electrons. The penetration is obvious since at 70keV O has an range of approximately 600 Å in Pt[21]. At $\psi = 10°$ the "perpendicular" energy is reduced to 2keV, where part of the beam is reflected at the surface and the penetration is shallow.

For the interpretation of the KLL Auger emission we compare in Fig. 6 the spectra

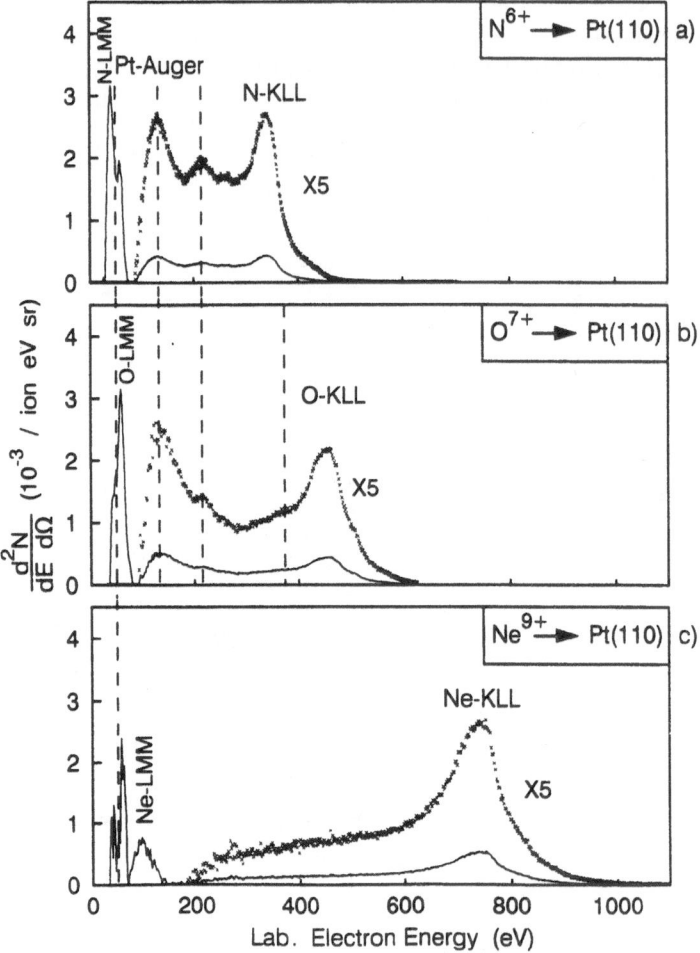

Figure 4. Electron energy spectra after background subtraction, correction for analyser transmission and acceptance.

generated by the impact of Ne^{9+} and Ne^{10+}. We note an additional emission at 850eV and weak features at 950eV. The emission at 790eV is about equal for both Ne^{9+} and Ne^{10+}. It is obvious that the additional emission is caused by the second hole in the K-shell of the projectile. The Ne^{10+} current was 25nA in the experiment. Similar spectra have been reported for the interaction of N^{7+} with a Ni(110) target[22]. In that experiments with slow ions the high energy shoulders are peaks which are identified as projectile KLM and KMN transitions. The broadening of the KLL peak is due to change in the screening when going from the H-like to the naked projectile. We note that target Auger emission is not excited by the Ne^{10+}, even though this ion hits the solid with 150keV.

DISCUSSION

The target Auger electron emission is identified as such on the basis of the independence in the projectile used. Furthermore the spectral position does not depend on a change of the angle between the incoming beam and the position of the spec-

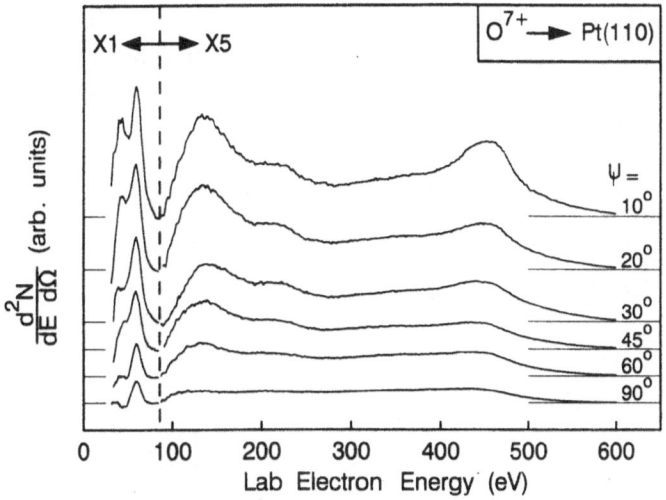

Figure 5. Electron energy spectra of 70keV O^{7+}, $\Theta = 120°$. The peaks below 90eV are target lines at 40eV and projectile LMM lines at 70eV. The broad peak at 135eV and 220eV are target Auger lines, the high energy peak is the O-KLL Auger electron emission.

trometer. The electrons are not emitted from a moving source. The missing of the spectral features in question in the Ne^{9+} and Ne^{10+} spectra is evidence for a vacancy transfer mechanism. Any excitation of ionization of the target by a close collision can be excluded. A transfer ionisation mechanism would also cause a shift of the spectral position depending on the projectile used[23], so we can exclude this mechanism as well.

We identify the target Auger transitions tentatively from standard electron excited Auger transitions as being NNV in the 370eV and 220eV region and NOO around 135eV. Unambiguous identification is afforded by calculating the lines using the atomic structure code of Cowan[24]. They are $N_{23}N_{45}V$ (370eV), $N_{45}N_{67}V$ (220eV) and $N_{45}O_1O_{23}$ (135eV). The shift of these energy levels due to the interaction with the O-K-shell is shown in Fig. 7. These curves are calculated diabatic potentials as a function of the internuclear separation R, i. e.

$$E_M^m(R) = E_M^m(\infty) - Q_N(R)/R. \tag{1}$$

The binding energy $E_M^m(R)$ of an electron m belonging to the center M is lowered by the Coulomb interaction with the center N. Following the model calculations of Stolterfoht[25] the effective nuclear charge of N is given by

$$Q_N(R) = Z_N^n \exp(-\alpha_0 R) \tag{2}$$

The screening constant α_0 is approximated by

$$\alpha_0 = s(\alpha_{MN})^t \tag{3}$$

with the "velocity" parameters $\alpha_M = \sqrt{2|E_M|}$ and $\alpha_N = \sqrt{2|E_N|}$ which add up to $\alpha_{MN} = \frac{1}{2}(\alpha_M + \alpha_N)$. The values s and t have been calculated previously[25], here we use $s = 0.65$ and $t = 0.5$. The effective nuclear charge Z_N^n in Eqn. 2 is estimated for a hydrogenic atom from $E^n = Z^2/2n^2$ which yields

$$Z_N^n = n\sqrt{2|E_N|} = n\alpha_N \tag{4}$$

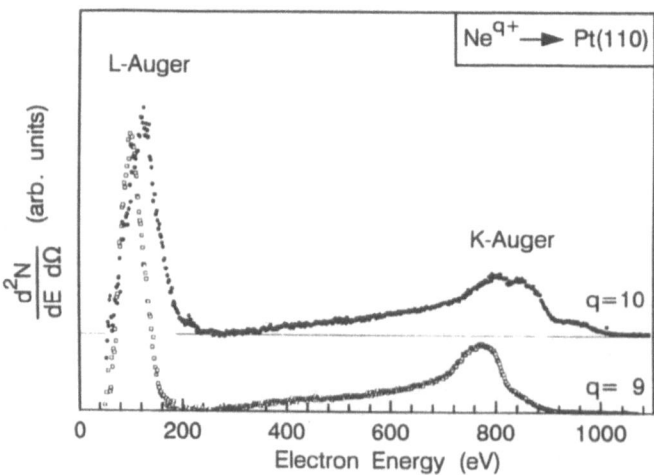

Figure 6. Comparison of the electron spectra from 90keV Ne^{9+} (H-like) and 150keV Ne^{10+} (naked), $\psi = 10°$ and $\Theta = 40°$. Note the broadening of the KLL peak due to the 2nd K-shell hole. The high energy tail is due to KLM emission. The background due to kinetic electron emission has been subtracted. The spectra are transformed into the projectiles rest frame.

As a result of these estimates we find curve crossings for the Pt-N_1-, -N_2- and -N_3-shells with the K-shells of Nitrogen and Oxygen (Fig. 7). There is no possible crossing with the Ne-K-state. This result gives an immediate qualitative interpretation of part of the experimental results, i. e. the target Auger emission at 370eV and the missing of it in case of Ne^{9+} and Ne^{10+} impact. The occurence of the vacancy transfer depends essentially on the binding energy of the K shell of the projectile with respect to the inner shell, here the N-shell, of the target at infinite separation. The Coulomb interaction forces projectile K-states to cross target states which are initially below the K-state. If the projectile K-state is initially below the target states in question no curve crossing will occur.

For each curve crossing the transition probability is then estimated using the Landau-Zener formula. Details of the calculations will be published elsewhere[20]. In case of the nitrogen platinum interaction we find probabilities in the angle of incidence dependent ($\psi = 10°$ - $90°$) range of 0.31–0.04 for the transfer of the nitrogen K-vacancy to the platinum N_{1-3}-states within the top 5 layers of the solid. Target Auger electrons ejected from deeper lying crystal layers are not detectable because of their limited escape depth. For the oxygen K-platinum N_{1-2} interaction the corresponding range of probabilities is 0.68–0.12. The probabilities estimated qualitatively account for the experimental relative intensities. The vacancy transfer from the Pt-N_{1-3}-states to the Pt-$N_{4,5}$-states giving rise to the emission at 220eV and 135eV is besides to a $N_{23}N_{45}V$ Coster-Kronig transition resulting in electron emission at 370eV (Fig. 4b) possibly due to a vacancy sharing between the neighbouring subshells[26].

Target Auger emission was also observed in N^{6+} and N^{7+} interacting with a Au(110) target[27,28]. For Au also the N-states are the recipients of the primary vacancy. It was

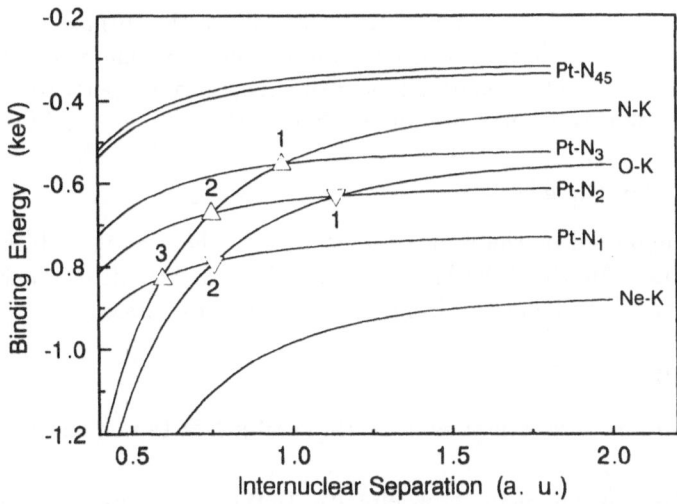

Figure 7. Model calculation of the electron binding energy of the Pt-N-shells (N_1, N_2, N_3 and N_{45}) when interacting with the K-shells of the projectiles N, O and Ne. The triangles mark "curve crossings" where Landau-Zener charge exchange occurs.

proposed that the vacancy transfer occurs via a promotion of the projectile K-vacancy. However it needs rather small internuclear distances for a sufficient promotion to occur. Our estimates show no promotional effects within the limits set by the distance of closest approach for a head-on collision, which is e. g. 0.13 a. u. for 60keV Ne^+ on Au. The vacancy transfer mechanism proposed here affords certainly a higher probability simply because the curve crossings are found in a range between 0.5 and 1.2 a. u. (Fig. 7).

A detailed discussion of the projectile KLL Auger electron emission will be given elsewhere[29]. The calculations with the Cowan Code reproduce the observed spectra when using the occupation of the L and M shells as "fit" parameters. Most important is the occupation of the L-shells. We find and average occupation of 5 electrons in the L-shells during the time of the KLL-Auger emission.

Finally from the observation of the decreasing intensity of the KLL emission with increasing grazing angle (Fig. 5) we can estimate the survival time t_s of the projectile K-vacancy within the solid. Assuming that all K-vacancies survive beyond the escape depth of the electrons of 7Å (see e. g. ref. 30) the time is $t_s = 8x10^{-16}$s. This is a lower limit which sets an upper limit for the transition rate of $\Gamma_s = 1.25x10^{15}$s.

Another time estimate is obtained from the impact at $\psi = 10°$. A MARLOWE trajectory analysis shows that the majority of the ions follow straight line trajectories down to a vertical depth of 7Å, i. e. the trajectory length is 40Å. For that length the 70keV oxygen needs $4.5·10^{-15}$s and obviously manages to emit KLL-electrons. In a recent study of the time dependence of relaxation processes of multi-charged ions[31] it was concluded that the "gas phase" value for the lifetime of K vacancies in N^+ ions of $1.0x10^{-14}$s is likely to be independent of the environment of the N atom. This assumption is supported by the experimental results of the ORNL group[16,31]. We may assume a similar time for the O-KLL transition, that the time window for the capture of the M and L electrons inside the solid is very narrow indeed. The coupling of the projectile KLL emission to the target NNV emission is additional experimental evidence for the relaxation processes. It is interesting to note that the relaxation is faster inside than outside. The filling of the outer shells of e. g. N^{7+} starts at about 20 a. u. above the surface, at $\psi = 10°$ the N^{7+} can travel about 60Å before hitting the plane of the surface ion cores. But the emission of KLL from above the surface is small compared to the emission from the bulk[16]. This observation seems to point at a solid state effect, i. e. an influence of the electron density and the dynamical properties of the solid state electrons. The solid state effect presumably affects the filling of the M and L shell.

ACKNOWLEDGEMENT

This work is supported by the Bundesministerium für Forschung und Technologie and the Stifterverband für Deutsche Wissenschaft. We thank J. Bleck-Neuhaus, Bremen, R. Morgenstern and L. Folkerts, Groningen for helpful discussions.

REFERENCES

1. W. Grotrian, Naturwissenschaften 27:214 (1939); I. S. Bowen and B. Edlén, Nature 143:374 (1939).

2. F. Bombarda, R. Giamella, E. Källne, G. J. Tallents, F. Bely-Duban, P. Faucher, M. Cornille, J. Duban and A. H. Gabriel, Phys. Rev. A37:504 (1988).

3. R. Geller, B. Jacquot and M. Pontonnier, Rev. Sci. Instr. 56:1505 (1985).

4. M. A. Levine, R. E. Marrs, J. N. Bardsley, P. Beiersdorfer, C. L. Bennet, M. H. Chen, T. Cowan, D. Dietrich, J. R. Henderson, D. A. Knapp, A. Osterheld, B. M. Penetrante, M. B. Schneider and J. H. Scofield, Nucl. Instr. Meth. B43:431 (1989).

5. H. D. Hagstrum and G. E. Becker, Phys. Rev. B8:107 (1973).

6. H. D. Hagstrum, Phys. Rev. 96:336 (1954).

7. P. A. Zeijlmans van Emmichoven, P.A.A.F. Wouters and A. Niehaus, Surf. Sci. 195:115 (1988).

8. S. Schippers, S. Oelschig, W. Heiland, L. Folkerts, R. Morgenstern, P. Eeken, I. F. Urazgil'din and A. Niehaus, Surf. Sci. 257:289 (1991).

9. R. L. Erickson and D. P. Smith, Phys. Rev. Letters 34:297 (1975).

10. J. C. Tully, Phys. Rev. B16:4324 (1977).

11. U. A. Arifov, L. M. Kishinevskii, E. S. Mukhamadiev and E. S. Parilis, Sov. Phys. Tech. Phys. 18:118 (1973).

12. P. Varga, Comments Atomic Mol. Phys. 23:111 (1989).

13. H. P. Winter, Z. Phys. Suppl. D21:129 (1991).

14. H. J. Andrä in "Electronic and Atomic Collisions" eds. W. R. MacGillivray, I. E. McCarthy and M. C. Standage, IOP Publ. Ltd, Bristol, p. 89 (1992).

15. A. Itoh, T. Schneider, G. Schiwietz, Z. Roller, H. Platten, G. Nolte, D. Schneider and N. Stolterfoht, J. Phys. B16:3965 (1983).

16. F. W. Meyer, S. H. Overbury, C. C. Havener, P. A. Zeijlmans van Emmichoven and D. M. Zehner, Phys. Rev. Lett. 67:723 (1991).

17. P. A. Zeijlmans van Emmichoven, C. C. Havener and F. W. Meyer Phys. Rev. A43: 1405 (1991).

18. R. Baragiola, in "Inelastic Particle Surface Collisions", eds. E. Taglauer and W. Heiland, Springer Series in Chem. Phys. 17:38 (1981).

19. S. T. de Zwart, A. G. Drentje, A. L. Boers and R. Morgenstern, Surf. Sci. 217:298 (1989).

20. S. Schippers, S. Hustedt, W. Heiland, R. Köhrbrück, J. Bleck-Neuhaus, J. Kemmler, D. Lecler and N. Stolterfoht, Phys. Rev. A46, in print (1992).

21. J. Ziegler, TRIM-Version 5.4 (1989).

22. J. Das, L. Folkerts and R. Morgenstern, Phys. Rev. A45:4669 (1992).

23. A. Niehaus, Comments At. Mol. Phys. 9:153 (1980).

24. R. D. Cowan, "The Theory of Atomic Structure and Spectra", Univ. of Cal. Press, Berkeley (1981).

25. N. Stolterfoht in "Progress in Atomic spectroscopy", eds. H. J. Beyer and H. Kleinpoppen, Plenum Press, New York, Part D: 415 (1987).

26. N. Stolterfoht, Physics Reports 146:315 (1987).

27. D. M. Zehner, S. H. Overbury, C. C. Havener, F. W. Meyer and W. Heiland, Surf. Sci. 178:359 (1986).

28. F. W. Meyer, C. C. Havener, S. H. Overbury, K. J. Reed, K. J. Snowdon and D. M. Zehner, J. de Physique (Paris) 50:C1-263 (1989).

29. S. Schippers, S. Hustedt, W. Heiland, R. Köhrbrück, J. Bleck-Neuhaus, J. Kemmler, D. Lecler and N. Stolterfoht, submitted to Nucl. Instr. Methods B; R. Köhrbrück et al., to be published.

30. M. P. Seah and W. A. Dench, Surf. and Interf. Anal. 1:2 (1976).

31. S. H. Overbury, F. W. Meyer and M. T. Robinson, Nucl. Instr. Meth. B67:126 (1992).

STUDIES ON SLOW PARTICLE-INDUCED ELECTRON EMISSION
FROM CLEAN METAL SURFACES BY MEANS OF
ELECTRON EMISSION STATISTICS

Friedrich Aumayr and Hannspeter Winter

Institut für Allgemeine Physik, Technische Universität Wien
Wiedner Hauptstr. 8-10, A-1040 Wien, Austria

ABSTRACT

A report is given on recent progress in studying slow ion-induced electron emission from clean gold, for which via experimental determination of the related electron emission statistics e.g. rather precise measurements of electron emission yields and a separation of (as long as remaining mutually independent) potential and kinetic contributions to the latter can be obtained.

We discuss the shape of electron statistics for slow singly charged ion-induced kinetic emission as well as features of potential emission due to impact of slow singly, doubly and multiply charged ions. In particular we have determined electron emission statistics and precise total electron yields for normal impact of slow multicharged N-, Ne-, Ar- and I ions on clean polycrystalline gold at impact velocities $1 \cdot 10^4 < v_p < 15 \cdot 10^4$ m/s. From the impact velocity dependence of the yields and the shapes of electron multiplicities two different contributions to electron emission have been identified, one being believed to result from autoionization before surface impact and the other from faster autoionization just upon surface impact. The latter part seems to be directly related to the number of electrons carried by the neutralized projectiles ("hollow atoms") in high Rydberg states just before their surface impact.

1. INTRODUCTION

Bombardment of the surface of a solid, in particular a clean metal, by slow neutral or ionized atoms or molecules may cause emission of electrons due to transfer of potential energy (potential emission/PE) and/or kinetic energy (kinetic emission/KE) from the projectile onto target electrons. Such processes are of both fundamental interest and considerable practical importance and have therefore been under scrutiny for a long time. General features of PE have been explained by Hagstrum (1954), and a recent review on this field has been given by Varga and Winter (1992). The status of knowledge on KE has recently been reviewed by Hasselkamp (1992). Both processes depend rather critically on target surface conditions, which is of considerable relevance for corresponding experimental investigations.

As a general distinction, PE results from electronic transitions between projectile and surface already <u>before</u> an impact has taken place, whereas KE can only be initiated <u>after</u> the projectile has made its close contact with the surface. In a more detailed view, PE arises from Auger-type processes involving time intervals of the order of 10^{-14} s, which is comparable to the flight time of a relatively slow ($v \leq 10^5$ m/s $\cong$ 60 eV/amu) particle within the distance for its probable electronic interaction with a metal surface (for singly charged ions typically some 10^{-10} m, cf. Hagstrum 1954, growing with the ion charge state q for multicharged ions, see Varga and Winter 1992). Consequently, even rather slow multicharged ions may not completely become neutralized and deexcited until their surface impact, because this would involve a relatively large number of electronic transitions. Therefore, PE is expected to become the more efficient the lower the impact energy, without an impact energy threshold.

On the other hand, KE is related to the stopping power of projectiles within the uppermost atomic layers of a solid and appears as the result of collision cascades which usually last not longer than 10^{-12} s. Only electrons from this surface-near region, which have been given a minimum kinetic energy of the order of 10 eV, may escape across the surface barrier into vacuum. Consequently, KE is subject to an impact energy threshold. From these simple considerations we may conclude that only for electron emission induced by rather slow particles the PE and KE processes can be regarded as mutually independent.

In the case of multicharged ions ("MCI" - Z^{q+}), electrons are captured resonantly from states near the Fermi edge of a metal surface within a critical distance d_c, which depends on the wavefunctions overlap of the surface density-of-states (S-DOS) with empty projectile states. Consequently, d_c increases for higher projectile charge q and/or lower surface work function W_ϕ. Once within this critical region, a slow MCI with impact velocity $v_p \ll 1$ a.u. will be rapidly further neutralized according to a characteristic time t_n related to the Fermi velocity v_F, which for metals is of the order of 1 a.u. (Ashcroft and Mermin, 1976). The then developing multiply excited ("hollow") atoms are subject to resonant ionization (RI) as well as autoionization (AI), which both together with the ongoing resonant neutralization (RN) will determine the projectile's electronic population until it hits the surface (Burgdörfer et al., 1991; Andrä et al., 1992). As a result of AI, slow electrons should be emitted from projectiles (Arifov et al., 1973) the more efficiently the lower the impact velocity (Delaunay et al., 1987a). What exactly happens to the electrons still bound to a projectile in highly excited states at the moment of surface impact has not been studied in detail so far neither theoretically nor experimentally. Sufficiently slow projectiles can be reflected in the repulsive planar surface potential (Andrä et al., 1992), whereas faster ones will penetrate into the solid to undergo various de-excitation processes until their complete neutralization and stopping.

The present work is based on recent experimental studies involving a new method (cf. sect. 2) which in contrast to the so far common techniques to investigate particle induced electron emission can distinguish among their different mutually independent contributions. This is of special interest in view to the relative importance of potential and kinetic emission for particular collision systems. We determine the probabilities for emission of a given number n = 0, 1, 2, etc. electrons due to impact of individual projectiles onto the surface, which results in the statistics of ion-induced electron emission ("ES"). The latter contains rather detailed informations on the involved electron emission processes, as demonstrated lateron for clean polycrystalline gold bombarded by slow singly (sect. 3), doubly (sect. 4) and multiply charged ions (sect. 5), respectively. Electron emission due to impact of MCI has so far been studied only by analyzing the related total electron emission yields (Hagstrum, 1954; Arifov et al., 1973; Delaunay et al., 1987a) and energy distributions (Hagstrum, 1954; Arifov et al., 1973; Delaunay et al., 1987b; de Zwart, 1987; de Zwart et al., 1989; Folkerts et al., 1990; Zeijlmans v.Emmichoven et al., 1991; Meyer et al., 1991a,b; Andrä et al., 1992), charge state composition of scattered projectiles (de Zwart et al., 1985; Winter, 1992) and emission of soft X-rays (Donets, 1985; Briand et al., 1990, 1991; Schultz et al., 1991; Andrä et al., 1992). The observed electron energy distributions are dominated by low energy continua (E_e < 30 eV) with comparably small contributions also by fast Auger electrons (Delaunay et al., 1987b; de Zwart, 1987; de Zwart et al., 1989; Folkerts et al., 1990; Zeijlmans van Emmichoven et al., 1991; Meyer et al., 1991a,b; Andrä et al., 1992; McDonald et al., 1992) which are now believed to originate mainly from below the surface (Meyer et al., 1991a,b). Several reviews on slow MCI-surface interaction (Varga, 1989; Winter, 1991; Andrä et al., 1992) and related semiclassical (Burgdörfer et al., 1991) and classical (Bardsley and Penetrante, 1991) theories have recently been published and quantum-mechanical calculations have been performed on the involved RN and AI processes (Wille, 1991, 1992; Vaeck and Hansen, 1991).

2. EXPERIMENTAL TECHNIQUES

Ion-induced electron emission from solid surfaces is usually being studied by determination of the total electron yield (i.e. the mean number of electrons emitted per projectile particle, conventionally being denoted by γ) and/or the ejected electron energy distribution dN_e/dE_e (c.f. figs. 1 a,b). However, both informations are only of qualitative value insofar as they do not provide a straightforward access to the mechanisms responsible for electron emission. In this situation it is of interest to measure the distribution of numbers of electrons resulting from individual emission events. This socalled emission statistics (ES) gives the set of probabilities W_n for ejection of a given number of n electrons (Hofer and Littmark, 1976) and is directly related to the total emission yield γ.

$$\gamma = \sum_{n=0}^{\infty} n \cdot W_n \; ; \quad \sum_{n=0}^{\infty} W_n = 1 \qquad (1)$$

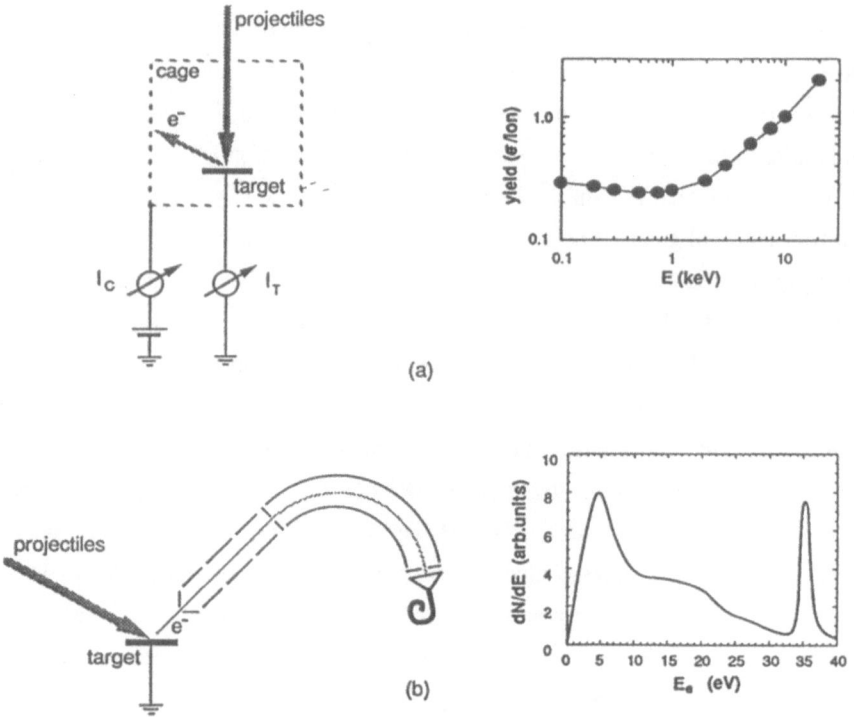

Figure 1. (a) Measurement of total electron yields. (b) Spectroscopy of ejected electron energy distributions

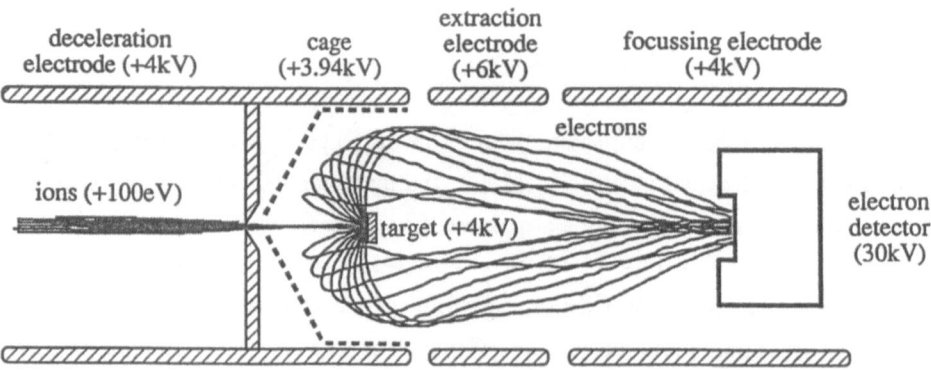

Figure 2. Setup for measuring ion-induced ES for impact of slow ions on clean polycrystalline gold (from Lakits et al., 1990).

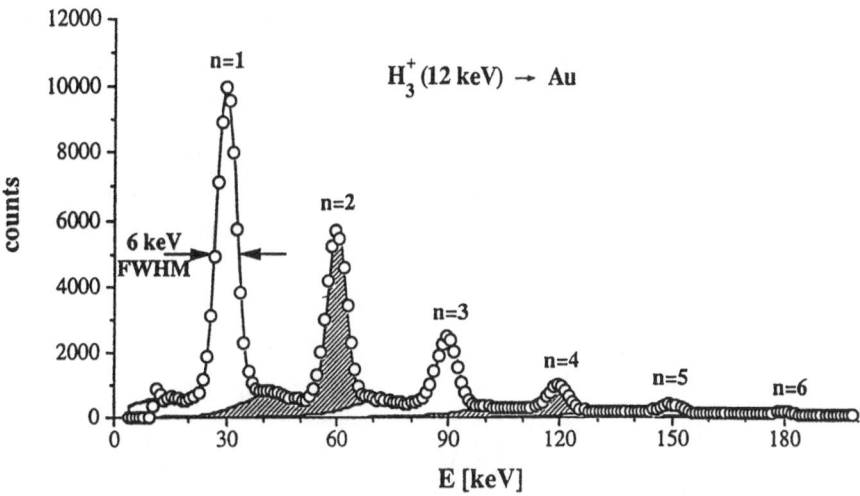

Figure 3. Typical uncorrected ES pulse height spectrum for impact of 12 keV H^{3+} ions on clean Au, and results of fitting procedure (from Lakits and Winter 1990). Note that here in contrast to fig. 2 the effective electron acceleration energy has been 30 keV, since the target was kept at ground potential.

Lakits et al. (1989, 1990), Aumayr et al. (1991) have described how such ES can be precisely determined for ionized as well as neutral projectiles. Fig. 2 shows a sketch of their apparatus for low ion impact energies.

Primary ions of interest are directed into the statistical detector setup in an UHV environment, where they can be decelerated just before hitting a sputter-cleaned poly-crystalline gold target which is situated inside a highly transparent conical electrode (Lakits et al. 1990). The latter assures backward deflection of the emitted electrons toward an acceleration lens system into a solid state detector which is situated on a positive potential (typically 25 kV) with respect to the target in the geometrical shadow of the latter. The emitted electrons thus hit the detector surface with typically 25 keV energy. Within the time resolution of the detector and subsequent electronics (typically >> 1 ns), emission of a group of n electrons due to impact of an individual projectile cannot be resolved into n single events but will rather be registered as one single detector pulse corresponding to the n-fold energy of one single 25 keV electron. This gives rise to raw pulse height distributions as presented in fig. 3, from which the relative values for W_n can be derived after proper corrections for detector resolution and background contributions (Aumayr et al. 1991). In further consequence, absolute values for W_n (including for n = 0) are being obtained by means of equ. 1, if the corresponding value for γ is already known. Otherwise, the latter can be determined with the present setup in a way described by Lakits et al. (1990), which permits rather precise total electron yield measurements even for γ values of less than 10^{-3} electrons/projectile. For ions in higher charge states (cf. fig. 4), the total yield does become so large that the quantity W_o can be safely neglected and γ can therefore be evaluated exclusively from the relative ES measurements according to equs. (1).

For the MCI studies presented in chapter 5 a recoil ion source (Mann, 1986) pumped by fast (3 - 11 MeV/amu) heavy ion beams from the GSI UNILAC accelerator delivered MCI fluxes of e.g. 100/s for Ar^{16+} or 10^4/s Ar^{10+} at the target surface. After their extraction with several hundred volts and charge-to-mass separation in a 180° magnet the recoil ions ($N^{q+}/q\leq6$, $Ne^{q+}/q\leq10$, $Ar^{q+}/q\leq16$, $I^{q+}/q\leq25$) were guided toward the ES detector.

Their small initial kinetic energy permitted deceleration to rather low nominal target impact energies $E \geq (2\pm1)\cdot q$ eV, by means of a four-cylinder lens in front of the target

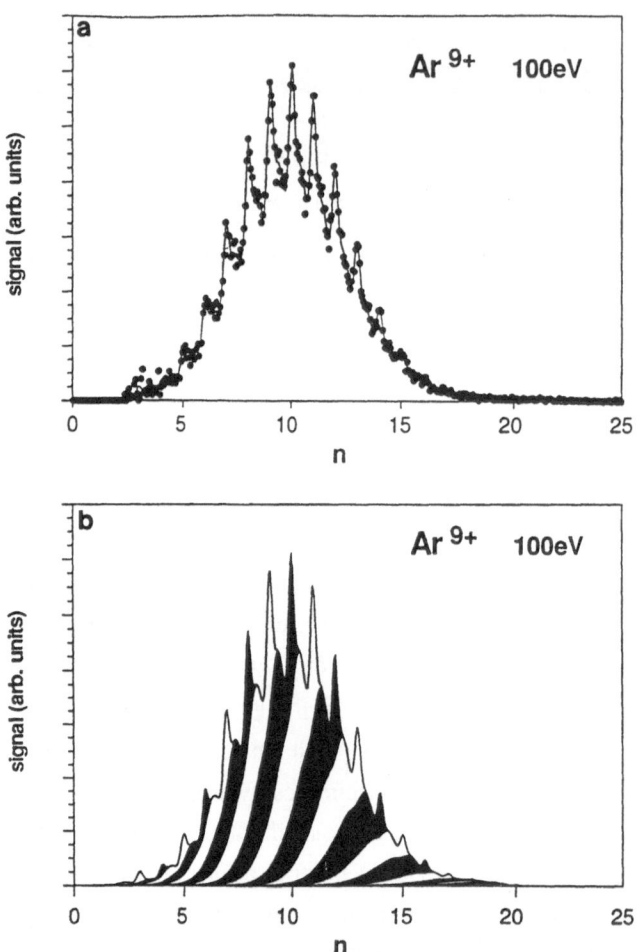

Figure 4. (a) Comparison between the measured electron emission statistics for impact of 100 eV Ar^{9+} ions on Au (full circles) and the results of a fit (solid line). Data from Kurz et al., 1992.
(b) Individual contributions to the fitted data including electron backscattering.
Contributions to even numbered peaks have been shaded for better visuality.

surface, cf. fig. 2. The final part of the MCI beam line and the ES detector assembly including the target (again atomically clean polycrystalline gold, sputter-cleaned by means of a built-in Ar^+ ion gun) was kept in UHV at a base pressure of typically 10^{-8} Pa during all measurements.

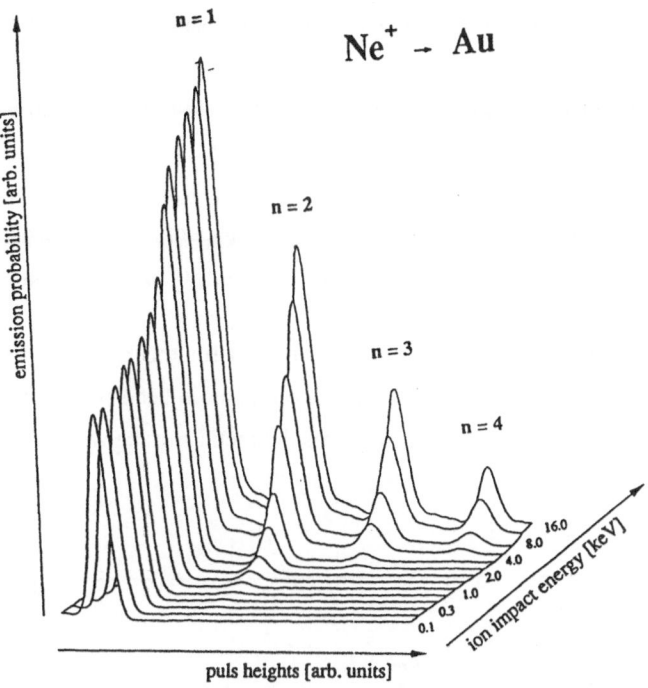

Figure 5. Raw experimental data of ES for impact of Ne$^+$ on clean polycrystalline gold vs. impact energy (from Winter et al. 1991).

Raw ES data (as shown in fig. 4 for 100 eV Ar^{9+}) had to be evaluated according to the procedures described by Aumayr et al. (1991), to obtain the emission probabilities W_n.

3. SINGLY CHARGED IONS

Raw ES data for impact of Ne$^+$ ions on clean gold are shown in fig. 5, with the impact energy varied from 100 eV up to 16 keV. Note that Ne$^+$ has no long-lived highly excited state (Winter 1982), because of which only Ne$^+$ ground state ions have been involved.

At the lowest impact energy (E ≤ 15 eV/amu) apparenly only one electron is ejected, whereas at higher E also emission of 2, 3, etc. electrons is gradually coming up (Lakits et al. 1990). Consequently, as long as KE is not possible, only PE can contribute to the total electron yield by ejecting one electron, although the transferrable potential energy (W ≤ 21,6 eV) would in principle suffice to emit up to three electrons. The here demonstrated results are supported by similar measurements for impact of He$^+$ (see fig. 6), Ar$^+$ (Töglhofer et al., 1992) and other singly charged ions on clean gold (Lakits et al. 1990). There seems to exist a fundamental difference between PE- and KE related ES. PE involves reasonably well defined transitions between electronic states of the target surface and the approaching projectiles, and therefore gives rise to emission of a correspondingly limited number of electrons. KE, on the other hand, involves dissipation of projectile kinetic energy among a relatively large number of target electrons which are facing comparable chances of being ejected. The number of these electrons increases with transferred kinetic energy and thus with the projectile velocity, whereas the potential energy carried by the approaching ion is transferred via Auger type electronic transitions to a single electron only.

In the following we demonstrate, that such measurements can be extended toward rather small impact energies. For example, consider impact of He[+] at low impact energy (cf. fig. 6). To identify the exact location of the KE threshold, the course of γ vs. impact velocity E as shown in fig. 6a is not quite significant. However, in fig. 6b the measured ES probabilities W_n for emission of $n \leq 3$ electrons have been plotted. The KE threshold impact energy (or at least its upper value) is now identified with the appearance of a second peak (i.e. $W_2 \neq 0$) in a straightforward manner, since the PE causes emission of one electron only (see above discussion).

According to classical dynamics, an upper limit for the KE threshold is principally given at that impact velocity where kinetic energy transfer in head-on collisions of projectiles with the quasi-free metal electrons just surpasses the metal surface work function W_ϕ.

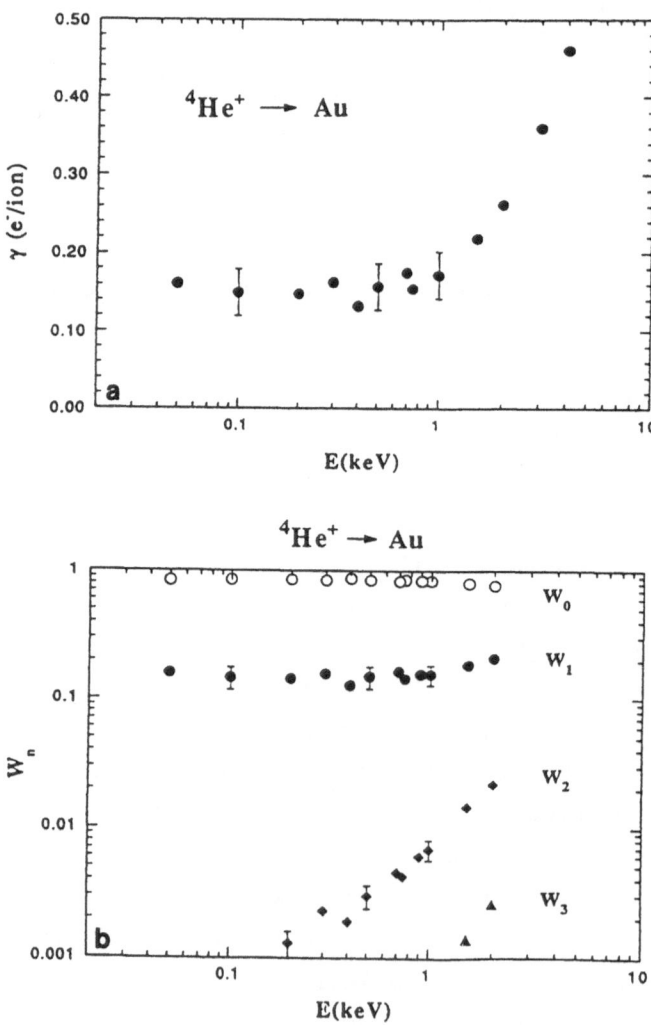

Figure 6. (a) Total electron emission yields for impact of He[+] ground state ions on clean gold vs. impact energy.
(b) Emission probabilities W_n for impact of He[+] on clean gold vs. impact energy.
Data from Töglhofer et al., 1992.

This threshold impact energy would be about 300 eV/amu for a clean gold surface (Alonso et al. 1986, Lakits et al. 1990) and thus considerably higher than the here determined KE threshold of less than 50 eV/amu for He+, cf. figs. 6b and 7. The precise origin of kinetic electron emission at such rather low impact velocities is not yet well understood, but may be caused by autoionisation of highly excited quasimolecules which are transiently formed in close encounters of projectile ions with single surface atoms. Similar measurements have also been carried out with other singly charged ions (Lakits et al.; 1990; Winter et al., 1991; Töglhofer et äl., 1992).

Our finding that PE induced by singly charged ion impact on Au can eject at most one electron, and the apparance of more electrons with increasing impact energy, which we thus attribute to the onset of KE, can now be used to separate PE and KE contributions to electron emission as follows. Let P_n and K_n be the individual probabilities for emission of n electrons by PE and KE, respectively. For the measured probabilities W_n the set of equations

$$W_n = \sum_{i=0}^{n} P_i \cdot K_{n-i} \tag{2}$$

can be solved for P_n and K_n under the assumptions that (i) $P_n = 0$ for $n \geq 2$ and (ii) $K_n = 0$ as long as W_{n+1} is negligibly small. For example, in this way our W_n data for He+ on Au have been separated into KE - and PE contributions, the results being shown in fig. 7.

Furthermore, it is of interest to consider the shape of the measured ES for KE, since they are often approximated by a Poissonian distribution for the corresponding total electron yield γ, i.e.

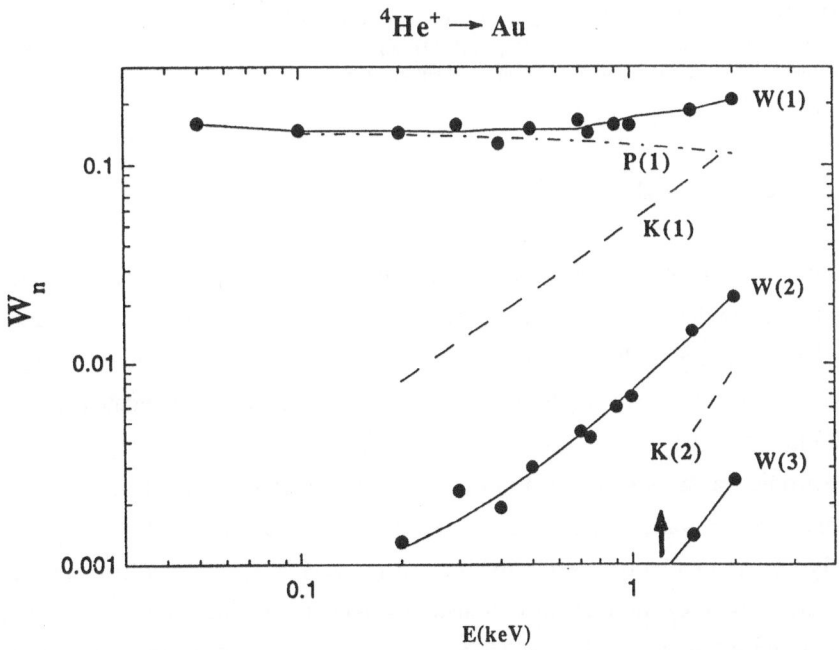

Figure 7. Contributions of PE and KE to electron emission statistics (c.f. text). Data from Töglhofer et al., 1992. Arrow marks "classical" upper limit for KE threshold at about 300 eV/amu (cf. text).

$$P_n(\gamma) = \frac{\gamma^n}{n!} \cdot e^{-\gamma} \quad \rightarrow \quad \gamma/(n+1) = P_{n+1}(\gamma)/P_n(\gamma) \tag{3}$$

Equs. (3) permit a straightforward comparison with the ratios of actually measured ES W_{n+1}/W_n, as shown in fig. 8 (from Ohya et al. 1992). The apparent clear deviations from the Poissonian shape could be related via Monte Carlo simulations to contributions from kinetic electron emission due to backscattered projectiles and/or recoiling target atoms.

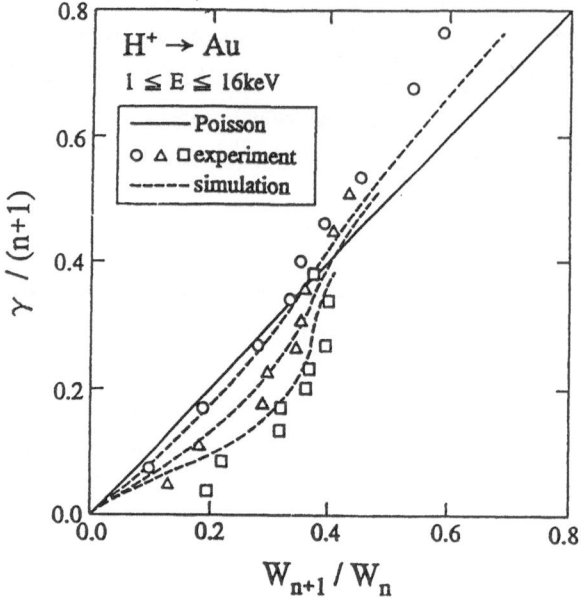

Figure 8. Course of $\gamma/(n+1)$ vs. $W_{n+1}(\gamma)/W_n(\gamma)$ for (1 - 16 keV) H^+ impact on clean gold (circles, triangles and squares refer to $n = 1, 2$ and 3, respectively), together with Monte Carlo - simulated data. The straight solid line indicates a Poissonian ES (from Ohya et al., 1992).

4. DOUBLY CHARGED IONS

Potential emission may become rather effective for impact of higher charged ions on a metal surface, due to the enhanced potential energy content of such projectiles (Delaunay et al. 1987a). This can already be demonstrated for impact of slow doubly charged noble gas ions on clean gold. In this context, ES measurements are very useful to elucidate the interplay of various electronic transitions governing such PE processes. According to Hagstrum (1954), electronic transitions involved in PE processes can engage one or more electrons. The only one-electron transition of interest is resonance neutralisation ("RN"), which does not directly result in electron emission but often acts as precursor for subsequent electron-emitting two- or more-electron transitions. Auger deexcitation ("AD") can cause electron emission if the involved deexcitation energy is larger than the metal surface work function W_ϕ. Auger neutralisation ("AN") may give rise to electron emission if the potential energy change in the related neutralisation step surpasses $2 \cdot W_\phi$, and autoionisation ("AI") will take place after formation of a doubly (multiply) excited projectile particle via double RN transitions ("DRN"). The last process has first been identified by Hagstrum and Becker (1973) for impact of He^{2+} on Ni single crystal surfaces.

Varga et al. 1982 have measured electron energy distributions for impact of the doubly charged rare gas ions $Ne^{2+} \div Xe^{2+}$ on clean W. They identified for both Ne^{2+} and Ar^{2+} a group of very slow electrons which they related to two successive deexcitation steps via two excited X^+ states on the way from X^{2+} toward the X^{+0} ground state along the path

$$X^{2+} \to RN \to X^{+*} \to AD(e^-) \to X^{+*'} \to AD(e^-) \to X^{+0} \tag{4a}$$

or along the alternative path

$$X^{2+} \to AN(e^-) \to X^{+*} \to AD(e^-) \to X^{+0} \tag{4b}$$

For actual occurence of electron-emission in the steps labelled by "(e^-)", sufficiently large deexcitation energies have to be made available (see above), as can indeed be found for the collision systems Ne^{2+}, Ar^{2+} - W (see also Varga 1988,1989).

However, in electron energy distributions measured by Wouters et al. (1989) for keV grazing collisions of He^{2+}, Ne^{2+} and Ar^{2+} on clean Cu, no evidence for transitions as described by equs. (4) has been found. These authors assumed that only the following transitions according to equs. (5), which each can give rise to one electron at most, are taking place in the $X^{2+} \to X^{+0}$ neutralisation steps.

$$X^{2+} \to DRN \to X^{0**} \to AI(e^-) \to X^{+0} \tag{5a}$$
$$\text{or}$$
$$X^{2+} \to AN(e^-) \to X^{+0} \tag{5b}$$

To clarify this situation, we have measured ES for impact of He^{2+}, Ne^{2+} and Ar^{2+} on clean gold. At 100 eV impact energy, for all three collision systems an exclusive PE situation is given (see figs. 5, 6). Emission of up to three electrons can unambiguously be seen for both He^{2+} (c.f. fig. 9) and Ne^{2+}, but not for Ar^{2+} (note that the third electron can be emitted via AN of X^{+0} to the X^0 ground state after conclusion of the transitions given by equs. 4). For impact of He^{2+} the dependence of W_n vs. ion impact energy is shown in fig. 9. With falling E both W_2 and W_3 go through a minimum and then rise again toward the lowest impact velocities covered by these measurements. This behaviour should be caused by improving probabilities for the transitions described by equs. (4) in comparison to those covered by equs. (5).

Further evidence for this behaviour is found in other trends for the three projectile species He^{2+}, Ne^{2+} and Ar^{2+}. For He^{2+} neutralisation, the first deexcitation step according to equs. (4) can involve at most 13,6 eV and for both Ne^{2+} (deexcitation into Ne^+ $2s2p^6$ ^2S) and Ar^{2+} (deexcitation into Ar^+ $3s3p^6$ ^2S) at most 14,1 eV (cf. Varga et al. 1982).

While these first deexcitation steps are almost equally large for all three ion species, binding energies of the related X^{+0} ground states differ strongly from each other (54.4 eV for He; 41.0 eV for Ne and 27.6 eV for Ar, respectively). In first approximation, the transition probabilities for electron emitting processes described by equs. (5) should be the larger, the smaller the involved X^{2+} - X^{+0} energy steps. Keeping this in mind, we can explain the decreasing related probabilities W_3 by the competition from processes following equs. (5), which become increasingly more important when changing the projectile species from He^{2+} via Ne^{2+} to Ar^{2+}.

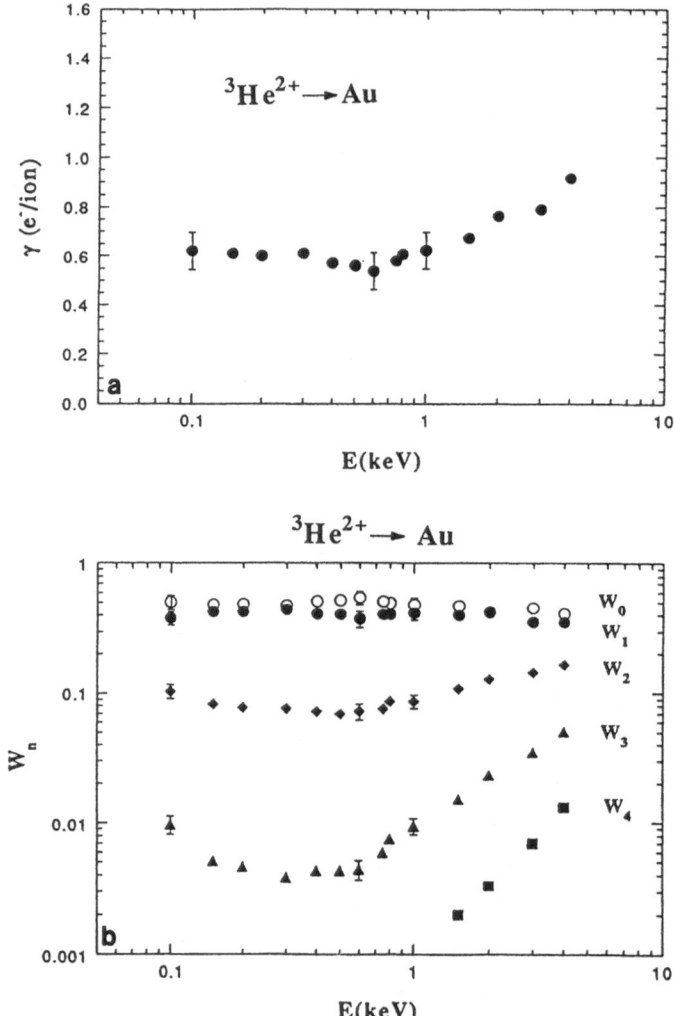

Figure 9. (a) Total electron emission yields for impact of He^{2+} ions on clean gold vs. impact energy. (b) Emission probabilities W_n for impact of He^{2+} on clean gold vs. impact energy. Data from Töglhofer et al., 1992.

Whereas Varga et al. (1982) identified a slow electron group also for Ar^{2+} - W collisions, we could not find the equivalent, i.e. emission of up to three electrons for impact of Ar^{2+} on Au. This may be caused by the difference in the surface-density of states ("S-DOS") for both target species. Polycrystalline tungsten features a surface work function

$W_\phi = 4.5$ eV and a large S-DOS just below the Fermi level (Feuerbacher and Christensen 1974), whereas for gold $W_\phi = 5.1$ eV and the S-DOS is much less pronounced in the vicinity of the Fermi level. These differences could explain a relative decrease of the probabilities for electron emission due to the first deexcitation processes described by equs. (4) in comparison with the electron emission probabilities covered by equs. (5).

Recently, electron energy spectra have been measured by Brenten et al. (1992) for very slow (< 10 eV) He^{2+} on clean and cesiated W. They show a peak at a few eV electron energy which they ascribed to electrons resulting from processes in accordance with equs. (4).

5. MULTIPLY CHARGED IONS

In this chapter we will deal only with the first leg of an MCI's journey from initial RN until its close contact with a clean metal surface, by studying the emission of slow electrons ($E_e \leq 60$ eV) during the projectile flight time $t_f = d_c/v_p$ toward the surface, to shed more light on the sequence of formation and decay of the transiently produced, multiply excited ("hollow") atoms.

To this purpose, we have determined ES for various MCI species. This method offers distinct advantages over the usual measurements of total electron yields from the currents of primary ions and emitted electrons, namely a much higher sensitivity relaxing the primary MCI current requirements dramatically, rather precise absolute electron yields directly available from the essentially relative ES measurements and, most notably, completely new information related to the electron emission multiplicity in such processes.

Raw ES spectra as measured for impact of 100 eV Ar^{q+} ions on clean Au are shown in figs. 4a and 10. From these measurements total electron yields for $N^{q+}/q\leq6$, $Ne^{q+}/q\leq10$, $Ar^{q+}/q\leq16$ and $I^{q+}/q\leq25$ projectile ions have been determined as described in chapter 2 and presented in figs. 11 and 12.

Fig. 11 shows how with increasing v_p the electron yields for impact of Ar^{q+} ions on Au first gradually decrease and then level off towards an apparently velocity-independent value γ_∞. The velocity-dependent part of the yield ($\gamma - \gamma_\infty$) originates presumably from AI of the projectiles on their way to the surface.

Calculating d_c according to Burgdörfer et al. (1991), which gives almost identical results as a corresponding classical treatment (Bardsley and Penetrante 1991), and taking $W_\phi = 5.1$ eV for clean gold, we obtain (atomic units are used unless otherwise stated)

$$d_c = \frac{1}{2 \cdot W_\phi} \sqrt{8 \cdot q + 2} \cong 0.4 \cdot \sqrt{q} \quad (nm) \tag{6}$$

At rather low impact energy, eq. (6) delivers typical projectile flight times $t_f \approx 10^{-13}$ s, and from the velocity-dependent parts of γ (cf. fig. 11) we thus obtain "mean apparent AI rates" of typically 10^{14} s^{-1}. If successive electron emission events would involve the same AI rates, ($\gamma - \gamma_\infty$) should scale like $1/v_p$.

However, fig.11 demonstrates that instead a relation

$$\gamma \approx \text{const.} \cdot v_p^{-0.5} + \gamma_\infty \qquad (7)$$

fits the measured total yields much better, which finding can be related to mean lifetimes linearly increasing for subsequent AI events (i.e. AI rates gradually decreasing with shrinking distance of the projectiles from the surface).

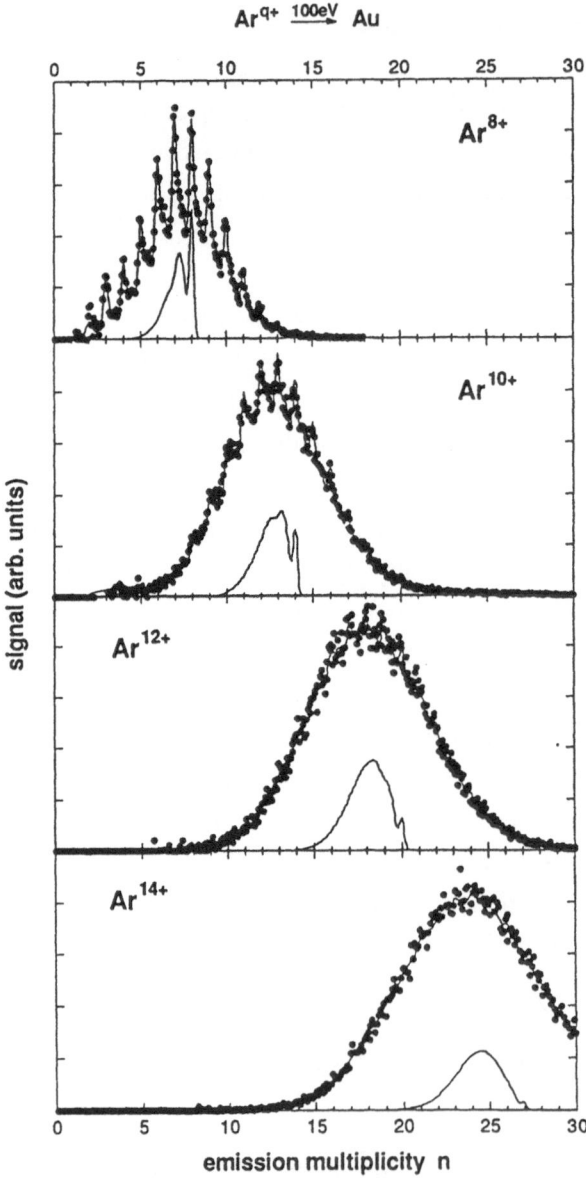

Figure 10. Measured pulse height spectra for impact of 100 eV Ar^{q+} in different charge states on clean polycrystalline gold. Measured data were fitted (solid line) for deconvolution into individual multiplicities. The shaded inserts show the energy deposition distribution for the respectively most probable number of emitted electrons (from Kurz et al., 1992).

We explain such a behaviour of the involved apparent AI rates as follows. On its way toward the surface the particle becomes subject to both RI and AI, but only the latter can cause the observed electron emission. It has been shown by Andrä et al. (1992) that because of "screening dynamics" RN populates projectile Rydberg states with increasingly lower principal quantum numbers as the particle approaches the surface. Therefore, the AI

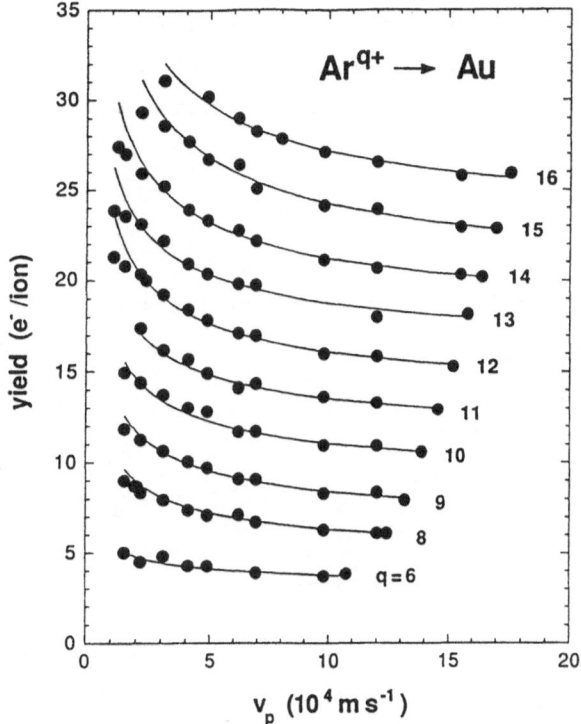

Figure 11. Total electron yields vs. impact velocity as derived from the measured ES, for impact of Ar^{q+} on clean polycrystalline gold (q = 6 + 16). The solid lines are fits according to eq. (7). Note the clear deviation from these fits at low v_p which is ascribed to image charge acceleration of projectiles (cf. text; from Kurz et al., 1992).

processes should also become slower when occuring nearer to the surface, considering the decreasing chance for the availability of empty lower projectile states permitting the relatively most probable AI transitions just into vacuum (Vaeck and Hansen 1991) and the fact that the competing RI will become gradually faster when closer to the surface. Fig. 11 shows, in addition, at the lowest impact velocities a deviation from the γ vs. v_p dependence as described by eq. (7), which can be related to a gain in projectile impact energy due to image charge attraction. Before its first RN at the distance d_c (cf. eq. 6), an ion Z^{q+} has already gained an additional kinetic energy $E_{q,im}$ according to

$$E_{q,im} = \frac{q^2}{4 \cdot d_c} \cong 0.9 \cdot q^{3/2} \quad (eV) \tag{8}$$

Fitting a $q^{3/2}$ dependence to experimental data from Winter (1992) for Ar^{q+} ($q \leq 6$) yields

$$E_{q,im} \approx 1.2 \cdot q^{3/2} \quad (eV) \tag{9}$$

Equ. (9) can be reproduced quite well by taking into account additional projectile acceleration inside d_c during the stepwise decreasing ion charge (Burgdörfer et al., 1991). Equ. (9) can be utilized to calculate the exact primary ion impact velocity in the low - v_p region. Apparently it makes no sense to aim for lower impact velocities than the absolute limit set by eq. (9), e.g. ca. 77 eV or $1.9 \cdot 10^4$ m/s for Ar^{16+}, irrespective of the chosen scattering geometry.

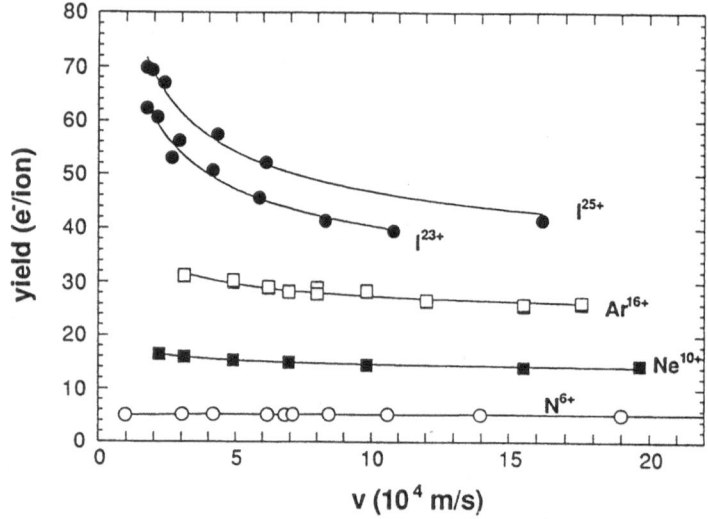

Figure 12. Total electron yields vs. impact velocity as derived from the measured ES, for impact of various MCI on clean polycrystalline gold (data from Kurz et al., to be published). The solid lines are fits according to eq. (7).

Turning now to the observed electron multiplicities, it is easily understood that the ES should follow a Poissonian distribution if the slow electron emission were related to mutually independent, equally fast AI processes. In fig. 13 measured ES have been plotted for Ar^{12+} in three different impact energies. Gaussian distributions fit these ES obviously quite well, whereas Poissonians for the same mean values γ are clearly too broad, as demonstrated for the 500 eV case.

This implies that the processes responsible for electron emission involve less randomness than one had to expect for fully independent emission of the individual electrons. The velocity-independent contribution γ_∞ (see above) cannot result only from "peeling-off" the $\leq q$ electrons still bound within highly excited states as the projectile reaches the surface (Burgdörfer et al., 1991), because for higher ion charge states γ_∞ gets clearly larger than q ($\gamma_\infty \approx 12$ for q = 12, but $\gamma_\infty \approx 21$ for q = 16, cf. fig. 11 and equs. 10). Kinetic electron emission cannot contribute by more than 0.5 electrons/ion at $v_p \leq 2 \cdot 10^5$ m/s (Lakits et al., 1990; Winter et al., 1991). We therefore propose some "ultimate", rather fast autoionization

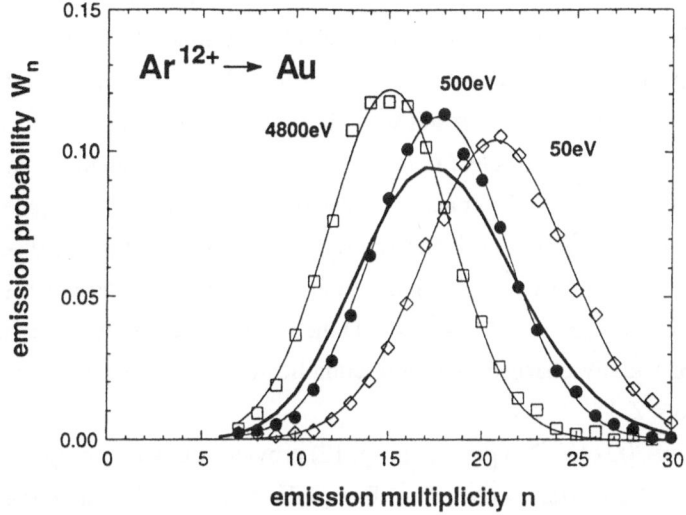

Figure 13. Electron emission statistics (ES) as derived by unfolding the raw pulse height spectra (cf. fig.10) for impact of Ar^{12+} with different impact energies on clean polycrystalline gold. The dashed lines are fitted Gaussian distributions. For the 500 eV data a Poissonian (solid line) with appropriate mean value γ has been added (from Kurz et al., 1992).

and/or Auger neutralization occuring close to the surface where the still populated highly excited projectile states overlap completely with the filled metal DOS. These processes should be fast enough to stay practically independent of the MCI impact velocity within its here considered limits.

For impact of Ar^{q+} (q = 8 ÷ 16) we found the following linear dependences for both γ and γ_∞ vs. ion charge q

$$\begin{aligned}
\gamma &= 2.97 \cdot q - 15.4 && \text{(for } E_{kin} = 100 \text{ eV)} \\
\gamma &= 2.67 \cdot q - 15.0 && \text{(for } E_{kin} = 1 \text{ keV)} \\
\gamma_\infty &= 2.10 \cdot q - 12.9 &&
\end{aligned} \tag{10}$$

Earlier studies (Arifov et al., 1973; Delaunay et al., 1987a; de Zwart, 1987) of MCI-related total electron yields demonstrated that for Ar^{q+} up to $q = 8$ the total yield γ remains directly proportional to the MCI's total potential energy, but for $q \geq 9$ the onset of inner shell vacancy formation in the projectile causes a marked levelling off from this behaviour. According to rels. (10) the increase of the contributions γ_∞ with q, which we have just ascribed to some "ultimate" AN/AI, corresponds to a direct proportionality of the yield increments with the number of electrons still carried in Rydberg states of the projectile upon its surface impact.

The arguments presented above are supported by our results for other projectile species, in particular N^{6+}, Ne^{9+} and I^{25+}. For bombardment with the first species we found almost no variation of the total electron yield ($\gamma \approx 5$) with v_p, which points to a complete domination of RI over AI and thus a rather effective suppression of electron emission $\gamma - \gamma_\infty$ during the approach of N^{6+} toward the surface. Consequently, the total electron yield measured for slow N^{6+} impact seems to be exclusively due to the above introduced "ultimate" AI/AN processes. This rather interesting result strongly suggests a reassessment of so far proposed models both for the filling of electronic states prior to the inner shell vacancy-related fast Auger electron emission (Andrä et al., 1992) and on the AI processes thought responsible for the slow electron emission (Vaeck and Hansen, 1991). For Ne^{9+} we also observed a very weak dependence of γ on v_p. In the light of our above presented arguments both findings suggest that for hydrogen-like MCI ($q = Z-1$) the AI processes are more strongly supressed by RI than for equally charged, but more complicately structured projectiles (i.e. for ion charge states $q < Z-1$).

The case of 200 eV I^{25+} ($\gamma \approx 70$; cf. fig. 12) provides our so far most extreme example for pure potential emission, where according to the respective ES more than 85 electrons could be ejected by a single MCI impact. In contrast to N^{6+} here the AI seems to be much less inhibited by RI, and the velocity-dependent contribution $\gamma - \gamma_\infty \approx 40$ electrons/ion is obviously a major part of the total yield. RN of the Ni-like I^{25+} ions cannot produce transient inner shell vacancies, which seems to increase the electron emission yield as a result of AI processes.

Roughness of the target surface might influence the total electron yield only via the γ vs. v_p dependence, but should not change the shape of a power law like eq. (7), which apparently applies quite well to all cases here studied (cf. fig. 11).

6. SUMMARY AND OUTLOOK

In the present paper we have demonstrated recent progress in studying slow ion-induced electron emission from clean metal surfaces, which has been gained from the measurement of the statistics of emitted electrons (ES). Such ES measurements permit separation of respectively potential and kinetic contributions to the total yield, and show their fundamental difference quite clearly.

A new, upper limit for the threshold impact velocity for onset of kinetic emission can now be precisely determined, and it has also become possible to study PE at impact energies above that apparent KE threshold. With our new ES technique we have also studied potential emission induced by doubly and multiply charged ions.

For impact of highly charged ions the ES method is especially well suited because of its sensitivity and informations supplementary to the ones already obtained by techniques so far applied in this field. The impact energy dependences of electron yields and emission statistics observed with MCI suggest that the former ones are composed of at least two parts. A first contribution is generated during the projectile's flight towards the surface and increases with decreasing impact velocity in a way that reflects a competition between autoionization and resonance ionization of the hollow projectile atoms being formed near the surface. The result of this competition should depend rather sensibly on both the target S-DOS and the electronic structure of the neutralized projectiles. A second, apparently impact velocity-independent contribution to the total yield is ascribed to rather fast multiple autoionization/Auger neutralization processes occuring just upon surface impact. This contribution seems to be directly proportional to the number of electrons still bound within highly excited projectile states at the moment of projectile impact.

All here presented results provide new insights in processes induced by slow ion – metal surface interaction and are particularly important for electron emission due to impact of MCI projectiles and the related fast Auger electron- and X-ray photon emission phenomena.

ACKNOWLEDGMENTS

This work has been supported by Austrian Fonds zur Förderung der wissenschaftlichen Forschung under project no. P8315TEC, and by Kommission zur Koordination der Kernfusionsforschung at the Austrian Academy of Sciences. Contributions by Mr. H. Kurz and Mrs. K. Töglhofer are gratefully acknowledged.

REFERENCES

Alonso E V, Alurralde M A, and Baragiola R A 1986 Surf.Sci. **166** L155

Andrä H J, Simionovici A, Lamy T, Brenac A, Lamboley G, Pesnelle A, Andriamonje S, Fleury A, Bonnefoy M, Chassevent M, and Bonnet JJ 1992 in Electronic and Atomic Collisions, eds. W.R. MacGillivray, I.E. McCarty and M.C. Standage (IOP Publ. Ltd./Adam Hilger, Bristol) p. 89

Arifov U A, Kishinevskii L M, Mukhamadiev E S, and Parilis E S 1973 Zh.Tekh.Fiz.**43** 181 (Sov.Phys.-Tech.Phys. **18**,118(1973))

Ashcroft N W and Mermin N D 1976 in Solid State Physics (CBS Publ. Asia Ltd., Philadelphia)

Aumayr F, Lakits G, and Winter HP 1991 Appl. Surface Sci. **47** 139

Bardsley J N and Penetrante B M 1991 Comm.At.Mol.Phys. **27**,43

Brenten H, Müller H, and Kempter V 1992 Z.Phys.D **22** 563

Briand J P, de Billy L, Charles P, Essabaa S, Briand P, Geller R, Desclaux J P, Bliman S, and Ristori C 1990 Phys.Rev.Letters **65** 159

Briand J P, de Billy L, Charles P, Essabaa S, Briand P, Geller R, Desclaux J P, Bliman S, and Ristori C 1991 Phys.Rev. A **43** 565

Burgdörfer J, Lerner P and Meyer F W 1991 Phys.Rev.A **44** 5674

Delaunay M, Fehringer M, Geller R, Hitz D, Varga P and Winter HP 1987a Phys.Rev. B **35** 4232

Delaunay M, Fehringer M, Geller R, Hitz D, Varga P and Winter HP 1987b Europhys.Letters **4** 377

Donets E D 1985 Nucl.Instrum.Meth.Phys.Res. B **9** 522

Feuerbacher B and Christensen N E 1974 Phys.Rev. B **10** 2373

Folkerts L and Morgenstern R 1990 Europhys. Letters **13** 377

Hagstrum H D 1954 Phys.Rev. **96** 325; 336

Hagstrum H D and Becker G E 1973 Phys.Rev. B **8** 107

Hasselkamp D 1992 in <u>Particle Induced Electron Emission II</u>, Springer Tracts in Modern Physics **123**, Berlin

Hofer W O and Littmark U 1976 Nucl.Instrum.Meth. **138** 67

Kurz H, Töglhofer K, Winter HP, Aumayr F, and Mann R 1992, Phys.Rev.Letters **69** 1140

Lakits G, Aumayr F, and Winter HP 1989 Rev.Sci.Instrum. **60**,3151

Lakits G, Aumayr F, Heim M and Winter HP 1990 Phys.Rev. A **42** 5780

Lakits G and Winter HP 1990 Nucl.Instrum.Meth.Phys.Res. B **48** 597

Mann R 1986 Z.Phys. D **3** 85

McDonald J W, Schneider D, Clark M W, and Dewitt D 1992 Phys.Rev.Letters **68** 2297

52 F W, Overbury S H, Havener C C, Zeijlmans van Emmichoven P A, and Zehner D M
    1991a Phys.Rev. Letters **67** 723

Meyer F W, Overbury S H, Havener C C, Zeijlmans van Emmichoven P A, Burgdörfer J, and Zehner D M
    1991b Phys.Rev. A **44** 7214

Ohya K, Aumayr F, and Winter HP 1992 Phys.Rev. B **46** 3101

Schultz M, Cocke C L, Hagmann S, Stöckli M, and Schmidt-Böcking H 1991 Phys.Rev. A **44** 1653

Töglhofer K, Aumayr F, and Winter HP 1992 Surface Science (in print)

Vaeck N, and Hansen J E 1991 J.Phys.B:At.Mol.Opt.Phys. **24** L469

Varga P, Hofer W, and Winter HP 1982 Surface Sci. **117** 142

Varga P 1988 in <u>Electronic and Atomic Collisions,</u> ed H.B. Gilbody et al, Elsevier, p. 793

Varga P 1989 Comm.At.Mol.Phys. **23** 111

Varga P, and Winter HP 1992 in <u>Particle Induced Electron Emission II</u>, Springer Tracts in Modern Physics
    123, Berlin

Wille U 1991 Z.Phys. D **21** S353

Wille U 1992 Phys.Rev. A **45** 3004

Winter HP 1982 Rev.Sci.Instrum. **53** 1163

Winter HP, Aumayr F and Lakits G 1991 Nucl.Instrum.Meth.Phys.Res. B **58** 301

Winter HP 1991 Z.Phys.D **21** S129

Winter H 1992 Europhys.Letters **18** 207

Wouters P A A F, Zeijlmans van Emmichoven P A and Niehaus A 1989 Surface Sci. **211/212** 249

Zeijlmans van Emmichoven P A, Havener C C, and Meyer F W 1991 Phys.Rev. A **43**,1404

de Zwart S T, Fried T, Jellen U, Boers A L, and Drentje A G 1985 J.Phys.B:At.Mol.Phys.**18** L623

de Zwart S T 1987 PhD Thesis, University Groningen

de Zwart S T, Drentje A G, Boers A L and Morgenstern R 1989 Surface Sci. **217** 298

Z_1 AND Z_2 OSCILLATIONS IN THE ENERGY LOSS OF SLOW IONS:
INHOMOGENEOUS ELECTRON GAS MODELS

J. Calera-Rubio[1], A. Gras-Marti[1] and N.R. Arista[2]

[1]Departament de Física Aplicada
Universitat d' Alacant
E-03080 Alacant, Spain

[2]Instituto Balseiro and Centro Atómico Bariloche (CNEA)
CC 439, RA-8400 Bariloche, Argentina

ABSTRACT

We evaluate the inelastic energy loss of slowly moving ions in solids taking into account the non-homogeneous electron density distributions of real solids. Both Z1 (ion atomic number) and Z2 (target atomic number) oscillations in the stopping power are calculated using a self-consistent (non-perturbative) description of the screening of the ion charge in an inhomogeneous electron gas, representing the core and the outer electrons of target atoms. The Z1 oscillations in the straggling parameter are also investigated.

The assumptions entering the electron gas model and the corresponding limitations will be outlined.

1. INTRODUCTION

The low-velocity inelastic stopping power of ions in solids provides an excellent bench for the discussion of theoretical models and experimental data on the interaction of ions with matter[1]. At low projectile velocities, atomic-physics models and solid-state descriptions have been employed in the theoretical analysis. The agreement between various approximations and the scarce experimental data is usually within a factor of two or better. Experiments exist both for crystalline and policrystalline materials, i.e., for energetic ions moving in either channeling or in random trajectories. Oscillations with Z_1 (ion atomic number) and Z_2 (target atom atomic number) of the stopping power and the straggling parameter have been measured, and their calculation has been attempted using different models.

The motivations for the present study are threefold. Firstly, there is a challenge to introduce an accurate description of the ion cores in electron gas models of solids[2]. This need is felt more strongly in the study of inelastic processes in solids, when excitations are intense and localized. Secondly, it was suggested[3] that a description of inner shells in an electron gas calculation could be done by simulating them as a denser electron gas. Thirdly, we would like to clarify some discrepancies between the predictions of the more accurate theoretical models and measurements of Z_1 oscillations in the stopping power, for various ions moving in random conditions in nearly-free electron systems, like in Al.

We shall briefly discuss here Z_1 and Z_2 oscillations in the inelastic stopping power, as well as the straggling parameter, of slow ions in solids. Rather than attempting a detailed comparison of our theoretical predictions with experimental data and with other calculations, which is done elsewhere[4], we shall outline the main assumptions entering the new model proposed, and its limitations, as well as the main characteristics of the predictions that it can give rise to.

2. ELEMENTS OF THE THEORY

The models that are used to describe low-energy inelastic stopping powers are based on linear-response theory[5,6] or on statistical concepts[7,8]. A different recent approach makes use of pseudopotential theory[9], but is limited to ions of the same atomic species as the target atoms.

The oscillatory Z_1 dependence of the energy loss of ions channeled in crystalline solids has been described using the transport cross section (TCS) approach[10] (to calculate the average momentum transfer to valence electrons scattered in the field of the ion moving along the crystal) and also by means of many-body formulations like density functional theory (DFT). The results emerging from the TCS-DFT model, for slow ions embedded in a uniform electron gas, provide a consistent approach to calculate the energy loss of channeled ions at a very good level of approximation[11,12].

However, for ions in random-incidence conditions, like in the penetration of polycrystalline Al foils or amorphous C foils, the comparison of the theoretical predictions with experimental data becomes worse. It is particularly striking that the approach seems to fail in the case of a nearly-free-electron metal like Al while, instead, it provides very accurate results for ions channeled in semiconductors[12]. The discrepancy has been explained[4] in terms of the different contributions to the stopping from intermediate-shell electrons in the case of particle motion in random *vs.* channeling conditions.

Let us briefly describe the TCS-DFT model of stopping calculations both for homogeneous and nonhomogeneous electron gases.

Ion slowing down as a scattering process

The stopping power for low-velocity ions ($v \ll v_F$, where v_F is the Fermi velocity) moving in a *homogeneous* electron gas of density n or one-electron parameter r_s (with $n=[4\pi r_s^3/3]^{-1}$) can be written in terms of the TCS which is expanded in terms of the phase shifts δ_l for the scattering of electrons at the Fermi surface[13],

$$S \equiv - \frac{dE}{dx} = \frac{3v}{v_F r_s^3} \sum_l (l+1) \sin^2\left[\delta_l(v_F) - \delta_{l+1}(v_F)\right]. \tag{1}$$

(Atomic units, $e = m = \hbar = 1$, will be used throughout).

The DFT theory employed in the calculation of the phase shifts[14] is only applicable for low-velocity ions, because the phase shifts δ_l are calculated for a spherically symmetric central potential. However, the limit of validity of this model, $v \ll v_F$, is not so severe when it is applied to intermediate or inner-shell electrons. Apart from the fact that low-velocity stopping powers have been experimentally observed to vary proportionally with v up to rather high velocities (up to $v \approx v_o$, the Bohr velocity)[15,16], there is another consideration coming from our description of the electron density in the stopping medium. Since $v_F \sim r_s^{-1}$, then the range of validity of the calculations for the nonhomogeneous electron density effects on the energy loss will be extended due to the larger v_F values corresponding to inner electrons.

DFT model for the homogeneous electron gas

The DFT model[13] provides self-consistent calculations of TCS and friction coefficients of slowly-moving charged particles in solids. The DFT treatment of the electron gas accounts for bound state occupation in the impurity charge (the projectile) as well as for the scattering of target electrons in the screened field of the projectile.

Oscillations in the stopping power as a function of projectile atomic number, as predicted by TCS-DFT calculations, are due to interferences (scattering and resonance effects) between wavefunctions of electrons of different angular momentum (with $\Delta l = \pm 1$) which are scattered in the field of the impurity[12]. The minima in the Z_1 oscillations are due to the formation of closed shells in the ion. Values around maxima are due to resonances in electron—ion scattering: for increasing Z_1, bound and free states decrease in energy and when one of these states crosses the Fermi level it gives rise to a fast change in a phase-shift, and correspondingly to a maximum of the transport cross section.

For high electron densities (small r_s) screening is so strong that bound states cannot be easily formed. On the contrary, for a dilute gas bound states approach orbital (atomic) nature, and the atomic character of scattering is recovered.

Many DFT schemes exist[13,17], with various degrees of sophistication. In the following we consider an approximate procedure to calculate the phase shifts for scattering, and the ensuing stopping parameters, based on a very simple physical model.

DFT "emulation"

This simplified approach, as a substitute of full DFT calculations, was pioneered by Cherubini and Ventura[18]. The non-linear calculation of the transport cross sections and friction coefficients starts from quantum-mechanical phase-shift calculations of the scattering of electrons with Fermi velocity v_F, in the screened potential of a stationary ion in the solid. For instance, a simple Yukawa-type potential, $V(r) = (Z_1/r)\exp(-\alpha r)$, with an adjustable screening parameter α, is used to represent the scattering potential. It has been shown[18] that the use of this, or similar forms, for the screened potential, yields TCS results that compare quite well with the full density- functional calculations provided that the screening parameter α in the potential is adjusted so as to satisfy the Friedel sum rule (FSR). This condition expresses that the medium screens out completely the impurity charge by displacing the same number of electrons.

We apply the self-consistent model just described to an impurity in a non-homogeneous electron gas which is described through a variable r_s parameter representing the actual electron—density profile in real solids: in this way we incorporate also the effect of a realistic electron distribution of the solid in the evaluation of the average inelastic energy loss of the moving ion.

Then, for a given ion atomic number Z_1 and for a fixed r_s value, we solve the usual radial Schrödinger equation by direct numerical integration, until the wavefunction approaches the asymptotic limit (for $r \to \infty$), $u_l(r) \longrightarrow A_l \sin(kr - \frac{1}{2}\pi l + \delta_l)$. In this way one determines phase shifts $\delta_l(Z_1, r_s)$ for electrons at the Fermi surface and for some initial assumption on the value of α. We repeat this procedure by varying α until we find the value for which the phase shifts satisfy the FSR. (The FSR is automatically satisfied in full DFT calculations). This method provides a simple way to adjust in a self-consistent manner the screening parameter and the scattering phase shifts for given values of Z_1 and r_s.

Technically, this procedure emulates a full DFT calculation, but, very importantly from a practical standpoint, it is computationally much simpler. The values of the phase-shifts obtained using this method[4] compare very well with those calculated by Puska and Nieminen using the full DFT[14].

Using in Eq.(1) the values of $\delta_l(Z_1, r_s)$ obtained in this manner we calculate self-consistent stopping power values $S(Z_1, r_s)$ for a homogeneous electron gas of arbitrary density.

LDA treatment for the inhomogeneous electron gas

The local density approximation (LDA)[5,6,19] assumes that the properties at a given point of an inhomogeneous electron distribution with given local density are equal to those of a homogeneous electron gas having the same density.

To represent the case of non-homogeneous electron distributions in real solids we consider an electron gas of varying density $n(\vec{r})$ and the corresponding $r_s(\vec{r})$ parameter. Then, also the quantities v_F, α and δ_l will now depend on the position $\vec{r}$ in the electron cloud of the target atom. To obtain the values of α and the ensuing δ_l, we repeat the self-consistent evaluation based on the application of the FSR, as described before, with a fine mesh of r_s values for each value of Z_1.

For simplicity, spherical electron distributions in each atomic cell are assumed. Screening of the ion charge is adjusted locally, so that the FSR is satisfied at every point of the ion's trajectory.

The combined LDA-DFT model incorporates then an adiabatic adjustment of the local effective charge of the projectile: as the projectile moves through regions of varying electron density, its effective charge state should change. This is represented by the filling of electron shells of the ion in the nonuniform electron gas.

We obtain the average energy loss of the ion by integrating over the whole atomic density profile $n(\vec{r})$, following the prescription of the local-density approximation[6,19],

$$<S> = 4\pi N_a \int_0^{r_a} r^2 dr\, S(r), \qquad (2)$$

where $S(r)$ is given by Eq.(1) —using the locally-adjusted values of $\alpha_l(Z_1,r_s)$ and $\delta_l(Z_1,r_s)$—, r_a is the atomic-cell radius, and N_a the number of atoms per unit volume.

In all LDA treatments of the electron gas[5,6,19] one considers electrons to be free. There is a binding force effect, however, included indirectly (through the local r_s values) when one averages over inhomogeneous electron densities calculated for real solids.

We have implicitly assumed that the incident ion does not greatly perturb the electron distribution of the target atoms[20]. This approximation becomes worse the larger the ion atomic number. A fully self-consistent calculation of the electron densities of the target atoms in the presence of the ion appears to be a difficult task.

Further assumptions

The existence of a minimum distance of approach r_{min} in the collision of the ion and a target atom has been neglected in the calculations. More properly, the lower limit of integration in Eq.(2) should be r_{min}. Although it may be cumbersome in practice, this approximation can be corrected for by using the procedure followed by Ventura [20]. By taking $r_{min} = 0$ in Eq.(2) we are providing an upper estimate for the stopping power due to the nonhomogeneous character of the electron density of solids. On the other hand, if one includes r_{min} effects, the strict proportionality of low-velocity stopping with the ion speed no longer holds[20]. It can be seen[4] that small values of r do not contribute significantly to the stopping, therefore the approximation of neglecting r_{min} is a reasonable one.

3. SOME APPLICATIONS

Detailed discussions of the predictions of the model will be reported elsewhere[21]. Here we shall briefly mention some results concerning the contributions of inner electrons to Z_1 and Z_2 oscillations in the stopping power, and to the calculation of the straggling parameter.

We have shown in ref. [4] that the predictions of DFT are very well reproduced by the procedure described above, that combines the use of a Yukawa potential and the constraint imposed by the FSR. The predicted stopping power for homogeneous electron gases of various densities is shown in Fig. 1a. We see in Fig. 1b that the maxima and minima in the Z_1 oscillations are shifted to larger values of the charge Z_1 when it is embedded in electron gases of increasing density. This is due to the fact

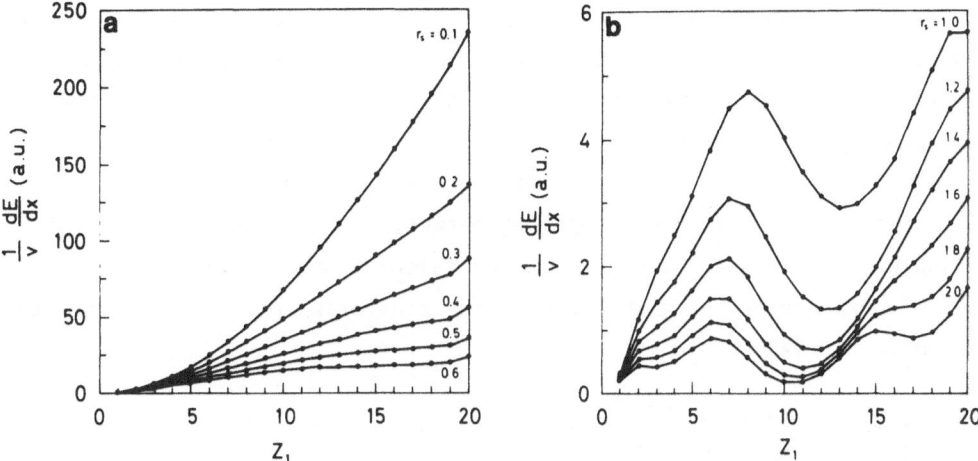

Figure 1. Stopping power of ions with atomic number Z_1 in electron gases of different densities, Eq. (1). The parameter on the curves is r_s. Results for typical electron densities corresponding to inner electrons and to valence electrons in real targets are shown. Note de different y-scale used in a) and b).

that as screening increases with decreasing r_s, it takes a stronger nuclear charge to bind extra electrons.

The behavior in Fig. 1 also suggests a possible explanation for the disagreement of stopping power calculations, based on the DFT for a homogeneous electron gas, with the experimental data for the stopping of ions in random orientations in Al: in the application of Eq. (2) one integrates over the effects of higher electron densities in the solid as they are probed by the ions, so that the corresponding contribution to the stopping power becomes larger, and the Z_1 dependence is modified.

Our calculations[4] agree very well with the observed Z_1-oscillations in the range $1 < Z_1 < 20$. It will be of interest to extend these calculations to higher Z_1 values (namely, $Z_1 \sim 50$), where DFT results for a homogeneous electron gas[23] predict a rather strong dip in the Z_1 oscillations, while experiments[22] with ions of velocity $0.8\ v_0$ in C and Al yield fairly constant results, followed by a rise for larger Z_1. We expect that the results for non-homogeneous distributions will smear the oscillations, as in the lower Z_1 range[4], and will improve the agreement with experiments.

Straggling

It has been found experimentally[24] that the Z_1-oscillations of the straggling parameter and of the stopping power appear to be out of phase, for ions with $9 \le Z_1 \le 21$ and energy of 16 keV/u, traversing carbon foils. The explanation advanced by the authors relies on some internal electron-promotion mechanism.

One would expect that intermediate-shell electrons will contribute significantly also to the straggling parameter W, because large energy transfers are weighted strongly in the calculations of this quantity.

The expression for W in terms of the scattering phase shifts was derived in ref.[25],

$$W = \int d\sigma\ T^2 = \frac{3\pi v^2}{4\sqrt{2}} \sum_{l} \sum_{m} (2l+1)\ (2m+1) \times$$
$$\{1 - \cos(2\delta_l) - \cos(2\delta_m) + \cos[2(\delta_l - \delta_m)]\}\ J_{lm} \qquad (3)$$

with the coefficients J_{lm} given therein[25].

Application of the procedure mentioned in section 2 to the calculation of this parameter for a homogeneous electron gas corresponding to the valence electron density in Al, and for the nonhomogeneous electron distribution of metallic Al, yield the results of Fig.2. Somewhat surprisingly, we observe no difference in the positions of the peaks in the Z_1-oscillations of the straggling parameter for the homogeneous and inhomogeneous case.

Z_2 oscillations in the stopping power

Z_2 oscillations in the stopping power and the straggling parameter of low-velocity ions in solids have been measured and calculated theoretically using various methods[26]. These oscillations simply reflect the oscillatory structure of the target electron density.

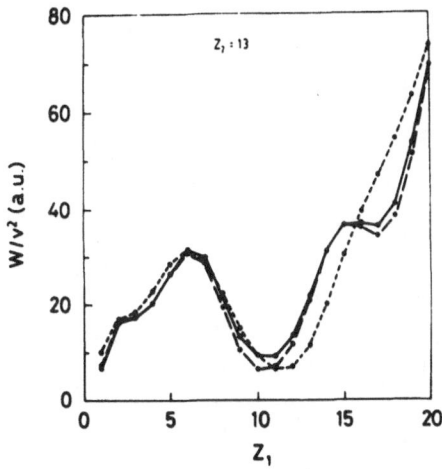

Figure 2. Straggling parameter of ions in Al as a function of Z_1. The dots are the theoretical calculations. The curves have been drawn to guide the eye. Short–dashed line: DFT for a uniform electron gas, $r_s = 2$, from ref.[25]. Long–dashed line: Yukawa potential plus FSR for a uniform electron gas, $r_s = 2$. Solid line: Yukawa potential plus FSR for a non-uniform electron gas corresponding to the density of Al.

Because of the computational difficulties in generating all phase shifts for arbitrary electron densities, the DFT formalism has not been used so far to make comparisons with experimental measurements of Z_2 oscillations in the stopping power of projectiles with given atomic number. This problem is circumvented with the procedure described above. Fig.3 shows the results of our calculation for a homogeneous and a nonhomogeneous electron density model. For the set of experimental data shown in the figure, the predictions of the inhomogeneous model overall agree better with experiments than the homogeneous electron gas model; but discrepancies are still present for some values of Z_2. One should keep in mind, however, that the velocities used in the measurements are not always small.

4. CONCLUSIONS

Our model provides an explanation of the distinct Z_1-dependences observed for the friction coefficient of slow ions, both in channeling and in random incidence, as due to the different electronic density profiles sampled by the ions in each case.

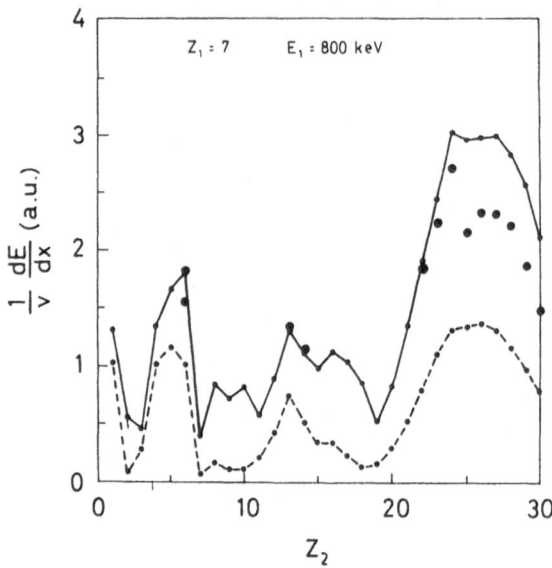

Figure 3. Stopping power of 800 keV N$^+$ ions in targets of atomic number Z_2. The experimental data (large dots) are taken from ref. [26]. The small dots are the theoretical calculations. The curves have been drawn to guide the eye and give the results of various models for the description of the target. Short—dashed line: Yukawa potential plus FSR for a uniform electron gas. Solid line: Yukawa potential plus FSR for a non-uniform electron gas.

We provide here evidence that inner electrons play an important role in the process of energy loss of slow ions in solids, and its inclusion is essential to understand the differences in ion-solid interactions for channeled and non-channeled trajectories.

The DFT emulation procedure outlined here is a first step towards a more consistent treatment of the inhomogeneous electron densities in real solids. Since the DFT is currently considered as one of the most powerful approaches to describe the behavior of impurities in solids, it is of interest to note that, according to the results presented here, the discrepancies between previous DFT calculations and experimental stopping power results should not be considered a failure of the DFT; in fact our results suggest that, by making an extension of current self-consistent DFT calculations to treat non-homogeneous electron distributions, the DFT could be further applied to produce *ab initio* calculations for slow ions in random incidence conditions.

On the other hand, the model presented here may be considered as an alternative to the standard DFT approach. It is strictly based on the use of the Friedel sum rule in order to achieve a self-consistent adjustment of screening parameters and scattering phase shifts for all r_s values. Our model retains the essential features of the non-linear stopping power description in terms of scattering phase shifts, and compares quantitatively well with DFT calculations. Because of its greater simplicity, the present approach may become very useful to investigate other non-linear phenomena of low-velocity particle-solid interactions[13,23].

The question remains open of an alternative microscopic mechanism for the observed energy losses of slow ions moving in a dense electron gas. Conceptually, the easiest way to incorporate the ion cores into an electron gas model may be to recognize that there is a gap between the last filled level in the core and the first empty level[3]. Now, if the energy levels are assumed rigid throughout the collision, direct scattering across this gap cannot occur in low-velocity interactions, because of the velocity threshold for excitation of internal electrons. Yet, it might occur if another electron, or the colliding system as a whole, would participate by providing some of the energy-momentum. However, no specific mechanism for this has been proposed yet.

Another possibility of inner-shell excitation is via electron-promotion mechanisms. These are usually not treated in the electron gas picture of energy loss. By this mechanism, some energy levels in the dynamically perturbed system will cross the gap and cause excitations above the gap. It has been shown that this mechanism can be very effective in providing the conditions for excitations of inner shells in the low-velocity quasi adiabatic limit[27]. The incorporation of these effects in the description of the energy loss of slow ions in solids, may also contribute to the understanding of the physical mechanisms underlying Firsov's[8] and other low-velocity theoretical models.

Acknowledgements

This project is supported in part by the Spanish DGICYT (project number PS89-0065) and the Consejo Nacional de Investigaciones Científicas y Técnicas (Argentina). A NATO grant is also acknowledged (CRG 901013). N.R.A. is visiting professor at the Universitat d'Alacant under the program PROPI of the Generalitat Valenciana. We thank Prof. F. Salvat for his advice in calculating the scattering phase shifts and for graciously providing a copy of his code. The Acciones Integradas program between Spain and Italy facilitated discussions with Prof. R.A. Baragiola. We also thank Vicente Esteve for his technical assistance.

REFERENCES

1. *Interaction of Charged Particles with Solids and Surfaces*, Ed. by A. Gras-Martí, H.M. Urbassek, N.R. Arista, and F. Flores (Plenum, New York, 1991), ASI-series, Vol. B271.

2. R.A. Baragiola, page 701 in ref. [1].

3. K.B. Winterbon, private communication to AGM.

4. N.R. Arista, J. Calera-Rubio, and A. Gras-Martí (submitted).

5. J. Lindhard and A. Winther, Kgl. Danske Videnskab. Selskab, Mat.-Fys. Medd. **34**, No.4 (1964).

6. J. Lindhard and M. Scharff, Kgl. Danske Videnskab. Selskab, Mat.-Fys. Medd. **27**, No. 15 (1953); E. Bonderup and P. Hveplund, Phys. Rev. **A4** (1971) 562.

7. J. Lindhard and M. Scharff, Phys. Rev. **124** (1961) 128.

8. O.B. Firsov, Zh. Eksp. Theor. Fiz. **36** (1959) 1517; Sov. Phys.-JETP **9**, (1959) 1076.

9. G. Falcone and Z. Sroubek, Phys. Rev. **B39** (1989) 1999; also, page 593 in ref. [1].

10. J.S. Briggs and A.P. Pathak, J. Phys. C **6**, (1973) L153; *ibid.* **7**, (1974) 1929.

11. P.M. Echenique, R.M. Nieminen and R.H. Ritchie, Solid State Commun. **37**, (1981) 779.

12. P.M. Echenique, R.M. Nieminen, J.C. Ashley and R.H. Ritchie, Phys. Rev. **A33**, (1986) 897.

13. P.M. Echenique and M.E. Uranga, page 39 in ref. [1].

14. M.J. Puska and R.M. Nieminen, Phys. Rev. B **27**, (1983) 6121.

15. A. Mann and W. Brandt, Phys. Rev. **B24**, (1981) 4999.

16. *Sakharov's rule* applies here also: the range of validity of a formula is usually larger than one expects on theoretical grounds. (R.H. Ritchie, private communication to AGM).

17. T. Ziegler, Chem. Rev. **91**, (1991) 651; M.J. Puska and R.M. Nieminen, Phys. Rev. **B43**, (1991) 12221.

18. A. Cherubini and A. Ventura, Lett. Nuovo Cim. **44**, (1985) 503.

19. R.E. Johnson and M. Inokuti, Comments At. Mol. Phys. **14**, (1983) 19.

20. A. Ventura, Il Nuovo Cim. **10D**, (1988) 43.

21. J. Calera-Rubio, A. Gras-Martí, and N.R. Arista, to be published.

22. W.N. Lennard, H. Geissel, D.P. Jackson and D. Phillips, Nucl. Instrum. Meth. **B13** (1986) 127; W.N. Lennard and H. Geissel, page 347 in ref. [1].

23. P.M. Echenique, F. Flores and R.H. Ritchie, Solid State Physics **43**, (1990) 229.

24. W.N. Lennard, H. Geissel, D. Phillips and D.P. Jackson, Phys. Rev. Lett. **57** (1986) 318.

25. J.C. Ashley, A. Gras-Marti and P.M. Echenique, Phys. Rev. **A34** (1986) 2495.

26. D.J. Land, J.G. Brennan, D.G. Simons and M.D. Brown, Phys. Rev. **A16**, (1977) 492.

27. W. Brandt and R. Laubert, Phys. Rev. Lett. **24**, (1970) 1037.

ELECTRON-SHELL EFFECTS IN PARTICLE-INDUCED KINETIC ELECTRON EMISSION FROM SOLIDS

Mona M. Ferguson[1] and Wolfgang O. Hofer[2]

[1] Daimler-Benz AG
Forschungszentrum Ulm
D-7900 Ulm, Germany

[2] Forschungszentrum Jülich, IPP[*]
D-5170 Jülich, Germany

Introduction

Ion-induced electron emission from solids is in general a complex interplay of electron excitation in the solid, transport of energetic electrons towards the surface, and emission of electrons across the surface barrier. This complexity is reduced in the case of metal targets, for which a separation into consecutive processes is possible. This is a consequence of the short mean free path length of electrons in metals - which, in turn, is a consequence of high free-electron densities. The electron escape depth is accordingly small too: it is of the order of the lattice constant at the free-electron densities typical of metals (10^{23} cm^{-3}). In the projectile-energy regime where kinetic emission prevails, the electron escape depth is much smaller than the excitation range along the projectile's track in the metal target. Most emitted electrons will, therefore, have received their energy via an electronic transport process from the location of excitation to the surface. High densities of free electrons effectively decouple the primary excitation from the emission proper.

An electron-collision cascade propagating in the free-electron system is generally assumed to mediate the transport of energy and momentum, and the electronic work function is generally chosen as the surface barrier which internally excited electrons must overcome.

It is an experimental fact that almost all experimentally accessible quantities are controlled by electron transport and the exit through the surface. The primary excitation is scarcely recognizable. For exploring the very genesis of ion-induced electron emission this is an obstacle, but remarkable and convenient similarity properties follow from this predominance

---

[*] EURATOM Association

Ionization of Solids by Heavy Particles, Edited by
R.A. Baragiola, Plenum Press, New York, 1993

of transport and emission. Most quantities measured outside the solid hardly depend on the projectile species since characteristics of the projectile enter predominantly through the primary ionization:

The **angular distribution** of emitted electrons is of a cosine shape for incident electrons, protons, and heavy ions; and

the **energy distribution** shows for all projectiles the typical collision cascade and surface-refraction effects.

A similar universality can be stated for the functional dependencies of the electron yield on the angle of incidence and the energy of the projectiles.

Moreover - and even more striking - this universality holds not only for variation of the projectile, it also applies for variations of the target, i.e. it holds for all metal targets. This is a consequence of the fact that for all ordinary metals, both the free electron density and the work function are, respectively, of the same order of magnitude ($n_e = 10^{23}$ cm^{-3}, $\Phi_w = 3 - 6$ eV).

Convenient as these similarity properties may be, they constitute an obstacle to investigations of the primary ionization process and, thus, to our understanding of the underlying physics. It is perhaps interesting to note that a similar situation exists for the emission of atoms, i.e. in sputtering - and this problem has the same roots, see e.g. Hofer (1990). For electron emission, however, there is one notable exception to this masking of the primary excitation: The influence of the electron shell on the electron yield. The total electron yield shows non-monotonic variations when, for a given target, the atomic number of the projectile is varied. With reference to a similar appearance in electronic stopping, this effect is often referred to as the Z_1-oscillation of the electron yield.

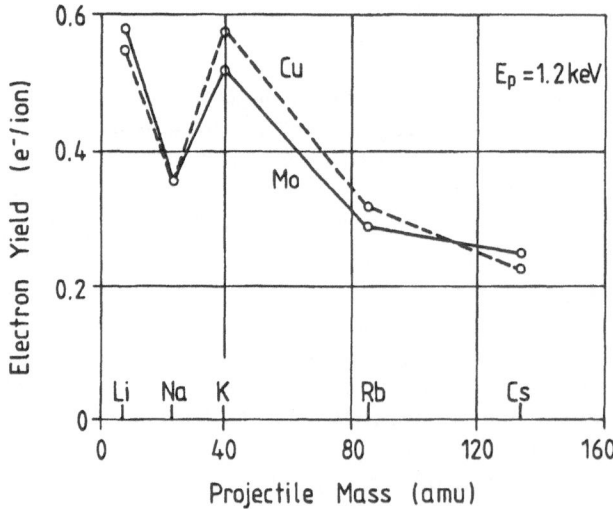

Fig. 1. Dependence of the electron yield from Cu and Mo surfaces upon 1.2 keV alkali-ion bombardment. From Ploch (1951).

The effect of the projectile's atomic number on the electron yield was discovered and correctly interpreted by Ploch (1951). In measurements on (moderately clean) Mo and Cu surfaces upon impact of alkali-metal ions, Ploch had found a non-monotonic yield dependence on the atomic number of the projectile, Fig. 1. He was aware that direct free-electron excitation by the heavy ions chosen in his experiment was a sub-threshold mechanism for electron emission across a surface barrier of several electron volts (Cu, Mo: Φ_w=4.5 eV). The excitation of electrons *bound in the ion cores* of the metal targets thus remained as the alternative. Excitation of bound electrons by heavy projectiles had been invoked earlier by other workers in electron emission but no mechanism for electronic excitation in low-energy atom / atom collisions was specified.

Ploch applied to solids the model of ionization of atoms in low-energy atom/ atom collisions developed by Weizel and Beeck (1932). This mechanism had originally been proposed for ionization in gases. It assumes electron promotion during that phase of the collision in which the electron clouds of the collision partners interpenetrate. During this time, the electronic system can be regarded as that of a transient molecule. In the two-centre force field of the nuclei, symmetry and parity requirements - and Pauli's exclusion principle - lead to splitting of atomic orbitals and promotion of molecular orbitals to higher principal quantum numbers. This part of the Weizel-Beeck model - i.e. the *excitation* part - is based on Hund's theory of molecular binding[*].

Regarding the *de-excitation / ionization* part of the model, Weizel and Beeck noted that during the separation phase of the colliding atoms there is a finite probability for promoted electrons to remain for a short time in the respective atomic orbitals. If the return to the ground state occurs in a nonradiative way - and this is the more probable de-excitation channel for outer electron shells - electrons are set free.

These are the electrons which may initiate electron collision cascades within the metal target. Of course, they may also be registered as "Auger electrons" outside the surface if the collision occurs within the escape depth of the solid. Investigations of ion-induced Auger electron emission thus provide most important information on the primary excitation process, cf. e.g. Baragiola (1982), Bonanno et al.(1990), Thomas (1983), Wittmaack (1979).

The Weizel-Beeck mechanism of ionization in low-velocity ($v_p < v_B \approx 2 \times 10^8$ cm/s) atom/atom collisions appears to have fallen into oblivion for more than 30 years. It was rediscovered by Fano and Lichten in 1965; cf. Briggs (1988), Hofer (1990). Barat and Lichten (1972) substantiated this model by introducing correlation diagrammes in atomic collision physics, see also Lichten (1980).

Ploch's work constitutes a milestone in the research on ion-induced electron emission. Not only was he able to provide a fully satisfactory explanation for an important effect he had the genius and fortune to discover, he also realized the enormous applicational aspect of his finding: an atomic-number dependent electron yield gives rise to an atomic-number dependent sensitivity of particle detectors relying on ion-to-electron conversion. Such high-sensitivity detectors are in operation, for instance, in all modern mass spectrometers.

So important did Ploch consider this implication that he published a warning in the form of a preliminary short note. How appropriate this step was becomes evident from the development in the understanding of the shell-effect in electron emission: for about three decades all investigations in this field were initiated by particle-detector problems, and atomic-number dependent detector sensitivity was their prime concern, see Fehn (1974, 1976), Staudenmaier et al. (1976), Rudat and Morrison (1978) and the references in these publications to earlier work. Progress in the understanding of the physics of shell effects in electron emission was meager. Ploch's work was either disregarded or unknown.

Experimental

Systematic investigations on the physics of the electron-shell effect on electron emission require a fine scan over a large range of the atomic number of the projectile. A universal ion

---

[*]The Hund-Mulliken model of chemical binding by molecular-orbitals.

source in conjunction with a mass filter is indispensable for this purpose. A secondary ion mass spectrometer analyzing the ion flux of a sputter target would appear to be an ideal solution - in particular when the yield of emitted electrons is determined via their emission statistics. This method *requires* low ion currents. Low ion currents are a characteristic of sputter-ion sources. This is often regarded as an instrumental shortcoming but has no negative influence here.

In order to measure the statistics of electrons emitted from solids, ion-electron converters are commonly employed. As such studies are the subject of a special section of these proceedings, we will abstain from a further description of the method; see also Thum (1979), Thum and Hofer (1979, 1984), Ferguson (1987).

Results and Discussion

A great number of data on the non-monotonic electron yield dependence became available with the wide-spread use of mass spectrometers in chemical analysis. These data are

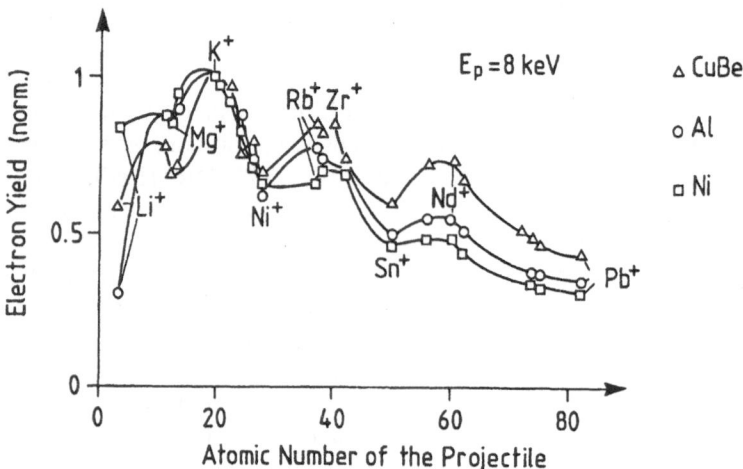

Fig. 2. Relative electron yields from CuBe, Al, and Ni dynodes upon bombardment with different singly-charged positive ions. From Fehn(1976).

mostly *relative* yields as they were deduced from the response of conventional particle detectors in the spectrometer. What is worse, all these measurements were carried out on *gas-covered* surfaces. Results obtained with these targets turned out to obscure rather than deepen the insight into the mechanisms effective in electron emission:
There was, firstly, the observation that the oscillations were independent of the chemical composition of the electron-emitting surface. In measurements on CuBe, Al, and Ni targets, Fehn (1976) noted, for instance, that only the amplitude of the oscillation was affected, the phase remaining unaltered, Fig. 2. More specifically, Fehn found in evaluations of both his own measurements and literature data that the inert-gas ions lay mostly in the maxima, while earth-alkali ions gave yields around the minima.

Such results cannot be reconciled with the excitation model of Weizel-Beeck-Fano-Lichten. According to this model, the excitation cross-section - and, thus, the electron yield - depends on a multitude of parameters, such as

- symmetry, spin, and angular momentum of the orbitals involved,
- matching of energy levels of electrons bound in the solid's and projectiles' ion cores;
- and whatever one might find to produce non-binding orbitals and thus electron promotion.

Of course, both particles involved, the projectile and the target atom, determine the excitation probability and the energy of the excited electrons. This contradicts the observation of target-independent phases of the oscillation, as well as the dominance by any one group of elements in the periodic table over another.

And there was, secondly, the hasty conclusion that the oscillations would follow those of the

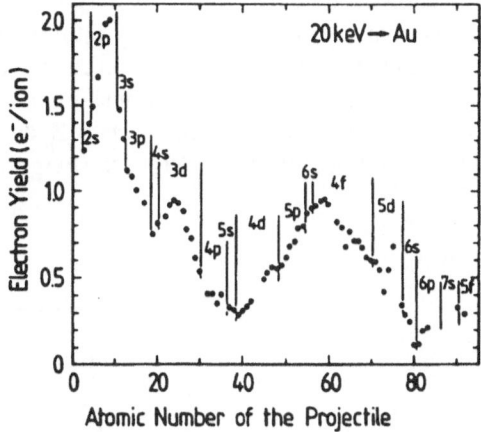

Fig. 3. Atomic-number dependence of the electron yield for bombardment of a clean Au-surface with singly-charged positive ions of 20 keV energy. From Thum and Hofer (1984).

electronic stopping power. Such a finding would support an approximation made in almost all theoretical treatments, namely of a proportionality between the yield and the electronic stopping power. This approximation is justified for H and He ions above about 10 keV, as has been shown by the groups in Gießen and Bariloche (Hasselkamp et al., 1980 -1987, Baragiola et al., 1979, 1980). It has no firm physical basis, however, for *heavy* ions ($Z_p > 2$): For heavy ions at velocities below the Bohr velocity, electronic stopping is controlled by collisions with conduction electrons. In electron emission, these collisions are sub-threshold processes. As was outlined above, electron emission by heavy projectiles is controlled by excitation of *bound* electrons.

In their work on clean Au surfaces, Thum and Hofer (1979, 1984) were able to resolve some of these inconsistencies. In short, their main findings were, Fig. 3:
- the oscillation in the yield is far more pronounced for emission from pure metal surfaces than from any of the gas-covered surfaces reported before; it is also
- much richer in detail - indicating characteristics of the *projectile's* electronic shell as well;

- the yields for cluster and molecule bombardment follow the same dependency, Thum (1979), Thum and Hofer(1980);
- the yield-oscillations show no resemblance to previous data on impure surfaces, in particular
- the maxima and minima are correlated
    - neither with certain groups of elements in the periodic table;
    - nor in a clear-cut way with electronic-level matching conditions;
    - nor with electronic stopping.
- contamination of the surface by air strongly affects the γ versus Z_p-plot, with a tendency for the yields at the minima to become more enhanced than those at the maxima.

These results raised new questions, however. Firstly, there was still the question to settle of whether or not the oscillations depend on the atomic number of the *target*.
Secondly, the inability to detect yield maxima for electronic matching conditions was felt embarrassing - particularly in view of the clear-cut correlations observable in ion-induced x-ray emission[*]. Measurements on targets other than Au - especially of lower atomic number - promised information on both points. Finally, the question was posed what role the *density of states* (DOS) in the conduction band of the target plays. The DOS is involved both in the excitation process - when electrons are promoted directly into the conduction band (Joyes, 1973) -, and in the Auger de-excitation process - if the non-radiative transition involves the conduction band; in the Weizel-Beeck conception the transition is an intra-atomic one.

The crucial task, therefore, was to investigate the γ versus Z_p dependence on clean surfaces other than Au. This was carried out by M.M. Ferguson (Ferguson, 1987; Ferguson and Hofer, 1989).

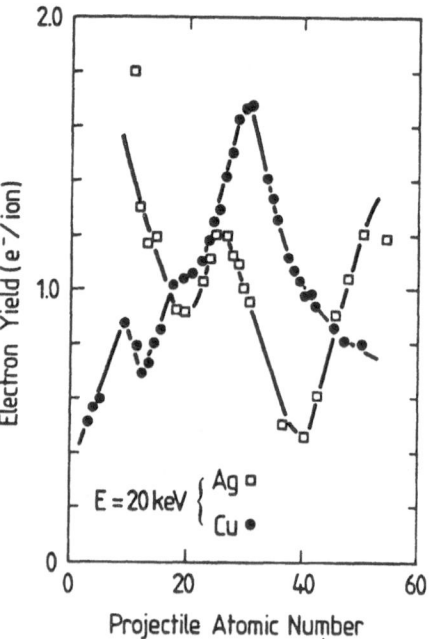

Fig. 4. Dependence of the electron yield of Cu and Ag upon bombardment with 20 keV ions. Ag and Au (Fig. 3) show largely the same yield dependence on Z_p. From Ferguson and Hofer (1989).

---

[*]Albeit at ion energies at least an order of magnitude higher than those used in studies of shell effects in electron emission.

Firstly, the question of whether the oscillations are dependent on the chemical element of the target was anwered in the affirmative, Fig.4. Clearly, there is no universal function for the dependence of the yield on the atomic number of the projectile - in agreement with the postulated excitation mechanism.

That there is nothing like a universal dependence of the yield on the atomic number of the target becomes evident also by a careful evaluation of the data available in the literature (Ferguson, 1987), see Fig. 5.

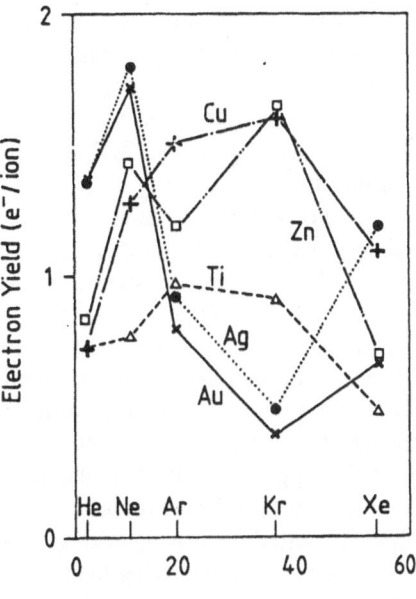

Fig. 5. Electron yields from different metals bombarded by 20 keV inert-gas ions. Evaluation of data from Zalm and Beckers (1985) by Ferguson (1989).

The density of states in the conduction band does not seem to play a major role. The overall form of the DOS is very similar for Cu and Ag, but these metals show quite different yield dependencies as seen in Fig. 4. The DOS of the divalent palladium, on the other hand, is clearly different from that of the monovalent inert metals Cu, Ag, and Au. Nevertheless, Pd shows the same yield dependency as Ag and Au, Ferguson and Hofer (1989).

The question remains whether or not there is a correlation between the maximum electron yields and orbital-matching conditions. In the picture developed from inner-shell excitation, orbital matching is fulfilled for equal binding energies. Thus, especially for the most obvious case of perfectly symmetric collisions, i.e. at $Z_p = Z_t$, there should be maxima in the electron yield for equal binding energies of projectile and target.

In this generality this is not the case here. Although yield maxima could be identified for the symmetric collisions with Ti and Cu targets, and it was possible to associate some of the smaller maxima in the γ vs. Z_p dependence for these, as well as for targets of Ag, Pd and Au with comparable electron binding energies, this correlation is neither compelling nor does it exist for the symmetric collisions with the medium-heavy to heavy targets Pd, Ag and Au.

Why don't we see what we are so well-acquainted with from inner-shell excitation? The following two remarks may be pertinent to this point:

It should be appreciated that **projectiles may also become electronically excited** in the collision and contribute thereby to electron emission. Should this be the case to any significant fraction of the yield, breaking of any one-to-one correlation with electronic-level matching would be an almost necessary consequence; comparisons with x-ray data[*] from the target are then irrelevant. In an approach where only *target* atoms are considered for excitation, it is the excitation cross section - and only the cross section - that is changed when the projectile species is varied. The amount of released energy per excitation remains constant since the energy levels involved are the same, namely that of the target. By varying the projectile species one samples with the projectiles particular energy levels set up by the target atoms. If, however, the projectile is subjected to excitation also, then not only the excitation cross section varies with Z_p but also the excitable electron orbitals. Therefore, in a variation of the atomic number of the projectile, an increase in the yield may be due to an enhanced cross section or a higher energy of the primary excited electrons. This complicates the structures in the $\gamma(Z_p)$-dependence and may well be a reason for the breaking of a correspondence with orbital matching.

In view of the data obtained with Au shown in Fig. 3, an excitation of the projectile is indeed suggested. Contrary to Harrison et al. (1965), who took the view that it is primarily projectile excitation that is responsible for electron emission, the present authors do not think that it will generally play a leading role. One should not discard it from the start, however.

Moreover, the data discussed here do not pertain to inner-shell excitation. All measurements on atomic-number effects in electron emission from solid targets were carried out in the low-energy regime. Primarily **outer shells are affected here**. These electron states are broadened due to solid state effects, so promotion along well defined molecular orbitals is hardly a fitting picture - in particular with metals, where excitation into the conduction band is possible, Joyes (1973), Sroubek (this conference), Fig. 6.

It is suggested dispensing with attempts to find electronic levels which match in the two collision partners and declaring them as having relative maxima in the excitation cross section. Energy level matching is not the only parameter influencing the excitation cross section - especially not when the outer electron shells play the decisive role. There are plenty of examples for this in the literature. One of those is the observation of enhanced Si 2p vacancy creation in *asymmetric* collisions (with either the projectile ion or certain components in composite targets), cf. Wittmaack (1979), Thomas (1983), and Bonanno et al. (1990) with references therein, as well as the presentation at this meeting.

That criteria other than electronic level matching determine electron emission in low-energy atom / atom collisions is evident also in the electron spectra from gas targets, e.g. Andersen et al. (1977, 1978); the spin state of the collision system and the symmetry of the orbitals involved appear to play the decisive role here. Moreover, the absence of solid-state effects, as well as of cascade multiplication with its smearing-out of the discrete energies characteristic of the ionization process, makes the unambiguous interpretation of the primary ionization step possible. Still, even for gas targets our understanding of the physics of ionization in atom / atom collisions has not yet reached a state where electron spectra can be predicted in a quantitative way. Before this state has been reached, it seems hopeless to expect quantitative information on the individual steps in electron emission from solids. A sensible strategy would, therefore, appear to be to try to understand these processes first with solid

---

[*]As with all spectroscopic data characteristic of the target. This argument holds true for all projectile energies.

state effects eliminated. Only when our understanding of outer-shell ionization has reached a state such that the emission lines can be predicted correctly, can many-particle effects in solids be included with a reasonable expectation of success.

Finally, some examples of the influence of exposure to reactive gases are presented and compared with emission from "technical" dynode surfaces used in particle-detector design,

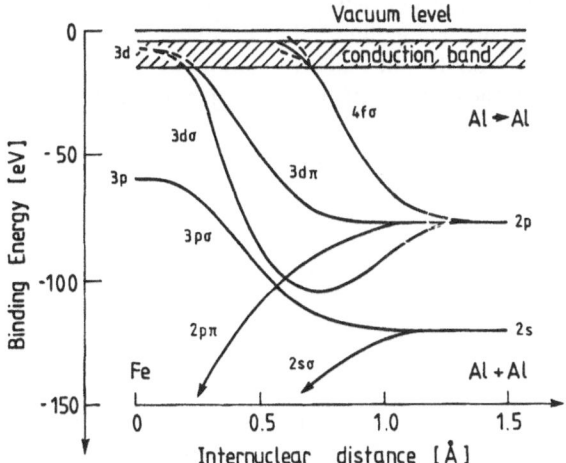

Fig. 6. Promotion of 2p electrons of an Al atom colliding with a second one in an Al metal target. At sufficiently close internuclear distances, the $4f\sigma$, $3d\pi$, and $3d\sigma$ molecular orbitals couple with the conduction band. Depending on whether or not an electron is picked up during the separation phase of the collision, the system may be left with an ion and a hot electron. This mechanism was first suggested for secondary-ion generation, see Joyes (1973).

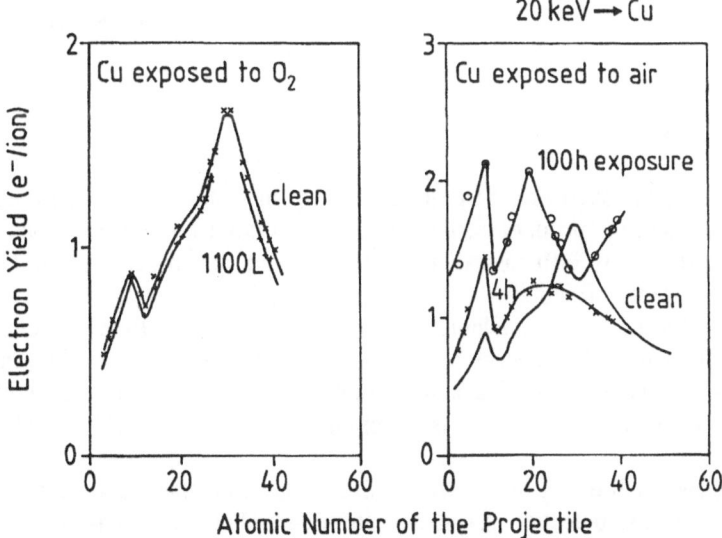

Fig. 7. Effect on the electron yield of exposure of copper to oxygen and air. From Ferguson (1988).

Fig. 7, 8. It is obvious from Fig. 7 that the exposure of Cu (Ag and Au behave very similar) to pure O_2 has a very much smaller effect than is often intuitively assumed. It should be noted, furthermore, that the yield is *reduced* over the whole range of ion species investigated here. Apparently, oxygen is merely adsobed on the surface and influences electron emission only by way of the surface barrier (work function). Neither in O_2-exposure nor in exposure to other reactive gases such as N_2 or H_2 has additional emission from the adsorbed layer been registered; the Z_p-dependencies are essentially parallel to those for the clean surface.

The situation is different for exposure to mixtures of gases such as air. Here, the yield dependence changes gradually and assumes after prolonged exposure a shape quite similar to that for surfaces prepared under atmospheric conditions (Fig.8). This is in keeping with Fehn's observation of a target-independent $\gamma(Z_p)$-dependence discussed in the introduction.

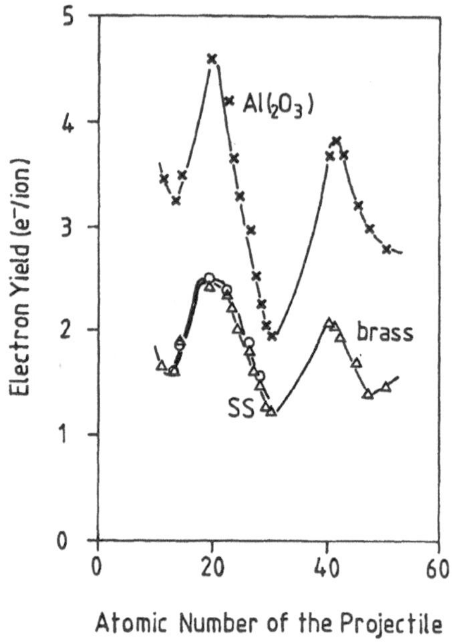

Fig. 8. Dependence of the electron yield from stainless steel, brass, and aluminum upon bombardment with 20 keV positive ions. From Ferguson (1988).

CONCLUSIONS

In summary, the shell effect in electron emission from solids is the only phenomenon where the physics of the primary excitation in the *bulk* comes into prominence. In most emission characteristics, any influence of the primary excitation process is masked by electron transport and surface refraction.

Qualitatively, the results available so far are in keeping with Ploch's model. A great deal more information from low-energy atom/atom collisions as well as on the influence of the band structure on the excitation/deexcitation process is needed, before a *quantitative* model can be expected to work. Presently it is not possible to predict the behavior of the yield oscillations.

The electron-shell effect on electron emission has a disquieting aspect in particle detection as it renders the detection sensitivity atomic-number dependent. Fortunately, surface treatments that are usually applied to enhance electron emission tend to smear out the amplitudes of the oscillations and thus alleviate the detector discrimination problem.

REFERENCES

Alonso EV, Baragiola RA, Ferrón J, Jakas MM, Oliva-Florio A. (1980). Z_1 Dependence of Ion-induced Electron Emission from Aluminum. Phys. Rev. B **22**, 80-87.

Andersen N, Østgaard Olsen J. (1977). Autoionising Neon levels Excited in 5-10 keV Collisions with Li and Na. J. Phys.B**10**, L719-722; ibid. 101-110; ibid **11** (1978) 3951.

Baragiola RA, Alonso EV, Oliva-Florio A. (1979b). Electron Emission from Clean Metal Surfaces Induced by Low-energy Light Ions. Phys. Rev. B **19**, 121-129.

Baragiola RA. (1982). Principles and Applications of Ion-Induced Auger Electron Emission from Solids. Radiat. Eff. **61**, 47-72.

Barat M, Lichten W. (1972). Extension of the Electron-Promotion Model to Asymmetric Atomic Collisions. Phys. Rev. A**6**, 211-229.

Bonanno A, Xu F, Camarca M, Siciliano R, Oliva A. (1990). Angle-resolved Auger study of 10 keV Ar^+-ion-induced Si LMM atomic lines. Phys. Rev. B**41**, 12590-98.

Briggs J. (1988). Fast Ion-Atom Collisions in: Proc. 15-th Int. Conf. *Electronic and Atomic Collisions* (ICACS) (Ed.: HB Gilbody et al.), Elsevier Sci. Pub., p. 13-20.

Fano U, Lichten W. (1965). Interpretation of Ar^+ - Ar Collisions at 50 keV. Phys. Rev. Lett. **14**, 627-629.

Fehn U. (1976). Variance of Ion-Electron Coefficients with Atomic Number of Impacting Ions. Int. Mass Spec. Ion Phys. **21**, 1-14.

Ferguson MM. (1987). Ioneninduzierte kinetische Elektronenemission an reinen und gasbedeckten Metalloberflächen. Thesis Techn. Universität Wien, Austria. Also published as report Jül-2201 (1988), ISSN 0366-0885.

Ferguson MM, Hofer WO. (1989). On the Z_1 Dependence of Ion-induced Kinetic Electron Emission. Radiat. Eff. Def. Solids. **109**, 273-280.

Harrison Jr. DE, Carlston CE, Magnuson GD. (1965). Kinetic Emission of Electrons from Monocrystalline Targets. Phys. Rev. **139**, A737-745.

Hasselkamp D, Scharmann A. (1982). The Ion-Induced Low Energy Electron Spectrum from Aluminum. Surface Sci. **119**, L388-L392.

Hasselkamp D, Hippler S, Scharmann A. (1987). Ion-induced Secondary Electron Spectra from Clean Metal Surfaces. Nucl Instr. Meth. Phys. Res. B**18**, 561-565.

Hasselkamp D, Hippler S, Scharmann A. (1987). Electronic Processes Induced by High Energy H^+, H_2^+, and H_3^+ Ions: A Scaling Relation. Z. Physik D**6**, 269-274.

Hasselkamp D, Lang KG, Scharmann A, Stiller N. (1981). Ion Induced Electron Emission from Metal Surfaces. Nucl. Instrum. Methods **180**, 349-356.

Hasselkamp D, Scharmann A, Stiller N. (1980). Ion Induced Secondary Electron Emission as a Probe for Adsorbed Oxygen on Tungsten. Nucl. Instrum.Methods **168**, 579-583.

Hofer WO. (1990). Ion-Induced Electron Emission from Solids. Scanning Microscopy, Suppl.**4**, 265-310.

Joyes P. (1973). Theoretical Models in Secondary Ionic Emission. Rad. Effects **19**, 235-241.

Lichten W. (1980). Molecular Orbital Model of Atomic Collisions. J. Phys. Chem. **84**, 2102-2116.

Ploch W. (1951). Massenabhängigkeit der Elektronenauslösung durch isotope Ionen. Z. Physik **130**, 174-195. See also the short communication in Z. Naturforschg. **5a** (1950) 570-571.

Rudat MA, Morrison GH. (1978). Detector Discrimination in SIMS: Ion-to-Electron Converter Yield Factors for Positive Ions. Int. J. Mass Spectr. IonPhys. **27**, 249-261.

Staudenmaier G, Hofer WO, Liebl H. (1976). Cluster Induced Secondary Electron Emission. Int. J. Mass Spectr. Ion Phys. **21**, 103-112.

Thomas EW. (1983). Inelastic Surface Collisions. Appl. Atomic Coll. Phys. **4**, 299-326.

Thum F. (1979). Sekundärelektronenemission durch Ionenbeschuß von Goldoberflächen. Thesis Techn. Univers. Munich, 1979; also Report IPP 9/29, Garching, Germany.

Thum F, Hofer WO. (1979). No Enhanced Electron Emission from High-Density Atomic Collision Cascades in Metals. Surface Sci. **90**, 331-338.

Thum F, Hofer WO. (1980). Ion-Induced Secondary Electron Emission by Heterogeneous Cluster Ions. Sympos. on Atomic and Surface Physics (SASP '80), Ed. E. Lindinger et al., Univ.Innsbruck, A. 19-21.

Thum F, Hofer WO. (1984). Z_1-Oscillations in Ion-Induced Kinetic Electron Emission. Nucl. Instr. Meth. Phys. Res. B2, 531-535.

Weizel W, Beeck O. (1932). Ionisierung und Anregung durch Ionenstoß. Z. Physik **76**, 250-259.

Wittmaack K. (1979). Characteristics of Ion-induced Silicon L-shell Auger Spectra. Surface Sci. **85**, 69-76.

Zalm PC, Beckers LJ. (1985). Ion-Induced Secondary Electron Emission from Cu and Zn. Surface Sci. **152/153**, 135-141; see also Philips J. Res. **39** (1984) 61-76.

SYMMETRIC AND ASYMMETRIC COLLISIONS IN AUGER ELECTRON EMISSION FROM Al, Mg, Si, AND Mg_xAl_{1-x} TARGETS INDUCED BY LOW keV ION BOMBARDMENT

A. Bonanno, P. Zoccali, N. Mandarino, A. Oliva, and F. Xu

Dipartimento di Fisica, Università della Calabria,
87036 Arcavacata di Rende, Cosenza, Italy
and INFM, unità di Cosenza

I. INTRODUCTION

Auger electron spectra of Mg, Al and Si obtained by Ar^+ bombardment on solid surfaces show some narrow line structures (~ 1 eV wide) and an underlying broad feature which are assigned to atomic-like $L_{23}MM$ and bulk-like $L_{23}VV$ transitions respectively[1]. Hole in the inner L-shell is produced in collisions between projectile and target atoms (asymmetric or p-t collisions) or between two alike target atoms in the collisional cascade (symmetric or t-t collision). The adiabatic approximation is usually employed in calculating the evolution of the electronic states of the quasi-molecule formed transiently during the collision as a function of the internuclear distance. Starting from these adiabatic curves Barat and Lichten proposed a "diabatizing" mechanism allowing the otherwise avoided crossing whenever the degeneracy allows the overlap of two molecular orbitals (MO)[2], thus one can attempt to follow the most probable path of one electron during a collisional event. For L-shell excitation occurring in the lighter partners in asymmetric collisions and in one or both atoms in symmetric collisions, Fano and Lichten proposed the electron promotion along the steeply rising, diabatic $4f\sigma$ MO[3].

Ionization of Solids by Heavy Particles, Edited by
R.A. Baragiola, Plenum Press, New York, 1993

The understanding of the role played by symmetric and asymmetric collisions in excitation of target core electrons is of fundamental importance in studies of ion induced Auger electron emission. In spite of the great effort put forward in this field some controversies still remain in both experimental and computer simulation investigations. Hou et al[4]. by using the MARLOWE code for 5 keV $Ar^+ \rightarrow Al$ found that the contribution of asymmetric collisions is 40% of the total Auger emission. Shapiro and Fine[5] by employing a modified molecular dynamics sputtering code Sput2, instead, claimed that all core excited atoms result from asymmetric collisions. For the same conditions Grizzi and Baragiola[6] evaluated an asymmetric contribution of 20% for the atomic peak. These latest authors also found that for higher energies the p-t collisions become much more important (70% at 15 keV).

In this paper we report a study on Auger electron emission from pure Mg, Al, and Si targets and from MgAl alloy samples under noble gas ion bombardment. The aim is to investigate the line shape and intensity variation of both the LMM and L^2MM atomic peaks as a function of the projectile energy, the asymmetry of the collisional system and the incidence angle.

II. EXPERIMENTAL

The experiments were conducted in a UHV chamber with a base pressure of 5×10^{-10} Torr. During ion gun operation it rose to 5×10^{-9}. Mechanically polished samples of high purity polycrystalline Mg, Al, Si and $Mg_x Al_{1-x}$ alloys (x= 0.25, 0.50, 0.75) were mounted on a dedicated sample manipulator with movements along three axes and rotation around the z axis. The chamber was equipped with an electron gun, a differentially pumped ion gun and two hemispherical energy analyzers, all lying in the same XY plane. The first analyzer, with an angular acceptance of ~2⁰, was mounted on a rotatable goniometer ~10 cm away from the sample for angle resolved study. The second analyzer was situated very close to the sample surface (semi-aperture angle ~25⁰) at a fixed angle of 70⁰ from the ion beam and 30⁰ from the electron beam. The ion gun discharge voltage was of 50 V and this condition ensured a contamination of <3% of Ar^{2+} ions in the primary beam which was not mass analyzed. Good homogeneity of alloy samples was confirmed by scanning electron microscopy measurements and cleaning was performed in situ with low energy ion sputtering. The surface Al concentration after sputtering for a long time was found to be 0.80, 0.56, 0.31 as determined with electron stimulated Auger spectroscopy for sample with nominal composition of 0.75, 0.50, 0.25, nearly independent on primary energy.

III. RESULTS AND DISCUSSIONS

III-A L²MM AUGER TRANSITIONS

In Fig. 1 we present three sets of angle resolved doubly core excited L²MM Auger electron spectra obtained by 15 keV Ar⁺ bombardment at $\theta_i =$ 40⁰ from the surface normal on Mg, Al and Si targets. All the spectra have

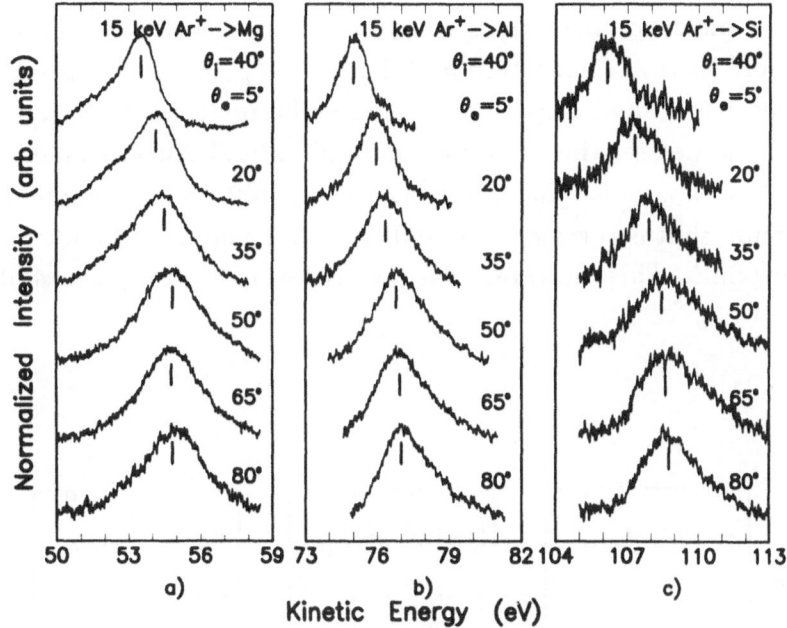

Fig.1 Angle resolved, background subtracted and intensity normalized L²MM Auger peaks induced by 15 keV Ar⁺ bombardment along $\theta_i =$ 40⁰ relative to the surface normal on: a) Mg, b) Al and c) Si targets.

been secondary electron background subtracted and normalized to the same height to emphasize the changes in line shape.

Previous studies have shown that a double L-shell excitation can be produced only in asymmetric encounters and the Auger decay takes place in the vacuum since the hole transfer mechanism in the subsequent collisional events prevents the deexcitation in the solid. Consequently, these excited atoms result essentially from the primary asymmetric collisions occurring at the top-most surface layer without suffering any further encounters[7]. In a recent study on the main Al L²MM peak we showed that the line shape changes for a set of detection angle resolved spectra with a fixed incidence angle can be well accounted for by a simple primary Ar-Al binary collision model[8].

The creation of a double core vacancy requires that the two colliding partners reach a minimal approach distance smaller than a critical value. This automatically implies the existence of a threshold transferred kinetic energy E^c_{trans} and a critical scattering angle θ^c. Because of the rapid decrease of the cross section for small impact parameters, the great majority of the observed excitation events occur at E^c_{trans} and θ^c, so that the maximum energy shift of the L²MM peak is observed at the detection angle corresponding to θ^c. Further, the incidence angle can strongly affect the L²MM yield, especially in the threshold energy region[9,10]. Indeed, for small E_p, the Al atoms are excited in collisions with a small impact parameter and a small scattering angle, thus a large θ_i is needed for the detection of atomic L²MM Auger signal.

From Fig. 1 we note that the peak energy shift and the angle at which the maximum shift is revealed are smaller for Mg than for Al and smaller for Al than for Si. This indicates that the critical minimal approach distance

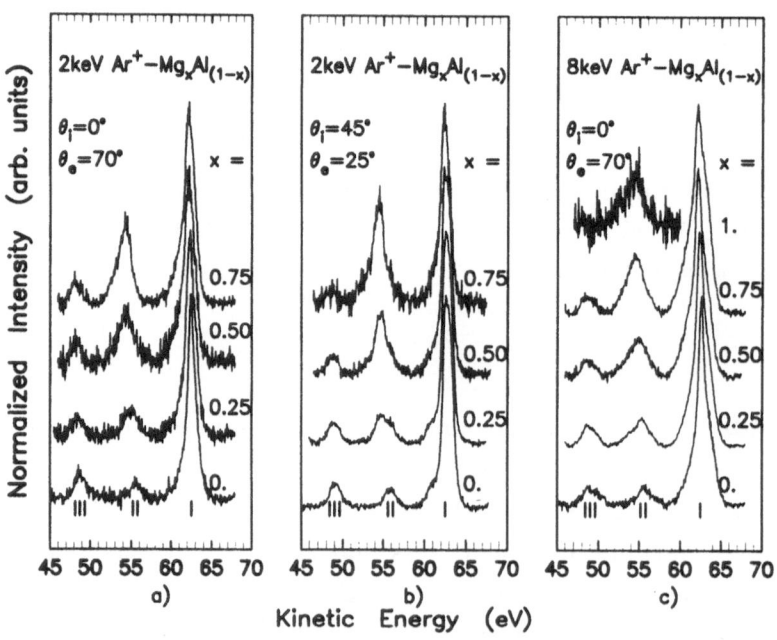

Fig.2 Al LMM and Mg L²MM normalized atomic peaks for Ar⁺ bombardment with: a) $E_p=2$ keV and $\theta_i=0^0$, b) $E_p=2$ keV and $\theta_i=45^0$, c) $E_p=8$ keV and $\theta_i=0^0$ on Mg_xAl_{1-x} alloy samples.

for double core excitation in the lighter partner is smaller in Ar-Si than in Ar-Al and in Ar-Mg and is in agreement with the correlation diagrams[11] and with the yield measurements[12], as well as with the fact that for Ar^+ incidence along the surface normal the L^2MM signals can be observed only for Mg but not for Al and Si.

On the other hand, it is expected that the inner excitation in a given element should depend also on the collisional partner. To quantitatively evaluate this aspect for Mg we studied the intensity and line shape variation of the Mg L^2MM peak for Ar bombarded Mg and MgAl alloy samples. Its overlap with an Al LMM peak, however, renders a detailed line shape analysis quite difficult. In Fig. 2 we present three sets of atomic Al Auger spectra in the energy range of 45-70 eV for 2 keV and 8 keV Ar^+ ion impact along normal and off-normal incidence angles for different Mg concentration. All these spectra were recorded with the fixed hemispherical analyzer and both the secondary electron background and the bulk LVV contribution have been subtracted.

We first note that Al-I and Al-III peaks are both gradually and rigidly shifted toward low kinetic energy as the Mg concentration is increased presumably because of a reduction in the surface work function (4.28 eV for Al and 3.66 eV for Mg, Ref. 13). The spectra show that for $E_p = 2$ keV the widths of both the Al-I and -III atomic lines remain unaltered as θ_i is changed from 0^0 to 45^0 and that the L^2MM Mg peak is absent in pure Mg even for off-normal incidence. This suggests that excitation in Ar-Mg encounters are not possible at this low energy. On the other hand, as the Mg concentration is increased, a new atomic peak, due to the Mg L^2MM transition, appears at low energy side of Al-II and grows in relative intensity and finally merges to completely overlap with, and dominate on, the Al contribution. As the Ar^+ primary energy is raised to 8 keV the increase of the Doppler shift and broadening renders the two peaks no longer separable. Furthermore, their total intensity is also greatly enhanced indicating that the excitation threshold energy is lower for Mg in Mg-Al than for Al in Al-Al.

In order to obtain a quantitative evaluation of the relative importance of the Ar-Mg and Al-Mg collisions in Mg L^2-excitations we extracted the Mg L^2MM intensity from the total peak area. It is important to point out that the ratio between Al-I and Al-III intensities results independent both on sample composition and on Ar^+ primary energy and so does that between various Mg LMM features. In fact, the relative intensities of LMM atomic peaks should depend only on the relative transition probabilities of different deexcitation channels but not on the excitation mechanism. By using the constant intensity ratio of Al-III and Al-II obtained for pure Al we subtracted the Al LMM contribution. The so obtained results, normalized to the surface Mg concentrations, are plotted in Fig. 3 for two different Ar[+] incidence angles together with those for pure Mg. It can be immediately noticed that the Mg L[2]MM yield is about an order larger in Mg-Al alloy than in pure Mg, indicating the dominant contribution from Al-Mg collisions. Further, in the low E_p range the increase of this intensity as a function of E_p is much greater for pure Mg than for alloy samples suggesting a lower threshold energy for Al-Mg than for Ar-Mg.

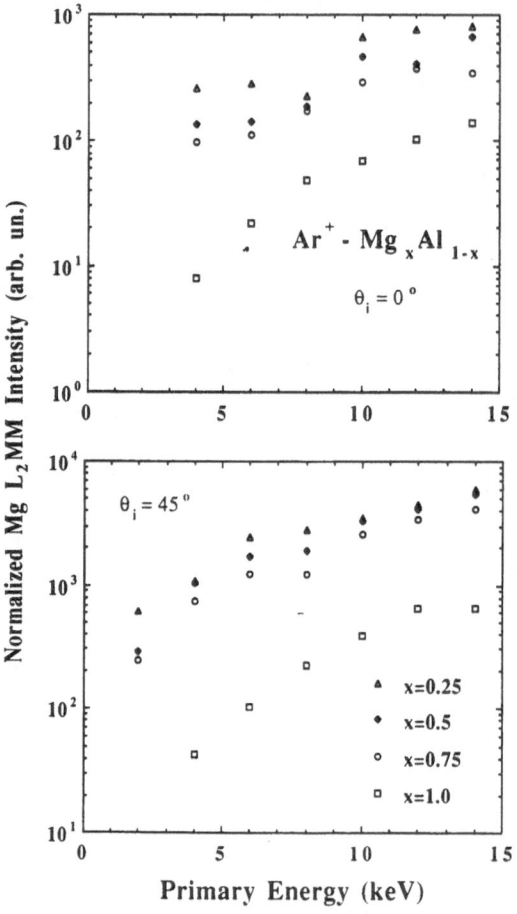

Fig.3 Mg L[2]MM Auger yield of MgAl alloys normalized to the Mg surface concentration and of pure Mg for: a) $\theta_i = 0^0$,   b) $\theta_i = 45^0$.

In a previous paper[7] we demonstrated that the L-shell vacancy transfer from one doubly excited Mg atom to a surface Al atom in a Mg-Al encounter can account for the observed Al double core excitation. We point out that the intensity of Al L[2]MM peak in the alloys is about an order lower than that for the Mg L[2]MM peak, suggesting that the vacancy transfer process has a quite small cross section.

Now let's come to discuss the LMM transitions in Al. The spectra in Fig. 2 show that, for $E_p = 2$ keV, the main peak is symmetric and its width remains independent on the sample composition. For $E_p = 8$ keV and $\theta_i = 0°$, a shoulder appears on the high energy side for pure Al and the overall line width increases with increasing Mg concentration in the alloys. We note that for off-normal incidence (not shown) the variation in the line shape of all the examined sample is much greater than in the 2 keV case. A similar trend is observed in all spectra induced by Ar^+ ions for $E_p > 4$ keV.

For $\theta_i = 0°$, the primary collisions can not directly contribute to the detected atomic peaks, we thus assign this shoulder essentially to the Al atoms excited in Ar-Al cascade collisions, faster than those excited in Al-Al ones[4]. We underline that for $E_p > 4$ keV the relative intensity of this structure increases gradually because of the increasing importance of the asymmetric collisions.

For alloy samples, the greater is the Mg concentration, the more frequent are the Al-Mg collisions, which produce, however, only excitation in Mg. On the other side, the relative weights of Ar-Al and Al-Al collisions in the Al 2p excitation are linearly and quadratically proportional to the Al concentration respectively. Given that the threshold transferred energies E_{tran}^c is higher for Ar-Al than for Al-Al, the Doppler broadening should be an increasing function of Mg concentration.

In panels a and b of the Fig. 4 we show angle resolved Auger spectra induced by Ar^+, Kr^+ and Xe^+ ions of various energies impinging at $\theta_i = 40°$ on an Al surface. All the peaks are relative to the main LMM transition of an Al neutral atom and present a clear broadening in the higher energy side. It is interesting to notice that for the 5 keV Ar^+ bombardment the peak is very similar to that obtained by Kr^+, Xe^+ and Ne^+ for both $\theta_e = 50°$ and $\theta_e = 80°$ and that the peak shape does not change with the primary energy in these later cases. In panels c and d spectra of the same transition peak obtained by Ar^+ impact at different E_p along $\theta_i = 60°$ are shown for two different observation angles ($\theta_e = 50°$ and $\theta_e = 80°$). The peak width in both panels changes greatly and a new shoulder appears at higher energy side. We attribute this structure to excitation in the primary Ar-Al encounters, which can produce fast moving excited Al atoms. This conclusion is supported by the fact that the observed Doppler shift of about 2.3 eV is very similar to that observed for L^2MM peak[8], certainly due to a primary Ar-Al collision. Further, the Ar^+ energy at which the shoulder appears is greater than 5 keV where the Ar-Al contribution to Al 2p excitations becomes important[6] and the position of such a structure does not change with the primary energy because most of the excited atoms have a velocity corresponding to that of the threshold value. Nevertheless, the weight of asymmetric collisions is strongly dependent both on primary energy and on incidence angle. The separation of this structure from the main LMM peak is much smaller than in the case of Ar-Si where the shifted component has been assigned to the primary Ar-Si collisions[14]. Indeed, the transferred kinetic energy is greater in Ar-Si than in Ar-Al.

We also point out that the total asymmetric contribution to LMM Auger transition can not be quantitatively estimated from the shoulder intensity[9] because the Ar projectiles can excite the Al or Si atoms also in cascade collisions for which the direction of the outgoing Al or Si atoms is not well defined so that only a peak broadening can be observed.

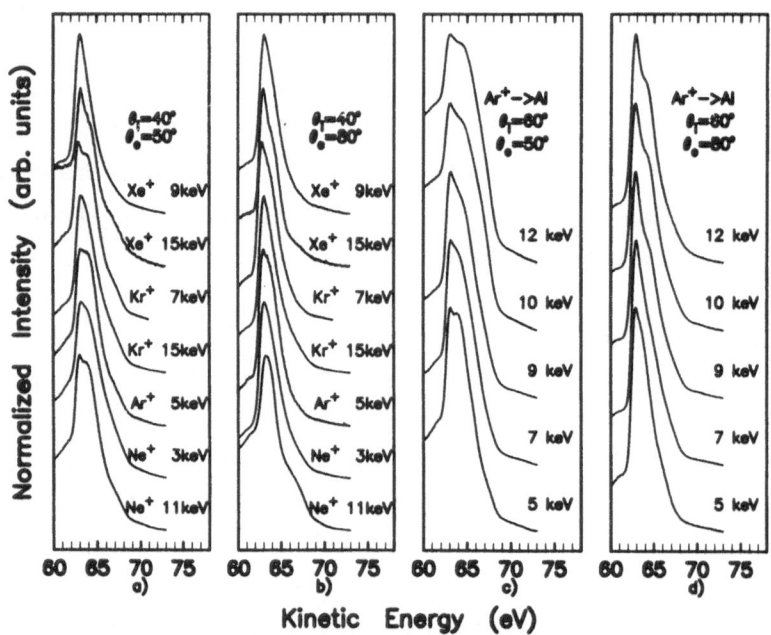

Fig.4 Angle resolved and intensity normalized main Al LMM peaks for Ar^+ ion impact for: a) $\theta_i = 60^0$ and $\theta_e = 50^0$, b)$\theta_i = 60^0$ and $\theta_e = 80^0$; and for various noble gas ion bombardment for: c) $\theta_i = 40^0$ and $\theta_e = 50^0$, d) $\theta_i = 40^0$ and $\theta_e = 80^0$.

IV. CONCLUSION

In conclusion, we have shown that in a collisional event with Ar^+, a Mg atom can be excited more likely than an Al or Si one and the excitation threshold energy is smaller in an Al-Mg encounter than in an Ar-Mg one. The asymmetric contribution to excitation depends not only on the the primary energy but also on the direction of the impinging Ar^+ ions.

ACKNOWLEDGMENTS

We thank V. Fabio and E. Li Preti for technical collaboration. This work was partially supported by CNR under contract of "Progetto Finalizzato Materiali Speciali per Tecnologie Avanzate".

REFERENCES

1. G. Zampieri and R. Baragiola, Phys. Rev. **29**, 1480 (1984).
2. M. Barat and W. Lichten, Phys. Rev. **A6**, 211 (1973).
3. V. Fano and W. Lichten, Phys. Rev. Lett. **14**, 627 (1965).
4. M. Hou, C. Benazeth and N. Benazeth, Phys. Rev. **A36**, 591 (1987).
5. M. H. Shapiro, J. Fine, Nucl. Instr. and Meth. **B44**, 43 (1989).
6. O. Grizzi, R. Baragiola, Phys. Rev. **A35**, 135 (1987).
7. F. Xu, M. Camarca, A. Oliva, N. Mandarino, P. Zoccali and A. Bonanno, Surf. Sci. **247**, 13 (1991).
8. A. Bonanno, M. Camarca and F. Xu, Surf. Sci. Lett. **264**, L 213 (1992).
9. R. A. Baragiola, L. Nair and T. E. Madey, Nucl. Instr. and Meth. **B58**, 322 (1991).
10. S. Valeri and R.Tonini, Surf. Sci. **220**, 407 (1989).
11. A. Bonanno, P. Zoccali, N. Mandarino, P. Calaminici, A. Oliva, F. Xu and N. Russo, unpublished.
12. F. Xu, F. Ascione and A. Bonanno, to be published.
13. Handbook of Chemistry and Physics, 65th Ed., CRC Press (1985).
14. A. Bonanno, F. Xu, M. Camarca, R. Siciliano and A. Oliva, Phys. Rev. **B41**, 12590 (1990).

ELECTRON EMISSION FROM SILICON

INDUCED BY BOMBARDMENT WITH OXYGEN IONS

Ewa A Maydell

Department of Metallurgy and Engineering Materials
University of Strathclyde,
Glasgow G1 1XN, Scotland, U.K.

INTRODUCTION

A multitechnique SIMS-Auger instrument, described in detail elsewhere (Maydell et al, 1992), comprises of a complete Auger electron spectrometer incorporated into a quadrupole based secondary-ion mass spectrometer. The instrument uses ion-induced electron emission for 'viewing' the sample surface. The intensity of low-energy electrons generated by the primary ion-beam, which scans the sample surface at TV frequency, is displayed on a TV monitor in a way similar to the acquiring of a secondary-electron image with a scanning electron microscope. The intensity of electron emission depends on the degree of oxidation of the surface, as well as on the nature and energy of the primary ions.

The electron emission spectra from magnesium, aluminium and silicon excited by Cs^+ and Ga^+ ion-beams have been studied and are found (Maydell, 1992) to resemble those generated by bombardment with noble-gas ions (Viel et al, 1976; Legg et al, 1980; Baragiola et al, 1982). Using a 360° rotation sample manipulator, we have examined the effect of angle of incidence of a Cs^+ ion beam on the intensity of atomic-like Auger electron peaks (Maydell and Fabian, 1992; Maydell, 1992a) as well as their dependence on the ion-beam energy (Maydell, 1992).

Ion-induced Auger electron spectra from surfaces covered with native oxides differ from those obtained from surfaces subjected to ion irradiation. Baragiola et al (1991), Xu and Bonnano (1992) and Xu et al (1992) all used oxygen background pressure to modify argon-induced atomic-like Auger electron spectra (peak shapes and relative intensities) for silicon and aluminium in an attempt to elucidate the relative importance of symmetric and asymmetric collisions in excitation of target atoms. In our work, we find that a prolonged ion bombardment is necessary to obtain a stable (high) intensity of atomic-like Auger emission; the oxygen concentration must be reduced to a

Ionization of Solids by Heavy Particles, Edited by
R.A. Baragiola, Plenum Press, New York, 1993

level not only undetectable by conventional AES, but to levels such that the O^- secondary-ion signal decreases to a value of around 10^{-4} or 10^{-5} times its intensity (normally high) for oxidised metal surfaces.

Argon-ion induced electron emission from oxides fabricated in-situ by implantation of a high dose of oxygen ions into silicon and aluminium has been found to show no observable atomic-like Auger emission (Maydell, 1992b). This also disagrees with the result reported by Thomas et al (1980), who observed no difference between the 200keV Ar^+ ion induced Auger emission from pure aluminium and from its oxide.

The electron emission generated by oxygen irradiation of silicon and aluminum has been studied recently by Wittmaack (1991), who examined the 'total' electron emission (making measurements of the number of electrons per ion) at nearly normal incidence - for ion energies ranging from 2keV/atom to 10keV/atom - but did not attempt electron energy spectrometry.

In the present work, the electron emission generated by oxygen bombardment of silicon at angles of incidence ranging from 0° to 75° and with ion energies from 0.75keV/atom to 10keV/atom has been studied. Argon-ion induced electron-emission spectra were also examined for the same angles of incidence. The results are compared and correlated with the secondary ion emission measured (using the same instrument) by quadrupole mass spectrometry and with conventional electron-excited AES for silicon bombarded with oxygen ions at different angles of incidence.

EXPERIMENTAL

Figure 1 shows schematically the geometrical arrangement of primary ion-beam, incident electron beam, electron analyzer acceptance cone, and secondary-ion extraction column. The primary beam, the electron energy analyzer and secondary-ion acceptance column are all in one vertical plane; with the primary ion beam horizontal, the electron spectrometer 36° below the primary ion-beam, and the secondary-ion column 45° above the primary ion-beam. The sample is mounted on a manipulator at the centre of a spherical analyzer chamber, and can be rotated (through 360°) about a horizontal axis perpendicular to the plane defined by the primary ion-beam and the directions of secondary-ion detection and electron detection. The Auger electron gun is mounted in a horizontal plane at an angle of 45° to the primary ion-beam.

Oxygen and argon ions are produced with a VG duoplasmatron cold-cathode ion source. For argon ions the source is fed with a mixture of $\sim$10% argon in oxygen. The ions are accelerated to energies of up to 10keV. The minimum energy attainable, while providing an ion current sufficient for ion-induced measurements, is 1.5keV. The ion-beam contains at least two charge states of oxygen: O_2^+ and O^+, with a possible admixture of O_2^{++} ions. The ion-species are separated using a Wien filter; but O_2^{++} and O^+ are indistinguishable because of their identical mass-to-charge ratio (they have, however, the same energy per oxygen atom). The energy range attainable for oxygen ion species is from 0.75keV/atom (O_2^+ at 1.5keV) to 10keV/atom (accelerating either O_2^{++} or O^+ by 10kV).

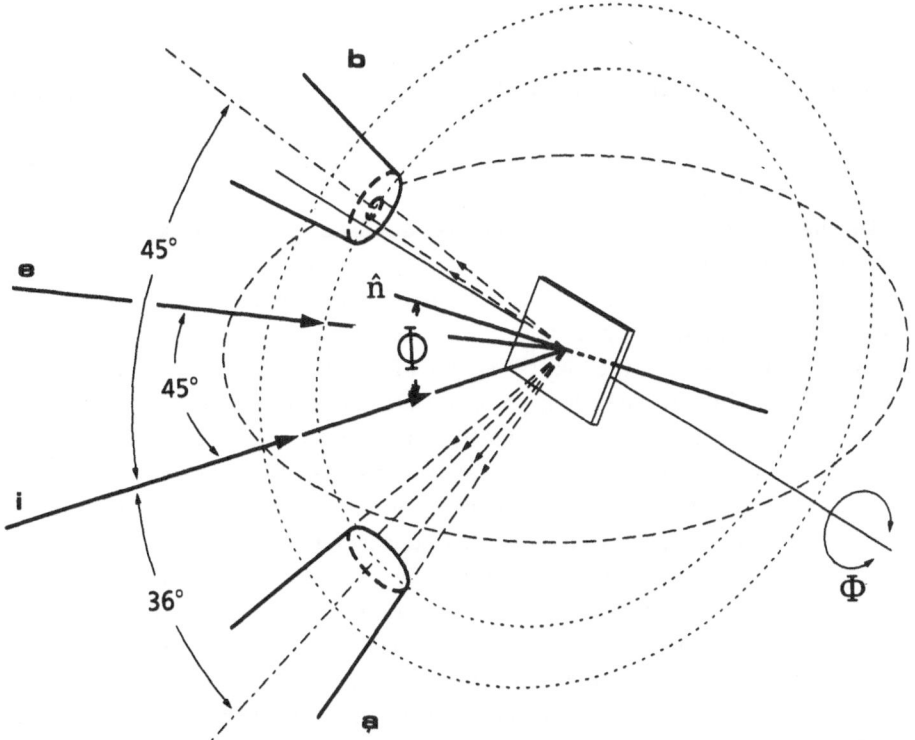

Figure 1. Geometrical arrangement for ion-induced Auger electron emission and SIMS measurements: **i** - ion beam; **a** - electron energy analyzer; **e** - electron beam; **b** - secondary-ion extraction column; **0** - angle of ion-beam incidence to the sample normal **n**.

The electron energy spectrometer (a VG CLAM 100) is a 180° hemispherical analyzer. In the present work it was operated with constant energy retard ratio. This mode of operation gives constant $\Delta E/E$, where ΔE is energy resolution and E is the energy of electrons. Using a retard ratio of 4 gives $\Delta E/E = 0.5\%$; ie for electron energies of 100eV or less, $E \lesssim 0.5eV$. Electron spectra were recorded as N(E) vs E, by pulse counting in 0.5eV wide channels.

The selected oxygen ion-beam (O_2^+ or O^+) was raster scanned at TV rate over an area of the sample which, for normal incidence to the sample surface, is ~ 350μm x 250μm, while for the ion-beam at an angle of incidence of ϕ is rectangular, ~ 350μm x 350μm/cosϕ. The maximum area irradiated at 75° incidence, 350μm x 650μm, is smaller than the field of electron detection by the electron analyzer (approximately 1mm x 1mm).

The sample for the present work was a WAKER (100) single-crystal, boron-doped p-type silicon, of resistivity 10-30 Ohmcm. Two types of experiment were performed: the first to clarify the influence of oxygen presence on argon-ion induced Auger emission, the second to study oxygen-ion induced electron emission from silicon. The oxygen pressure in the analyzer during all measurements was $(0.8-2.0)\times10^{-9}$ mbar.

In the first experiment 10keV O_2^+ ions were implanted (in-situ) in an area ~ 350μm x 250μm of a silicon surface at normal incidence. The implanted dose of ~ $7 \times 10^{17} cm^{-2}$, was sufficient to form an oxygen-saturated layer of SiO_2 thicker than 400A. A 5keV argon ion-beam was then scanned over a central area (~175µm x 125µm) within the oxidized rectangle, while a sequence of ion-induced electron spectra was recorded in a similar manner to dynamic SIMS depth profiling.

In the second experiment a silicon surface was irradiated with oxygen ions while electron spectra were being measured. In order to establish a constant oxygen concentration in the bombarded surface (ie to produce a dynamic equilibrium), the sample was first irradiated for 10-15mins with the selected oxygen species (using currents of 500nA of O_2^+ at 10keV, down to 20nA for O_2^{2+} at 2keV), and then an ion-induced electron spectrum was measured, followed by an electron-induced Auger spectrum; the latter gave a measure of the oxygen concentration at the sample surface.

For quantification of the intensity of an atomic-like Auger peak in the spectrum, the peak-to-peak (positive to negative) 'excursion' in the differentiated spectrum (dE/dN) was multiplied by the peak-width; the latter being measured as the energy difference between the positive and negative excursions. The second differential of the electron spectrum (d^2N/dE^2) provided an exact measure of the peak energies.

The secondary-ion emission was measured without extraction field, so that only the ions emitted in the direction of the secondary-ion extraction column were 'accepted', without distortion of the secondary-ion spacial distribution. Measurement of a secondary-ion mass spectrum with a quadrupole mass-analyzer, as in the present instrument, requires the sample to be biased in order to match the mean secondary-ion energy with the 'window' of the energy filter preceding the analyzer. A bias of 20V was used, corresponding to an electric field of ~$1Vcm^{-1}$, which does not affect the secondary ion spacial distribution. Measurements of +ve and -ve secondary-ion spectra generated from silicon by primary O_2^+ ions at angles of incidence of 0°, 25°, 45° and 65° were made for correlation with surface oxygen content, as well as to assist with the interpretation of oxygen-ion induced electron emission.

RESULTS

Figure 2 shows the 5keV argon-ion induced electron spectra obtained for silicon dioxide and silicon, and the 3keV electron excited Auger spectrum for silicon dioxide. The ion-induced electron spectrum for silicon dioxide shows no atomic-like Si LMM Auger peaks. Only after the oxide has been etched away do the atomic-like Auger peaks appear, characteristic of the argon-induced Auger spectrum from silicon. The electron-excited spectrum for the oxide shows two broad Si LVV Auger peaks, at energies of 72eV and 58eV; these are typical of SiO_2. Note, that the energy scales in figures 2 and 3 are off-set by ~ +5eV, due to electron spectrometer calibration.

Figure 3 shows the electron spectra generated by irradiation of silicon with 10keV O_2^+ ions, at angles of incidence (to surface normal) of 0°, 25°, 45°, 65° and 75°. The spectra for 0° and 25° ion-beam incidence show a high intensity of low-energy electron emission. Those for 45°, 65° and 75° show, in addition to the low-energy peak, Si LMM atomic-like Auger peaks super-imposed on a broad bulk solid emission background. Using the Whaley and Thomas (1984) scheme of identification, the peak marked 'I' identifies with excited neutral atoms (Si°*) while those labelled 'II' and 'III' result from the Auger decay of excited ions (Si⁺*). The peak, labelled 'IV' is assigned to Auger decay in a double **2p**-hole excited Si⁺ ion, while 'i' and 'ii' are the excited-ion and excited-neutral Auger peaks, postulated but not resolved in their measurement by Whaley and Thomas (1984). These peaks were detected at oblique angles of argon ion-beam incidence, combined with glancing angle electron detection, by De Ferrariis et al (1986), Bonanno et al (1990 and

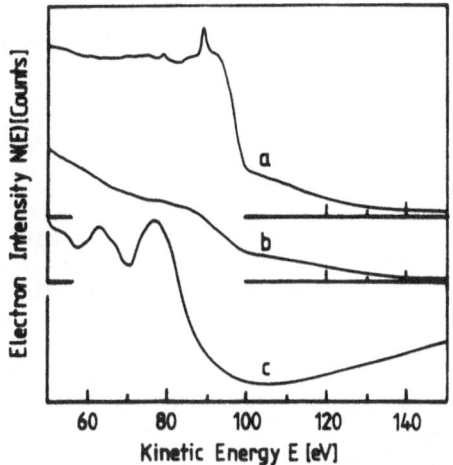

Figure 2. (a) and (b): electron spectra induced with normal-incidence 5keV argon ions, from a) silicon and **b**) SiO$_2$ (synthesized by implantation at normal incidence of 10keV O$_2^+$ ions), (c): Si LVV Auger spectrum for stoichiometric SiO$_2$ on silicon, excited with 3keV electrons.

1990a) and by Oliva et al (1991), and also in our work using oblique caesium-ion bombardment (Maydell and Fabian, 1992; Maydell, 1992a).

All of the spectra in Figure 3 show a broad feature between 20 and 40eV, with its maximum at an energy which depends on the angle of ion-beam incidence (with its corresponding angle of electron detection). Similar features have been observed in the Cs⁺ and Ar⁺ ion-induced electron spectra, from magnesium, aluminum and silicon (Maydell, 1992) measured using instrumental geometries that permit the 'acceptance' of negative ions into the electron analyzer. When using electron spectrometers operating with retarding fields, without special (magnetic trim) arrangements, negative ions

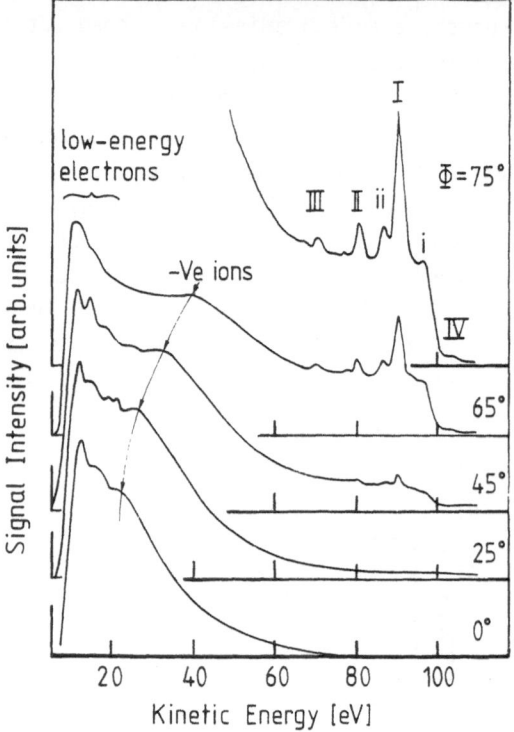

Figure 3. Electron spectra induced by irradiation of silicon with a 10keV O_2^+ ion-beam incident at $0°$, $25°$, $45°$, $65°$ and $75°$. For identification of peaks **I-IV, i** and **ii** see text.

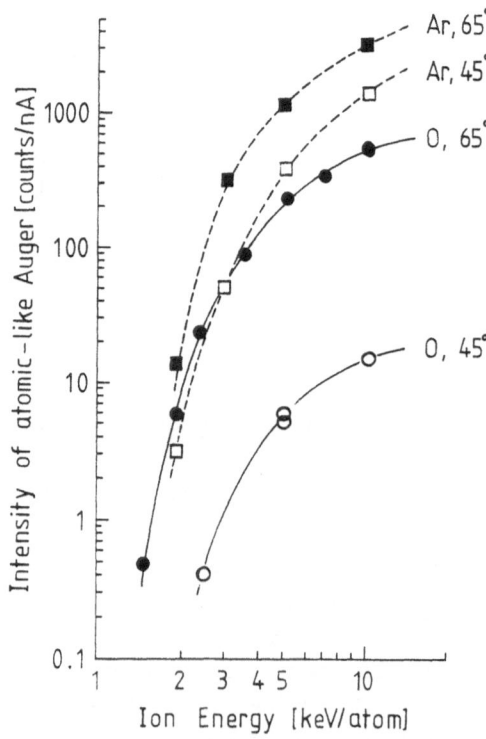

Figure 4. Dependence of ion-induced Auger electron emission from the Si^{O*} excited neutral (peak I) on the primary ion energy for oxygen and for argon ion bombardment of silicon at $45°$ and $65°$. [The drawn lines are for guiding the eye].

are indistinguishable from electrons with the same energy. The energy
dependence of the excited neutral Si^O* (peak I) intensity for 45° and 65°
oxygen ion-beam incidence is shown in Figure 4. The other main atomic-like
excited-ion peaks (II and III) follow the Si^O* dependence on energy for all
the angles of incidence studied; for clarity they are omitted from figure 4.
The energy dependence of the argon-ion induced atomic-like Si^O* Auger peak is
shown for the same angles of incidence.

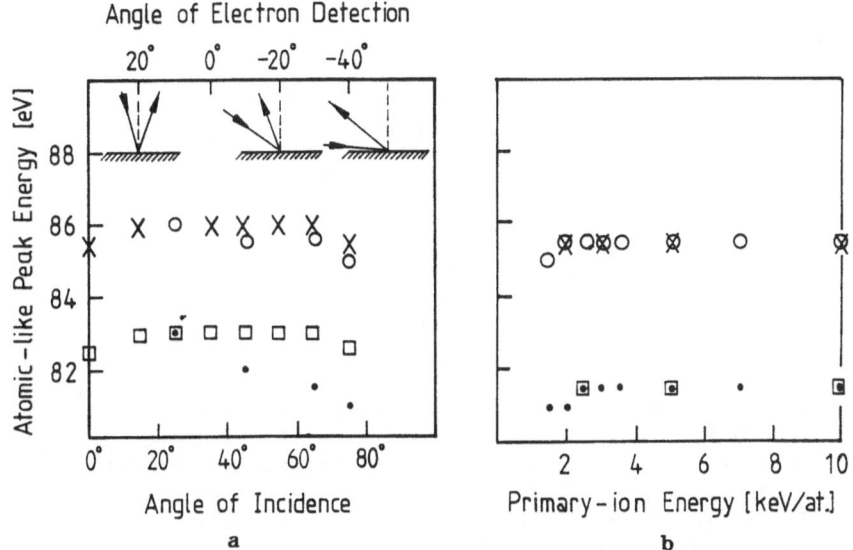

Figure 5. Dependence of the measured energy of the ion induced atomic-like
peaks on (**a**) angle of primary-beam incidence and (**b**) primary-beam energy.
Primary oxygen ions: (**O**) – peak **I**, (**●**) – peak **ii**.
Primary argon ions: (**X**) – peak **I**, (**◻**) – peak **ii**.

Energies of the principal atomic-like peaks (I-IV) shift slightly with
angle of oxygen ion-beam incidence (and corresponding angle of electron
detection). Figure 5a shows the dependence for the measured peak energies.
Peak I is detected at an energy of 86.0eV for 25° incidence, and at 85.5eV
for 65° incidence; a shift of -0.5eV. Peak 'ii', detected at 83eV for 25°
incidence, shifts by ∼ -1.5eV for 65° incidence. The energies of all the
peaks are almost independent of the oxygen-ion energy; only at the lowest
primary-ion energies is a difference of ∼ -0.5eV observed. Figure 5b shows
the peak I and peak ii energies measured using oxygen-ions at 65° incidence.
Peak I is detected at energy 85.5eV with oxygen ions of energy 2-10keV/atom
and at 85.0eV with oxygen ions of 1.5keV/atom. Peak 'ii', on the other hand,
is detected at a constant energy of 81.5eV with oxygen ions of 2.5-10keV per
atom, and at 81.0eV with oxygen ions of 1.5-2keV/atom.

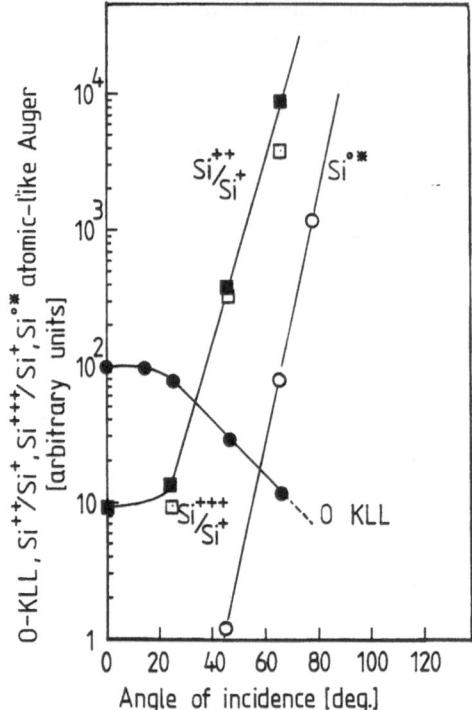

Figure 6. Dependence on the angle of incidence of:
(O) Si^{o*} (peak I) induced by 10keV O_2^+ bombardment of
silicon; (●) O **KLL** Auger electron signal excited with
3keV electrons for silicon implanted with 10keV O_2^+
to saturation at any given angle of incidence;
(■) Si^{++}/Si^+ and (□) Si^{+++}/Si^+ relative intensities of
multiply charged secondary ions generated by bombardment
of silicon with 10keV O_2^+.

Figure 6 shows the dependence of O KLL Auger signal excited with 3keV
electrons immediately after measurement of each of the oxygen-ion induced
electron spectra. The intensity of the O KLL Auger signal for 25° and normal
oxygen-beam incidence are virtually identical. However, the O KLL signal
decreases by a factor of ten for silicon irradiated with oxygen at 65°
incidence, corresponding to an implanted oxygen concentration of ~ 7%.

Positive and negative secondary-ion mass-spectra generated by primary
oxygen ions at normal and at 65° incidence show large variations in the
intensities of molecular-ion species relative to the Si^+ atomic ion. Table I
summarises the ratios of the signals that show the strongest variation with
the angle of primary ion beam incidence to that of Si^+ secondary-ion. The
ratio $Si_2O^+:Si^+$ increases ~ 160 times and $Si_2O_2^+:Si^+$ ~ 80 times for 65°
oxygen incidence, while the ratio $SiO^+:Si^+$ decreases by ~ 50%.

Table 1. Relative yields, Y(O), of +ve secondary ions measured using oxygen primary ions incident at angles $\phi = 0°$ and $65°$ to the sample normal. The yields are normalised to the intensity of the Si^+ ion.

	Si^{++}	Si^{+++}	SiO^+	Si_2^+	Si_2O^+	$Si_2O_2^+$
$Y(0°)$	4.4×10^{-5}	ND	2.6×10^{-2}	1.4×10^{-3}	1.4×10^{-3}	1.2×10^{-4}
$Y(65°)$	3.3×10^{-2}	5×10^{-4}	1.3×10^{-2}	5.5×10^{-2}	2.3×10^{-1}	9.3×10^{-3}
$Y(65°)/Y(0°)$	750	--	0.5	40	164	77

Table 2. Relative yields, Y(O), of -ve secondary-ions measured using oxygen primary ions incident at angles, $\phi = 0°$ and $65°$ to the sample normal. The yields are normalised to the intensity of the Si^- ion.

	O^-	SiO_2^-	SiO_3^-
$Y(0°)$	80	0.32	6.8×10^{-2}
$Y(65°)$	7.9	4.0	1.1
$Y(65°)/Y(0°)$	0.1	12	16

The negative secondary-ion mass spectra show a higher intensity ratio, $SiO_n^-:Si^-$ for oblique incidence, while $O^-:Si^-$ is ten times that for the normal incidence. The ratios, $SiO_n^-:Si^-$, for normal and for 65° oxygen-beam incidence are summarized in Table 2.

DISCUSSION

Primary-Beam Angle of Incidence

An important consideration in the present study is the low concentration of implanted oxygen appearing at high (oblique) angles of primary-beam incidence. Clark et al (1987) implanted low-energy oxygen ions into silicon at various angles of incidence and observed the formation of a suboxide of silicon at oblique incidence. They studied the surface composition (ex-situ) by both Auger-electron and photo-electron spectroscopies and concluded that, by contrast to the stoichiometric SiO_2 formed on bombardment with oxygen ions at normal incidence, the suboxide (SiO_x, $x < 2$) is formed under bombardment at angles of incidence $> 40^O$.

Computer simulations, using TRIM code (Webb, 1991), predict for 5keV per atom of oxygen impinging on a silicon surface, an increase in reflection coefficient from $\sim 3\%$ for normal incidence up to $\sim 25\%$ for 65° incidence. Preferential sputtering of oxygen from the surface can also be expected because of the efficient kinetic energy transfer for colliding atom-species of equal mass. However, these two effects taken together can only extend the time required for the stable-phase (steady-state) composition of silicon dioxide. Oxygen is a reactive ion and its interaction with silicon is not purely ballistic. The lower concentrations of implanted oxygen when oblique angles of bombardment are used is not therefore simply the result of oxygen preferential sputtering or of the increased reflection of oxygen ions competing with their penetration. To account for the observations, an additional mechanism or mechanisms must be considered. One such could be an increase in the probability of formation of molecular oxygen-rich secondary ions.

The observed intensities of secondary ions (by SIMS) for normal incidence bombardment with oxygen ions indicate the formation of silicon dioxide by implantation of oxygen. Enhancements of Si^+, SiO^+ and O^- occur and the results are consistent with the accepted 'bond breaking' mechanism of secondary-ion formation for ionic solids (Sroubek, 1988). Ionic 'fragments' of the oxide survive departure from the surface and, in the absence of 'tunnelling neutralisation' (Yu and Lang,1986) are observed with increased intensities; a well-known effect of positive secondary-ion enhancement for oxides and oxygen-flooded or oxygen-bombarded metallic surfaces. With oblique oxygen-beam incidence, the relative intensities of the positive secondary ions: Si_2O^+ and $Si_2O_2^+$, as well as the negative secondary ions: SiO_2^- and SiO_3^-, are enhanced by factors of up to x160, causing a significant depletion of oxygen at the surface of silicon.

Interestingly, the ratio of Si^{++} to Si^+ observed at oblique incidence shows a massive (x750) enhancement over normal incidence, while the Si^{+++} ion, undetectable at normal incidence, is seen only at oblique incidence.

Multiply charged silicon-ion species relate to Auger auto-ionisation processes occurring in excited silicon ions outside the solid surface. They are observed only with large angles of oxygen incidence, for which Auger decay of excited atoms and ions outside the solid surface is also evident. On the other hand, the atomic-like Auger decay of excited neutrals giving rise to peak I, and leading to the formation of singly charged Si^+ ions, is not reflected in the absolute intensity of the Si^+ signal. This is a result of the change of mechanism of ionisation from 'bond-breaking', for bombardment with oxygen ions normal to the sample surface, which form implanted stoichiometric oxide, to 'tunnelling neutralisation' followed by auto-ionization, for bombardment and oxygen-ion implantation at angles $> 40^o$. Sputtered ions can seemingly be neutralised by electron tunnelling as they leave a silicon surface containing a net 30% oxygen concentration. This we can deduce from the observed emission of atomic-like Auger electrons from excited neutrals outside the solid. The same conclusion can be reached about the 'take-up' of valence electrons by tunnelling when exited ions leave the surface and the observation of atomic-like Auger emission from these excited ions. Target atoms sputtered by 'bond-breaking' from an ionic solid, on the other hand, leave as ions; no charge-exchange with the solid surface takes place.

In addition to asymmetric collisions between projectile ions and target atoms (IT collisions), symmetric collisions between target atoms (TT collisions) occur. In IT or TT collisions with sufficiently small impact parameter (ie 'close' collisions), **2p** electrons are promoted by formation of molecular orbitals (Barat and Lichten, 1972; Fano and Lichten, 1965). For the IT or TT collisions the only difference between a pure metal and its oxide is to be found in the density of metal (target) atoms; these are more 'dispersed' in the oxide. The density of silicon atoms in silicon dioxide is 40% of that in pure silicon. This leads to relative probabilities between silicon and its oxide of 1:0.4 for suitable IT close collisions occurring and 1:0.16 for suitable TT collisions, for any given primary ion of given energy. The observed intensities of the atomic-like Auger emission from exited neutral silicon atoms provide an estimate of the probability for normal incidence oxygen-ion bombardment being $\sim 6 \times 10^{-4}$ of that measured with 65° primary-beam incidence. For the oxygen ion-beam incident at 65° we still observe some oxidation of silicon and therefore the true reduction of the atomic-like neutral silicon emission (peak I) must be higher. We can take as reference the intensity of peak I induced by 5keV argon ions (figure 4), this having approximately the same kinetic energy transfer to a silicon target-atom as for 10keV O_2^+ ions. Then, at 65° incidence, the reduction of the peak I intensity for 10keV O_2^+ at normal incidence is $\underline{1:4 \times 10^{-6}}$. An accurate measurement of this reduction would require an argon ion-beam free of any oxygen background (the latter being $1-2 \times 10^{-9}$ mbar in the present work).

The 40.000 times reduction in the intensity of the atomic-like Auger emission from silicon, induced by oxygen-ion bombardment at normal and nearly-normal incidence, can thus only be explained by an effect of the surface electronic structure. Tunnelling neutralisation of sputtered excited species cannot occur for a stoichiometric oxide. The ion-species sputtered from an ionic solid have an incomplete M-shell, and the probability of their de-excitation by Auger decay is substantially reduced. This is consistent with results reported by Dahl et al (1976) for gas phase ion-atom collisions. Gas-phase neutralisation of collisionally excited species cannot take place and a **2p** hole in an excited atom of a light element with additional vacancy or vacancies in its valence shell is not filled by direct Auger decay; instead, fluorescent soft x-ray emission with longer decay time, is observed.

Primary-Beam Energy

The observed 'shifts' in measured energies of oxygen induced atomic-like Auger peaks are of markedly different character from those observed for argon and caesium ion-induced emission (Maydell, 1992). Caesium and argon ion-bombardment at 36° incidence, with electron detection normal to the surface, produces positive shifts from $\sim$ +0.5 to $\sim$ +2.2eV in the measured energies for the atomic-like peaks with respect to the energies measured using glan-cing angle electron detection. We note in figure 5a, for argon-ion irradiation, that the 'shifted' energies remain almost constant, over a wide range of detection angles, $\pm 30^\circ$ centred around the surface normal. In the case of caesium-ion bombardment an additional workfunction shift in energy of all atomic-like peaks is observed ($\sim$ -1.7eV with respect to the energies measured under argon-ion irradiation) at all angles of incidence; this being an effect similar to that described by Zampieri and Baragiola (1984) for argon-ion bombarded aluminium on which sodium metal was adsorbed. Using argon-ion irradiation, the energy of peak I decreases by $\sim$ 0.5eV, for electron detection angles from -10° to $+30^\circ$, while the energy of peak 'ii' steadily decreases by $\sim$ 1.5eV for the same variation of detection angle.

The Doppler shifts observed for atomic-like Auger peaks using argon-ion bombardment (Maydell, 1992) show a pattern similar to that reported by Bonanno et al (1990). The variations in energy of these same peaks when measured using oxygen irradiation can therefore only be caused by changes of workfunction of silicon with oxygen concentration when implanted at the different angles of incidence.

Only a slight primary-ion energy dependence of the measured energy-shifts for the atomic-like Cs^+ and Ar^+ ion-induced Auger peaks is observed, as previously (Maydell, 1992), and the effect - if the shifts are simply Doppler in origin - is not fully understood. If the **2p**-excited sputtered species are ejected in asymmetric collisions between the primary ions and target atoms, they should have kinetic energies, and therefore exhibit Doppler shifts, that depend linearly on the primary-ion energy. This is not observed. For **2p**-excitations in the collision cascade, we examine the model described by Wilson and Webb (1986), in which the cascade will have a 'hot' anisotropic energetic centre surrounded by a 'cold' isotropic outer region. Thus, if symmetric collisions in the collision cascade are the source of the **2p**- excited sputtered species, only target atoms with high kinetic energy (those recoiling from 'head on' collisions) can be a source of these excited species, and (in contrast to a 'normal' isotropic collision cascade) the emission of highly excited species will be anisotropic - with kinetic energy transferred to the collision partner again depending linearly on the primary-ion energy.

Interpretation of ion-induced Auger electron spectra is bedeviled by the difficulty of distinguishing between processes caused by collisional excitations of atoms in the bulk and perturbations caused by interaction of the sputtered excited species with the surface as they leave. The electronic structure of the surface plays crucial role in modifying the electronic configurations of the excited species. While the surface electronic structure of most solid materials is well understood (Hagstrum, 1977), The surfaces under ion irradiation still elude description Corbett (1977); only when the various defect states resulting from the motion of atoms in a collision cascade are accurately modelled, will an adequate description of ion-induced excitations of atoms in the solid and of their de-excitation in vacuum be achieved.

We have to account for the absence of primary-ion dependence of the measured Doppler shifts in energies of Auger electrons ejected from the secondary ions and are led to question whether molecular secondary ions do not play a role, with Auger-electron emission from **2p**-excited fragment ion species. A more complete picture would be obtained with information from complementary de-excitation processes, such as those giving rise to UV-visible and soft x-ray emission (Cairns et al, 1979) associated with the filling of **2p** holes by valence shell electrons.

SUMMARY

It has been demonstrated that bombardment of silicon with oxygen ions causes emission of atomic-like Auger electrons from sputtered excited neutrals and ions once the implantation-effect of oxygen in the surface is reduced by the enhanced formation of oxygen-rich molecular ions. Stoichiometric oxides bombarded by oxygen ions, as with argon ions, cannot give rise to 'tunnelling neutralisation' of sputtered ion-species. This causes suppression of the atomic-like Auger emission. Correlations of oxygen-ion induced electron emission spectra with secondary-ion mass spectra supports both the accepted 'bond-breaking' and 'tunnelling neutralisation followed by autoionisation' mechanisms of secondary-ion formation.

The energies of atomic-like Auger peaks are found to depend on the geometry of the excitation but not on the primary-ion energy. Extending the studies of ion-solid interaction to radiative de-excitation processes (photon and soft x-ray emission) would enable more complete understanding of the role of molecular-ion species and surface interactions to be reached.

ACKNOWLEDGEMENT

The author wishes to thank Dr Derek Fabian for encouragement and helpful discussion and comment. Thanks are due also to Dr Hassan Bolouri at the University of Strathclyde, for technical assistance, and to Dr Asunta Bonanno of University of Calabria for making results available prior to publication. This work has been supported by the UK Science and Engineering Research Council.

REFERENCES

Baragiola R.A., Alonso E.V., and Raiti H.J.L., 1982, Phys. Rev. A25:1969.

Baragiola R., Nair L. and Madey T., 1991, Nucl. Instr. Meth in Phys. Res. B58:322.

Barat M. and Lichten W., 1972, Phys. Rev. A6:211.

Bonanno A., Xu F., Camarca M., Siciliano R. and Oliva A., 1990, Nucl. Instr. Meth. in Phys. Res. B48:371.

Bonanno A., Xu F., Camarca M., Siciliano R., and Oliva A., 1990, Phys. Rev. B41:12590.

Cairns J.A. and Holloway D.F. 1979, Nucl. Instr. Meth. 159:99.

Clark E.A. Dowsett M.G. Spiller G.D.T. Augustus P.D. Thomas G.R. and Sunderland I., 1988 Vacuum 38:937.

Corbett J.W., 1979, Surface Sci. 90:205.

Dahl P, Rodbro M, Hermann G, Fastrup B and Rudd M E (1976) J Phys B8, 1581.

De Ferrariis L. Grizzi O. Zampieri G.E. Alonso E.V. and Baragiola R., 1986, Surface Sci. Lett. 167:L175.

Fano U. and Lichten W., 1965, Phys. Rev. Lett.14:627.

Hagstrum H.D., 1977, Ch. I in Inelastic Ion-surface Collisons, Eds. N.H. Tolk, J.C. Tully and C.W. Heiland, Academic Press Inc., N.Y. San Francisco,

Legg K.O., Metz W.A. and Thomas W.A., 1980, Nucl.Instr.and Meth. 168:379.

Maydell E.A.,Fabian D.J. and Bolouri H., 1992 J.Phys.E (in print).

Maydell E.A., 1992 (in preparation - for Surface Sci.).

Maydell E.A. and Fabian D.J., 1992 Nucl.Instr. and Meth. in Physics Res. B67: 610.

Maydell E.A. 1992a, Surface and Interface Analysis 19:65.

Maydell E.A., 1992b (in preparation - for Surface Sci.).

Sroubek Z. 1988, SIMS IV Conf. Proc. Eds. Benninghoven A., Huber A.M. and Werner H.W., J.Wiley and Sons, Chichester, NY, Brisbane, Toronto, p 17- 24.

Viel L., Benazeth C. and Beanazeth N., 1976, Surface Sci. 54:635.

Webb R.P., 1991, University of Surrey, private information.

Wilson I.H. and Webb R.P. 1986, Invited Paper in Proc. of Third Int. Conf on Radiation Effects in Insulators, I.H.Wilson and R.P.Webb Eds., Gordon and Breach Science Publishers, p.765.

Whaley R. and Thomas E.W., 1980 J. Appl. Phys. 56:1505.

Wittmaack K. 1991, Nucl. Instr. and Meth. in Physics Res. B58:317.

Xu F. and Bonanno A., 1992, Surf. Sci. Lett. (in print).

Xu F., Mandarino N., Zoccali P. and Bonanno A, 1992, Phys. Rev. A (in print).

Yu M.L. and Lang N.D., 1986, Nucl. Instr. and Meth. in Physics Res. B14:403.

Zampieri G. and Baragiola R., 1984, Phys. Rev. B29:1480.

ELECTRON EJECTION INDUCED BY FAST PROJECTILES

G. Schiwietz

Hahn-Meitner-Institut Berlin GmbH
Glienicker Str. 100, 1000 Berlin 39
Germany

INTRODUCTION

In recent years different quantum-mechanical and classical collision models have been applied to the calculation of atomic cross sections. Most of these models may be applied for the prediction of the primary electron flux in solids, as far as insulators or innershells of conductors are concerned. From comparisons of experimental data and theoretical results one may derive the limits of perturbation theory or classical collision theories. It will be shown that there is some overlap of the corresponding ranges of validity if the number of observed quantities is kept low. However, only extensive coupled-channel calculations seem to cover the range of incident energies from a few eV up to a few hundred MeV. The contents of this paper will be restricted to the intermediate to high energy regime.

The connection between atomic cross sections and electron emission from foils will be demonstrated for 8 MeV/u U^{68+} incident on carbon foils. Results of a simple transport model based on semi-empirical cross section formulas for the initial electron flux will be compared to experimental data for this case. Furthermore, it will be shown that electron spectroscopy in fast (about 100 MeV) Ne^{q+} ($q = 7 \dots 10$) collisions with foils allows for a determination of the nuclear track potential in insulators.

ATOMIC ELECTRON EMISSION PROCESSES

Electron emission in ion-atom collisions has been investigated intensively during the last decades (ICPEAC, 1991). Our present understanding of single-electron processes seems to be quite complete even if multiply differential cross sections are concerned. For total electron capture, excitation or ionization cross sections, there are theoretical models which allow for highly accurate predictions (Fritsch and Lin, 1982, 1983; Winter, 1982).

However, only a few theoretical models are usually applied to the calculation of multiply differential cross sections. The ranges of validity of these models will be discussed in the following. Fig.1 serves to classify the approximations involved in the different models for ionization. The upper left box shows the exact quantum-mechanical solution of the dynamical many-body problem and each arrow stands for an additional approximation involved in the treatment. Most of the models are based on the impact-parameter picture, where the wavefunctions of the heavy particles (projectile and target) are replaced by trajectories of classical particles

Ionization of Solids by Heavy Particles, Edited by
R.A. Baragiola, Plenum Press, New York, 1993

(point-charges are used in most non-relativistic cases). The so-called eikonal transformation may be used to convert results from classical to quantum treatments of the projectile and vice versa (Salin, 1989; Flannery and MacCann, 1973; Willets and Wallace, 1968).

Most of the models use certain types of independent electron approximations where electron-correlation effects are ignored (McGuire, 1991). Often the initial-state frozen-core approximation is sufficient to account for the influence of the electron-electron interaction in single-electron processes (Schiwietz, 1988, 1990a).

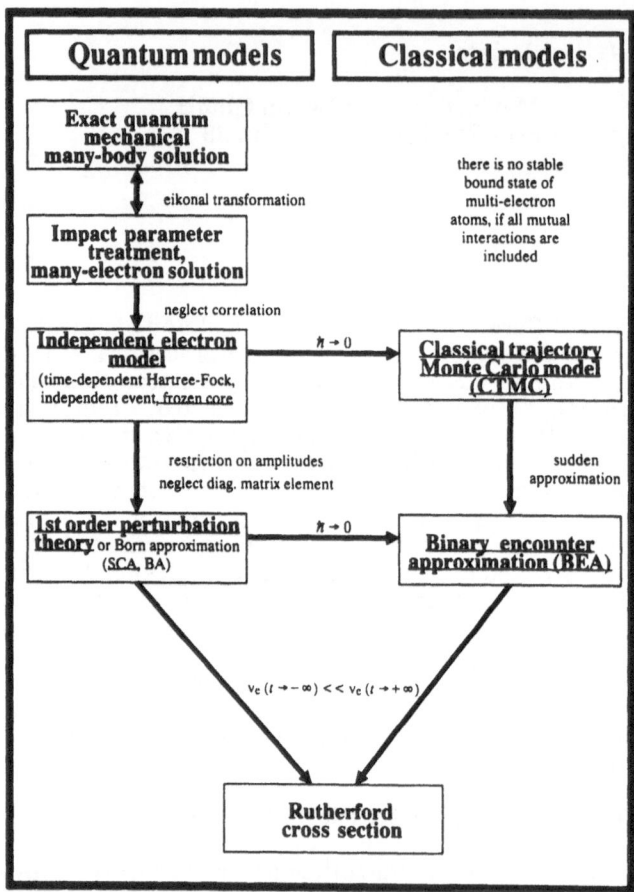

Fig.1. Atomic collision models and corresponding approximations.

Some progress in the description of multi-electron processes has been made in the independent event model, where complete relaxation of the intermediate (singly-excited) states is incorporated (Deb and Crothers, 1991; see also Hippler et al., 1984). If the interaction with the projectile may be considered a small perturbation, then the semi-classical approximation (SCA) or the equivalent three-body plane-wave Born approximation may be applied (Bates and Griffing, 1953; Bang and Hansteen, 1959). Similarly, if the target-electron interaction is weak compared to the projectile - electron interaction the strong - potential Born approximation (SPB) may be applied (Macek, 1992). At low velocities the influence of both nuclei on the time-evolution of the electronic wavefunction is of similar importance. In this case, however, perturbation theory may be applied for transitions between quasi-molecular orbitals. If neither of the two nuclei provides a strong perturbation, then the electron-emission cross sections converge to the Rutherford cross section (for bare projectiles). This is the case especially for high incident energies and electron-ejection energies somewhat below the maximum energy-transfer in a binary encounter. Boxes on the right side of the figure are related to classical mechanics only, which might be viewed as the limiting case for $\hbar \Rightarrow 0$. However, the binary-encounter approximation (BEA) may be derived from the Born approximation by applying additionally the impulse approximation (Vriens, 1969).

Singly differential ionization cross-sections $d\sigma/dE$ for bare-ion - atom collisions may be subdivided into four partial cross sections, corresponding to different electron-production mechanisms, according to

$$\frac{d\sigma}{dE} = \frac{d\sigma_H}{dE} + \frac{d\sigma_S}{dE} + \frac{d\sigma_C}{dE} + \frac{d\sigma_M}{dE}$$

The partial cross section $d\sigma_M/dE$ corresponds to quasi-molecular ionization processes (Wille and Hippler, 1986), as e.g. direct ionization in the united atom limit or during molecular promotion. In both cases electrons are emitted from a moving source, namely the center-of-charge. These mechanisms are important, whenever the projectile velocity is small compared to the mean of orbital velocity of the electrons under consideration. However, in this work only those cases will be discussed where quasi-molecular processes are of minor importance.

The partial cross section $d\sigma_C/dE$ corresponds to electron capture into the projectile continuum. The physical picture for this process is very similar to the electron capture into bound projectile-states, but electrons are ejected with a velocity near to the projectile velocity. This process may be described within the framework of electron-capture theories (Macek, 1992). At high incident energies a small part of the initial state velocity distribution matches with the projectile velocity and only this part leads to electron capture into bound or continuum states of the projectile. Hence, the capture cross sections are of minor importance (proportional to v_p^{-11}) at high incident velocities v_p. However, at intermediate energies the contribution to the total ionization cross section might be as high as about 20%.

The partial cross sections $d\sigma_H/dE$ and $d\sigma_S/dE$ stand for hard and soft collisions, corresponding to large and small momentum transfers between the projectile and the ejected electron. Large momentum transfers occur predominantly at small impact parameters. Both contributions $d\sigma_H/dE$ and $d\sigma_S/dE$ may be calculated within the quantum mechanical plane-wave Born-approximation (PWBA) or the equivalent semi-classical approximation (SCA). Both theories base on the first-order perturbation theory and may be applied in the case of a small perturbation, e.g. fast projectiles with small nuclear charge.

The binary-encounter approximation (BEA) includes only contributions from a perturbative treatment of hard collisions. If the velocity distribution of the initial state is represented by the the mean orbital velocity, the partial cross-section may be written (in atomic units) for high incident energies as

$$\frac{d\sigma_H}{dE}\left(E\right) = 2\pi \frac{Z_p^2}{v_p^2} \sum_j n_j \left(\Delta E_j^{-2} + \frac{4}{3} I_j \; \Delta E_j^{-3}\right) \quad with \; \Delta E_j = I_j + E \quad ,$$

where n_j is the number of electrons in the subshell j with binding energy I_j (Vriens, 1969; Stabler, 1964). This expression is only valid for electron energies below $2v_p^2$ (v_p is the projectile velocity). The difference between results from the PWBA and the BEA corresponds to the soft-collision contribution and may be parametrized (in atomic units) as (Schiwietz et al., 1990)

$$\frac{d\sigma_S}{dE}\left(E\right) = Z_p^2 \sum_j n_j \, c \, (I_j, v_p) \, \Delta E_j^{-2.9 + I_j^{0.08}} \quad ,$$

where $c(I_j, v_p)$ is a target dependent quantity. It should be emphasized, that this partial cross section does not appear in classical collision theories and reflects the long-range nature of the Coulomb interaction. As can be seen from the above expressions the hard as well as the soft collisions contributions are proportional to the squared projectile charge (Z_p^2). This is a results of first-order perturbation theory. Thus, it holds true only for small perturbation, as e.g. large impact parameters, small projectile charges and high projectile speeds. Since the hard collision contribution corresponds to small impact parameters (large momentum transfers), it may be strongly affected by effects, which go beyond the first-order perturbation theory. It is noted, that the different partial cross sections as discussed in this section can not be clearly distinguished, especially in models which go beyond first order perturbation theory. The soft-collisions contribution may be influenced by couplings to non-dipole transitions (hard collisions), the capture contribution is not orthogonal to the hard-collisions contribution, and all partial cross sections may be influenced by molecular effects.

Fig. 2 displays absolute singly-differential electron-ejection cross sections for a He target (Platten et al., 1987 and Rudd et al., 1979) divided by the squared projectile charge (Z_p^2) in comparison with different theoretical results. For high electron energies all theoretical results are in reasonable agreement with the experimental data. The classical-trajectory Monte Carlo (CTMC) results do also exhibit the enhancement of the Ne data in comparison with the proton data. This is due to the fact, that the CTMC goes beyond perturbation theory similar as coupled-channel or time-dependent Hartree-Fock calculations. For light projectiles (protons) the influence of the so-called binding effect becomes important. High energy electron emission takes place at small internuclear distances and the corresponding enhancement of the initial -state binding energy leads to a reduc-

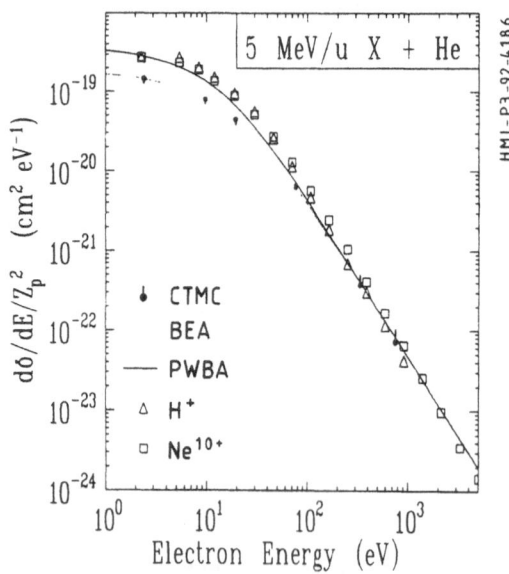

Fig. 2. Absolute singly differential cross sections as function of the electron energy for 5 MeV/u protons and bare Ne ions incident on He. CTMC results are circles (H+ results) with bars (the end of each bar points to the Ne10+ result).

tion of the electron-emission cross sections. For heavier projectiles additional multi-step processes come into play which broaden the spectral distribution. Thus, low energy electron emission is suppressed and the cross section for fast electrons is enhanced. Furthermore, for low electron energies only quantum theories may be applied, since classical theories (BEA and CTMC) fail in the description of the long-range part of the dipole interaction (Boesten et al., 1975).

The different ionization mechanisms discussed before determine also the angular distribution of ejected electrons. It should be kept in mind, that all of the electron production mechanisms have a non-zero intensity at any ejection angle or energy. However, each of the production mechanisms correspond to at least one maximum in the energy or angular distribution of ejected electrons. The contribution due to electron capture to the projectile continuum has a pronounced maximum (ECC or so-called cusp peak) at an ejection angle of $\Theta_e = 0°$ and for electron velocities near to the projectile velocity (Berry, 1985). The hard- and especially the soft-collisions contribution correspond to a maximum at zero energy (soft-collisions peak). Thus, there are two maxima corresponding to the velocities of the charged projectile and target nucleus. A third structure in the energy and angular distribution of ejected electrons appears at an energy of about

$$E_e = 2\,v_p^2 \cos^2(\Theta_e)$$

and corresponds to the hard collisions. The momentum transfer to the ionized electron is high enough, so that the problem may be treated as a binary-encounter process. The relation between ejection angle and electron energy, as well as the cross section follow the Rutherford cross section for an ion incident on an electron at rest. At lower electron energies the initial electron momentum becomes comparable to the momentum transfer from the projectile and the so-called binary-encounter peak becomes broadened. Consequently, at low electron energies the binary encounter peak can not be distinguished from the soft-collisions peak, since the width of the binary encounter peak reaches 90° and more.

Fig. 3 displays the ratio of doubly-differential cross sections for 5 MeV/u Ne[10+] on He (Platten et al., 1987) divided by the corresponding Z_p^2-scaled proton values (Rudd et al., 1979) for three ejection angles as function of the electron energy. Perturbative approaches predict a ratio of one, independent of electron ejection-angle or energy. As can be seen from the figure for fast electrons forward ejection is enhanced by a factor of 4 and the backward ejection is reduced by a factor of 5. At intermediate angles the ratios are close to one. Two sets of theoretical results (CTMC, CDW-EIS) are also displayed in the figure and both show similar tendencies as the experimental data. As discussed above the CTMC includes all interactions between both nuclei and the active electron (Abrines and Percival, 1966). The problems connected with the inadequate treatment of slow electrons (underestimated dipole contribution) cancel partly for the ratios plotted in Fig.3. The CDW-EIS results go beyond perturbation theory in that two-center continuum wavefunctions are incorporated in this model (Fainstein and Rivarola, 1987). Thus, the agreement of these results with the experimental values points to the conclusion that continuum-continuum interactions yield a forward focussing of ejected electrons. In a first step an electron is transferred into a target-continuum state and in a second step it is attracted by the projectile and focussed towards 0°. This so-called two-center effect (Stolterfoht et al., 1987) may be viewed as the onset of the cusp peak, where an infinite number of projectile-nucleus electron interactions is necessary to describe electron capture in a target-centered basis (Schiwietz and Grande, 1992). It is emphasized that similar deviations from perturbation theory have been found also for 5 MeV/u C[6+], O[8+], 25 MeV/u Mo[40+], and 6 MeV/u U[38+] on He and Ar targets. The ratios for He and Ar are nearly identical and scale with Z_p/v_p, which is to be expected for such a redistribution effect in the continuum (Schneider et al., 1989). The deviations from perturbation theory reach about an order of magnitude for 6 MeV/u U[38+] ions.

IONIZATION AT SMALL IMPACT PARAMETERS

Figs. 4 and 5 show coincidence data for ionization of He by 100 and 300 keV protons at small impact parameters. Time-of-flight analyzers were used for detection of electrons and recoil ions. The projectile scattering angles were measured by means of a fast position-sensitive double-channelplate device. Details of the experimental setup and the data analysis may be found in Skogvall and Schiwietz (1990). The curves labeled SCA and AO represent results of the impact-parameter Born approximation and a coupled-channel calculation (using target-centered basis states) respectively. At the incident energy of 300 keV, there

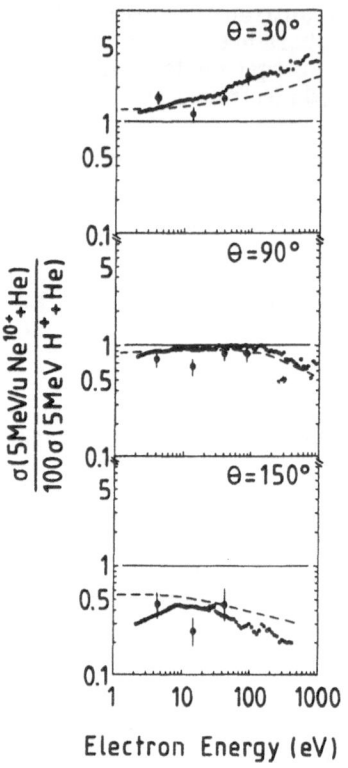

Fig.3. Ratio of doubly differential cross sections for bare 5 MeV / u Ne ions and protons divided by 100. Experimental results: Open cirles; CTMC results: closed cirlces with error bars; CDW-EIS results: dashed lines; perturbation theory solid lines.

is very good agreement between experimental data, SCA and AO results for the energy distribution. At 100 keV, however, the SCA results overestimate the experimental results by a factor of three for electron energies around 150 eV. At this incident energy electron capture, excitation and ionization are of similar importance. An electron may be ionized in a first step and thrown back to bound states of the target or the projectile in a second step. Such multi-step processes are accounted for in AO or CTMC calculations but not in a first-order treatment.

In Fig. 4a results of two different CTMC calculations are also displayed. One calculation was performed using a hydrogen like target potential with an effective charge (Z_{eff}) of 1.68. In the sCTMC calculation, however, the interaction of the active electron with the residual target core is given by a self-consistent field for the initial state (Montemayor and Schiwietz, 1989). Hence, the resulting initial-state momentum distribution is very similar to the quantum-mechanical Hartree-Fock distribution. The sCTMC results seem to converge to the experimental data for high electron energies, whereas the standard-CTMC results are on the average closer to the measured data. It is emphasized that the selection of small impact parameters tests not only high momentum components of the initial-state momentum distribution but also the coordinate-space distribution near the target nucleus. This coordinate-space distribution is in general unrealistic for micro-canonical statistical ensembles which are the starting point of most CTMC calculations. The classical mean orbital radius calculated for a hydrogen like target potential ($Z_{eff}=1.68$), however, is closer to the exact quantum-mechanical result than the sCTMC initial mean orbital radius. Thus, the classical momentum and coordinate space relation leads to the deviations between experimental data and CTMC results at small impact parameters. These deviations are strongly reduced when no impact-parameter selection is performed.

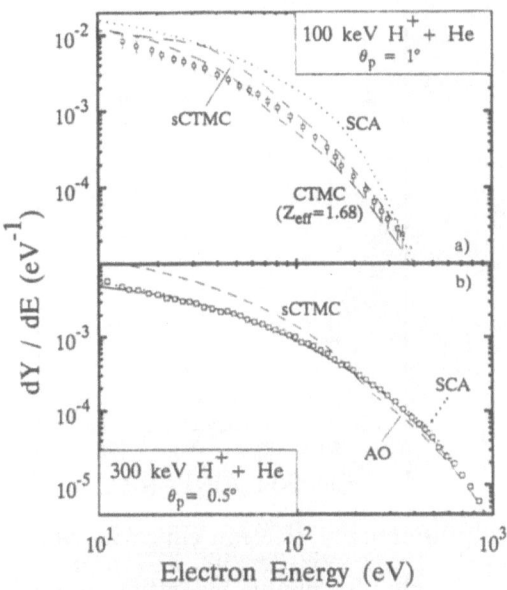

Fig.4. Absolute singly differential electron emission yields as function of the electron energy for proton He collisions at small impact parameters: a) at 100 keV, b) at 300 keV. Experimental results: open squares.

Fig. 5 displays the angular distribution of ejected electrons in 300 keV H+ + He collisions at small impact parameters (Skogvall and Schiwietz, 1990). Theoretical results displayed in this figure correspond to the models described above. The dCTMC results, however, incorporate also a dynamic electron-electron interaction and go somewhat beyond the sCTMC approximation. The dCTMC results in Fig. 5a lie somewhat above the experimental data due to the classical momentum and coordinate-space relation. The SCA results predict a curve for the angular distribution, which is too flat in comparison with the data. The AO results, however, are in good agreement with the experimental results. The reason for the deviation between first-order (SCA) and AO results is the same redistribution as reported before for non-coincident electron spectra. Since the AO model includes all couplings between continuum states, it is able to quantitatively reproduce the forward focussing of ejected electrons.

In the double-ionization case, as displayed in Fig. 5b, all theoretical approaches yield order-of-magnitude deviations from the measured data. The dCTMC results are too high because of the classical phase-space distribution discussed above. The quantum results are too high, since both rely on the initial-state frozen-core approximation. Thus, relaxation effects and the difference between first and second ionization potential are neglected in the SCA and AO model. Correspondingly, the ionization probability for the "second" electron is overestimated. It is emphasized, that the shape of the angular distribution differs also significantly from the one for single ionization in contradiction to the independent-electron approximation. The reason for this difference is the mutual repulsion of both electrons in the two-electron continuum, which is not accounted for in any of the above theories. Most likely, both electrons are initially ejected

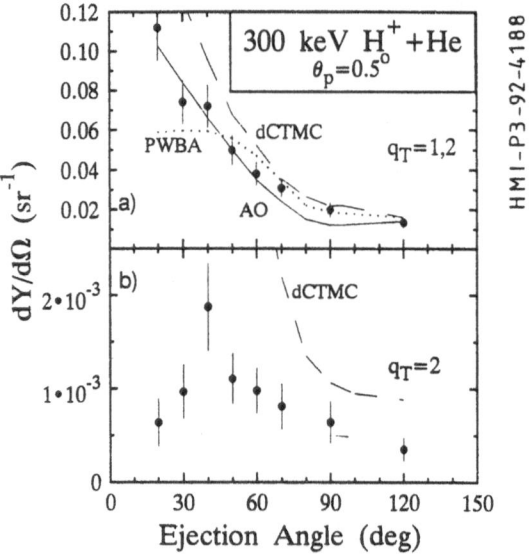

Fig.5. Absolute singly differential electron emission yields (ejection angles are measured relative to the beam axis) for 300 keV proton He collisions at small impact parameters: a) Single plus double ionization, b) double ionization contribution.

towards 0° and then deflected off each other. It is noted, that non of the current collision theories provides the full solution for the double-ionization process independent of incident energy. From the experience with independent particle calculations one may estimate that 10^4 to 10^5 basis states are necessary to perform rigorous coupled-channel calculations (using two-electron wavefunctions) that yield multiply differential double-ionization cross sections.

Table 1 summarizes the comparison between experimental data and theoretical results with respect to the single-ionization process. The ranges of validity are given for quantum-mechanical perturbation theory and for classical models in general. The CTMC is taken as a reference for classical models, since it fully incorporates all electrostatic interactions between the particles and the use of classical mechanics for the electronic motion is the only approximation involved (in the case of $H^+ + H$ collisions). As can be seen from the table, the classical approximation is reasonable for intermediate energies around the cross section maximum. In this region, perturbation theory fails and is valid only for very high incident energies (> 2 MeV H^+), if multiply-differential cross sections are

TABLE 1. Range of validity for classical collision theories (CTMC) and for first order perturbation theory (SCA) as a function of the initial-state binding energy I, the projectile velocity v_p and the projectile charge q. The indicated range stands for deviations of less than 30% from the experimental data. This table contains partly results of Manson et al. (1975).

Cross section type		CTMC: $\sqrt{I}/v_p$	SCA: q/v_p (for q = 1*)
σ		0.1 ... 0.55	< 0.6
$\dfrac{d\sigma}{dE}$		as above	< 0.25
$\dfrac{d\sigma}{d\Omega}$		as above	< 0.25
$\dfrac{d^2\sigma}{d\Omega\,dE}$		as above	< 0.1
P	(b = 0)	0.3 ... 1	< 0.45
$\dfrac{dP}{dE}$	(b = 0)	#	< 0.35
$\dfrac{dP}{d\Omega}$	(b = 0)	0.3 ... 1	< 0.22
$\dfrac{d^2P}{d\Omega\,dE}$	(b = 0)	#	< 0.15

*   for heavier projectiles the upper limits of q/v have to be multiplied by 2.
\#   the uncertainties exceed 30% in the whole investigated energy region.

considered. For perturbation theory the range of validity depends strongly on the number of observables and for total cross sections there is even some overlap with the range of validity of classical collision theories. Especially for those regions of ejection angles and electron energies where perturbation theory (SCA or PWBA) fails the so-called strong-potential Born approximation (SPB) may be applied (Macek, 1992). Thus, a combination of PWBA (or SCA) and SPB may be used to extend the range of validity down to a few hundred keV/u. It is emphasized that the single-center AO calculations yield good agreement with multiply differential experimental data down to about 100 or 200 keV/u incident energy. Two-center AO calculations seem to have no limited range of validity since the influence of electron capture and the formation of quasi-molecular orbitals (for static potentials) is fully taken into account (Grande and Schiwietz, 1992; Fritsch and Lin, 1982, 1983). Such calculations, however, are very time consuming even in comparison with single-center AO calculations. Furthermore it is noted, that coupled-channel calculations may also be performed with quasi-molecular orbitals as basis set (Sroubek, 1992) and the corresponding range of validity extends from hyper-thermal energies to energies somewhat below the cross section maximum.

ION-INDUCED DELTA-ELECTRON EJECTION FROM FOILS

A rigorous theoretical description of δ-(or secondary) electron ejection from solid matter requires a quantitative treatment of many different types of electronic transitions. In the case of amorphous or polycrystalline materials transport theory may be applied for the computation of electronic motion under the influence of such different perturbations. At very low incident velocities ($<$ 1 keV) electron ejection takes place via different types of di-electronic transitions at or in front of the surface. These di-electronic electron-emission processes are referred to as potential emission, since the projectile's potential energy determines the number of ejected electrons (for projectile charge-states below about 10) per incident ion. There is, however, a lack of realistic theoretical predictions for absolute transition probabilities and for the ion-induced perturbation of the surface band - structure. This precludes ab-initio calculations for slow highly-charged ions until now (Niehaus, 1992). Such calculations would also require a detailed understanding of charge-exchange processes in ion-surface interactions, which is subject to many recent investigations (Wille, 1992). At incident energies above a few keV/u the so-called kinetic electron emission dominates the electron production. Kinetic emission includes all electron production mechanisms, which vanish in the limit of low incident energies. For low energy electrons in solids there are many mechanisms which influence the transport to the surface, such as plasmon creation and decay, exciton production and electron-phonon coupling.

In this work we will focus our attention to the interaction of fast highly charged projectiles with thin foil targets. In all cases the foil thickness is small enough to ensure that the incident particles lose only a minor part of their incident energy. Furthermore, using stripper foils, incident charge states near the mean equilibrium charge were selected to reduce non-equilibrium effects. Thus, the target induces only minor changes of the projectile state. This allows for a more selective investigation of target dependent effects via electron spectroscopy.

In order to describe the spectra of ejected electrons theoretically we developed a transport model using the SELAS approximation (separation of energy loss and angular straggling). Details of the transport theoretical derivation can be found elsewhere.(Schiwietz et al., 1990, Tougaard and Sigmund, 1982; Schou, 1980) In the following we will give an overview of the basic ingredients of the model. We distinguished between four different electron production mechanisms:

i) The emission of surface electrons was described by semi-empirical atomic ionization cross sections and the surface potential as well as the solid state binding energies were explicitly taken into account.

ii) The production of target electrons due to the interaction with the projectile nucleus and bound projectile electrons was described by semi-empirical atomic ionization cross sections, and the transport of electrons from inside the solid to the surface was calculated in the SELAS approximation. The semi-empirical ionization cross sections are quite reliable for fast highly charged ions and include effects which go beyond first-order perturbation theory.(Schiwietz et al., 1990) Solid-state effects were taken into account for the electron production (energy levels) and for the transport calculation (plasmon excitation, surface potential).

iii) The production and transport of projectile electrons was treated in a fashion similar to that described in (ii), except that the differential ionization cross sections were transformed from the projectile frame to the laboratory frame of reference. Since the projectile population numbers are the only unknown quantities in this calculation, their determination could follow a fit to the experimental data. It should be noted that the electron production mechanisms (i) and (ii) are also slightly influenced by the projectile state populations. Furthermore it is noted that lattice effects such as, e.g. the resonant coherent projectile ionization (Datz et al., 1978) which might show up in the spectrum of ejected electrons, were not included in the present treatment. This is expected to be a reasonable simplification since we are not dealing with single crystals or channeling in this work.

iv) The contribution due to cascade electrons (electrons produced by successive electron-electron collisions) was calculated by iterating the transport equations for the different generations of cascade electrons. Semi-empirical electron-atom collision cross sections were used to describe the production of the cascade electrons.

The majority of all ejected electrons are both created and observed in the forward direction. Thus we may neglect the path length straggling for fixed production and emission angles (β_0 and β) with respect to the incident beam. Electrons produced inside the solid (mechanisms ii, iii and partially iv) will undergo multiple collisions during transport to the surface. The mostly high electron energies ($>> 100$ eV) investigated in this work correspond to electron production deep inside the solid and the multiple collision character of the electron transport destroys the interdependence between angular scattering and energy loss. This justifies the Separation of Energy Loss and Angular Scattering (SELAS approximation) in the so-called propagator function. Furthermore, we neglect the energy straggling and describe the energy loss mechanisms simply by a stopping power $S(E)$. This should be a reasonable approximation for fast electrons. Under the above assumptions the electron emission yield for the mechanisms ii) and iii) becomes

$$d^2P/d\Omega dE\,(E, \beta) = |\cos(\beta)/S(E)| \int dE_0 \int d\beta_0 \sin(\beta_0)\, F(ER_0, \beta_0)\, M(E_0, \beta_0, E, \beta).$$

$F(E_0, \beta_0)$ is taken as the product of the atomic ionization cross section and the target density. $M(E_0, \beta_0, E, \beta)$ is an analytic expression for the angular scattering probability. It preserves unitary and includes multiple scattering effects.

Fig. 6 displays experimental and theoretical electron spectra for three electron ejection angles in 8 MeV/u U + C collisions. At non-zero electron ejection angles the spectra are decreasing with energy. At 0° there is a pronounced peak (at about 4.4 keV) due to capture and loss of electrons into the projectile continuum, the so-called cusp peak. The intensity of this peak depends strongly on the population of excited projectile states. A fit of the SELAS results to the experimental data, with

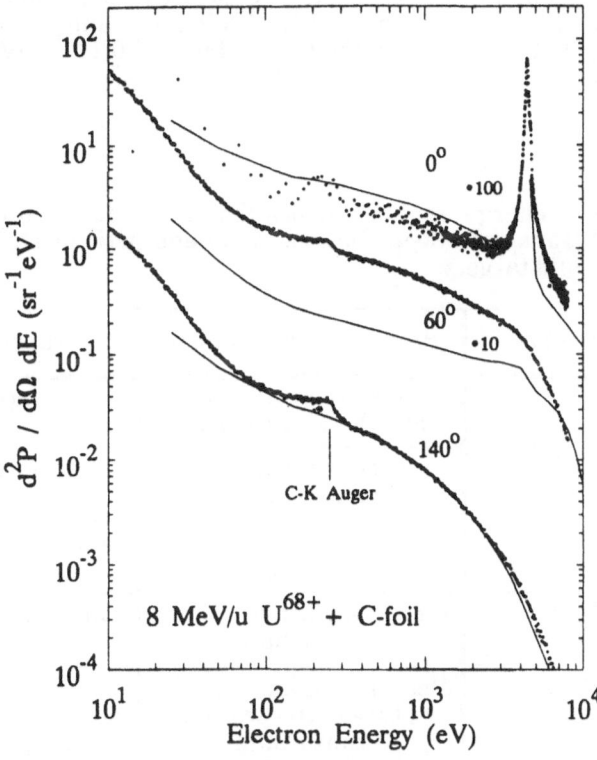

Fig.6. Electron energy spectra taken at three ejection angles for 8 MeV/u U^{68+} ions normally incident on a 44 µg/cm^2 carbon foil. The solid lines represent the SELAS results.

the projectile Rydberg-state population number as a free parameter, allows for determination of this population number *inside* the foil. This method, however, seems to be reliable only for few-electron ions (Schiwietz, 1990b; Schneider et al., 1992). In the experimental spectra, there is another structure at about 260 eV superimposed on the δ-electron continuum. This structure is due to target K-Auger electron emission.

The theoretical δ-electron results are seen to agree well with the experimental data for backward ejection and for electron energies above 4000 eV. However, for slow electrons ejected at large forward angles there are discrepancies between experimental and theoretical results which reach a factor of six. These deviations for 60° are most likely due to the SELAS approximation used in the model. The average path length for an electron in a solid depends only on the detection angle in this model and not on the initial ejection angle. Thus, the path length for electrons created with velocity vectors parallel to the surface is strongly underestimated and a more refined treatment would increase the theoretical results for low energies and detection angles near 90°. It is emphasized, that about 2500 electrons are emitted per incident U ion. At such high incident energies (3.5 to 8.5 MeV/u) the forward ejection yield exceeds the backward yield by a factor of two to three and about 30% of all ejected electrons are fast (>100 eV) (Schiwietz, 1990b; Schneider et al., 1992). This is a clear deviation from the approximate forward/backward symmetry and the dominance of low energy electrons (> >90%) stated for lower incident energies (<about 1 MeV/u). (Rösler, 1991; HP. Winter, 1991; H. Rothard et al., 1991) Furthermore, it is noted that the mean electron energy for about 5 MeV/u heavy ions is 800 eV in the case of forward ejection and 250 eV for backward ejection respectively. The total electron yields, however, seem to scale with the electronic stopping-power of the ions as reported previously for protons as well as for fast and heavy ions.(Hasselkamp, 1991; Rothard et al. 1991) The so-called specific yields (ratio of the total electron yield and the electronic stopping power) of 0.59 ± 0.24 Å/eV for 3.5 and 8 MeV/u U ions (Schneider et al., 1992) and 0.25 ± 0.03 Å/eV for 5 and 8.5 MeV/u Ne ions (Schiwietz et al., 1990) are close to the value of 0.31 ± 0.14 Å/eV determined by Rothard et al.(1991).

TABLE 2. Electron energy regions associated with the dominant production mechanisms and transport effects for ejection of energetic electrons by 8 MeV/u ions penetrating solid targets.

Electron Energy	Production Mechanism	Transport Effects
0 .... 50 eV	electron collision cascades at the surface	surface potential
15 ... 300 eV	direct ionization of surface atoms	surface potential
100 eV ... $4*E_p/M_p$	direct ionization of bulk atoms	energy loss and angular straggling
$>4*E_p/M_p$ #	direct ionization of inner shells (bulk)	energy loss and angular straggling
E_p/M_p at 0° #	electron loss and electron capture to the projectile continuum	energy loss and angular straggling and Coulomb focussing / defocusing

\#   The ratio E_p/M_p has to be calculated in units of eV/m_0, where m_0 is the electron mass.

Table 2 gives the energy regions corresponding to certain electron production mechanisms as extracted from our model calculations. As a general tendency low energy electrons stem from outer shells of surface atoms (or the conduction band) and at higher electron energies inner shells of deep-layer atoms gain importance. It is noted, that the present model is not valid for the ejection of low energetic electrons (<50 eV), since the used energy loss function is a very crude approximation for this energy regime and diffusion-type motion is not accounted for. Moreover, electron ejection due to the decay of bulk and surface plasmons is not incorporated. This is crucial for the prediction of electron spectra at low incident energies (Rösler, 1991) or when the mean production depth of ejected conduction-band electrons is needed (Rau, 1992). However, the table should be valid for electron energies above about 50 eV. The Coulomb focussing/defocusing effects mentioned in the table for cusp electrons arise only in transport models that include the post-collision interaction with the projectile ion (Burgdörfer and Bottcher, 1988).

NUCLEAR TRACK POTENTIALS IN INSULATORS

For heavy ions at high incident energies it is well known that the formation of a transient electrostatic track potential has to be considered to understand plasma effects in solid-state particle detectors or the plastic deformation of metallic glasses. (Klaumünzer et al., 1986; Johnson, 1992) There is no easy way to access the important quantities, such as maximum track potential or recombination time, for most materials. However, because of the increased track radius (Schou, 1992) and suppressed electron-hole recombination processes insulators might be ideally suited to determine the maximum track potential from the spectrum of ejected electrons. These electrons created at or near track serve as a probe and the track potential may be extracted from a shift of the energy spectrum.

In the following we present results for target K-Auger spectra produced during (normal incident) penetration of polypropylene foils by 100 MeV Ne^{9+} ions. (Schiwietz et al., 1992) In order to avoid macroscopic charging of the foils or foil temperatures in excess of 50°C polypropylene (PP) targets of 2μm thickness were used with an aluminum coating (20 μg/cm^2) on one side of the foil. For most insulators this technique would also lead to macroscopic charging of the samples (Cazaux, 1992) but the resistivity of PP is known to drop by several orders of magnitude during irradiation. The absence of macroscopic charging was checked, by measuring a constant total electron ejection yield (ratio of target and beam current times mean projectile charge) as function of irradiation time and beam current (0.3 and 2 particle nA). Thus, recombination occurs along the tracks and the electron spectra will be influenced by single tracks only. It is interesting to note that the total electron ejection yield (33.5 $\pm$ 3) for PP foils (with one aluminum surface) at small fluences is equal to the yield determined for carbon foils (34.5 $\pm$ 1). At higher fluences ($>10^{15}$ ions/cm^2) a 10% reduction of the PP electron yield was found, which might be due to the chemical changes of the samples. At this point it should be emphasized, that the chemical processes leading to the so-called carbonization of the samples are of minor importance for the short-time processes during the penetration of the solid by a fast heavy ion.

Fig.7 displays three electron spectra taken at 120° with respect to the beam for the passage of 100 MeV Ne^{9+} ions through a carbon foil and a polypropylene foil (with an aluminum film at the beam exit side) at two different fluences. In order to reduce the temperature load and to achieve a high accuracy for the determination of the ion fluence, a custom made target wobbler was used. The motion of a target frame was synchronized with the energy scan of the electrostatic electron spectrometer. Thus, all points of an electron spectrum correspond to the same fluence and to the same number of incident ions. The PP foil spectra for low fluences are dominated by the carbon K-Auger peak at 180 eV. This peak is shifted towards higher energies when the fluence is increased. In the case of carbonfoils, or PP foils at fluences above $2 * 10^{15}$ ions/cm^2 the peak energy is similar to atomic Auger energies. The low energy tail of the peak corresponds to emission of Auger electrons from up to about 20 Å below the surface. The deeper the electrons are produced inside the solid the more energy they loose on their way to the surface. In auxiliary measurements it was verified that the PP Auger energy shift depends on whether electrons from the entrance or the exit side are detected (the aluminum coatings of the foils were always on the opposite side), which is another indication for the absence of macroscopic charging of the samples. Consequently, the shift of the Auger peak can be assigned to the influence of a single track, in the case of low fluences.

The relative position of the PP K-Auger structure was determined by a fit to the corresponding carbon foil spectra. (Schiwietz et al., 1992) Asymptotic energy shifts for zero fluence are 68 ± 4 eV at 120° and 35 ± 3 eV at forward angles. The

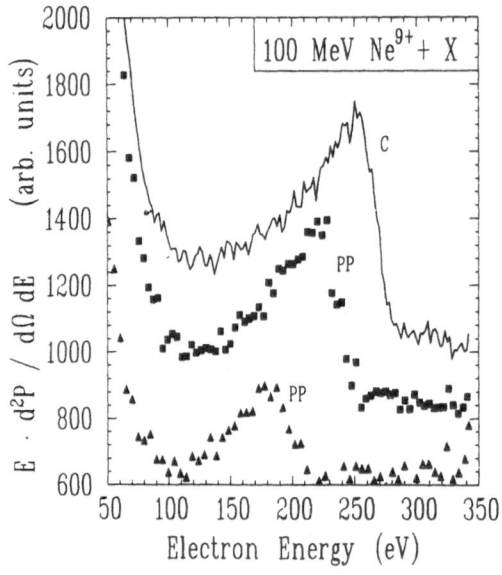

Fig.7. Electron energy spectra taken at an ejection angle of 120° for 100 MeV Ne^{9+} ions on thin target foils. Solid line : 20 µg/cm^2 C; squares : 1.55 µm polypropylene at a fluence of $1.1 \cdot 10^{15}$ ions /cm^2; triangles: 1.55 µm polypropylene at a fluence of $3.6 \cdot 10^{13}$ ions /cm^2.

asymptotic value for 0°-Auger electron ejection in a 170 MeV Ne^{7+} + mylar experiment was 32 ± 4 eV. The energy shift tends to zero in the limit of high fluences. The lowest measured shift is 1 ± 0.5 eV at $3.1 * 10^{15}$ ions/cm^2. At this fluence the ratio of H to C atoms is reduced by a factor of about 2.5 compared to C_3H_6. However, the changed chemical environment effects the carbon K-Auger peak energy for a certain transition by only about 5 to 10 eV. Therefore, we conclude that the transition from an insulator to a conducting material is responsible for the fluence dependence of the PP spectra.

In an insulator, recombination processes are strongly suppressed. Thus, for an inner-shell excited carbon atom inside a PP environment there is a high probability that its charge state is conserved during the decay time. The direct carbon L-shell ionization probability is about 0.48 for a single electron in 100 MeV

Ne^{9+} +C collisions. It follows that the most probable K-Auger states in an insulator are 1s2s^2, 1s2s2p and 1s2p^2. The Auger energy for the transition from 1s2s2p to 1s^2 is 22.6 eV lower than the energy of the corresponding diagram lines. If the initial state is given by a four-electron configuration, the Auger energy is reduced by only about 10 eV. In a conductor, recombination can take place before the Auger transition. This results in a population of the 1s2s^{2}2p^2 initial state and leads to an energy shift with respect to an insulator. In total one may estimate a shift of less than 30 eV (the most probable energy shift is 23 eV) due to suppressed recombination and the presence of either hydrogen or carbon atoms near the Auger atom (chemical shift). The maximum observed shift is too large to be justified by the effects discussed above. Thus, the origin of the remaining shift of about 45 eV must be explained.

Using a simple Monte Carlo approach for the production and transport of electrons we calculated the motion of electrons inside the target for the investigated collision system (Schiwietz et al.,1992). The track potential $V_T(\vec{R})$ was computed from the spatial charge density of electronic vacancies $\rho_v(\vec{r})$ for all bands and from the charge density of stopped electrons $\rho_s(\vec{r})$ via

$$V_T(\vec{R}) = \int dr \frac{\rho_v(\vec{r}) - \rho_s(\vec{r})}{|\vec{R} - \vec{r}|}; \quad for |\vec{R} - \vec{r}| > 1.5 \text{ Å} .$$

The computed maximum potential inside a PP foil of 10,000 Å is about 35 eV. This result is a accord with unpublished data by A. Akkerman (1992). Since carbon Auger electrons may escape the surface only from a depth of up to 20 Å, an average Auger energy shift of 20 eV may be extracted from the results of the calculation. Our theoretical estimate does neither incorporate quantum mechanical three-body cross sections for the electron production nor collective excitations in the electron energy-loss or the (self-) interaction of directly ionized electrons with the track potential. Estimates for these effects show that an Auger energy shift between 15 and 60 eV is consistent with a track potential.

From the above discussion it becomes clear that the measured maximum Auger energy shift is mainly due to the track potential (about 45 eV) and that the fluence dependence of the shift is correlated to recombination processes in the carbonized environment of a track. Since the solid is completely carbonized in the limit of high fluences, the PP Auger spectra become nearly identical to the corresponding carbon foil spectra. In this limit recombination takes place within less than 100 a.u., i.e., before the Auger decay. Therefore the carbon K-Auger energies from solid carbon, irradiated PP at high fluences and singly ionized atomic carbon are equal to within less than 5 eV.

CONCLUSIONS

In summary, our present understanding of electron ejection for fast ions interacting with solid and gaseous targets has been reviewed. Emphasis is given to investigations of multiply differential electron ejection cross sections and probabilities. Especially electron spectroscopy provides a sensitive tool to determine the ranges of validity of different ion-atom collision theories. From comparisons with experimental data it becomes clear that classical collision theories may be applied for intermediate incident energies (about 100 to 300 keV/u) and only as long as no impact parameter selection is performed. This limited range of validity is due to failures in the description of quasi-molecular effects, of long-range dipole interactions and of the initial phase-space distribution. Perturbation theory, on the other hand, is valid only for fast light projectiles (> 2 MeV protons). At lower incident energies the binding effect and different types of multi-step processes come into play. The binding effect leads to a reduction of high energy electrons and is accounted for in perturbed stationary state theories. The dominant types of multi-step processes (at not too low projectile velocities) lead to a forward bending of ejected electrons, which is taken into account in CDW-EIS calculations, and to a flattening of the electron energy spectra. Only atomic-orbital (AO) coupled-channel calculations include all above mentioned effects and seem to be accurate in the whole range of incident energies.

The single ionization mechanisms are in most cases even quantitatively well understood, whereas our knowledge about the double-ionization process is still poor. This represents a challenge to theoreticians, that have to account for several new effects (dynamic screening, shake-off, TS-1 and additional correlation terms) in their models. It should be noted here that the treatment of multi-electron processes is usually easier, since mean field effects dominate the transitions. On the experimental side, however, there is a need for sophisticated coincidence techniques to identify the different basic two-electron processes.

In the case of electron ejection from foils only fast heavy ions (about 5 MeV/u) have been considered in this work. Results of a simple electron-transport model (SELAS) are in reasonable agreement with our experimental data. The remaining discrepancies, however, call for improved transport models which should also not rely on perturbation theory. It is emphasized, that the electron spectra taken for fast projectiles differ significantly from those taken previously for intermediate incident energies. The mean electron energies are much higher (about 800 eV in the case of forward ejection) and the ratios of forward to backward emission (about 3) are approximately twice the ratio determined for lower incident energies (1 MeV/u). A corresponding forward/backward asymmetry of Auger-electron intensities due to electron-electron collision cascades has also been observed for high incident energies (Schiwietz et al., 1988). An investigation of target Auger-electron spectra from polypropylene foils has shown significant Auger energy-shifts (about 70 eV) as a function of fluence. It was verified that this shift was not due to macroscopic charging-up effects. A contribution of 45 eV could be assigned to the slowing down of Auger electrons in the presence of the nuclear track potential. This value agrees with theoretical estimates for the maximum track potential near the surface. More refined models are, however, needed to relate this result to the plasma effect in solid-state particle detectors or to the ion-induced plastic deformation of materials.

ACKNOWLEDGEMENTS

We are much indebted to P.L. Grande for helpful comments on the present work.

REFERENCES

R. Abrines and I.C. Percival, Proc. Phys. Soc. **88**, 861-83 (1966)

A. Akkerman, private communication (1992), see also the contribution in these proceedings

J. Bang and J.M. Hansteen, Kgl. Dan. Vidensk. Selsk. Mat. Fys. Medd. **31**, No.13 (1959)

D.R. Bates and G. Griffing, Proc. Phys. Soc. **A66**, 961 (1953)

S.D. Berry, A. Glass, I.A. Sellin, K.-O. Groeneveld, D. Hofmann, L.H. Andersen, M. Breinig, S.B. Elston, P. Engar, M.M. Schauer, N. Stolterfoht, H. Schmidt-Böcking, G. Nolte, and G. Schiwietz, Phys. Rev. **A31**, 1392-98 (1985)

L.G.J. Boesten, T.F.M .Bonsen and D. Banks, J. Phys. **B8**, 628-637 (1975)

J. Burgdörfer and C. Bottcher, Phys. Rev. Lett. **61**, 2917 (1988)

J. Cazaux, see contribution in these proceedings (1992)

S. Datz, C.D.Moak, O.H. Crawford, H.F. Krause, P.F. Dittner, J. Gomezdel Campo, J.A. Biggerstaff, P.D. Miller, P. Hvelplund and H. Knudsen, Phys. Rev. Lett. **40**, 843 (1978)

N.C. Deb and D.S.F. Crothers, J.Phys. **B24**, 2359-65 (1991)ICPEAC 1991, International Conference on the "Physics of Electronic and Atomic Collisions", Brisbane, Australia 1991, ed. by W.R. MacGillivray, I.E. McCarthy and M.C.Standage (Adam Hilger Publishing, Bristol, 1992)

P.D. Fainstein and R. Rivarola, J. Phys. **B20**, 1285 (1987)

M.R. Flannery and K.J. MacCann, Phys. Rev. **A8**, 2915 (1973)

W. Fritsch and C.D. Lin, J. Phys. **B15**, 1255-1268 (1982)

W. Fritsch and C.D. Lin, Phys. Rev. **A27**, 3361-3364 (1983)

P.L. Grande and G. Schiwietz, submitted for publication (1992)

J.H. McGuire, Advances in Atomic, Molecular and Optical Physics, Vol. 29 217-315 (1991)

D. Hasselkamp, "Particle Induced Electron Emission II", Springer tracts of modern physics, Vol. 123, (Springer, Berlin, 1991)

R. Hippler, J. Bossler, and H.O. Lutz; J. Phys. **B17**, 2453 (1984)

R.E. Johnson, see contribution in these proceedings (1992)

S. Klaumünzer, Ming-dong Hou and G. Schumacher, Phys. Rev. Lett. **57**, 850 (1986)

J. Macek, see contribution in these proceedings (1992)

S.T. Manson, L.H. Toburen, D.H. Madison, and N. Stolterfoht; Phys. Rev. **A12**, 60 (1975)

V. Montemayor and G. Schiwietz, J. Phys. **B22**, 2555-65 (1989)

A. Niehaus, see contribution in these proceedings (1992)

H. Platten, G. Schiwietz, T. Schneider, D. Schneider, W. Zeitz, K. Musiol, T. Zouros, R. Kowallik, and N. Stolterfoht, Proceedings of the XV. Conf. of the Phys. of Electronic and Atomic Collisions, ed. by J. Geddes et al., p. 437 (Brighton, 1987); H. Platten, PhD Thesis, Freie Universität Berlin, 1986

C. Rau, see contribution in these proceedings (1992)

M. Rösler and W. Brauer, "Particle Induced Electron Emission I", Springer tracts of modern physics, **122**, (Springer, Berlin, 1991), see also the contribution in these proceedings

H. Rothard, K.-O. Groeneveld, and J. Kemmler, "Particle Induced Electron Emission II", Springer tracts of modern physics, **123**, (Springer, Berlin, 1991), see also the contribution in these proceedings

M. E. Rudd, L.H. Toburen, and N. Stolterfoht, Atomic Data and Nuclear Data Tables **23**, 405 (1979)

A. Salin, J. Phys. **B22**, 3901-3914 (1989)

G. Schiwietz, Phys. Rev. **A37**, 370-376 (1988)

G. Schiwietz, Phys. Rev. **A42**, 296-306 (1990a)

G. Schiwietz, Radiation Effects and Defects in Solids 112, 195-200 (1990b)

G. Schiwietz and P.L. Grande, Nucl. Instr. and Meth. **B69**, 10 (1992)

G. Schiwietz, D. Schneider, J.P. Biersack, N. Stolterfoht, D. Fink, A. Mattis, B. Skogvall, H .Altevogt, V. Montemayor, and U. Stettner, Phys. Rev. Lett. **61**, **2677 (1988)**

G. Schiwietz, J.P. Biersack, D. Schneider, N. Stolterfoht, D. Fink, V. Montemayor, and B. Skogvall, Phys. Rev. **A41**, 6262-71 (1990)

G. Schiwietz, P.L. Grande, B. Skogvall, J.P. Biersack, R. Köhrbrück, K. Sommer, A. Schmoldt, P. Goppelt, I. Kádár, S. Ricz, and U. Stettner, Phys. Rev. Lett., in press (1992)

D. Schneider, D. DeWitt, A.S. Schlachter, R.E. Olson, W.G. Graham, J.R.M owat, R.D. DuBois, D.H. Loyd, V. Montemayor, and G. Schiwietz, Phys. Rev. **A40**, 2971-75 (1989), and references therein

D. Schneider, G. Schiwietz, and D. DeWitt, submitted for publication (1992)

J. Schou, Phys. Rev. **B22**, 2141 (1980), see also the contribution in these proceedings (1992)

B. Skogvall and G. Schiwietz, Phys. Rev. Lett. **65**, 3265 (1990) and references therein; ibid Phys. Rev. A in print.

Z. Sroubek, see contribution in these proceedings (1992)

R.C. Stabler, Phys. Rev. **133**, A1268 (1964)

N. Stolterfoht, D. Schneider, J. Tanis, H. Altevogt, A. Salin, P.D. Fainstein, R. Rivarola, J.P. Grandin, J.N. Scheurer, S. Andriamonje, D. Bertault, and J.F. Chemin, Europhys. Lett. 4, 899 (1987)

S. Tougaard and P. Sigmund, Phys. Rev. **B25**, 4452-66 (1982)

P. Varga and HP. Winter, "Particle Induced Electron Emission II", Springer tracts of modern physics, **123**, (Springer, Berlin, 1991), see also the contribution in these proceedings

L. Vriens, in Case Studies in Atomic Physics, edited by E.W. Mc. Daniel and
    M.R.C. Mc Dowell (North-Holland, Amsterdam, 1969), 1, p. 335.
U. Wille and R. Hippler, Phys. Rep. **132**, 129-260 (1986)
U. Wille, Phys. Rev. **A45**, 3004-24 (1992)
L. Willets and S. J. Wallace, Phys. Rev. **169**, 84, (1968)
T.G. Winter, Phys. Rev. **A25**, 697-712 (1982)

ELECTRON EMISSION FROM SWIFT HYDROGEN CLUSTER INTERACTION WITH THIN CARBON FOILS

Hermann Rothard[1], Jean-Paul Thomas[1], Joseph Remillieux[1],
Jean-Claude Poizat[1], Robert Kirsch[1], Karl-Ontjes Groeneveld[2],
Mireille Fallavier[1], and Denis Dauvergne[1]

[1]Institut de Physique Nucléaire de Lyon,
 IN2P3-CNRS / Univ. Claude Bernard, F-69622 Villeurbanne Cedex
[2]Inst. f. Kernphysik, J.W. Goethe Univ., D-6000 Frankfurt am Main 90

1. INTRODUCTION AND EXPERIMENT

Electron emission yields from solids are a possible indirect measure of the energy loss of charged particles[1,2,3]. Most of the theoretical models on particle induced kinetic electron emission from solids[1] consider the average yield γ of electrons per projectile to be proportional to the electronic stopping power S_e of the medium, i.e.

$$\gamma = \Lambda \, S_e \tag{1}$$

For protons and even for heavy ions, where tabulated stopping power values are available[4], it has been shown that electron yields are proportional to the energy deposited near the surface[2,5], but it is an open question whether such a simple relation is also valid for molecular ions[3] and, more generally, for clusters.

Recently, first studies of the energy loss of large (n>3) hydrogen clusters H_n^+ in thin carbon foils have been reported[6]. It is clear that for projectiles such as clusters, the structure of which is modified as they penetrate into a solid, much more information on collective effects is obtained when, rather than the total yield, yields of backward and forward emission from entrance and exit surfaces of a thin foil, respectively, are measured separately.

Therefore, we have measured electron emission yields from the entrance- (*backward* yield γ_B) and exit (*forward* yield γ_F) surfaces of thin carbon foils bombarded with swift (E/p = 20-600 keV/p) hydrogen clusters (H_n^+, n = 1-13) in

Ionization of Solids by Heavy Particles, Edited by
R.A. Baragiola, Plenum Press, New York, 1993

standard vacuum (p<1μTorr). As the results do not differ significantly for different targets (760, 1045 and 1100 Å thick, assuming that the density of carbon foils is 1.65 g/cm³), in the following we present averaged electron yield values.

The experimental setup is shown in. fig. 1. The hydrogen cluster beams are delivered by a 600 keV Cockroft-Walton accelerator equiped with a cryogenic cluster source[6]. The singly charged clusters are analysed according to their mass and energy by an electrostatic- and a magnetic analyser. The emitted electrons are collected by

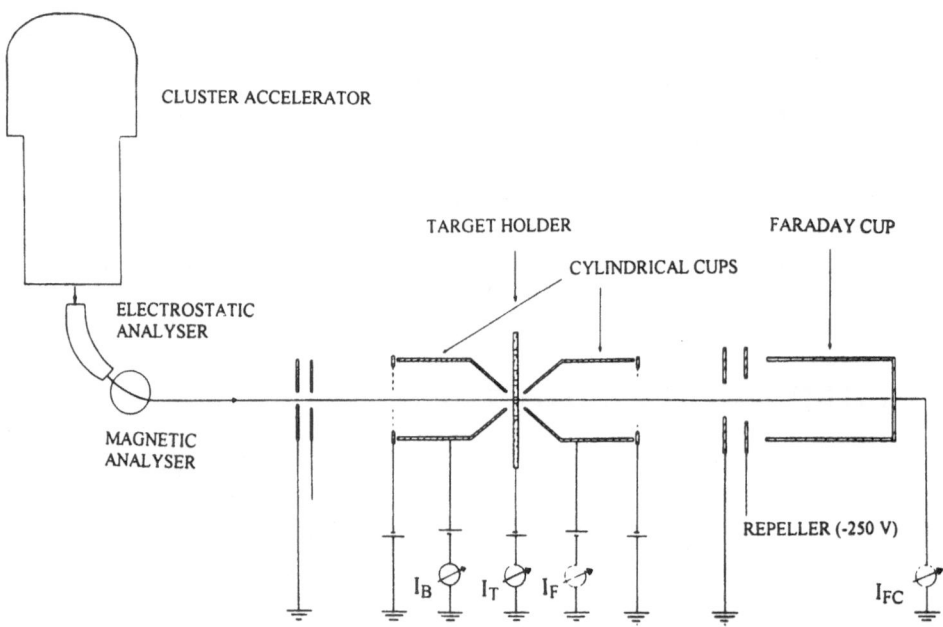

Figure 1. The experimental setup (see text).

two cylindrical metal cups up- and downstream the target foil (backward- and forward electron current I_B and I_F, resp.). Also, the target current (I_T) induced by the beam can been measured.

The electron yields can easily be calculated by measuring the cluster beam current I_{FC} (with target foils removed from the beam) with a Faraday cup. To ensure that nearly all of the emitted electrons are detected, the cylinders were held at a positive potential of +45 V, whereas the target itself as well as two grids in front of the cylinders were held at -45 V. This experimental arrangement as well as the data evaluation procedures have been described in detail elsewhere[3].

2. PROTONS: VELOCITY DEPENDENCE

Fig.2 shows the observed energy dependence of *total* electron yields, i.e. $\gamma_T = \gamma_B + \gamma_F$ (top), and the "material parameter", $\Lambda_T = \gamma_T/S_e$ (bottom), for *protons*. The stopping power S_e values are taken from Ziegler, Biersack, and Littmark[4]. We have included electron yield data for the high energy range (E = 0.4-9.5 MeV) from

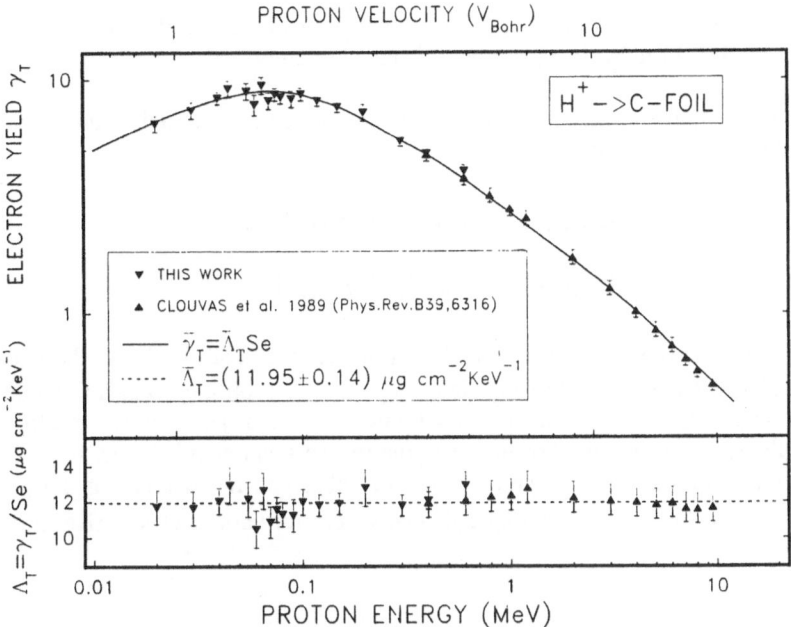

Figure 2. Energy dependence of the total electron yield γ_T (top) and of the "material parameter" $\Lambda_T = \gamma_T/S_e$ (bottom) for protons. The stopping power (S_e) values are taken from Ziegler, Biersack and Littmark[4]. The dashed line shows the mean value of the "material parameter" as indicated in the figure, and the full line shows the corresponding total electron yield.

Clouvas et al.[5] It is important to note that both data sets have been obtained under similar vacuum conditions and with thin carbon foils produced in the same laboratory.

Over the wide energy range of 20 keV $\leq$ E $\leq$ 9.5 MeV the electron yields and the stopping power are found to be proportional within a few percent! Thus, eq.(1) is valid at least for *proton* impact.

3. HYDROGEN CLUSTERS: VELOCITY- AND SIZE DEPENDENCE

A common practice to quantify the behaviour of composite projectiles such as molecular or atomic clusters during interaction with matter is to refer to deviations from the behaviour of the single components. For example, to study the stopping power of matter for hydrogen clusters in comparison to the stopping power for protons of the same velocity, it is convenient to introduce the ratio

$$R(n,v) = S_e(H_n^+) / nS_e(H^+) \qquad (2)$$

Recent studies[6,7] have shown that the stopping power ratio of carbon for hydrogen clusters depends on both cluster velocity v and size n. If no collective effects are present, i.e. for individual protons, R is equal to unity.

In analogy to the stopping power ratios (eq.2) we define backward- and forward electron yield ratios

$$R_B(n,v) = \gamma_B(H_n^+) / n\gamma_B(H^+) \qquad (3a)$$

$$R_F(n,v) = \gamma_F(H_n^+) / n\gamma_F(H^+) \qquad (3b)$$

The experimental results are shown in figs. 3-5. Fig. 3 shows the *backward-* (left half) and *forward-* (right half) electron yields per proton (γ/n) as a function of the cluster size n. As can be seen from this figure, backward- and forward electron yields are of the same magnitude. For instance, the forward emission is higher by about 20% for incident protons. Figs. 5 and 6 show the observed ratios $R_B(v)$ and $R_F(v)$ as a function of the hydrogen cluster specific energy E/p for various cluster sizes $n = 2,5,7$.

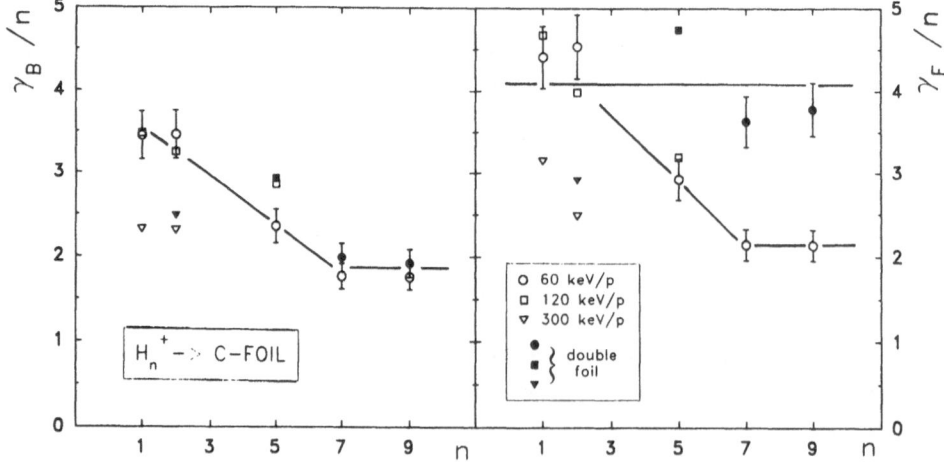

Figure 3. Backward- (left half) and forward- (right half) electron yields per proton (γ/n) as a function of the cluster size n for double foils (full symbols) and single foils (open symbols) for different energies of $E/p = 60$, 120 and 300 keV . The lines are drawn to guide the eye.

3.1 Backward Electron Yields

In the following, we compare the stopping power ratio R (eq.2) to the backward electron yield ratio R_B. R(v) has been found to increase with increasing velocity: Below a "critical velocity" v_{cr}, R(v) values below unity have been observed[6], whereas above v_{cr}, R(v) values larger than one have been reported[6], or predicted[7,8]. v_{cr} is approximately equal to the Fermi velocity of the nearly free conduction electrons (for glassy carbon foils, v_{cr} corresponds to E/p $\cong$ 50 keV/p). The magnitude of the deviation of R(v,n) from unity increases with the cluster size n, and has been shown[6] to reach a saturation value at n $\cong$ 9. Similar effects have been observed by measuring secondary electron yields induced by H^+, H_2^+ and H_3^+ molecular ions up to several MeV/u [1-3,9].

Like the stopping power ratio R, the backward electron yield ratio R_B increases with increasing specific energy E/p. However, a R_B-value larger than one is found only for E/p as high as at least 150 or 300 keV/u. Indeed, with thin foil targets, $R_B > 1$ has only been observed at higher energies of at least 600 keV/u[1-3,9]. These results are in agreement with backward electron yield data[10] for Au, Mo and stainless steel targets obtained with H_n^+ (n=1...9).

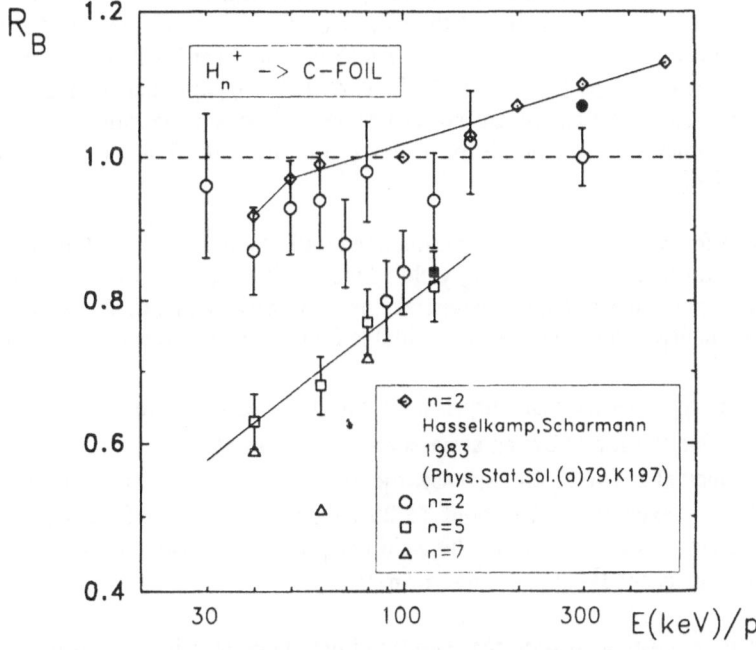

Figure 4. Specific energy (E/p) dependence of the backward electron yield ratio (eq.3a) for various cluster sizes (n=2, 5, 7). The lines are drawn to guide the eye. Full symbols correspond to a double foil (see text).

For comparison, in fig. 4 we have included R_B-values obtained for H_2^+ in ultrahigh vacuum with sputter-cleaned graphite targets by Hasselkamp and Scharmann[11]. It is interesting to note that in this case, the critical velocity $v_{cr}(R_B=1)$ is much lower (80 keV/u) than for the glassy carbon foil targets in standard vacuum conditions. Probably, this difference results from different target densities and structure, in particular close to the surfaces[1].

A more detailed comparison of the energy dependence of R and R_B indicates that the relation of hydrogen cluster induced backward electron yields and the stopping power of carbon foils (related to the *production* of electrons) seems to be more complicated than a simple proportionality (eq.1). Other processes which could lead to a reduction of electron yields must be taken into account. Possibly, the *transport* and/or the *transmission* of electrons through the surface are modified in the vicinity of a swarm of collectively interacting protons.

The magnitude of the molecular- or cluster effect as measured by the backward electron yield ratio R_B increases with increasing n (fig.3, left half, compare also fig.4). A saturation is reached[12] at $n \cong 9$, which is in agreement with the theoretical and experimental results[6] for the stopping power ratio R.

3.2 Forward Electron Yields

Due to Coulomb explosion and multiple scattering, the components of a cluster should have lost their correlation at the exit side of thick enough foils and should act like independent protons. The validity of this assumption can easily be checked with a double foil arrangement (two foils of about 400 Å are mounted with a spacing of 1mm). This assures that all protons are well separated when penetrating the second foil after having undergone Coulomb explosion in the first one. This is demonstrated in fig.3 (right half).

The forward yields per proton obtained with the double foils (full symbols) do not depend on n, whereas a strong yield reduction is observed for single foils (open symbols), in particular at large cluster sizes n=7 and n=9 (compare also figs. 4 and 5). Of course, the R_B-values obtained with double foils and single foils are the same.

From fig.5 we see that, surprisingly, forward electron yield ratios R_F are about 0.7 for n=5 and 0.9 for n=2 even at an energy as low as 40 keV/p after penetration of targets as thick as $\cong 950$ Å. In this case, the mean distance between the protons at the exit surfaces is expected to be about 15-20 Å[8]. As can be seen from fig.3, electron yield supressions of about 50% of the yield per proton are observed for n=7 and n=9. A saturation is reached[12] at about n=7 or n=9.

It is interesting to note that similar effects have also been reported with H_2^+ and H_3^+ traversing thin foils (C, Al, Ti, Ni, Cu) of about 500 Å thickness in ultrahigh vacuum[3].

4. OUTLOOK: TARGET THICKNESS DEPENDENCE

Apparently, as far as electron emission is concerned, the proton swarm keeps some correlation over unexpectedly large distances even at low projectile velocities. Therefore, it is necessary to study the target thickness dependence of the electron yields as well as that of the electron yield ratios (eq.3).

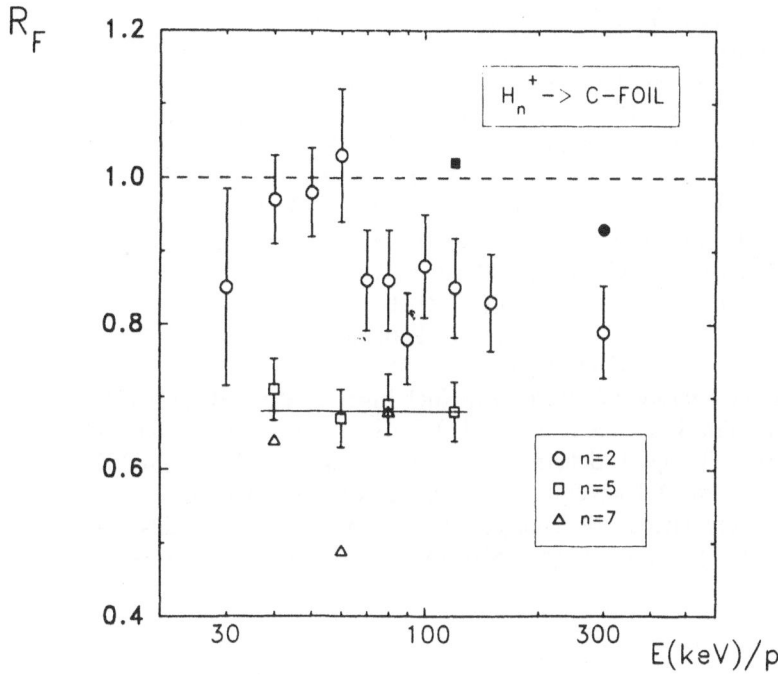

Figure 5. Specific energy (E/p) dependence of the forward electron yield ratio (eq.3b) for various cluster sizes (n=2, 5, 7). The line is drawn to guide the eye. Full symbols correspond to a double foil (see text).

Preliminary results with a variety of targets (in a thickness range between about 170 and 2800 Å) indicate that *forward* yield reductions of about 25% are still present with the thickest carbon target studied (for n > 5 evan at 40 keV/p).

Backward yields do not depend on the target thickness for targets thicker than about 300 Å. Further studies (also with thin copper- and gold foils) of electron *yields* and electron *energy distributions* are in progress.

ACKNOWLEDGEMENTS

We would like to thank Annick Billebaud for her assistance in performing the experiments and evaluating the data. Also, we thank the Lyon cluster accelerator staff for delivering the cluster beams. H.R. acknowledges a post-doc grant from Deutsche Forschungsgemeinschaft (DFG, Bonn).

REFERENCES

1. D. Hasselkamp, H. Rothard, K.O. Groeneveld, J. Kemmler, P. Varga, H. Winter
        "Particle induced Electron Emission" (G. Höhler, E.A. Niekisch, Eds.)
        Springer Tracts in Modern Physics 123 (1991)
2. H. Rothard, J. Schou and K.O. Groeneveld, Phys. Rev. A45 (1992) 1701
3. H. Rothard et al., Phys. Rev. B41 (1990) 3959
4. J.F. Ziegler, J.P. Biersack, U. Littmark, "The Stopping and Ranges of Ions
        in Matter", Pergamon, New York (1985)
5. A. Clouvas et al., Phys. Rev. B39 (1989) 6316
6. M. Ray, R. Kirsch, H.H. Mikkelsen, J.C. Poizat, J. Remillieux,
        Nucl. Instrum. Meth. B69 (1992) 133
7. I. Abril, M. Vicanek, A. Gras-Marti, N. R. Arista,
        Nucl. Instrum. Meth. B67 (1992) 56
8. M. Farizon, N.V. de Castro Faria, B. Farizon-Mazuy, M.J. Gaillard
        Phys. Rev. A45 (1992) 179
9. D.M. Suszcynsky, J.E. Borovsky, Nucl. Instrum. Meth. B53 (1991) 255
10. Y. Chanut, J. Martin, R. Salin, H.O. Moser, Surf. Science 106 (1981) 563
11. D. Hasselkamp, A. Scharmann, physica status solidi (a)79 (1983) K197
12. H. Rothard, D. Dauvergne, M. Fallavier, K.O. Groeneveld, R. Kirsch , J.-C.Poizat,
        J. Remillieux, J.-P. Thomas, Rad. Eff. and Defects in Solids
        (1992) , (Proceedings of SHIM 92, Bensheim, FRG, May 19-22)

STATISTICS OF HEAVY PARTICLE-
INDUCED ELECTRON EMISSION
FROM A FOIL

A.A. Kozochkina[1], V.B. Leonas[1], V.E. Fine[2]

[1]Institute for Problems in Mechanics
Russian Academy of Sciences, Moscow, Russia
[2]Joint Institute of Nuclear Research, Dubna, Russia

ABSTRACT

The multiple electron emission from a thin (~ 50 Å) carbon foil has been measured for impacting primary $H(H^+), He, O, S$ - atoms with energies in the range of 10-200 keV. The absolute probabilities $P_k, P_{k'}, P_{kk'}$ of the emission of $k, k'(= 0, 1 \ldots)$ numbers of electrons in forward (k), and/or backward (k') direction from the foil have been derived. Using the Monte Carlo approach the processes of internal secondary electron production and transport were simulated. The observed SEE features were successfully reproduced in computations. Thus, the key values of the main processes which determine the peculiarities of heavy particle-induced electron emission from solids can be extracted from their simulations.

I. INTRODUCTION

Secondary electron emission (SEE) resulting from bombardment of solids by fast atomic particles has been studied for a long time[1]. Nevertheless, its differential features i.e. the distribution in the number, energy and direction of secondary electrons emitted from a surface after impact of particles of given mass and energy, still cannot be safely predicted from theoretical models available now[2,3,4].

The unsatisfactory development of theory and the demand for detailed knowledge of the probability of multiple emission of secondary electron (MUSE) for practical applications has stimulated new experiments to study these effects[5,6].

The purpose of the present study is threefold:

- to perform the measurements of MUSE statistics,

- to develop a computational approach of calculate all measurable (along with non-measurable) SEE-features,

Ionization of Solids by Heavy Particles, Edited by
R.A. Baragiola, Plenum Press, New York, 1993

- to fit the calculations to the measurements. In this way the inverse problem solution, can be achieved by deriving the characteristics of main collisional processes that govern the secondary electron emission.

A well suited method for the first objective is the investigation of MUSE from thin foils ($\sim 1\mu g/cm^2$). In this case, energy dependent effects can be separated, because the energy loss of the impacting particle in the foil can be neglected. Moreover, the yields of secondary electrons in forward or backward direction or in both directions simultaneously can be measured directly. Also, the principal atomic process producing the electrons inside the material - the ionization by collision - can be studied directly, because additional processes with small cross sections ($Q < 10^{-17}cm^2$), like multiple ionization or ionization by recoil atoms, may be neglected.

Therefore, an interpretation of the experimental results employing transport models[4,8] is considerably simplified and a realistic, quantitative simulation of MUSE by Monte-Carlo calculations can be achieved.

In Section II the description of an experimental approach and the main parameters of the experiment are presented; Section III describes how to derive absolute values for partial probabilities from measured data and gives a short discussion of primary results of measurement. The following Section IV includes the discussion of MC simulation of MUSE; the comparison of computations and measurements are presented in Section V. The main results of the present study are summarized in the final Section.

II. EXPERIMENT

The experimental method is described in full detail in previous paper[9]. In short, a beam of ions of known mass and well determined, selectable energy between 10 keV and 200 keV is generated in a linear accelerator. Part of the ion beam is neutralized by the residual gas ($10^{-7}\ldots10^{-6}hPa$) on the way from the accelerator to the chamber. While the ions are deflected by electrostatic fields, the neutral particles ($10^3\,s^{-1}$) enter into the instrument (Fig.1) .

A $\sim$ 50 Å thickness carbon foil (supported by a grid of 83% transmissivity) is mounted perpendicular to the beam between two identical electrostatic mirrors (total mirror transmissivity 84%) (Fig.1). Electrons emitted from the foil are accelerated up to $\sim 1keV$, deflected and transported to the detectors D_1 (forward) and D_2 (backward). These specially developed microchannel-plate detectors[10] have rather tight pulse height distributions (FWHM=30 ÷ 40%). Thus, the output signals of D_1 and D_2 are proportional to the number of accepted electrons (compare Fig.2) . This allows to clearly separate the emission events corresponding to a particular number. After the penetration of the foil, the beam particles are detected by a third, commercial microchannel-plate detector D_3. The following standard electronics consisting of CAMAC modules (amplifiers, discriminators, coincidence logic) controlled by a microcomputer is shown in Fig.1 (an additional 8k-resolution pulse height analyzer system was used as well). With this electronic equipment, it is possible to record pulse height distributions simultaneously for both detectors as well as two-dimensional spectra in coincidence or anticoincidence between any two of the detectors with the third detector. With the beam intensity ($10^3\,s^{-1}$) used and the low intrinsic noise, accidental coincidences can be neglected in this system.

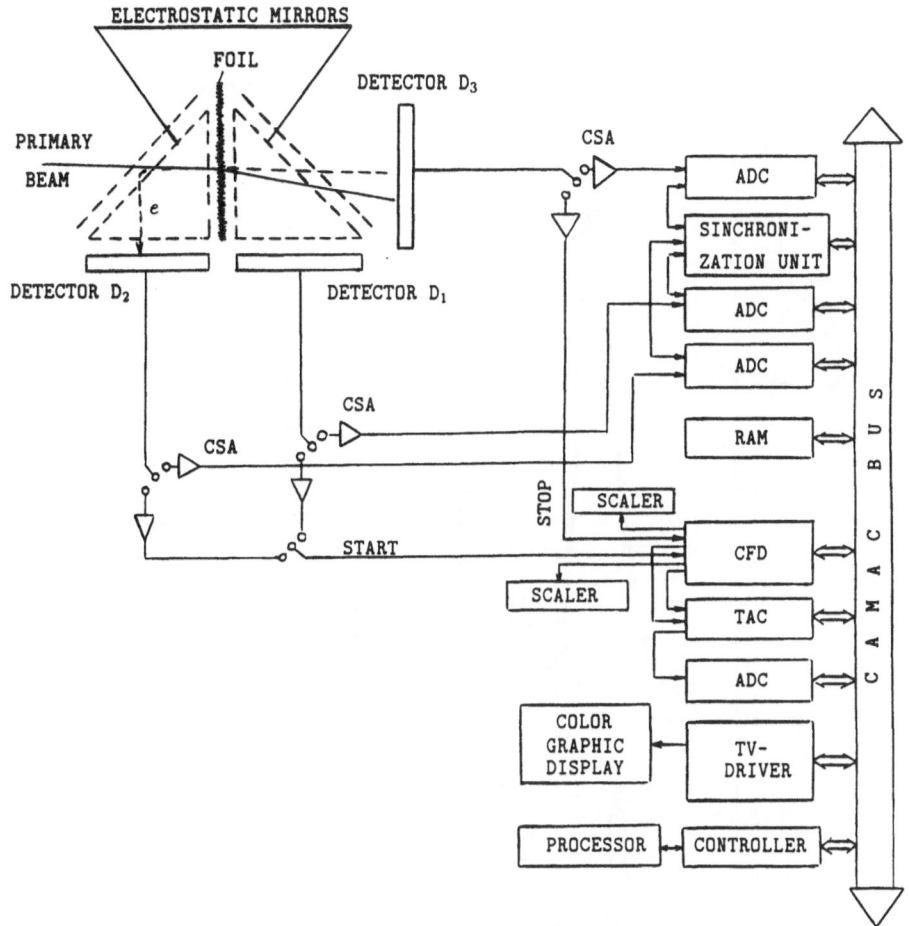

Figure 1. Schematic diagram of the experimental setup. Particles of the primary beam produce secondary electrons in the carbon foil; emitted electrons are accelerated and after deflection by the electrostatic mirrors are directed into the microchannel-plate detectors D_1 and D_2. The primary particle rates are recorded by the detector D_3. CSA-charge sensitive amplifier, ADC-analog-to-digital convertor, RAM-random access memory, CFD-constant fraction discriminator, TAC-time-to-amplitude convertor.

III. MEASUREMENTS AND ANALYSIS

The intensity of the primary beam and its stability have been controlled with a semiconductor detector, which periodically was moved into the beam in front of the measuring device. The beam was sufficiently stable so that its variations during the time of the measurement could be neglected. The absolute intensity I_0 can be determined[11] from the MUSE-measurement itself. Using the count rates I_1 (or I_2) and I_3 in the detectors D_1 (or D_2) and D_3, the coincidence rate $I_{1,3}$ (or $I_{2,3}$) between the two detectors, I_0 can be calculated from

$$I_0 = I_1 \times I_3 / I_{1,3}. \tag{1}$$

This rate was in good agreement with the one measured directly with the solid state detector, taking into account the finite transmissivity of the grids to be traversed by the beam (which had to be used to support the foil and in the electrostatic mirrors).

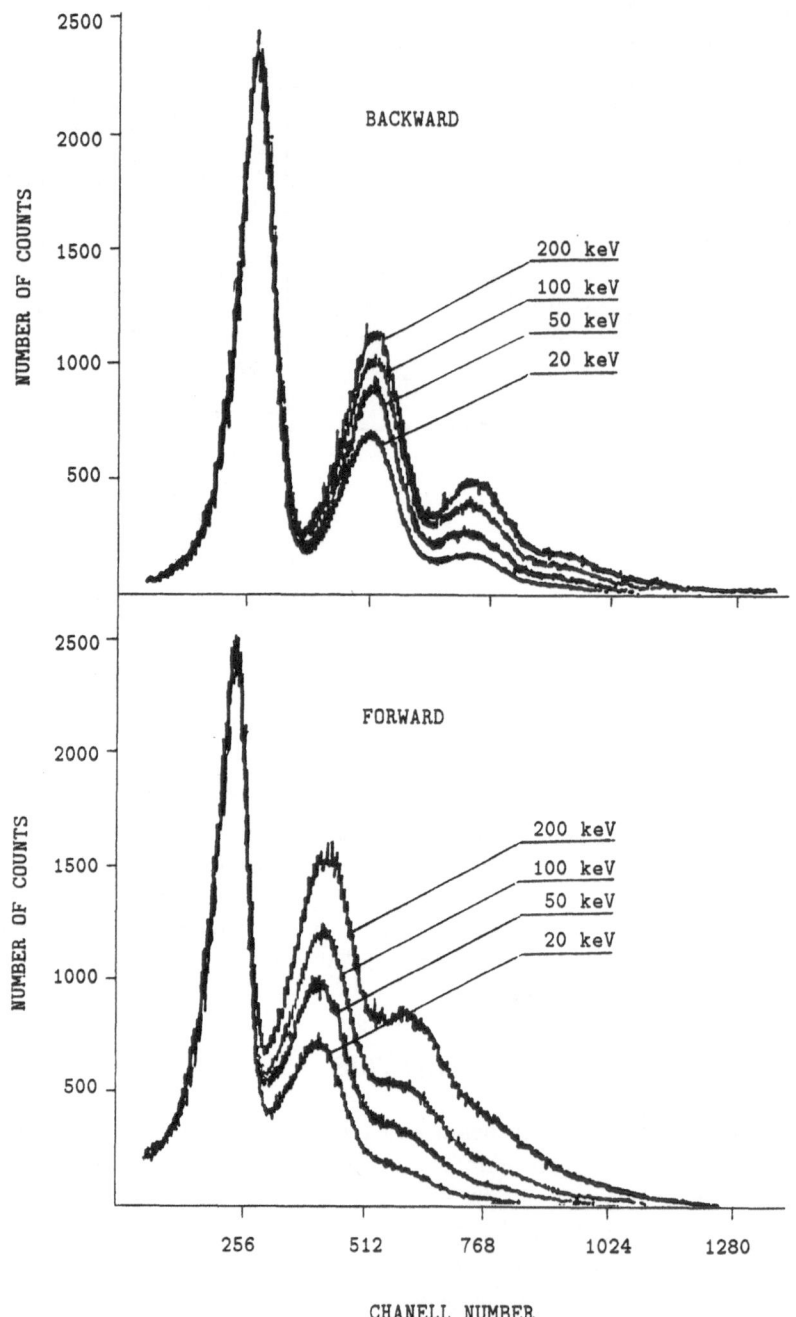

Figure 2. Measured pulse height distributions (output of detector D_1, forward direction and detector D_2, backward) for four different energies of the primary Helium particles. Resolvable peaks result from events with 1,2 or 3 incident electrons, but events with even more secondary electrons are registered.

Fig.2 shows one-dimensional pulse height spectra in detector D_1 (forward) and detector D_2 (backward direction) for four different energies of the primary He-projectiles. Three distinct peaks can be distinguished, resulting from the incidence of 1, 2 or 3 secondary electrons. Events with larger amplitudes indicate the simultaneous incidence of 4 or more electrons.

The spectra shown are normalized at the maximum of the first peak. From the larger numbers of counts in the second and third peak it is obvious that the probability for multiple electron events grows with the energy of the primary particle (with some preference of multiple electron events in forward direction).

Fig.3 visualizes a two-dimensional spectrum of measured coincident events in detec-

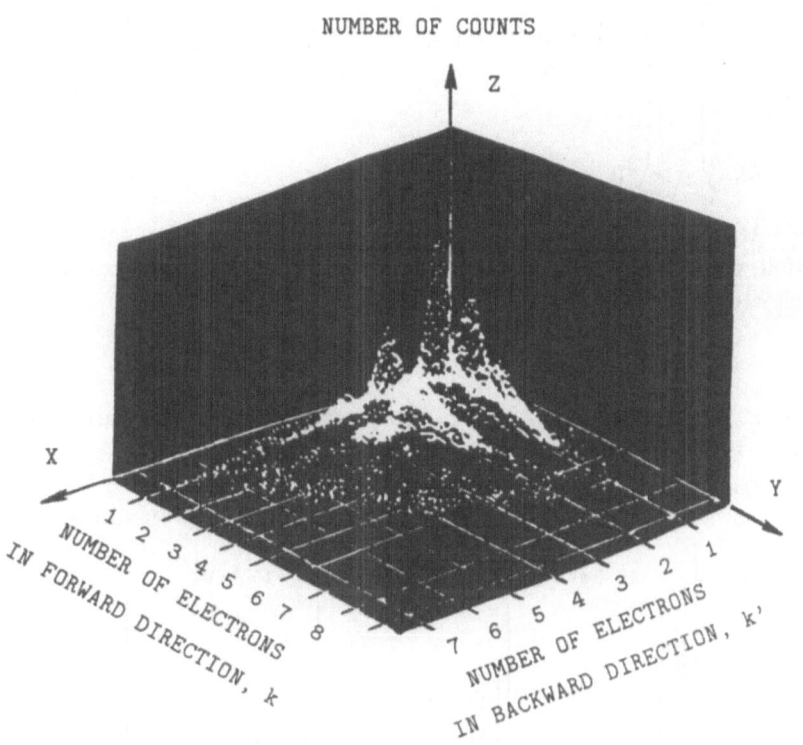

Figure 3. A two-dimensional spectrum of the coincident signals as measured with the detectors D_1 and D_2, showing the multiple electron emission to both sides of the Carbon-foil after penetration by a 200 keV Helium atom.

tors D_1 and D_2 resulting from 200 keV-He-projectiles. The peaks are clearly separated from up to three secondary electrons in either direction. The similar one- and two-dimensional spectra have been measured for energies of 20(30)keV, 50 keV, 100 keV and 200 keV for protons (H^+), Hydrogen (H), Helium (He), Oxygen (O) and Sulphur (S).

The individual peaks of the spectra have been fitted by Gaussian distributions in order to determine the real measured number of events M_k, ($M_{k'}$, $M_{kk'}$) in each peak. Here, $k, (k', kk') = 1, 2 \ldots$ denote the number of secondary electrons emitted in forward (backward, coincident in both) directions, respectively.

The number of events M_0, in which no secondary electrons ($k = 0, k' = 0$) have been produced, was deduced from the known intensity I_0 of the primary beam using the relationship

$$M_0 = M - \sum_{i=1} M_i \qquad (2)$$

(where $M = I_0 \times t$ is the total number of particles penetrated the foil during the measurement time t).

To find the absolute probabilities P_k, $P_{kk'}$, the PA-spectra of Figs.2,3 have to be deconvoluted. This numerical deconvolution procedure takes into account the real transmissivity of the grid system (æ is the total transparency of the grids passed by emitted electrons, μ is the efficiency of electron detection). This deconvolution procedure can be visualized by the following equation

$$
\begin{pmatrix} M_0 \\ M_1 \\ M_2 \\ . \\ . \\ M_n \end{pmatrix}
=
\begin{pmatrix}
1 & (1-\varepsilon) & (1-\varepsilon)^2 & \cdots & (1-\varepsilon)^n \\
0 & \varepsilon & 2(1-\varepsilon)\varepsilon & \cdots & C_{n-1}^k(1-\varepsilon)^{n-1}\varepsilon \\
0 & 0 & \varepsilon^2 & \cdots & C_{n-2}^k(1-\varepsilon)^{n-2}\varepsilon^2 \\
. & . & . & \cdots & . \\
. & . & . & \cdots & . \\
0 & 0 & 0 & \cdots & \varepsilon_n
\end{pmatrix}
\times
\begin{pmatrix} N_0 \\ N_1 \\ N_2 \\ . \\ . \\ N_n \end{pmatrix},
\qquad (3)
$$

where M_k and N_k correspond to the spectra measured by the detector and emitted by the foil, respectively. $\varepsilon = \mu æ$ is the probability for an electron emitted by the foil to be detected, $C_n^k = n!/(n-k)!k!$. Now

$$P_k = N_k/M, \qquad (4)$$

where M is the number of projectiles having traversed the foil during the spectrum acquisition time. The value of $P_{kk'}$ can be calculated accordingly from appropriate values of $N_{kk'}$ and M.

Fig.4 shows these probabilities P_k, $P_{k'}$ derived from the one-dimensional spectra

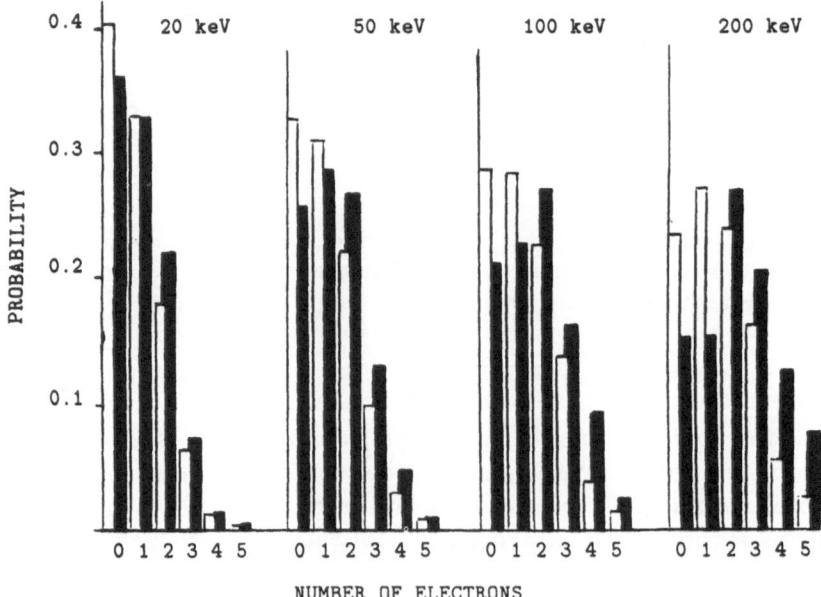

Figure 4. Derived probabilities of the emission of electrons in forward (black) and backward (white) direction induced by Helium atoms with different energies. At higher energies, the development of a directional anisotropy can be seen.

for four energies of primary Helium atoms in forward (black) and backward (white) directions. Probabilities $P_{kk'}$ for coincident emissions are shown in Fig.5 for 200 keV Helium atoms. In addition, in Tab.1 we list the values of P_k, $P_{k'}$ for all projectiles investigated. The following trends can be seen from the data:

1. The probability P_k, $P_{k'}$ - values for H - and H^+ - projectile are almost identical, differing by a few percent only within the energy range studied. This allows to conclude that the energy of the stripped electron is so low ($E_e \leq U$) that it can not escape from the foil.

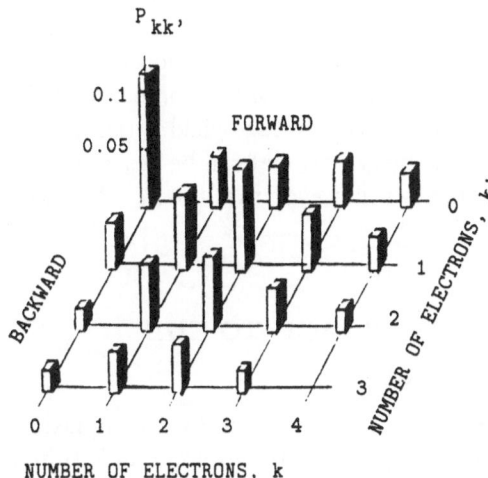

Figure 5. Derived $P_{kk'}$ probabilities of the simultaneously emission of k, k' electrons in both directions.

Table 1. The measured probabilities for forward (P_k) and backward ($P_{k'}$) emission (%)

E_B,keV	γ'	γ	$P_{0'}$	P_0	$P_{1'}$	P_1	$P_{2'}$	P_2	$P_{3'}$	P_3	$P_{4'}$	P_4
H 20	1,06	1,13	39,7	36,5	31,9	32,7	18,0	21,2	7,0	7,5	0,15	0,17
50	1,16	1,45	36,0	27,3	30,2	27,4	21,1	26,2	8,8	13,1	2,6	4,7
100	1,22	1,67	32,3	21,7	32,9	26,2	21,7	27,2	9,0	16,1	2,9	6,1
200	1,15	1,37	36,4	27,9	31,0	31,3	19,8	24,1	9,0	10,9	2,5	4,5
He 20	0,98	1,07	40,4	36,1	33,0	32,8	18,0	21,9	6,4	7,3	1,4	1,5
50	1,24	1,47	32,7	25,7	31,0	28,5	22,1	26,7	9,9	13,0	2,9	4,6
100	1,44	1,82	28,7	21,1	28,4	22,6	22,7	27,1	13,9	16,3	4,0	9,4
200	1,67	2,33	23,4	15,1	27,0	15,3	23,9	26,8	16,3	20,5	5,6	12,6
O 30	1,35	1,44	34.7	32,9	25,0	23,6	21,4	21,9	11,7	13,5	4,6	4,9
50	1,52	1,63	30,9	29,2	24,5	21,7	21,9	23,1	12,9	14,8	5,9	6,8
100	1,78	2,04	25,8	22,0	22,4	19,5	23,1	22,4	14,7	17,2	8,0	11,0
200	1,99	2,38	21,6	18,5	22,6	17,7	22,8	19,3	15,8	18,2	8,5	12,1
S 30	1,19	1,30	43,3	43,2	21,6	20,1	19,1	18,4	9,2	11,5	4,4	4,6
50	1,52	1,77	31,5	28,7	23,4	18,5	22,2	22,3	13,3	16,9	5,7	8,3
100	1,86	2,31	25,2	21,6	20,7	12,4	23,8	22,9	16,3	18,6	7,7	13,0
200	2,16	2,81	19,2	14,8	18,5	10,4	23,4	20,0	18,6	20,8	10,4	15,2

2. The probability $(I - P_{00})$ that at least one secondary electron is emitted, is of the order of 80% to 90% and increases with the energy of the primary particle.

3. As expected, the number of secondary electrons grows with the energy and the Z-value of the primary particle.

4. At lower energies, there is no significant preference for an emission in forward or backward direction. At higher ones, however, in cases where two or less secondary electrons are emitted, there is a slight preference for backward emission, while in cases with more than two secondary electrons, more electrons are emitted in forward than in backward directions.

The probability correlation coefficient $\rho_{kk'}$ and the total charge correlation coefficient $\hat{\rho}$ can be calculated according to the standard definition. The interesting result is that there is no significant correlation between emission events with $k, k' > 0$ as well as between charges ejected from both sides of the bombarded foil.

Assuming Poisson statistics, the average yield γ (an average number of secondary electrons per penetrating primary particle) can be described by $\gamma = -\ln P_0$. Investigation of the data of Tab. 1 shows, however, that the P_k, $P_{k'}$ found here are not consistent with Poisson statistics.

IV. MONTE CARLO APPROACH TO THE DESCRIPTION OF MUSE

Both the diversity of physical processes and the complexity of each one involved in the SEE-phenomenon make it difficult to develop a quantitative theory. At present a nearly complete understanding of the key processes that accompany energetic atomic particle penetration of matter has been obtained. Consequently, approaches to calculate the main characteristics of these processes were developed. Thus a complete numerical MC method opens the way to a quantitative description of MUSE. The experimental results shown above can be fitted using a MC simulation of the transport of both primary particles (fast heavy atoms) and the internal secondary electrons.

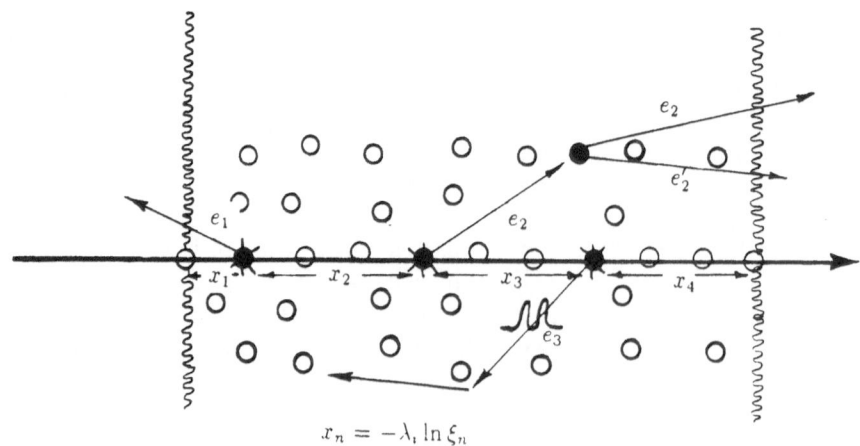

$$x_n = -\lambda_i \ln \xi_n$$

Figure 6. Qualitative picture of fast projectile penetration of a thin foil and internal secondary electron migration to the surface.

MC computations are seriously simplified in the case of bombardment of thin foils, because the trajectories of atoms can be taken as straight lines, and the energy loss can be neglected, only the projectile charge-state changes.

According to the widely accepted view to SEE of solids i.e., the three-step-process scheme, in our model we treat MUSE as a result of the consequent (and independent) steps:

1. The ionization of target atoms by the projectile.

2. Internal secondary electron migration to the solid-vacuum boundary including the possibility of cascade electron production.

3. Electron escape as a result of getting over the surface barrier of height U. The qualitative picture of these three steps is presented by Fig.6 .

The production of internal secondary electrons takes place after a path length $x_n (n = 1, 2, 3 \ldots)$. Its value is determined by the mean free path for ionization (λ_i) and can be found from

$$x_n = -\lambda_i \ln \xi_n \tag{5}$$

(ξ_n are uniformly distributed random numbers in the interval $\xi \in (0, 1)$). Due to the relation of λ_i to the ionization cross-section $Q_i(\lambda_i = (Q_i n_0)^{-1})$ which, in turn, depends on the projectile charge-state, the values of x_n can be simply generated by taking into account the real charge-state of the projectile inside the target. The internal secondary electrons are characterized by the position of their birth-place, the direction of ejection and their energy. The two latter parameters can be found from the distributions assumed. In a first approximation, these distributions were taken as independent and similar to those found experimentally by Rudd et al.[12] for ionization of gaseous targets.

Internal secondary electrons while migrating to the surface are subjected to interaction with target matter. This interaction can be subdivided[13] into processes of **binary** and **collective** interaction. The binary interactions include the elastic scattering by target atomic nuclei (cross-section Q_{el}), the electron-hole excitation due to interaction with target atom valence electrons (cross-section Q_{e-h}) and inner-shell electron ionization (cross-section Q_{e-i}). Two last processes are responsible for production of the tertiary and etc. electrons. The collective interaction corresponds to plasmon excitation (effective cross-section Q_{pl}). There is a specific common feature of all inelastic processes mentioned: all excitation functions $Q_{e-i}(E_e), Q_{e-h}(E_e), Q_{pl}(E_e)$ have either real or apparent energy thresholds. For $Q_{e-i}(E_e)$ the value of the threshold is at about 300 eV in the case of carbon. Thus this process can be excluded if we are interested in low-energy secondary electrons only.

For Q_{e-h}, Q_{pl} the effective thresholds (as will be seen later) are close to $E_e = 30 eV$. To take into account the stopping of low ($E_e < 30 eV$) energy electrons, it is useful to employ the scheme of continuous stopping[13]. According to this scheme, the energy of electrons after having travelled between two consecutive elastic collisions has to be lowered by the value $\Delta E = (dE_e/dx) \cdot x_{el}$.

Finally, the escape of an internal electron from the solid into vacuum was treated by assuming planar potential barrier. The escape conditions can be described by simple equations

$$\begin{aligned} E_e \cos^2 \theta &= E_{oe} \cos^2 \theta_0 - U \\ E_e \sin^2 \theta &= E_{oe} \sin^2 \theta_0. \end{aligned} \tag{6}$$

Here, E_e, E_{oe} denote the energies of escaped and internal electrons respectively, and θ, θ_0 – the polar angles (counted from the normal to the surface) for free and internal electrons.

The model described above is capable of being systematically improved. It is easily possible to include any new or more reliable data. The mean free path for ionization (λ_i) is determined by the value of the ionization cross-section. As a first approach, it can be estimated by dividing the foil thickness d by the measured total yield $\bar{\gamma} = \gamma + \gamma'$, i.e.

$$\lambda_i = d/\bar{\gamma}. \tag{7}$$

Ionization cross-section for singly- and doubly-charged ions (e.g. He^+ and He^{2+}) differ by a factor 2-3 as can be seen from the measurements[14]. In our MC simulations this factor was taken as a free parameter.

The energy spectrum of the internal electrons was described by the equation

$$d\sigma/dE_e = A(I + E_e)^{-a}, \tag{8}$$

following to Rudd et al.[12], here A is a normalizing factor, I – the electron binding energy, a – a spectrum rigidity factor ($a = 2$ in the case of ionization of gaseous targets by protons). The same description was used for various projectiles. This is justified by the close similarity of secondary electron energy spectra from graphite for H^+, Ar^+, Kr^+, Xe^+-ions[15].

The energy spectrum has to be truncated at the high-energy tail; the cutoff electron energy $E_{e\,max}$ is determined by the equation

$$E_{e\,max} = \alpha(m_e/m_p)E_B, \tag{9}$$

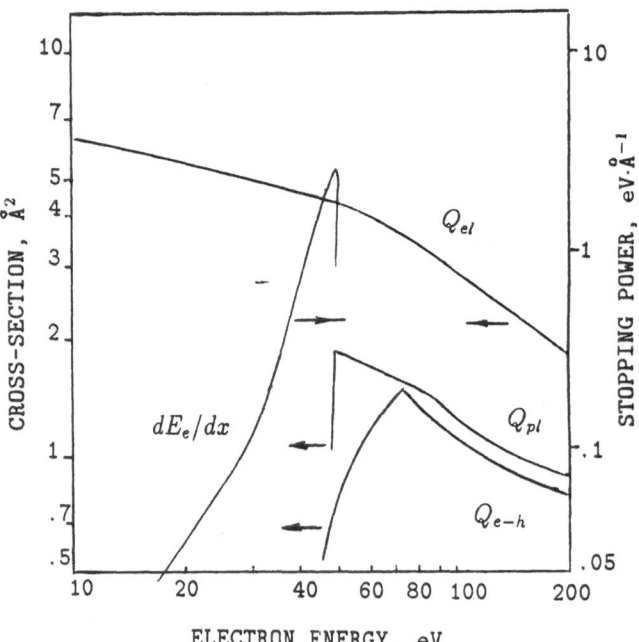

Figure 7. The energy dependencies of the cross-sections and of the stopping-power for internal electrons in carbon.

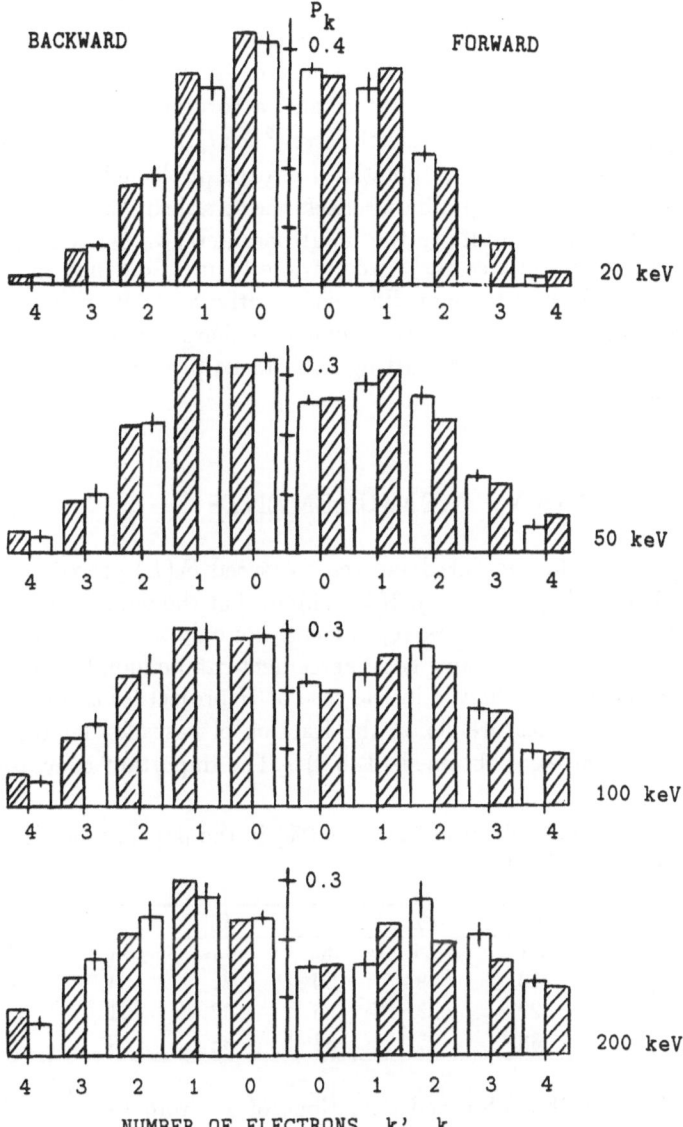

Figure 8. Comparison of measured and computed P_k, $P_{k'}$, values for He-projectiles of various energy. The accuracy of the computed values is better than that indicated by error bars for the measured P_k, $P_{k'}$, values.

with m_e, m_p -electron and projectile masses, E_B - projectile energy, α - factor which depends on the projectile mass.

The angular distribution of internal electrons was again assumed to be similar to the one found by Rudd et al.[12],

$$d\sigma/d\Omega = A_1 \exp(-A_2\theta_0) \cos(\theta_0 A_3) + A_4. \tag{10}$$

A_i are constants whose final values have to be determined by fitting of the measured yield asymmetries. The inelastic and elastic cross-sections used here are shown in Fig.7. They were calculated by Gibrekhterman[16] according to the recommendations of Walker[17] for Q_{el}, of Ritchie[18] for Q_{e-h}, of Ferrel[19] for Q_{pl}. The stopping-power for low energy electrons was found by extrapolation from the published data[16].

This large set of necessary input data reflects the complexity of the SEE-phenomenon. The sensitivity of the MC results to the input parameter values was investigated in a separate series of computations. It is worthwhile to mention that the number of input parameters (taking into account different sensitivity of MC results to them) does not exceed the number of the measurable emission characteristics. Thus, the "inverse problem" solution by means of MC fitting will indeed provide with the unique values of parameters involved.

MC CALCULATIONS VS. MEASUREMENTS

The comparison of the computed and the measured $P_k(E)$ probability distributions is presented in Fig.8 (He-projectiles). It is evident that the model (as discussed above) describes very well the measurements, reproducing not only the yields, but also statistics and asymmetry. Moreover, the same good agreement can be found for the probabilities $P_{kk'}$. Close agreement was achieved in the case of H-projectiles as well. By fitting the MC simulations to the measurements, the fraction of cascade electrons contributing to the partial yields P_k were obtained (Tab.2) . The important conclusion which can

Table 2. Cascade electron fractions (%) in the partial yields P_k, $P_{kk'}$ for *Helium*-projectile.

E,keV	P_1	$P_{1'}$	P_2	$P_{2'}$	P_3	$P_{3'}$	P_4	$P_{4'}$
20	2.2	2.0	3.4	2.5	4.3	3.6	6.9	5.1
50	13	10	18	20	22	24	28	26
100	15	22	23	26	32	31	33	35
200	13	38	25	38	33	39	40	46

be drawn for Tab.2 is that observed deviations of P_k from the Poisson distribution are explainable by the presence of cascade contributions added to the true secondary electrons. Our computations not only allow to fit our stimulations to the measured electron yields and the probability distributions but also to electron energy spectra. An example is shown in Fig.9, where we compare a fitted spectrum for an 80 ÅC-foil bombarded by 300 keV energy H-atoms to a spectrum obtained experimentally[7] with a massive graphite target bombarded by 500 keV protons. The possibility of reproducing exactly the shape of the measured spectrum confirms the adequacy of our MC stimulations and of the input data used.

In this respect, the angular distribution in turn are not so informative because the real target surface structure differs very much from the one used in the simulating

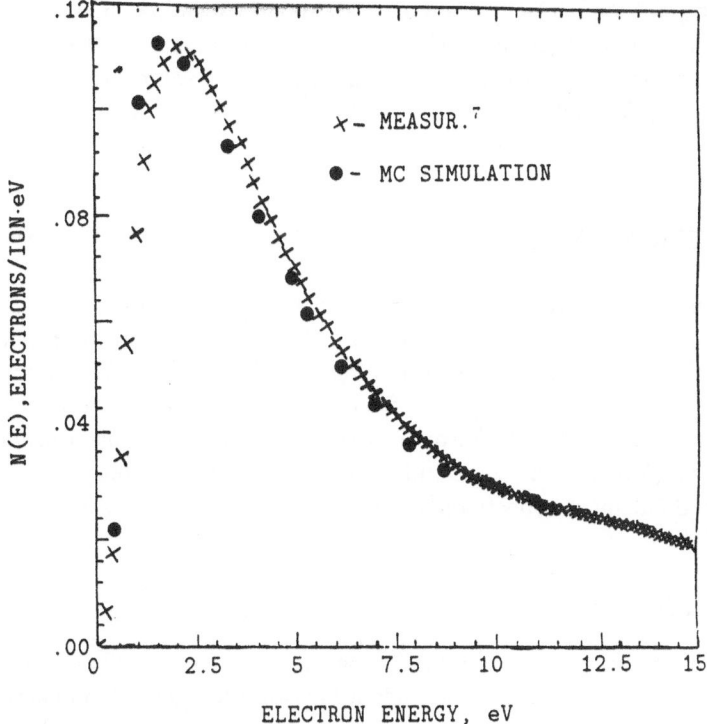

Figure 9. Comparison of a computed energy spectra to a spectrum obtained experimentally[7] (see the text for details).

computations. Concluding, we would like to point out that MC simulation of MUSE enables us to determine some unmeasurable or hardly measurable target characteristics such as the escape depth λ_e et cetera.

VI. SUMMARY AND CONCLUSIONS

The main results of the present study are the following:

1. In the present paper, we have reported on the development of a new experimental approach to investigate MUSE induced by heavy particle bombardment.

2. This approach was used to measure of MUSE statistics for C-foil bombarded by energetic (10-200 keV) protons (H^+) and atoms (H, He, O, S).

3. A Monte Carlo simulation code was developed and used to fit the observed MUSE-features. A close agreement of the computations and the measurements was achieved. This allowed to derive the values of some important quantities which characterize the collision processes in a carbon-foil.

4. It was demonstrated that MUSE-investigations can now be employed for an extraction of detailed information concerning the elementary collision processes in solids.

5. The successful application of our new approach to the particular case of carbon makes it desirable to extend MUSE studies to other target materials. In the such studies simultaneous measurement of as much observable properties as possible have to be performed.

ACKNOWLEDGMENTS

The measurements have been performed at the accelerators of the University of Moscow and the Max Planck Institute, Lindau. We wish to thank both teams, in particular Drs M.Witte and G.A.Iferov and B.Wilken, H.Sommer and W.Boker for their extremely helpful and continuous support. Stimulating discussions with Drs A.Akkerman and A.Gibrekhterman and M.Gruntman in the early phase of this study are appreciated. One of us (V.B.Leonas) gratefully acknowledges very valuable discussion with Dr.H.Rothard about most aspects of this paper.

REFERENCES

1. For comprehensive references on SEE investigations refer to "Particle induced electron emission",I, II Springer Tracts on Modern Physics, 123 (1991); 123 (1992).

2. E.J.Sternglass, Theory of secondary electron emission by high-speed ions,*Phys.Rev.* 108:1 (1957).

3. E.S.Parilis, L.M.Kishinevskii, On the theory of ion-electron emission, *Sov.Phys.Stat.Sol.* 3:885 (1960).

4. J.Schou, Transport theory for kinetic emission of secondary electrons from solids,*Phys.Rev.* B22:2141 (1980).

5. Y.Yamazaki, K.Kuroki, Studies of multiply emitted secondary electrons (MUSE) in charged-particle solid interactions, *Nucl.Instr.and Meth.* A262:118 (1987).

6. G.Lakits, F.Aumayr and H.Winter, Statistics of ion-induced electron emission from clean metal surface,*Rev.Sci.Instr.* 60: 3151 (1989).

7. S.Hippler, *Thesis*, Giessen, FRG (1988).

8. J.Schou, Secondary electron emission from solids by electron and photon bombardment, *Scanning Microscopy*, New York, 2: 607 (1988).

9. M.A.Gruntman, A.A.Kozochkina, V.B.Leonas, The instrument for investigation of statistics of secondary electron emission of thin foils, *Instrum.and Exper.Tech.*3:157 (1989).

10. A.A.Demchenkova, Parameters of assembly of microchannel plates with micrometer gaps, *Instrum.and Exper.Tech.*30:1182 (1987).

11. M.A.Gruntman, V.A.Morozov, H-atom detection and energy analysis by use of thin foils and TOF technique, *J.Phys.E.*15:1356 (1982).

12. W.Q.Cheng, M.E.Rudd, Y.-Y.Hsu, Angular and energy distribution of electrons from 7.5 - 150 keV proton collisions with oxygen and carbon dioxide, *Phys.Rev.A* 40:3599 (1989).

13. A.F.Akkerman, A.L.Gibrekhterman, Comparison of various Monte-Carlo schemes for simulation of low energy electron transport in a matter, *Nucl.Instr.and Meth.* B6 (1985).

14. P.P.DuBois, Coincidence techniques to study electron emission in ion-atom collisions, *Nucl.Instr.and Meth.* B11/12:120 (1985).

15. D.Hasselkamp, *Thesis*, Giessen, FRG (1985).

16. A.L.Gibrekhterman, (in russian) Etrin-package of programs for models of low energy electron transport in heterogeneous matter, *Preprint N83-19* of Kazakh.Academy HEP-Institute, Alma-Ata (1983).

17. D.Walker, Relativistic effects in low energy electron scattering from atoms, *Adv.Phys.* 20:257 (1971).

18. H.Ritchie, Interaction of charged particle with a degenerate Fermi-Dirak electron gas, *Phys.Rev.* 114:644 (1959).

19. R.A.Ferrel, Angular dependence of the characteristic energy loss of electrons passing trough metal foils, *Phys.Rev.* 101:554 (1956).

INCIDENT AND EXIT CHARGE STATE DEPENDENCE OF SECONDARY ELECTRON EMISSION FROM A CARBON FOIL BY THE PASSAGE OF SWIFT OXYGEN AND CARBON IONS

T. Azuma, Y. Yamazaki, K. Komaki, H. Watanabe

Institute of Physics, College of Arts and Sciences
University of Tokyo, Meguro, Tokyo 153, Japan

M. Sekiguchi

Institute for Nuclear Study, University of Tokyo
Tanashi, Tokyo 188, Japan

T. Hasegawa

Department of Engineering, Miyazaki University
Miyazaki, 889-21, Japan

T. Hattori

Research Lab. of Nuclear Reactors, Tokyo Institute of Technology
Meguro, Tokyo 152, Japan

K. Kuroki

National Research Institute of Police Science
Chiyoda, Tokyo 102, Japan

INTRODUCTION

A swift projectile ion travelling in solid targets loses its energy through the excitation of electrons and plasmons. The charge distribution of the projectile ion itself comes close to the state of equilibrium through the charge exchange process with target

Ionization of Solids by Heavy Particles, Edited by
R.A. Baragiola, Plenum Press, New York, 1993

atoms. Primary electrons produced in the direct ionization of target atoms by the projectile ion give birth to "secondary" electrons through a cascade process. Plasmons also decay into electrons [1]. A part of these electrons diffuse to the surface, overcome the surface potential, and escape from the surface. They are generally called "secondary electrons", whose energy is mostly below ~10 eV. The investigation of secondary electrons has a very long history and many review articles have been reported [2][3]. The parameters observed in the secondary electron emission have been usually an average number, the energy distribution the angular distribution, etc.

Recently we have developed a novel detection technique of multiply emitted secondary electrons (MUSE) [4-7] as well as other groups [8-15]. It enables us to detect not only the average number of MUSE but also the number distribution of them. It also enables us to do the measurements with a faint ion beam, and so free from damage or charging up of the target. In the present experiment the exit charges of projectile ions were analyzed by applying an electrostatic deflector, and these projectile ions were detected by a Position Sensitive Detector (PSD). Thus we can get new information, i.e., the number of secondary electrons on the condition that the projectile ions have specific incident and exit charges, by making use of the projectile ion with a specific incident charge and by taking a coincidence with a output signal of the PSD for the purpose of the discrimination of the exit charge. This is impossible by the conventional measurement techniques. We measured both the forward (the direction downstream to the beam) and backward (the direction upstream to the beam) emitted MUSE from a very thin ($\sim$ 80Å) carbon foil by the passage of 1MeV/u O, C with this technique. The present study was peformed from several view points.

First, we have been much interested in the energy loss of multiply charged ions or partially stripped ions in a non-equilibrium condition. The secondary electron yield is closely related to the stopping power. Therefore, it is an appropriate approach to investigate the effect of the incident and exit charges of the projectile ions by monitering MUSE.

The nuclear charge of a partially stripped ion is screened by bound electrons causing a reduced effective charge. According to the Bethe theory [15][16], the rate of the energy loss of a fully stripped ion is proportional to Z_p^2 (Z_p denotes the nuclear charge of the projectile ion). The effect of the screening of partially stripped ions has been taken into account by introducing Z_{eff} which was experimentally determined in place of Z_p.

However, the energy loss of a partially stripped ion is influenced not only by the excitation of target atoms but also by the excitation and ionization of the projectile ions [17], and it is also expected that the energy loss is affected by the electron loss and capture of the projectile ions through X-ray and electron emission. The effect of electron loss and capture by the projectiles is expected to be revealed in the secondary electron emission especially in the condition that the charge state of ions has not reached equilibrium. We adopted target foils so thin that the exit charge distribution of projectile ions has not reached the equilibrium state even at the exit surface. The data concerning the charge distribution of the projectiles in a non-equilibrium condition has been very limited so far. Recently measurements of the projectile charge dependence of the average yield of the secondary electron emission have begun to be reported [18] [19] [20]. In the present study we show the data of MUSE from a very thin foil.

Second, our data involve information on the electron escape depth in the low energy region. It is an important parameter to evaluate the attenuation of the signal intensity of SEM, AES or XPS. But the direct measurement of the electron escape depth is very difficult and available theoretical calculations are not very accurate.

The secondary electron originates near (i.e. less deeper than the escape depth) the entrance (or exit) surface. Thus it should reflect the effective charge Z_{eff} of the projectile near the surface, where the charge state of ions would be close to the incident (or exit) charge. If the target foil is very thin, the region from which the forward emitted secondary electron originates will have much overlap with that corresponding to the backward emitted electron. Then, the MUSE emitted in the forward and backward directions should be similar. Thus, we can obtain the electron escape depth and discuss the region from where the seconday electrons originate by moniterning the foil thickness dependence of the average number of MUSE.

Third, the number distribution of MUSE should inform us of the production process of secondary electrons like the cascade process, which is not noticeable in the average number of MUSE. Further, the coincidence technique gives us the correlation between the forward and backward emitted MUSEs. This information would also provide a clue to understand the mechanism of secondary electron production like the role of close vs. distant collisions.

EXPERIMENTAL

A detail of the detection method of MUSE was already described elsewhere [4] [5] [6] [7]. Oxygen and carbon ion beams were provided by the SF cyclotron at the Institute for Nuclear Study (INS), University of Tokyo. The projectile energy was

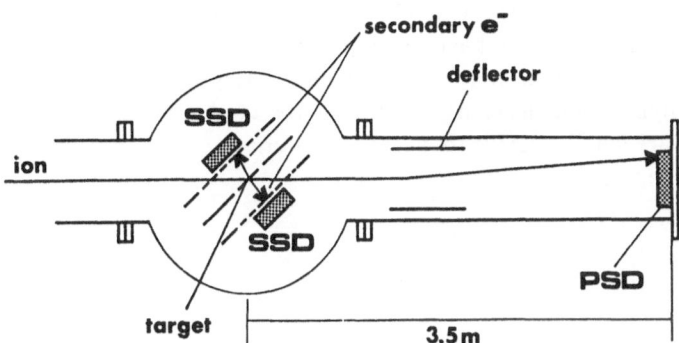

Figure.1 The experimental set up. SSD: solid state surface barrier detector for MUSE. PSD: position sensitive detector for the projectiles.

fixed at 1 MeV/u. In this energy region it is enough to take only the kinetic emission (KE) into account, and little contribution of the potential emission (PE) is involved. The available charge of the oxygen ion beam from the cyclotron was 2+ and ions with charges from 4+ to 8+ were obtained by passing the initial beam through a thin charge stripping carbon foil. Carbon ions with charges from 2+ to 6+ were obtained by the same technique. The beam was collimated (0.2mm x 0.2mm) and guided into an experimental chamber, which was kept at ~1x10^{-6}Torr. There, it passed through a thin carbon foil of 1.1μg/cm^2, which was tilted 45° to the beam direction. The thickness of the foil was ~80Å.

Emitted electrons from the carbon foil both in the forward and backward directions were accelerated to 15keV (the foil was set to -15kV with respect to earth potential), and detected by two solid state surface barrier detectors (SSD) (Canberra SPD100-12-100: the resolution is 12keV for α particle), each of which was placed at 20mm forward and backward position of the target foil, respectively, as shown in fig.1. The exit charge state of projectile ions was analyzed by an electrostatic deflector and detected by a PSD (Hamamatsu S2291).

The pulse height of the signals from the two SSDs for MUSE show the multiplicities of electrons which come into the detector simultaneously, because n electrons deposit n times as large energy as one electron does.

The counting gate of the two SSDs for MUSE was opened for several μs after the projectiles entered the PSD. These three kinds of output signals were amplified, digitized by an ADC, and stored in a personal computer in a list mode. The current of the beam was limited to several 100 cps in order to avoid sputtering and pulse pile-up in the ADC of the data-acquisition system. A part of the electrons which are injected in the SSD detector backscatter from its surface into vacuum. Because of this backscattering effect, the absolute value of the yield of MUSE is supposed to be reduced by roughly 10-15% . Therefore, the discussion below is limited in the relative intensities of MUSE.

RESULTS AND DISCUSSION

1. Exit Charge State Distribution of Projectile Ions

Tables 1 and 2 show the exit charge state distributions of oxygen and carbon projectiles, respectively; they have not reached equilibrium after passing through the 80Å carbon foil. The deviation from the equilibrated charge distribution is especially apparent for higher incident charges, which is reasonable because the electron capture cross sections are small relative to electron loss cross section in this energy region.

Table 1. The exit charge state (Q_{ex}) distribution [%] vs. the incident charge state (Q_{in}) for 1MeV/u oxygen incident on a 1.1μg/cm^2 carbon foil. -: the value was too small to be measured. *: the reported data [21]

		Q_{ex}					
		3+	4+	5+	6+	7+	8+
	2+	-	1.2	15.8	58.5	22.6	1.8
	4+	-	1.1	15.6	59.0	22.4	1.8
Q_{in}	5+	-	0.6	14.1	59.6	23.4	2.0
	6+	-	0.2	5.6	57.3	32.0	3.4
	7+	-	0.1	2.4	28.2	56.0	9.0
	8+	-	0.1	2.0	15.9	38.6	34.6
equi.*		0.02	0.93	11.4	51.7	31.4	4.52
		0.04	1.06	11.6	51.6	31.6	4.08

2. Data Evaluation of MUSE

Individual raw data sets consisting of output signals of the two SSDs and the PSD for the projectile ions with a specific incident charge Q_{in} were obtained. First of all, we took a coincidence between the pulse height spectra of the SSDs for the forward and backward emitted MUSE and the output signal of the PSD for the projectile ions of the specific exit charge, Q_{ex}. The incoming position of the projectile ions on the PSD depends on their exit charge, since the electrostatic deflector was placed between the target and the PSD as described already. Thus we can get the 2-dimensional spectra of the number distribution of MUSE simultaneously emitted in the forward and backward directions, $f(n_F, n_B; Q_{in}, Q_{ex})$ by bombardment with projectiles with a specific incident (Q_{in}) and exit (Q_{ex}) charge.

Table 2. The exit charge state (Q_{ex}) distribution [%] vs. the incident charge state (Q_{in}) for 1MeV/u carbon incident on a $1.1\mu g/cm^2$ carbon foil. *:the reported data [21]

		Q_{ex}			
		3+	4+	5+	6+
Q_{in}	2+	1.8	30.9	51.6	15.7
	3+	1.7	31.3	51.7	15.3
	4+	0.6	28.0	54.3	17.1
	5+	0.2	9.1	63.1	27.6
	6+	0.2	4.5	32.3	63.0
equi.*		1.56	26.9	52.8	18.8

It is assumed that the SSD pulse height spectrum consists of several peaks, which are representative of a multiplicity, i.e. the number of secondary electrons which are emitted simultaneously, and each of them follows the Gaussian distribution. However the width of the distribution increases for the larger multiplicities. In the present condition the multiplicity is too large to treat each peak separately due to the lack of resolution. Therefore, we analyzed the spectra without a decomposing procedure. Instead, we obtained the multiplicity, with the help of a calibration factor between the multiplicity and a channel number of multichannel analyzer which collected the output signals of the SSD. This was done using 1MeV/u He ion, where the spectra have a good resolution due to a small multiplicity [7]. The number distribution obtained through the above procedure i.e. $f(n_F, n_B; Q_{in}, Q_{ex})$ was normalized with respect to n_F and n_B.

$$\sum_{n_F} \sum_{n_B} f(n_F, n_B; Q_{in}, Q_{ex}) \equiv 1 \qquad (1)$$

A typical example is shown in Fig.2.

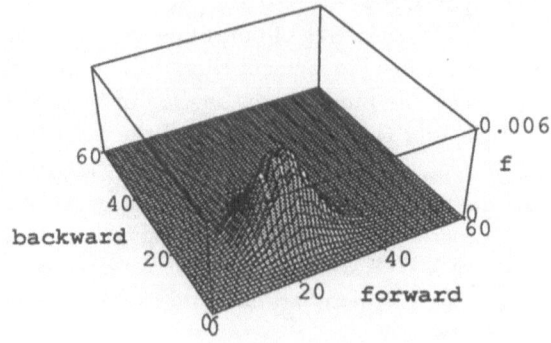

Figure.2 $f(n_F, n_B; Q_{in} = 6, Q_{ex} = 6)$ for 1MeV/u oxygen incident on a $1.1\mu g/cm^2$ carbon foil.

3. Average Number Distribution of MUSE

First of all, to get the average numbers emitted in the forward and backward directions, $\bar{n}_F(Q_{in}, Q_{ex})$, $\bar{n}_B(Q_{in}, Q_{ex})$, we calculated as follows.

$$\bar{n}_F(Q_{in}, Q_{ex}) = \sum_{n_B}\sum_{n_F} n_F \cdot f(n_F, n_B; Q_{in}, Q_{ex})$$

$$\bar{n}_B(Q_{in}, Q_{ex}) = \sum_{n_F}\sum_{n_B} n_B \cdot f(n_F, n_B; Q_{in}, Q_{ex}) \tag{2}$$

The results for oxygen bombardment are mapped in fig.3 and fig.4, respectively. Those for carbon bombardment are shown in fig.5 and fig.6.

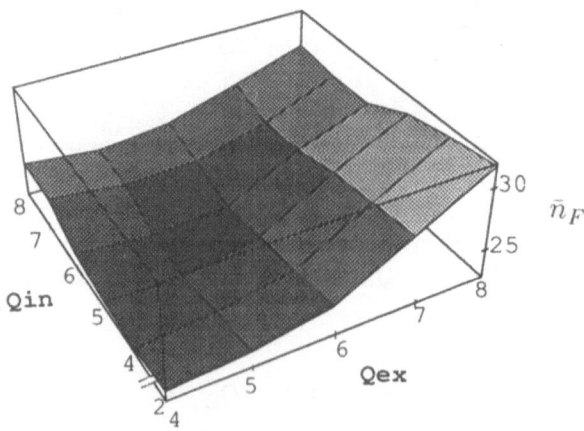

Figure.3 $\bar{n}_F(Q_{in}, Q_{ex})$ distribution for 1 MeV/u oxygen incident on a $1.1\mu g/cm^2$ carbon foil

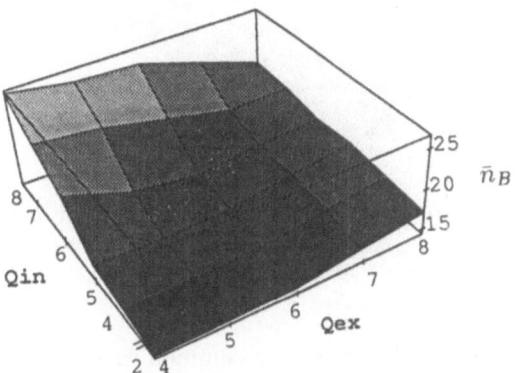

Figure.4 $\bar{n}_B(Q_{in}, Q_{ex})$ distribution for 1MeV/u oxygen incident on a $1.1\mu g/cm^2$ carbon foil

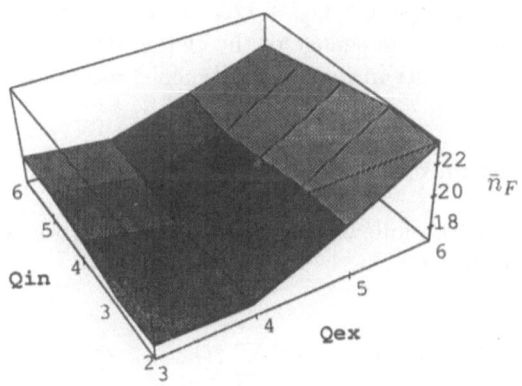

Figure.5 $\bar{n}_F(Q_{in}, Q_{ex})$ distribution for 1MeV/u carbon incident on a $1.1\mu g/cm^2$ carbon foil

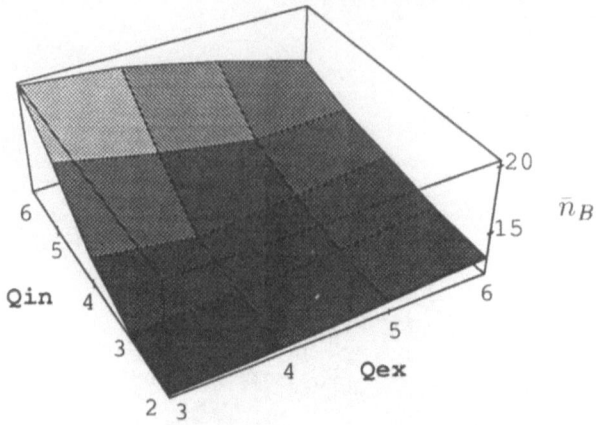

Figure.6 $\bar{n}_B(Q_{in}, Q_{ex})$ distribution for 1MeV/u carbon incident on a 1.1μg/cm^2 carbon foil

These spectra have remarkable features in common with oxygen and carbon.

1. $\bar{n}_F(Q_{in}, Q_{ex})$ correlates more strongly with Q_{ex} than with Q_{in}, and increases for increasing Q_{ex}. On the other hand, $\bar{n}_B(Q_{in}, Q_{ex})$ depends much on Q_{in}, and increases for increasing Q_{in}. This tendency is attributed to the fact that the majority of the secondary electrons originate near the exit (or entrance) surface, where the charge state of ions would be close to Q_{ex} (or Q_{in}), as already described. Furthermore, it implies that the escape depth of the electrons is less than 80Å, otherwise these dependences would not appear.

2. When projectile ions capture or lose more electrons, $\bar{n}_F(Q_{in}, Q_{ex})$ and $\bar{n}_B(Q_{in}, Q_{ex})$ have a tendency to be larger. That is, both $\bar{n}_F(Q_{in}, Q_{ex})$ and $\bar{n}_B(Q_{in}, Q_{ex})$ increase for increasing $| Q_{in} - Q_{ex} |$. This implies that a close collision which induces a large change of the charge state of the projectile, although a rare event, results in many simultaneous secondary electron emission per collision.

4. Number distribution of MUSE

Here, we introduce the following number distribution of MUSE emitted in the forward and backward directions.

$$
\begin{aligned}
f(n_F; Q_{in}, Q_{ex}) &= \sum_{n_B} f(n_F, n_B; Q_{in}, Q_{ex}) \\
f(n_B; Q_{in}, Q_{ex}) &= \sum_{n_F} f(n_F, n_B; Q_{in}, Q_{ex})
\end{aligned}
\tag{3}
$$

It has always been found that $f(n_F; Q_{in}, Q_{ex})$, or $f(n_B; Q_{in}, Q_{ex})$ has a wider distribution than the Poisson distribution with the same $\bar{n}_F(Q_{in}, Q_{ex})$ or $\bar{n}_B(Q_{in}, Q_{ex})$, regardless of the type of target, projectile or projectile energy [4-15]. The succesive and independent ionization of target atoms by the projectile results, in principle, in the Poisson distribution. What is the origin of the deviation? In the case of slow ($<$

1 a.u.) projectile ions this property was explained by the variation of the large angle scattering leading to backscattering of the incident ions or recoiling of the target atoms [22]. However, the swift projectiles with which we deal, do not suffer backscattering. Thereby, this is not a suitable explanation for the origin of the wider distribution in the present study.

One possibility is the cascading electron production process by the primary electrons. Now we consider a simplified model as follows.

$$P(N) \; = \; \sum_{n=0}^{\infty} p(n) \sum_{m_1 + m_2 + ... = N} q(m_1)q(m_2)...q(m_n) \tag{4}$$

$P(N)$: a probability to emit N electrons

$p(n)$: a probability to produce n primary ionized electrons per one projectile

$q(m)$: a probability to produce m "secondary" electrons per one primary ionized electron

If no cascading occurs i.e. $q(0) = 1$ and $q(m) = 0$ for $m = 1...\infty$, $P(N)$ exactly follows the Poisson distribution. If both $p(n)$ and $q(m)$ follow the Poisson distribution, $P(N)$ follows a wider distribution than the Poisson distribution. If $p(n)$ follows the Poisson distribution and $q(m)$ follows the binominal distribution, then $P(N)$ just follows the Poisson distribution. The actual distribution of the number of cascading "secondary" electrons, i.e., $q(m)$ would be dependent on the energy distribution of the primary electrons. Thus, the width of the distribution would be a good index of the cascading process.

In the present study the wider distribution was always obtained again regardless of Q_{in} and Q_{ex}. The charge dependent deviation from the Poisson distribution was examined. We introduced a specific value as a guiding index for the deviation, i.e. $d \equiv (variance)/(average)$. For instance, if both $p(n)$ and $q(m)$ follow the Poisson distribution, this value is larger than unity and is rewritten as follows.

$$\begin{aligned} d \equiv \frac{(variance)}{(average)} &= \frac{(N - \bar{N})^2}{\bar{N}} \\ &= \frac{\bar{n} \cdot \bar{m}(1 + \bar{m})}{\bar{n} \cdot \bar{m}} \\ &= 1 + \bar{m} \end{aligned} \tag{5}$$

The results are shown in Figs.7, 8, 9 and 10. It is clearly seen that there exists an apparent Q_{in} and Q_{ex} dependence of the width of the distribution. In general, $d(f(n_F; Q_{in}, Q_{ex}))$ has a stronger positive correlation with Q_{ex} than with Q_{in}, whereas $d(f(n_B; Q_{in}, Q_{ex}))$ depends much on Q_{in}.

5. Correlation between the forward and backward emitted MUSE

It is possible to get information on the correlation between the simultaneously emitted MUSE in the forward and backward directions. We employ the following correlation coefficient $\Delta f(n_F, n_B; Q_{in}, Q_{ex})$.

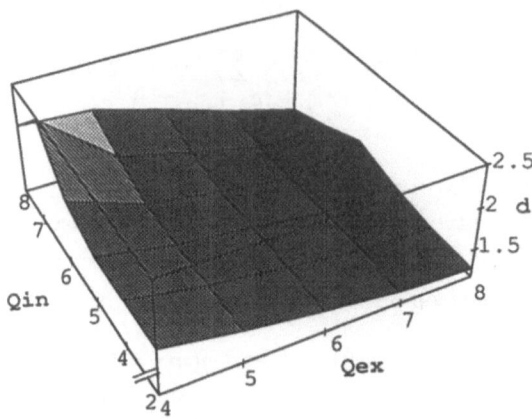

Figure.7 $d(f(n_F; Q_{in}, Q_{ex}))$ distribution for 1MeV/u oxygen incident on a $1.1\mu g/cm^2$ carbon foil

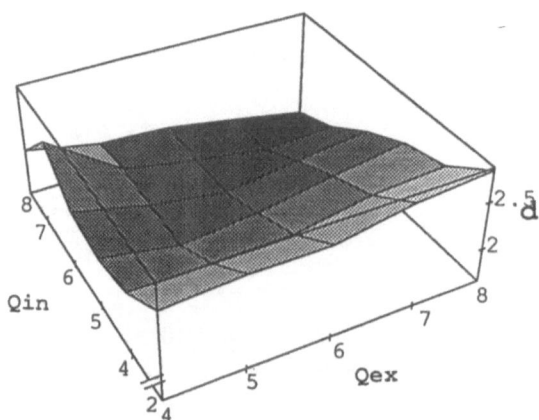

Figure.8 $d(f(n_B; Q_{in}, Q_{ex}))$ distribution for 1MeV/u oxygen incident on a $1.1\mu g/cm^2$ carbon foil

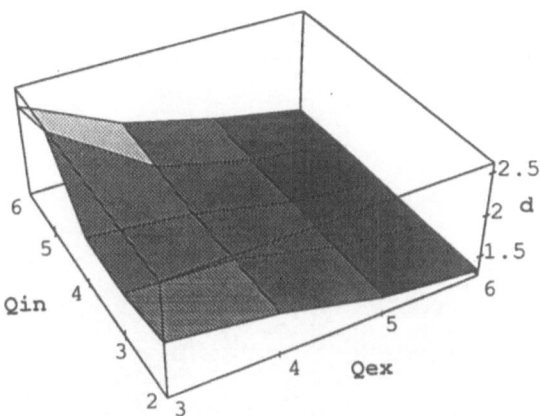

Figure.9 $d(f(n_F; Q_{in}, Q_{ex}))$ distribution for 1MeV/u carbon incident on a $1.1\mu g/cm^2$ carbon foil

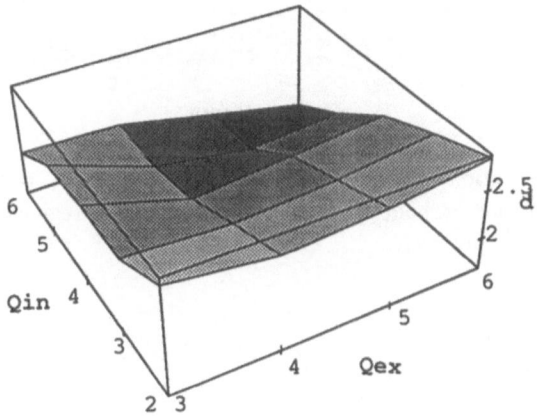

Figure.10 $d(f(n_B; Q_{in}, Q_{ex}))$ distribution for 1MeV/u carbon incident on a $1.1\mu g/cm^2$ carbon foil

$$\Delta f(n_F, n_B; Q_{in}, Q_{ex}) \equiv f(n_F, n_B; Q_{in}, Q_{ex})$$

$$- \sum_{n_F} f(n_F, n_B; Q_{in}, Q_{ex}) \sum_{n_B} f(n_F, n_B; Q_{in}, Q_{ex}) \quad (6)$$

$\sum_{n_F} \sum_{n_B} \Delta f(n_F, n_B; Q_{in}, Q_{ex}) = 0$ from the definition. When no correlation exists, $\Delta f(n_F, n_B; Q_{in}, Q_{ex}) = 0$ for any n_F and n_B.

Figure 11 shows the mapping of the correlation coefficient between the forward and backward emitted MUSE for 1MeV/u oxygen incident on a $1.1\mu g/cm^2$ carbon foil. The

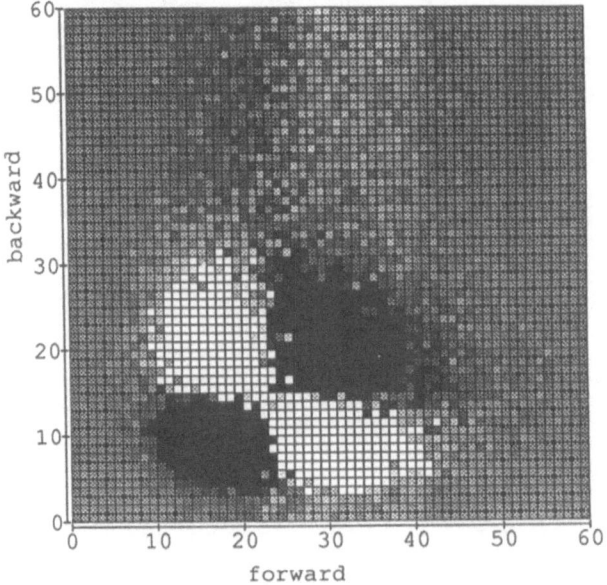

Figure.11 The correlation coefficient $\Delta f(n_F, n_B; Q_{in} = 6, Q_{ex} = 6)$ for 1MeV/u oxygen incident on a $1.1\mu g/cm^2$ carbon foil. The white and black regions represent positive and negative coefficients, respectively.

region of positive $\Delta f(n_F, n_B; Q_{in}, Q_{ex})$ was seen around the area of $n_F + n_B \cong 40$. It corresponds to a clear negative correlation; when $< f(n_F; Q_{in}, Q_{ex}) >$ is larger, $< f(n_B; Q_{in}, Q_{ex}) >$ is smaller at the same time, and vice versa. The origin of the negative correlation is not understood well yet, but it results from the condition that the total number of the secondary electrons is kept constant and they are divided into the both directions.

References

[1] M.S.Chung and T.E.Everhart, Phys.Rev. B15:4699(1977)

[2] W.O Hofer, Scannning Miroscopy Supplement 4:265(1990)

[3] "Particle Induced Electron Emission I and II", Springer,Berlin, (1991)

[4] Y.Yamazaki and K.Kuroki, Nucl. Instr. and Meth. A262:118(1987)

[5] K.Kuroki and Y.Yamazaki, Nucl. Instr. and Meth. B33:276(1988)

[6] Y.Yamazaki, in "High Energy Ion-Atom Collisions", D.Berenyi ed., Springer, Berlin (1988) p.322

[7] T.Azuma, Y.Yamazaki, K.Komaki, M.Sekiguchi,T.Hasegawa, T.Hattori and K.Kukori, Nucl. Instr. Meth. B67:636(1992)

[8] G.Lakits, F.Armayr and H.Winter, Rev. Sci. Instrum. 60:3151(1989)

[9] G.Lakits, A.Aumayr and H.Winter, Radiat. Effects and Defects in Solids 109:129(1989)

[10] G.Lakits and H.Winter, Nucl. Instr. and Meth. B48:597(1990)

[11] G.Lakits, A.Arnau and H.Winter, Phys.Rev. B42:15(1990)

[12] M.A.Grutman, A.A.Kozochkina and V.B.Leonas, JETP Lett. l51:22(1990)

[13] A.A.Kozochkina, V.B.Leonas and M.Witte, Nucl. Instr. and Meth. B62:51(1991)

[14] V.B.Leonas, Sov.Phys.Usp. 34:317(1991)

[15] H.A.Bethe, Ann.Phys. 5:325(1939)

[16] M.Inokuti, Rev.Mod.Phys. 43:2978(1971)

[17] Y.-K.Kim and K.-T.Cheng, Phys.Rev. A22:61(1980)

[18] P.Koschar, K.Kronenberger, A.Clouvas, M.Burkhard, W.Meckbach, O.Heil, J.Kemmler, H.Rothard, K.O.Groeneveld, R.Schramm and H.-D.Betz, Phys. Rev. A40:3632(1989)

[19] H.Rothard, K.Kronenberger, A.Clouvas, E.Veje, P.Lorenzen, N.Keller, J.Kemmler, W.Meckbach and K.O.Groeneveld, Phys. Rev. A41:2521(1990)

[20] H.Rothard, J.Schou, and K.O.Groeneveld, Phys. Rev. A45:1701(1990)

[21] K.Shima and T.Mikumo, Atom. Data and Nucl. Data Tables 34:357(1986)

[22] K.Ohya, F.Aumayr and H.Winter, Phys Rev. B46:3101(1992)

ELECTRON EMISSION PHENOMENA IN GRAZING COLLISIONS OF FAST IONS WITH SURFACES

H. Winter, G. Dierkes, A. Hegmann, J. Leuker, H.W. Ortjohann and R. Zimny

Institut für Kernphysik der Westfälischen Wilhelms-Universität Münster
Wilhelm-Klemm-Str. 9, D-4400 Münster, Germany

INTRODUCTION

The interaction of fast atoms and ions with a clean and flat surface of a metal under a grazing angle of incidence is characterized by two vastly different regimes with respect to the motion parallel and normal to the surface plane. In Fig.1 we give a simple sketch of the geometry of grazing surface collisions, where some relevant polar and azimuthal angles are indicated. The motion parallel to the surface proceeds with about the initial energy of the projectile E, whereas the energy for the normal motion is given by $E_y = E \sin^2\Phi$ (Φ = grazing angle of incidence or emergence). For $\Phi = 1°$ - a typical angle of incidence in our studies - we have $E_y = 3*10^{-4}$ E, so that for beams with keV-energies the normal motion proceeds with energies of about eV. This feature of grazing surface collisions has a number of consequences with respect to the interaction of the projectiles with the surface. Due to the low normal energies E_y the projectiles cannot overcome the repulsive potentials formed by the atoms of the topmost layer of the surface and will not penetrate into the bulk of the solid (for a perfect surface!). It is evident that the emission of electrons during ion impact with the solid is affected by the specific features of the grazing incidence collision geometry. We will discuss here some examples of our recent work in this respect.

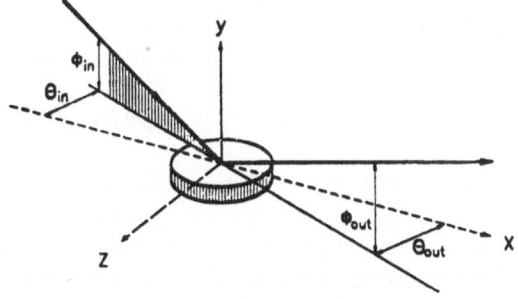

Figure 1. Sketch of the geometry for grazing ion surface scattering from a crystal

Ionization of Solids by Heavy Particles, Edited by
R.A. Baragiola, Plenum Press, New York, 1993

STUDIES OF ELECTRON EMISSION UNDER SURFACE CHANNELING CONDITIONS VIA MEASUREMENTS OF THE TARGET CURRENT

When fast ions are scattered from a surface of a metal under grazing angles of incidence, the current recorded at the target exceeds clearly the current of the incident projectile beam. This observation indicates that quite a few electrons are ejected during the scattering event. The number of emitted electrons per incident ion - the electron emission coefficient γ - is generally significantly larger for grazing incidence geometry than for impact of ions under normal incidence. As an example we refer to a study by Hasegawa et al.[1] with 0.5 to 2 MeV He$^+$-ions impinging on a SnTe(001)-surface, where γ is found to increase monotonically from about 3 for normal incidence to more than 200 for $\Phi_{in} < 1°$. The angular variation of the data is well approximated by a $1/\sin\Phi_{in}$-dependence, which reflects the length of the projectile trajectories within the escape depth for excited electrons from deeper layers of the solid. However, for grazing angles of incidence Φ_{in} the trajectories of the projectiles are affected by surface channeling phenomena, and a deviation of γ from the simple $1/\sin\Phi_{in}$-scaling is observed. These deviations become prominent below a critical angle for planar channeling at the surface plane of a crystal given by[2,3]

$$\Phi_{crit} = \left(2\pi \, n_s \, Z_1 \, Z_2 \, a_{TF} \, / \, E \right)^{1/2} \tag{1}$$

where n_s is the areal density of atoms in the topmost surface layer, Z_1 and Z_2 are the atomic numbers of incident ion and target atoms, respectively, a_{TF} is the Thomas-Fermi screening length, and E is the projectile energy. The azimuthal angle Θ shown in Fig.1 is referred to a low index direction <uvw> in the (hkl)-plane of the target surface. In case the angle incidence to a <uvw>-string $\Psi = (\Phi^2 + \Theta^2)^{1/2}$ is smaller than a critical angle[3,4]

$$\Psi_{crit} = \left(2 \, Z_1 \, Z_2 \, / \, E \, d_{<uvw>} \right)^{1/2} \tag{2}$$

where $d_{<uvw>}$ denotes the interatomic spacings in a string of atoms along <uvw>, axial channeling phenomena may become important.

Grazing ion surface collisions are run for $\Phi_{in} < \Phi_{crit}$ under planar channeling conditions at the topmost (hkl)-plane, and the projectiles are reflected specularly from the target. However, for azimuthal settings of the target, where the incident projectile beam is directed close to a <uvw>-direction, the concept of planar channeling breaks down, and the scattering is dominated by the axially symmetric potentials formed by strings of atoms at the surface. As a consequence axial surface channeling phenomena can be observed in the regime of planar surface channeling, where the transition between both regimes can be estimated from an analytical expression given by Bill et al.[5] and Varelas[6]. For interaction potentials with a Moliere-type screening[7] this transition is expected to proceed for an azimuthal angle

$$\Theta_{ap} = \left(\pi \, d_{ap}^2 \, Z_1 \, Z_2 \, n_s \, a_1 \, b_1 \, \exp\left(-b_1 y_{min}/a_{TF}\right) \, / \, 4 \, E \, a_{TF} \right)^{1/2} \tag{3}$$

where d_{ap} is the distance between adjacent <uvw>-strings in the (hkl)-plane, y_{min} is the distance of closest approach under planar channeling conditions, i.e. $\Theta = <$"random"$>$, and $a_i = (0.35, 0.55, 0.1)$, $b_i = (0.3, 1.2, 6)$ are the parameters for the Thomas-Fermi-Moliere potential[7]. In grazing ion surface collisions impact parameters are large and in good approximation only a_1, b_1 are of relevance.

A variation of the target current during an azimuthal rotation in grazing ion-surface scattering was noticed in studies on polarized light emission phenomena[8]. Thereafter detailed studies of this type have been performed for fast protons scattered from clean as well as contaminated Ni(110)-surfaces[9] and from a Si(001)-surface[10]. The Kyoto group[1] performed similar studies with He[+]-projectiles at MeV-energies interacting with SnTe(001)- and PbSe(001)-surfaces.

Our motivation to study the variation of the target current for an azimuthal rotation of a monocrystalline sample was based primarily on the need of a simple on-line method for the orientation of a low index direction in the surface plane relative to the axis of the projectile beam. Such a technique allows to perform reliable "random" adjustments between beam and atomic strings, where the projectiles are specularly reflected from the target under (semi-) planar surface channeling conditions. It turns out that a recording of the target current during an azimuthal rotation of the crystal provides indeed a simple and effective technique for an on-line orientation of monocrystalline samples with good precision.

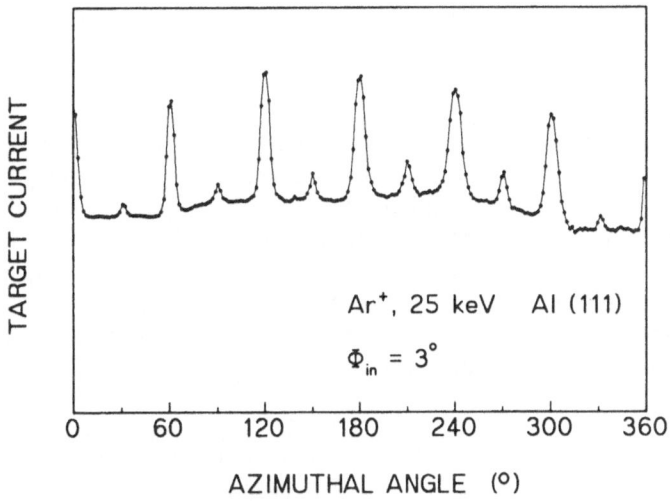

Figure 2. Dependence of the target current on the azimuthal angle of rotation of a clean Al(111)-surface for the scattering of 25 keV Ar[+]-ions under $\Phi_{in}= 3°$.

Since an important procedure in the preparation of clean surfaces of crystals with a low amount of defect structures consists in sputtering of the target under grazing angles of incidence (Φ_{in} up to some degree), the information on the structural dependence of electron emission phenomena is obtained as a "by-product" in the cycles of preparation. We clean generally our targets by sputtering with 25 keV Ar[+]-ions under continuous azimuthal rotation with respect to the surface normal[11]. Fig.2 shows a typical result obtained in the final state of preparation of an Al(111)-target. The (positive) target current is recorded for a complete rotation of the sample and for a continuous bombardment with an Ar[+]-beam of about 0.5 μA under $\Phi_{in}= 3°$. The data show a number of prominent peaks, which are separated by intervals of 60°. It is straightforward with the help of the sketch of the positions of atoms in the (111)-plane of a fcc-crystal (see Fig.3) to identify the directions, where the target current is enhanced, with an azimuthal adjustment of the low index <1$\bar{1}$0>-strings along the beam axis of the projectiles. The structures observed for the current reflect the symmetry of the (111)-plane.

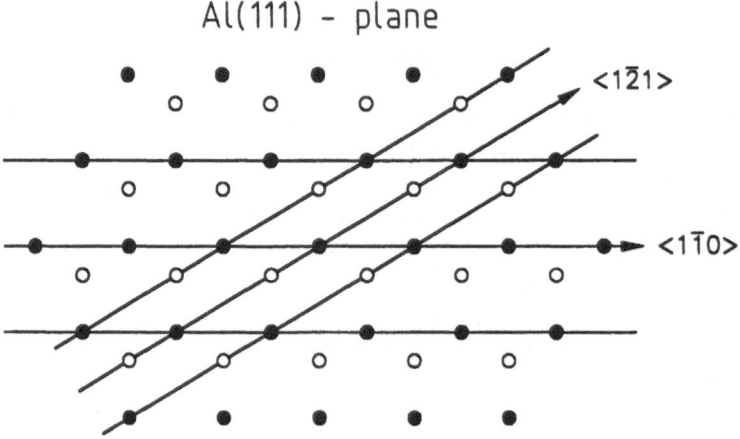

Figure 3. Sketch of the positions of atoms in the topmost layer (solid dots) of an unreconstructed Al(111)-surface. The open cirlces represent the positions of atoms in the second layer of the crystal. The solid lines indicate the directions of two low index axial channels in the surface plane.

Comparison of the current of incoming ions with the target current yield γ about 5 for a random azimuthal orientation, so that - despite an almost complete reflection and neutralization of the projectiles in the scattering process - the ion current plays only a minor role here; i.e. the signal is dominated by kinetic emission processes of electrons. From eq.(1) we deduce here a critical angle Φ_{crit}= 6.8°, so that the experiments are run for a random azimuthal orientation under planar channeling conditions with a distance of closest approach to the surface plane y_{min}= 1.3 a.u.

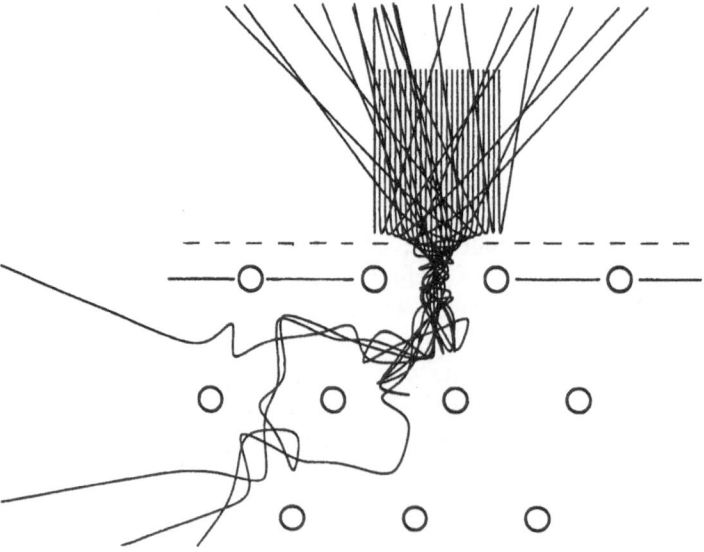

Figure 4a. Sketch of trajectories of 25 keV Ar-projectiles interacting with an Al(111)- -surface under Φ_{in}= 3° along the <110>-direction. The figure displays a projection of trajectories onto a plane normal to the surface plane and to <1$\bar{1}$0> (yz-plane in Fig.1).

Instead of a detailed description of the electron emission phenomena we will limit our discussion here to qualitative arguments based on the trajectories of projectiles during the collision with the crystal. For the sake of a simple presentation we ignore here any defect structures, thermal vibrations of lattice atoms, and reconstruction phenomena. In Fig.4a we show a projection of trajectories onto a plane normal to the (111)-surface and to the $\langle1\bar{1}0\rangle$-strings. The trajectories are calculated for axial continuum potentials[2,3] with a Moliere-type of screening[7]. A spatially constant flux of 25 keV Ar-projectiles is directed under $\Phi_{in}= 3^o$ ($E_y= 70$ eV) along $\langle1\bar{1}0\rangle$, i.e. $\Theta_{in}= 0^o$.

In contrast to planar channeling conditions, where all projectiles have the same y_{min} (dashed line in the figure), the axial channeling with respect to $\langle1\bar{1}0\rangle$-strings results in an enhanced flux of projectiles between neighbouring strings ("focusing") and a partial penetration of projectiles to deeper layers of the crystal. Then the longer trajectories and the higher densities of conduction electrons[12] in comparison to a planar channeling under a random alignment lead to increased probabilities for the excitation and emission of electrons.

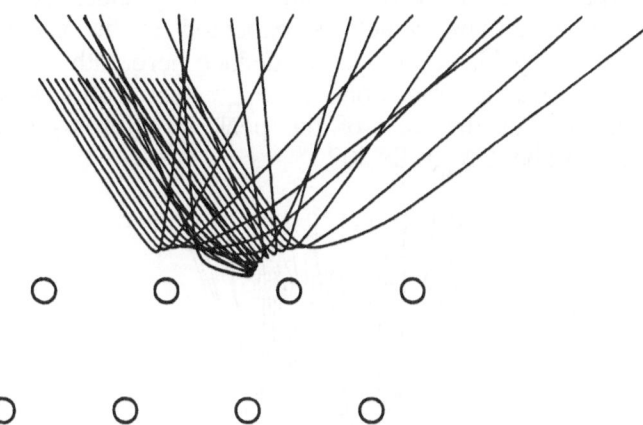

Figure 4b. Same as in Figure 4a, but $\Theta_{in}= 2^o$ with respect to the $\langle1\bar{1}0\rangle$-strings.

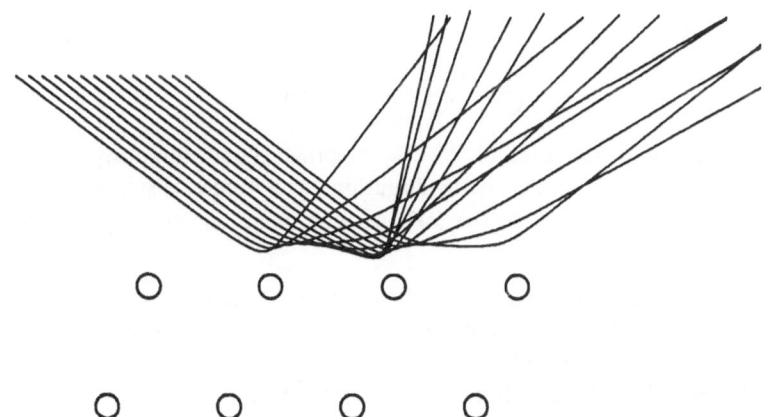

Figure 4c. Same as in Figure 4a, but $\Theta_{in}= 4^o$ with respect to the $\langle1\bar{1}0\rangle$-strings.

In Figs.4b, 4c, and 4d we display projections of trajectories for $\Theta_{in}= 2^o$, 4° and 8°, respectively, which indicate that the mean distance of closest approach to the surface and the interaction length of projectiles decrease with increasing deviation from the string direction. It is obvious that these features

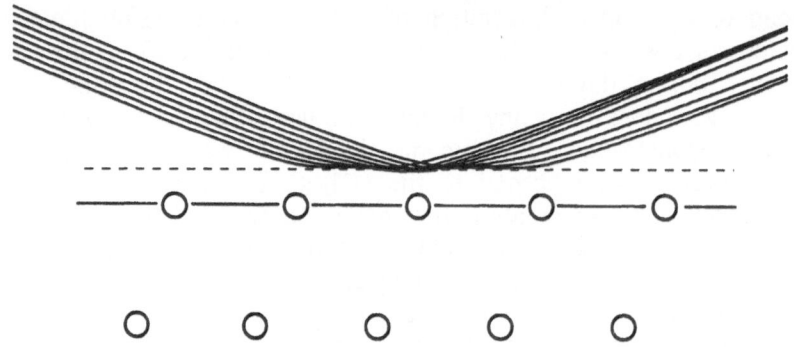

Figure 4d. Same as in Figure 4a, but $\Theta_{in}= 8°$ with respect to the <$1\bar{1}0$>-strings. The dashed line indicates the distance of closest approach under planar channeling conditions.

lead also to a reduced probability for the emission of electrons (here: reduction of the target current) as observed in the experiments. The sequence of Figs.4a - 4d illustrates the gradual transition from scattering under axial to that under planar channeling conditions, which is expected to be about completed a $\Theta_{in}= 8°$ as shown in Fig.4d. From eq.3 we deduce here with respect to an axial channeling along <$1\bar{1}0$> a transition at $\Theta_{ap}= 9.5°$. This agrees with our experimental findings for the tails of the peak structures displayed in Fig.2, which correspond to about $\Theta_{ap}= 8°-10°$.

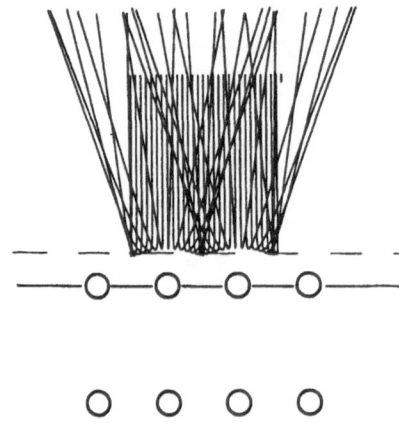

Figure 5a. Sketch of trajectories of 25 keV Ar-projectiles interacting with an Al(111)-surface under $\Phi_{in}= 3°$ along the <$2\bar{1}\bar{1}$>-direction. The figure displays a projection of trajectories onto a plane normal to the (111)-surface plane and to <$2\bar{1}\bar{1}$>.

The data displayed in Fig.2 do show aside from a poor modulation over the full rotation - caused by a slight misalignment of the target with respect to the axis of azimuthal rotation - clearly poorer variations of the current for other low index directions. This observation can be explained by the fact that for the energy of normal motion in this case ($E_y= 70$ eV) the separations between corresponding strings of atoms are too small for pronounced axial channeling effects. We illustrate this feature for the <$2\bar{1}\bar{1}$>-strings with help of Fig.5a, where the trajectories and the distances of closest approach for planar (dashed line) and axial channeling show minor differences only.

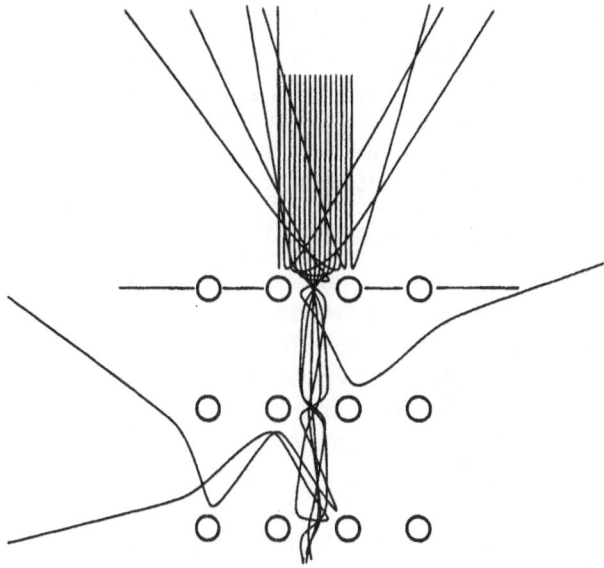

Figure 5b. Same as Fig.5a, but for 450 keV Ar-projectiles and $\Phi_{in}= 1.25°$.

For larger E_y, however, the projectiles can overcome the collective axial potentials formed by neighbouring strings of higher index directions ("axial hyperchanneling"). This is sketched in Fig.5b for 450 keV Ar^+-ions scattered from Al(111) under $\Phi_{in}= 1.25°$ ($E_y= 215$ eV) along $\langle 2\bar{1}\bar{1}\rangle$, where a focusing and a partial penetration of projectiles is evident. In Fig.6 we display the corresponding experimental data and find a pronounced increase of the target current (equivalent to an increase of emitted electrons) also for the $\langle 2\bar{1}\bar{1}\rangle$-directions. These peaks are less intense and narrower than for the $\langle 1\bar{1}0\rangle$-directions and appear at the expected positions between the prominent peaks obtained for an orientation along $\langle 1\bar{1}0\rangle$ (compare Fig.3). Indications for peaks due to higher index directions can be found also in the data.

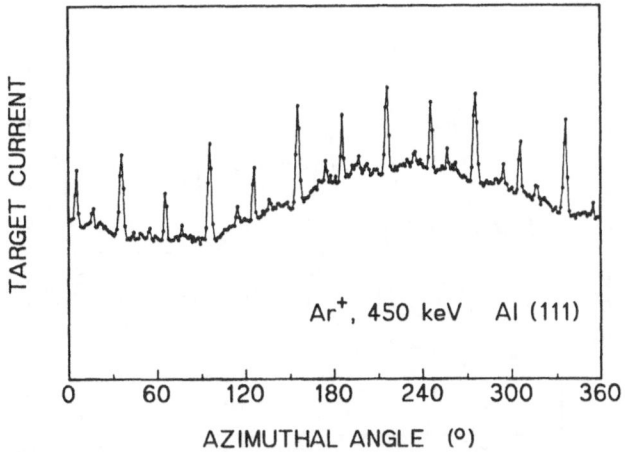

Figure 6. Dependence of the target current on the azimuthal angle of rotation of a clean Al(111)-surface for the scattering of 450 keV Ar^+-ions under $\Phi_{in}= 1.25°$.

Since the azimuthal angle for the transition from axial to planar channeling Θ_{ap} shows about an $E^{-1/2}$-dependence in eq.3, the widths of the peak structures are expected to decrease with increasing projectile energy E. This is evident from a direct comparison of the data obtained for 25 keV (Fig.2) and for 450 keV (Fig.6). From eq.3 we calculate here for the $\langle 1\bar{1}0\rangle$-direction $\Theta_{ap}= 3.9^{\circ}$ and for $\langle 2\bar{1}\bar{1}\rangle$ $\Theta_{ap}= 2.2^{\circ}$. This is consistent with the transitions observed in the experiments at 3.5°-4° for $\langle 1\bar{1}0\rangle$ and 1.7°-2.2° for $\langle 2\bar{1}\bar{1}\rangle$.

It should be noted that all structures in the data for the target current presented here depend sensitively on the state of preparation of the target surface. In the initial state of preparation peaked structures in the signals are difficult to identify. With increasing efforts and cycles of treatments of the crystal peaks grow out of an initially smooth dependence. This observation can be explained simply by the dominant role of the crystallographic structure at the surface plane on the presence of surface channeling effects. In this respect the dependence of the target current on the azimuthal angle reflects the defect structures at the surface plane. On a qualitative scale we make use of this feature as a simple analytical tool during the preparation of our samples for ion-surface scattering experiments[11].

In Fig.7 we present results obtained with 25 keV Ar$^+$-ions scattered from a W(100)-surface, where we have varied the energy of the normal motion E_y by the setting of the angle of incidence Φ_{in}. The data reflect clearly the different symmetry of the (100)-plane of a bcc-crystal in comparison to the (111)-plane used in the experiments discussed before. For Ar-W(100) we compute from eq.(1) a critical normal energy for planar surface channeling $E_{crit}= E\,\Phi_{crit}^2 = 1.0$ keV, and for $E = 25$ keV we have $\Phi_{crit}= 11.5^{\circ}$. This indicates the stronger interaction potentials for a tungsten target compared with aluminum.

The data obtained at $\Phi_{in}= 1.6^{\circ}$ ($E_y= 20$ eV) show slight indications of peaks for projectiles incident along the low index $\langle 100\rangle$- and $\langle 110\rangle$-strings. The strong interaction potentials lead here to a relatively large distance of closest approach $y_{min}= 2.2$ a.u., and the modifications of trajectories in a transition from planar

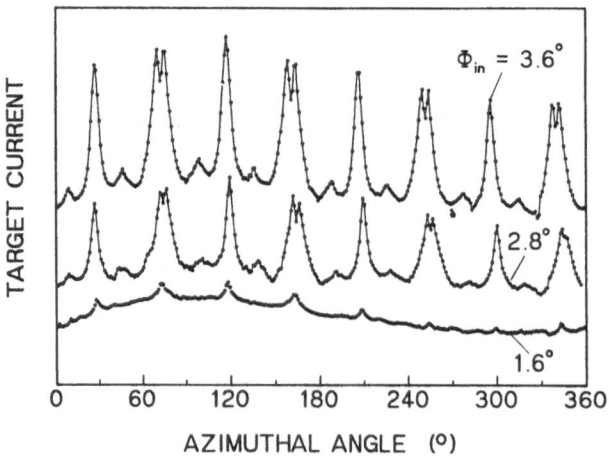

Figure 7. Dependence of the target current on the azimuthal angle of rotation of a clean W(100)-surface for the scattering of 25 keV Ar$^+$-ions under $\Phi_{in}= 1.6^{\circ}$, 2.8°, and 3.6°.

to axial surface channeling are small. This situation is changed with increasing energies for the normal motion, i.e. larger angles of incidence. The data for $\Phi_{in}= 2.8^{\circ}$ ($E_y= 60$ eV) and $\Phi_{in}= 3.6^{\circ}$ ($E_y= 100$ eV) show a pronounced increase of the target current for scattering along low index directions. This can be understood by the trajectories and the density of conduction electrons at a W(100)-surface[12]. The peaks for scattering along <100> show dips in the centers, which increase with Φ_{in}, whereas along the <110>-directions narrower and simple peak structures are observed. Qualitatively the dips observed along <100> can be understood by a focusing of flux of projectiles between <100>-strings in the topmost layer onto <100>-strings in the second layer of target atoms. For an exact orientation along <100> this focusing leads to a rather direct reflection of projectiles from the <100>-strings below the surface. A sketch of the projection of trajectories for this case is shown in Fig.8a.

For a slight deviation from the <100>-direction by $\Theta_{in}= 2^{\circ}$ the projectiles are focused on neighbouring <100>-strings in the surface plane. Then projectiles perform a bouncing motion between the strings (see Fig.8b), and the extended

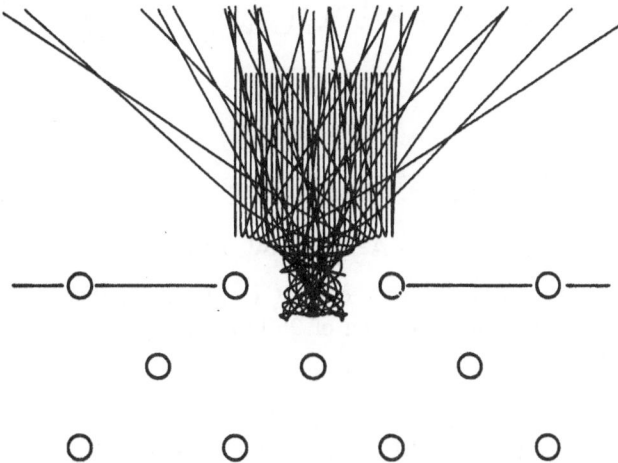

Figure 8a. Sketch of trajectories of 25 keV Ar-projectiles interacting with an W(100)-surface under $\Phi_{in}= 3.6^{\circ}$ along the <100>-direction. The figure displays a projection of trajectories into a plane normal to the (100)-surface plane and to <100>.

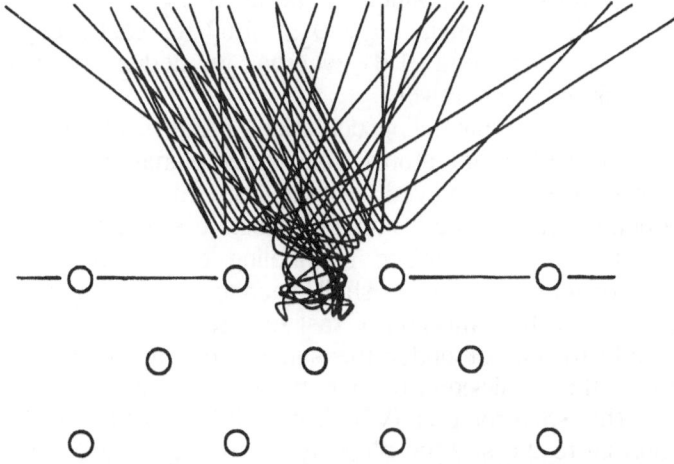

Figure 8b. Same as Fig.8a, but $\Theta_{in}= 2^{\circ}$ with respect to the <100>-direction.

lengths of trajectories lead to an enhanced electron emission in comparison to a perfect axial alignment. Along <110> the flux is focused between the (001)-planes (normal to the (100)-surface plane), and no scattering from strings below the surface plane can take place as sketched for a similar case in Fig.5b. In that case we have a simple peak structure as observed in the experiments.

The width of the structures in the target current for Ar-W(100)-scattering as shown in Fig.7 are reproduced rather well by eq.3. We calculate for Φ_{in}= 3.6° and scattering along <100> Θ_{ap}= 19.2° and along <110> Θ_{ap}= 13.6°. The experimental data yield about 16° and about 10°-12°.

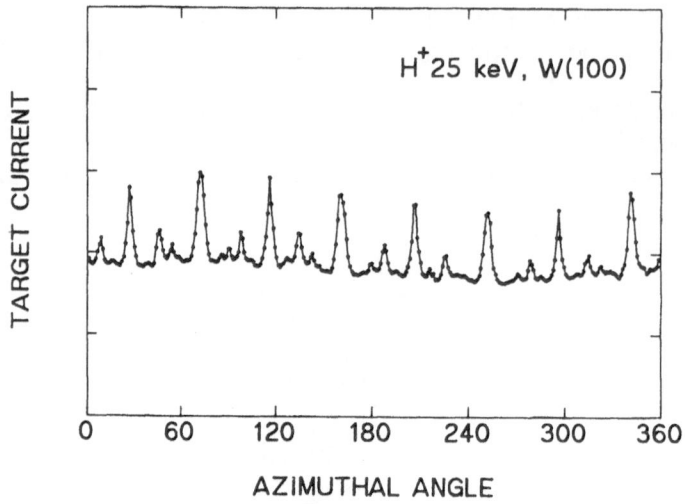

Figure 9. Dependence of the target current on the azimuthal angle of rotation of a clean W(100)-surface for the scattering of 25 keV protons under Φ_{in}= 1.6° (E_y= 21 eV).

In Fig.9 we present data obtained for the scattering of 25 keV protons scattered from the same W(100)-target under Φ_{in}= 1.6° (E_y= 20 eV). The weaker interaction potentials for H-W(100) in comparison to Ar-W(100) lead to closer distances of approach for projectiles to planes and strings formed by target atoms, so that the data show much clearer structures than the corresponding results obtained with Ar+-ions under the same Φ_{in} in Fig.7. The smaller widths of the peak structures are explained by the about $Z_1^{1/2}$-dependence in eq.3. We calculate for the data shown in Fig.9 Θ_{ap}= 7.6° for scattering along <110> and Θ_{ap}= 5.4° for scattering along <110>, whereas we deduce from the experimental data 7°-8° and 5°-6°, respectively.

Under axial surface channeling conditions the projectiles have usually larger distances to individual target atoms than under planar channeling (see e.g. the sketch of trajectories displayed in Fig.4a). Then processes, which have increased probabilities for small impact parameters, should be suppressed on a relative scale under axial surface channeling conditions. As an example we discuss the production of inner shell vacancies of target atoms and their subsequent decay via the emission of fast electrons in Auger-type processes.

In our study we detect under the same experimental conditions as for the previous case the emission of electrons with a hemispherical electron spectrometer. This spectrometer (VG CLAM 100) is positioned with the entrance lens about normal to the surface plane (y-axis in Fig.1) and is set to an electron energy of about 173 eV in order to register Auger-electrons emitted in the decay of N-shell vacancies of tungsten atoms. The data displayed in Fig.10 indicate a pronounced dip-structure. The curve can be considered as a rather perfect inversion of the data obtained for the target current, i.e. the total emission of electrons. In case the spectrometer is set to energies, where contributions of target Auger-electrons can be excluded (e.g. 77 eV), the intensity of the emitted electrons shows a comparable dependence on the azimuthal angle as the target current. Selection of a further Auger-line at about 48 eV yields again a dip-structure as observed in Fig.10. Similar observations have been reported for a Si(001)- surface by Pfandzelter[10] and for the Auger-emission from Ni(110) by Schuster and Varelas[14]. A direct comparison of the data obtained from the measurements of the target current and of the electron spectra indicate very clearly that kinetic emission of valence electrons dominates by far the ejection mechanisms of electrons in comparison to Auger-processes related to target atoms in grazing ion surface scattering.

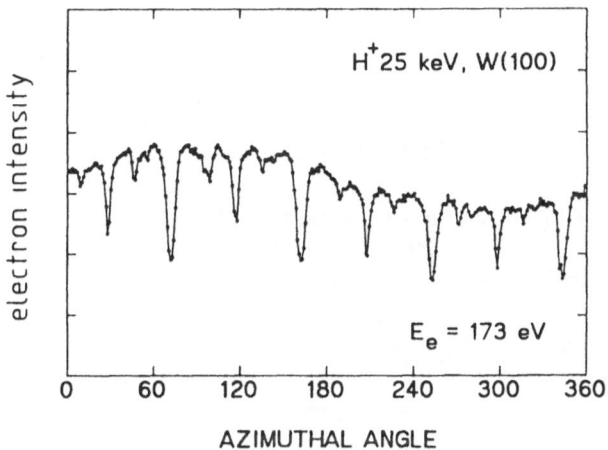

Figure 10. Dependence of the intensity of electrons emitted with an energy of 173 eV from a clean W(100)-surface during the scattering of 25 keV protons under $\Phi_{in} = 1.6°$.

Finally we note that the structures in the target current allow to identify a (hkl)-plane and to perform a precise on-line matching of low index directions <uvw> in the surface plane with respect to the direction of a projectile beam. We estimate the accuracy of this simply applicable technique to about some 0.1°. For larger angles of incidence, i.e. $\Phi_{in} > \Phi_{crit}$, axial and planar channeling effects are then characterized by a reduction of small impact parameter collisions with target atoms and an enhanced penetration along axial or planar channels into deep layers of the solid. This features result consequently in a reduced electron emission probability in the channeling regime and in the observation of dip-structures in the electron yields for variations of azimuthal and polar angles. In low energy ion scattering Feijen et al.[15] observed those structures for 7 keV H_2^+-ions scattered under e.g. $\Phi_{in} = 35°$ from Cu(001) and proposed from their findings a method to orient the surface of a crystal within 0.2°.

ENERGY DISTRIBUTIONS OF ELECTRONS EMITTED IN GRAZING ION-SURFACE COLLSIONS

With an electron spectrometer - an element of our instrumentation for surface analysis - we performed studies of the energy distributions of electrons emitted during the scattering of fast protons, He^+-, and He^{2+}-ions from an Al(111)- and a Fe(110)-surface. The spectrometer is a commercially available "Combined Lens and Analyzer Module" (CLAM 100), which accepts with an opening angle of about 3° electrons ejected from the surface in the yz-plane (see Fig.1) under a fixed angle of 15° with respect to the y-axis. Earth and stray magnetic field are compensated by sets of large Helmholtz coils to less than about 10 mGauss. The electron spectra are recorded with a constant analyzer energy, so that the transmission of the spectrometer shows only a small variation over the energy interval in our studies.

Electron spectra obtained with He^{2+}-projectiles scattered from Al(111)

We will discuss here some electron spectra obtained with He^{2+}-projectiles scattered under grazing angles of incidence from a clean Al(111)-surface. The cleanness of the surface is controlled by Auger spectroscopy. SPA-LEED-studies[16] show rather flat surfaces with terraces separated by monoatomic steps and average widths larger than 100 nm. These high quality surfaces are an essential prerequisite for grazing incidence surface collisions, where the defect structure of the target has a much larger effect on the scattering than for normal incidence. The well defined surface plane results here in trajectories which are characterized by a small angular spread for the scattered projectiles with FWHM polar widths ranging from 0.3^o to less than 1° for projectile energies from 2 keV to 25 keV and $\Phi_{in} < 3^o$.

Electron spectra for He^{2+}-projectiles scattered from metal surfaces under normal impact have been investigated in early studies by Hagstrum[17] and Hagstrum and Becker[18]. Recent studies have been reported by the group of Niehaus[19], Folkerts[20], and Schippers et al.[21]. The spectra show a broad low energy peak with a pronounced tail ranging up to a peak structure at about 35 eV due to the autoionization (AI) of doubly excited He I $2s^2$ ^1S and $2p^2$ ^1D. A detailed analysis and a simulation of the spectra obtained for a Pb(111)-surface has been performed by Zeijlmans van Emmichoven et al.[22].

In Fig.11 we show two spectra obtained for 10 keV He^{2+}-ions scattered from an Al(111)-surface under $\Phi_{in} = 2^o$ for an azimuthal orientation of the projectile beam along the $<1\bar{1}0>$-direction and $\Theta_{in} = 5^o$ with respect $<1\bar{1}0>$ ("random"), respectively. Aside from some uncertainties in the comparison of the intensities obtained in both spectra, we notice a clear difference in the low energy peak structures and their intensities relative to the tails and the autoionization peaks. The spectra indicate that the larger electron emission yield γ for an orientation along a low index direction - as observed by the target current (see previous section) - is due to an enhanced emission of low energy electrons. Electrons originating from Auger neutralization (AN) and autoionization processes are emitted in front of the surface plane, so that their intensities should depend only weakly on the arrangement of atoms at the target surface. The kinetic emission of electrons, however, is concentrated to a domain close to the surface plane and eventually to deeper layers of the crystal. This was the basic assumption in the previous section for the explanation of the pronounced structural dependences found for the target current and explains also this feature of our electron spectra shown in Fig.11.

A striking feature of the data for a scattering of the beam along <1$\bar{1}$0> is a peak at about 9 eV. We have studied the formation of this structure by a variation of the angle of incidence from Φ_{in}= 1° to 2.9° and find a relative decrease of this peak in comparison to a peak structure at an energy of about 3 eV (Fig.12). Such a low energy peak structure is observed also for normal impact of keV-ions[23] and is ascribed to the excitation of single conduction and core electrons. However, we note that because of the low energy the electrons are easily affected by residual magnetic and electric fields. Then it is likely that the form and position of this peak structure will be changed by small fields and a varying transmission of the spectrometer (see below).

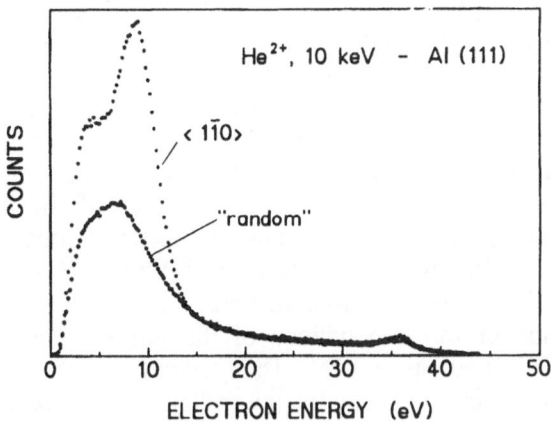

Figure 11. Electron spectrum obtained for 10 keV He^{2+}-ions impinging on a Al(111)-surface under Φ_{in}= 2° along <1$\bar{1}$0> and Θ_{in}= 5° with respect to <1$\bar{1}$0>.

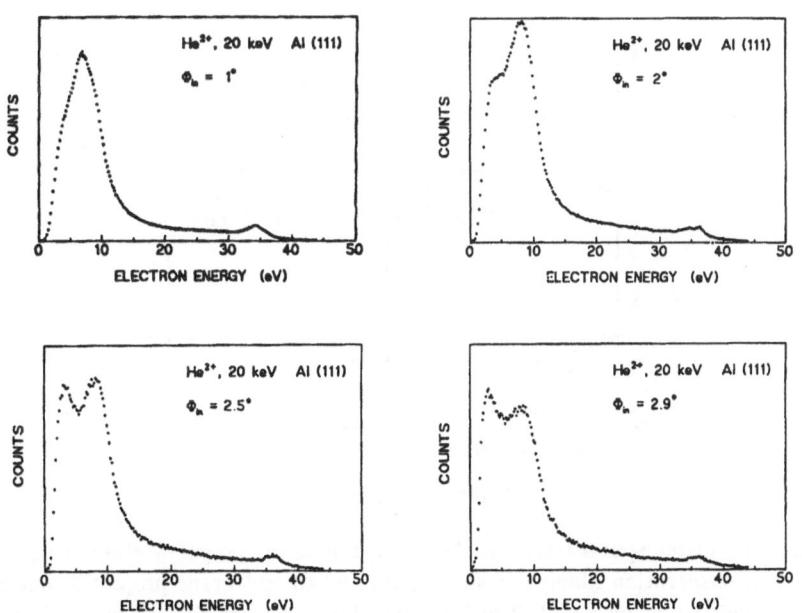

Figure 12. Same as Fig.11, but for a variation of the angle of incidence Φ_{in}.

The peak at an electron energy of about 9 eV is positioned between the energies for the single electron-hole pair decay of surface (ω_s- W = 6.3 eV) and bulk plasmons (ω_p- W = 10.8 eV) in aluminum and may be attributed to a superposition of both excitation modes. This interpretation is supported by our observation that this peak does not appear in experiments with surfaces of transition metals (e.g. Fe(110)), where pronounced plasmon peaks are absent in the excitation spectra. In addition, the overall structure does not depend on the nature of the projectile. Similar structures at this energy have been observed for electron scattering[24] and for normal impact of fast ions[25] on aluminum surfaces. Further investigations of this feature are in progress.

Another interesting aspect of the spectra shown in Figs.11 and 12 is related to the intensity of the peaks at about 35 eV due to autoionization of doubly excited helium atoms in front of the surface plane. These excited terms are populated via resonant electron transfer processes on the incident trajectory at relatively large distances from the surface. In Fig.13 we show a selection of electron spectra obtained by Folkerts[20] for He^{2+} impinging on a polycrystalline tungsten target under Φ_{in}= 15° and observation of electrons normal to the surface plane. The spectra show a decrease of the peak intensities with increasing projectile velocity. A similar decrease of the intensity can be found for a constant projectile energy and an increase of the angle of incidence[21]. From these experiments it is evident that the energy of the normal motion $E_y = E \sin^2\Phi_{in}$ is a crucial parameter in the understanding of these observations. This has been found out already by Hagstrum and Becker[18] for normal incidence scattering of He^{2+}-ions with a Ni(100)-surface.

An interpretation of the dependence of the intensity of the AI-peak on E_y has been given already in ref.18. The formation of the doubly excited He I $(2l)^2$-terms proceeds via resonant neutralization, where a competing Auger

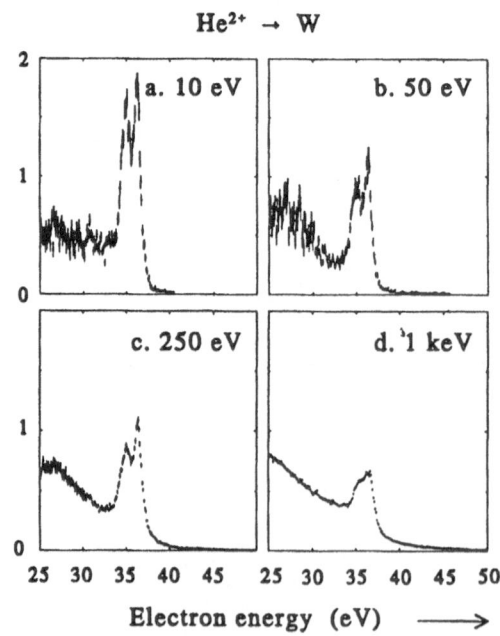

Figure 13. Electron spectra for He^{2+}-ions incident on polycrystalline tungsten under Φ_{in}= 15° and observation along the surface normal for different projectile energies. The data are obtained by Folkerts[17]. The corresponding energies for the normal motion are: 0.7 eV (10 eV), 3.3 eV (50 eV), 16.7 eV (250 eV), and 67.0 eV (1 keV).

neutralization channel may form He I 1s2s. The decay of the doubly excited terms takes place via autoionization - an intraatomic process associated with the emission of an electron of well defined energy (see spectra) - or via Auger deexcitation. Since the Auger processes are effective at clearly closer distances (typically about 3 - 4 a.u.) than the onset of resonant neutralization processes (about 10 - 15 a.u.) and AI-rates are basically independent of the distance from the surface, a AI-decay of He I $(2l)^2$ is favourable for long time scales; i.e. small energies for the normal motion E_y. At larger E_y the projectiles reach the close vicinity of the surface after the onset of the electron transfer much faster, and the formation and the decay of the doubly excited terms are dominated by the Auger mechanisms instead of autoionization.

We have studied the intensities of the AI-peaks for a variation of the projectile energy by a setting of the angle of incidence to $\Phi_{in}= 2^o$, which is a smaller angle than used in previous studies. The data displayed in Fig.14 indicate, as found in similiar studies of this type, a pronounced decrease of the intensity of the AI-peaks with increasing projectile energy (corresponding also to increasing E_y). However, a comparison with the data of Folkerts[20] in Fig.13 reveal an essential difference in the dependence on E_y. For low projectile energies the AI-peak heights relative to a background are about comparable in both data sets for about the same E_y: 50 eV data in Fig.13 ($E_y= 3.3$ eV) and 4 keV data in Fig.14 ($E_y= 4.8$ eV). But for larger projectile energies this situation is changed dramatically. The data for 250 eV at $\Phi_{in}= 15^o$ ($E_y= 16.7$ eV) and for 15 keV at $\Phi_{in}= 2^o$ ($E_y= 18.2$ eV) are obtained at about the same E_y. However, whereas for 250 eV projectiles the AI-peaks form a prominent structure in the selected spectrum, the corresponding peaks in our data for 15 keV are comparable with the "background" signal.

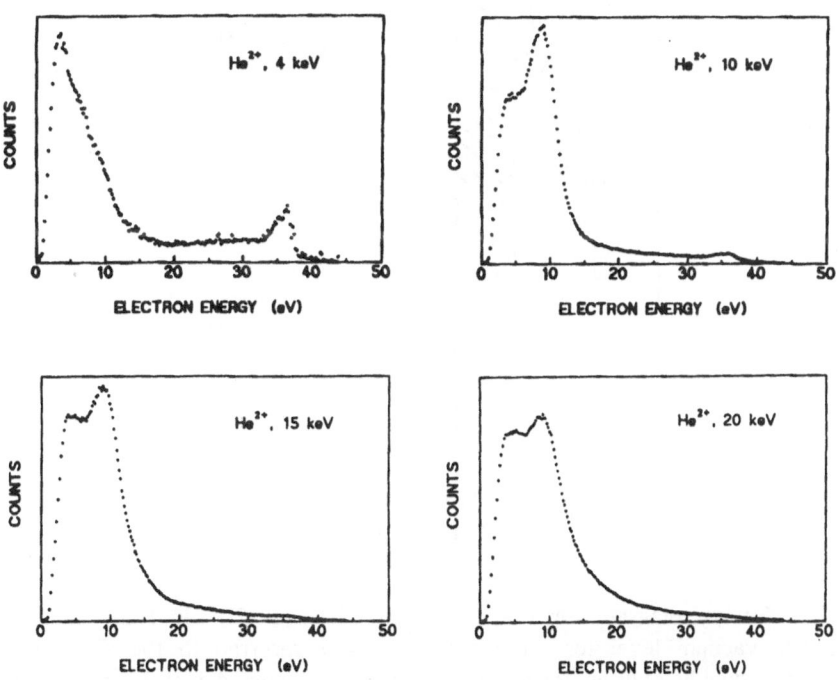

Figure 14. Electron spectra obtained during impact of He^{2+}-ions with different energies with a clean Al(111) under $\Phi_{in}= 2^o$. The corresponding energies for the normal motion are: 4.9 eV (4 keV), 12.2 eV (10 keV), 18.3 eV (15 keV), and 24.4 eV (20 keV).

Our observation implies that in addition to the energy of the normal motion E_y also the parallel motion with about the projectile energy has an effect on the emission of electrons via autoionization. A rather simple explanation would be the relative increase of the kinetic electron emission yield with increasing projectile energy, which results in an enhanced background for the AI-peak. Since the structure of the tails in the electron spectra does not vary significantly for the higher projectile energies, we conclude that this effect seems to play only a minor role here.

Therefore we consider an alternative mechanism, which is related to a specific feature of grazing ion-surface collisions. Extensive investigations of charge state distributions[26] have shown that charge transfer in this geometry is affected in a characteristic way by the effect of the energy (velocity) of the motion parallel with respect to the surface plane $E_x = E \cos^2\Phi_{in} \sim E$. In a simple picture one can illustrate the effect of the parallel motion, i.e. the Galilei-transformation between surface and moving atom, by a "Doppler-Fermi-Dirac distribution"[27] of electronic conduction band levels in the moving frame of the projectile. The parallel velocity effect on the distribution of occupied and unoccupied electronic levels in the metal is similiar to the result of a (very) high temperature for a "static" Fermi-Dirac distribution.

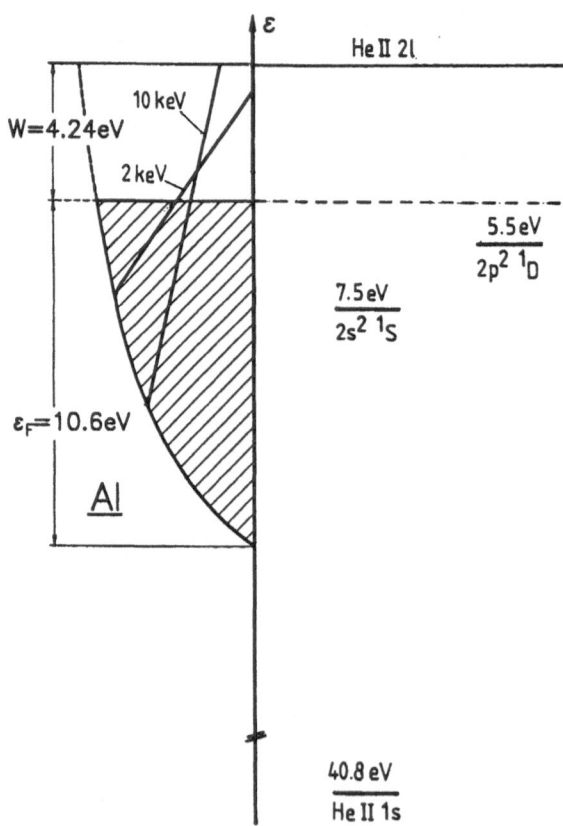

Figure 15. Energy diagram for the ionization of He I $(2l)^2$ in front of an Al(111)-surface. The vacuum level for the atomic terms is referred to the single ionization energy of the doubly excited terms. The density of occupied metal levels in the free electron metal is sketched for the static case (hatched area) and for finite energies of the parallel motion ("Doppler-Fermi-Dirac distribution"). The AI-peaks result from the ejection of an electron in the transition from He I $(2l)^2$ to He II 1s.

In Fig.15 we have sketched an energy diagram of the two autoionizing He I $(2l)^2$-terms and an aluminum metal, where the vacuum level for the atom is adjusted to the single ionization limit He II 2l. From this scheme we find that in the "static" case, i.e. for low projectile energies, the $(2l)^2$-terms are in resonance with occupied metal states only, so that these terms can not be depleted via resonant ionization. This situation changes with increasing projectile energy (better: velocity of the parallel motion), because the Doppler-effect brings an increasing number of unoccupied metal states into resonance with the $(2l)^2$-terms, opening up an effective competing channel for the depopulation of these terms. At a projectile energy of 10 keV (see "Doppler-Fermi-Dirac distribution" sketched in Fig.15) the contribution of resonant ionization is so large already that the autoionization plays only a minor role in comparison to the "static" case at low energies. This interpretation is supported by our observation that the intensities of the AI-peaks depend strongly on the workfunction of the metal. For e.g. Fe(110) (W = 5.0 eV) the AI-peaks are clearly weaker than for Al(111) (W = 4.24 eV) and show the expected stronger effect on the parallel motion. Corresponding data are presented in the nect section.

Electron spectra observed in collisions with a Fe(110)-surface

We will close our presentation of recent studies on electron emission phenomena in grazing ion-surface collisions with a discussion of results obtained with a Fe(110)-surface. In Fig. 16 we show a spectrum obtained with 10 keV He^{2+}-ions scattered under Φ_{in}= 2° and random azimuthal orientation from Fe(110). This spectrum deviates clearly from the results displayed in Fig.11 for the Al(111)-target. We find a low energy peak structure with a smaller width and a sharp maximum at about 3 eV. The relative extension of the tail of the peak structure towards larger energies is less pronounced here than for Al(111), and no indication for the AI-peaks around 35 eV can be found.

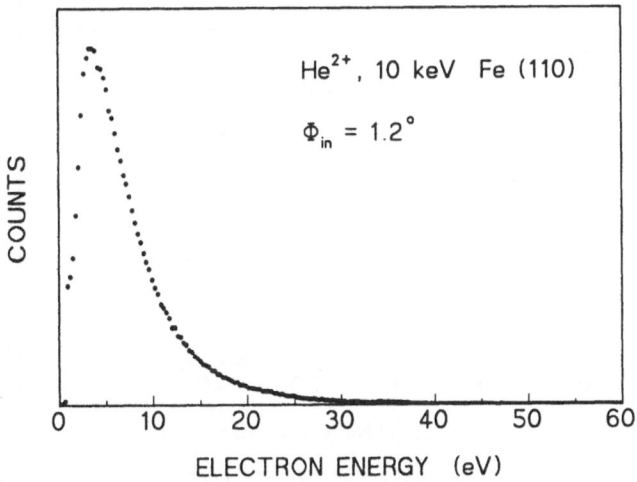

Figure 16. Electron spectrum observed for 10 keV He^{2+}-ions scattered from a Fe(110)-surface under Φ_{in}= 1.2° and random azimuthal orientation.

At lower projectile energies (e.g. 5 keV) the Al-peaks can be identified, however, the relative intensities are much smaller than for the Al(111)-target. This is consistent with our interpretation given above, since the workfunction of Fe(110) is larger than for Al(111), so that the kinematically induced resonance ionization process is more effective here.

In Fig.17 we show a spectrum obatained for 20 keV He$^+$-ions, which aside from a broader peak structure is comparable with the data in Fig.16. In the previous section we have pointed out already the sensitivity of the results obtained at low electron energies on residual fields and on the transmission of the spectrometer. With the setup used in these studies it was not possible to improve the experimental conditions in this respect, so that we are left with quite some uncertainties with respect to the shape of the spectra in the low electron energy domain. In this respect we state that we did not succeed here to find evidence for a complete suppression of kinetically emitted electrons and structured low energy spectra as claimed by Rau et al.[28] for 25 keV protons and He$^+$-ions impinging under comparable target and detector settings on Ni(110).

Interesting information is obtained from a study, where we biased the target by a low negative voltage V, which shifts all electron energies by $E_V = eV$ towards larger values and facilitates the handling of electrons with initially very low energies. In Fig.17 we show data for V = - 10 Volts, which are - aside from a trivial shift by 10 eV - clearly different than the data obtained with a target kept on ground potential. It is obvious that the low energy onset for the spectra obtained with the accelerated electrons is much steeper. This may indicate probable spurious effects on the low energy data for a grounded target. With an additional electrode in the vicinity of the biased target we tried to make sure that the steep increase of the spectra is not affected by electron focusing effects due to the additional electric field. Work on this subject is in progress.

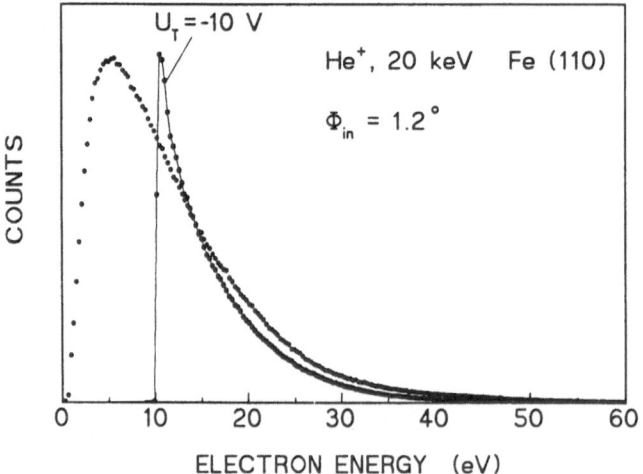

Figure 17. Electron spectra observed for 20 keV He$^+$-ions scattered from a Fe(110)-surface under Φ_{in}= 1.2° and random azimuthal orientation with the target kept on ground potential and biased by a voltage of - 10 Volts, respectively.

REFERENCES

1. M. Hasegawa, K. Kimura, Y. Fujii, M. Suzuki, Y. Susuki, and M. Mannami, Nucl. Instr. Meth. B33:334 (1988).
2. D.S. Gemmell, Rev. Mod.Phys. 46:129 (1974).
3. C. Erginsoy, Phys. Rev. Lett. 15:360 (1965).
4. J. Lindhard, Kgl. Dan. Vidensk. Selk. Mat. Fys. Medd. 34:No.14 (1965).
5. U. Bill, R. Sizmann, C. Varelas, and K.E. Rehm, Rad. Eff. 27:59 (1975).
6. C. Varelas, Habilitationsschrift, Universität München (1979).
7. G. Moliere, Z. Naturforschung A2:133 (1947).
8. H.J. Andrä, H. Winter, R. Fröhling, N. Kirchner, H.J. Plöhn, W. Wittmann, W. Graser, and C. Varelas, Nucl. Instr. Meth. 170:527 (1980).
9. H. Winter, Thesis, Universität München (1986) [author not identical with first author of this paper].
10. R. Pfandzelter, Nucl. Instr. Meth. B48:351 (1990).
11. H. Winter, Physics Reports, to be published.
12. K. Mednick and L. Kleinman, Phys. Rev. B22:5768 (1980).
13. M. Posternak, H. Krakauer, A.J. Freeman, and D.D. Koelling, Phys. Rev. B21:5601 (1980).
14. M. Schuster and C. Varelas, Surf. Sci. 134:195 (1983).
15. H.H.W. Feijen, L.K. Verhey, A.L. Boers, and E.P.Th.M. Suurmeijer, J. Phys. E6:1174 (1973).
16. U. Scheithauer, G. Meyer, and M. Henzler, Surf. Sci. 178:441 (1986).
17. H.D. Hagstrum, Phys. Rev. 122:83 (1961).
18. H.D. Hagstrum and G.E. Becker, Phys. B8:107 (1973).
19. P.A. Zeijlmans van Emmichoven and A. Niehaus, Comm. At. Mol. Phys. 24:65 (1990).
20. L. Folkerts, Thesis, Rijksuniversiteit Groningen (1992).
21. S. Schippers, S. Oelschig, W. Heiland, L. Folkerts, R. Morgenstern, P. Eeken, I.F. Urazgil′din, and A. Niehaus, Surf. Sci. 257:289 (1991).
22.. P.A. Zeijlmans van Emmichoven, P.A.A.F. Wouters, and A. Niehaus, Surf. Sci. 195:115 (1988).
23. D. Hasselkamp, Comm. At. Mol. Phys. 21:241 (1988).
24. J. Pillon, D. Roptin, and M. Cailler, Surf. Sci. 59:741 (1976).
25. D. Hasselkamp and A. Scharmann, Surf. Sci. 119:L388 (1982).
26. H. Winter and R. Zimny, in: "Coherence in atomic collisions physics", eds. H. J. Beyer et al., Plenum Press, New York, p.283 (1988).
27. D.M. Newns, Comm. Cond. Mat. Phys. 14:295 (1989).
28. C. Rau, K. Waters, and N. Chen, Phys. Rev. Lett. 64:1441 (1990).

ELECTRONIC ENERGY LOSS OF FAST PROTONS
IN PLANAR SURFACE CHANNELING

R. Pfandzelter and F. Stölzle

Sektion Physik
Universität München
Amalienstraße 54
8000 München 40
Germany

INTRODUCTION

The trajectories of fast ions which are incident at a grazing angle upon an atomically flat surface are determined by many consecutive small angle deflections due to collisions with the surface atoms. There is some correlation between the collisions, which gives rise to a governed motion. This motion can be described using the concept of continuum scattering: the ions do not feel individual surface atoms, but are steered by the planar continuum potential, which is obtained by averaging the individual screened Coulomb potentials along the plane. As a result, the energy in the motion perpendicular to the surface (transverse energy) is conserved, i.e. the angle of reflection equals the angle of incidence (specular reflection). Such a motion we may term *planar surface channeling* or *semiplanar channeling* due to its close relation to the well-known planar channeling in the bulk.

In planar surface channeling the interaction of the ions with the crystal atoms is essentially restricted to the topmost layer. This fact accounts for the extreme surface sensitivity of experimental techniques involving grazing-angle ion-surface scattering. Most promising in this connection is the (spin-polarized) electron emission spectroscopy (see the contributions of Rau[1] and Winter[2] in these proceedings), where medium-velocity ($v \approx v_0$, where v_0 is the Bohr velocity) light ions are grazingly scattered by a flat crystal surface and the spin- and angle-resolved energy distribution of emitted electrons is measured. This technique is successfully employed to retrieve information on the electronic and magnetic properties of the surface. The most important theoretical models of electron emission consider the total electron yield to be proportional to the electronic stopping power. It is thus indispensable for future progress in electron emission induced by planar surface channeled ions to know the electronic stopping power near the surface and its dependence on the distance and velocity of the ion.

Ionization of Solids by Heavy Particles, Edited by
R.A. Baragiola, Plenum Press, New York, 1993

In the following we report experiments on the electronic energy loss of medium-velocity ($0.76v_0 \leq v \leq 1.88v_0$) H^+ and H^0 in planar surface channeling. Experiments in the medium-velocity range are highly desired, because, apart from the above mentioned applications, this is a transition region for many available theories. Beyond that, analysis of energy loss spectra in planar surface channeling gives direct indication of the trajectories involved. This is of particular interest, since the ions' trajectories should be subject to the (dynamical) image interaction.

HIGHLY ORIENTED PYROLYTIC GRAPHITE SURFACES

The lateral length of interaction with the surface atoms in planar surface channeling of medium-velocity H^+ amounts to some ten nm for ordinary specular reflection. A small fraction of the incident projectiles is expected to describe more intricate trajectories due to the image attraction, where multiple reflections are involved[3]. The wavelengths of these skipping trajectories are estimated at about one hundred nm. Thus, experiments on grazing-angle scattering of medium-velocity H^+ require target surfaces with regions of atomic flatness extending over hundreds of nm. Polycrystalline highly oriented pyrolytic graphite (HOPG) surfaces have been found to meet this crucial prerequisite and thus provide a unique opportunity for studying planar surface channeling with considerable potential for a successful observation of skipping motion. Large-scale scanning tunneling microscopy (STM) images of freshly cleaved HOPG show a surface which is characterized by very flat terraces separated by small steps with heights between 0.66 and 3.01 nm[4]. The lateral dimensions of the terraces are on the order of several hundred nm. Atomic-scale images show the characteristic hexagonal lattice pattern of the unreconstructed surface, devoid of any defects[5].

Our HOPG samples ($12 \times 12 \times 1.5$ mm^3) have a Gaussian distributed mosaicity with $\sigma = 0.13° \pm 0.02°$ and $\sigma = 0.16° \pm 0.02°$, respectively. The azimuthal

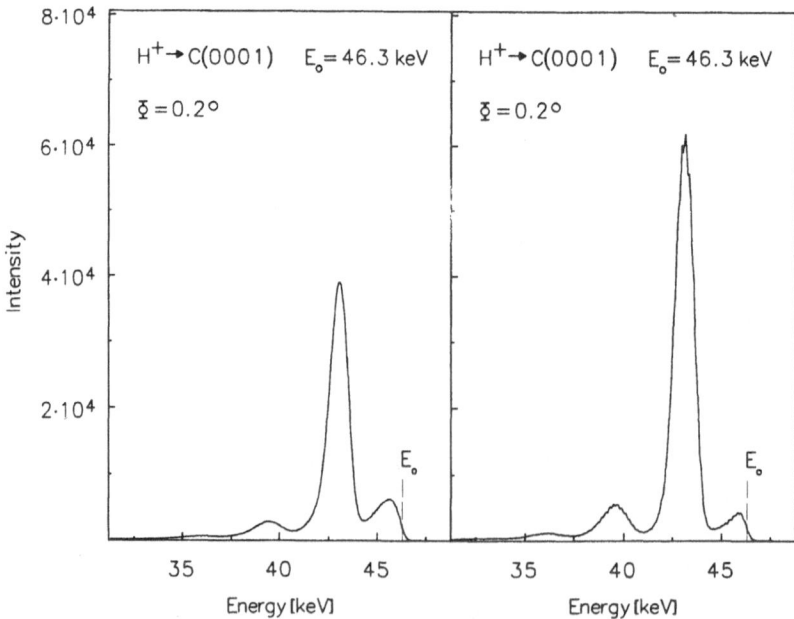

Figure 1. Energy spectra of H^+ specularly reflected from a graphite surface before annealing (left panel) and after annealing at 1000°C for 10 min (right panel).

orientations of the crystallites are randomly distributed as confirmed by low energy electron diffraction (LEED). The linear dimensions of the crystallites are on the order of 10 μm as obtained by stylus surface profiling. Each sample is cleaved in air immediately before it is mounted into the scattering chamber. The extreme low gas adsorption efficiency guarantees a clean surface devoid of any adsorbate atoms as confirmed by Auger electron spectroscopy (AES). Prior to the measurements, the samples are annealed at 1000°C for 10-120 min. This treatment yields an additional flattening effect (thermal surface smoothing), which is monitored by the gradual increase of the yield of specularly reflected H^+ (fig.1).

Fig.2 shows a typical STM image at large-scan size (1 μm $\times 1\mu$m) of our HOPG sample[6]. The extended flat terraces are separated by a sharp, well-defined boundary. Smaller-scan images reveal that these boundaries consist of steps or step-like structures with some nm in breadth.

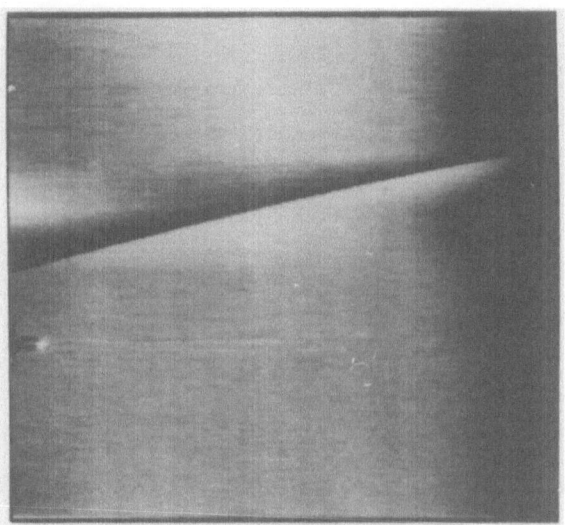

Figure 2. STM image of a 1μm $\times 1\mu$m area of our HOPG sample.

EXPERIMENTAL

A mass-selected, mono-energetic H^+ beam is incident upon the graphite surface at a grazing angle Φ. Diaphragms in the beam line collimate the beam to a size of 0.3×0.1 mm^2 (1.0×0.2 mm^2 for the 0.76 v_0 measurements) and a divergence of less than $\pm 0.03°$. The current densities are 1-10 nA/mm^2. The scattered H^+ are energy analyzed by a 127° cylindrical analyzer (ESA) with an energy resolution of $\Delta E/E = 10^{-2}$. The ESA is positioned at an angle Φ to the surface in the plane containing the surface normal and the incident beam. The angular acceptance range is $\Delta\Phi = \pm 0.12°$. The H^0 beam ($\pm 0.1°$ divergence) is generated by transmission of an H^+ beam through a thin self-supporting foil (3 μg/cm^2) and elimination of the H^+ part in the transmitted beam by electrostatical deflection. The experiments were performed in a UHV scattering chamber ($5 \cdot 10^{-10}$ mbar) equipped with a LEED/AES system and coupled to an ion accelerator via a differentially pumped beam line.

ENERGY LOSS SPECTRA

In fig.3 some energy spectra of planar surface channeled H^+ for an incident H^+ beam of energy $E_0 = 75.8$ keV ($v = 1.74\,v_0$) and different incidence angles Φ are reproduced. The spectra are dominated by a strong peak with nearly Gaussian distribution (the *most probable* energy loss deviates from the *mean* energy loss by less than 20 eV). This peak is caused by ordinary specular reflection, i.e. the ions approach the surface plane, suffer a series of correlated small-angle deflections and are reflected back. The energy loss of the main peak decreases monotoneously with increasing incidence angle Φ. This has been found to hold true for all projectile velocities used (fig.4). There are two counteracting effects determining the variation of the energy loss with increasing incidence angle: the decreasing length of interaction and the increasing electron density sampled. It is evident that the former effect predominates here. It has been further found that the energy loss does not depend on the charge state of the incident projectiles. The energy losses for H^+ (fig.4, triangles)

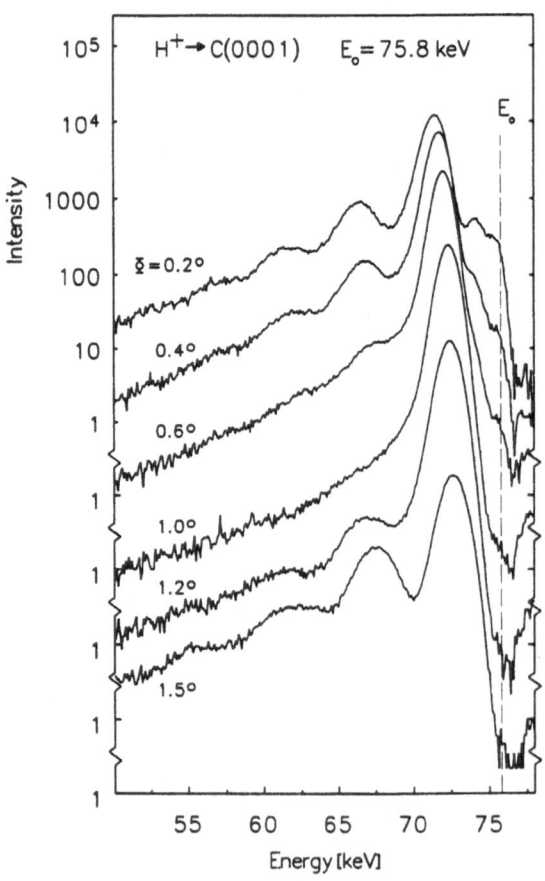

Figure 3. Energy spectra of H^+ specularly reflected from a graphite surface for an incident H^+ beam of energy $E_0 = 75.8$ keV ($v = 1.74\,v_0$) and different incidence angles Φ. For clarity, the origins have been displaced vertically.

and H^0 (fig.4, squares) are the same within experimental error for all incidence angles and velocities. Since both the planar surface potential and the stopping power are expected to depend on the projectile's charge state, it must be concluded that charge state equilibrium is established already at large distances from the surface, where energy loss is negligible.

Besides the main peak, the spectra at the smallest and largest incidence angles used exhibit at least two further discrete peaks at lower energies (i.e. larger energy losses). The peaks are separated by nearly equal energy spacings. The peaks at the smallest incidence angles can be reconciled with skipping motion type of trajectories, which are characterized by multiple approaches of the ions to the surface plane[7,8]. Each subsequent peak represents an additional vibration in the surface potential well, which is composed of the repulsive planar continuum potential and the attractive dynamical image potential. Trapping and detrapping can only occur when transverse energy conservation is violated. One reason may be the insufficiency of the continuum scattering approximation: in planar channeling, there is no correlation in the *lateral* component of the impact parameter in successive collisions, only in the *transverse* component. This causes intrinsic fluctuations in the transverse energy. Another reason one can envisage are charge exchange processes, which occur frequently in the vicinity of the surface[9]. This is supported by the observation that trapping into the surface potential well is slightly more effective, when H^0 are used as projectiles, since the image attraction is missing on the incident trajectory (fig.5). With increasing incidence angle Φ, the skipping peaks grow weaker and eventually disappear at an angle $\Phi \approx 0.8°$. The energy losses of the main peak, the second and the third peak are in a ratio of nearly 1:2:3, as has been predicted theoretically for skipping motion[10].

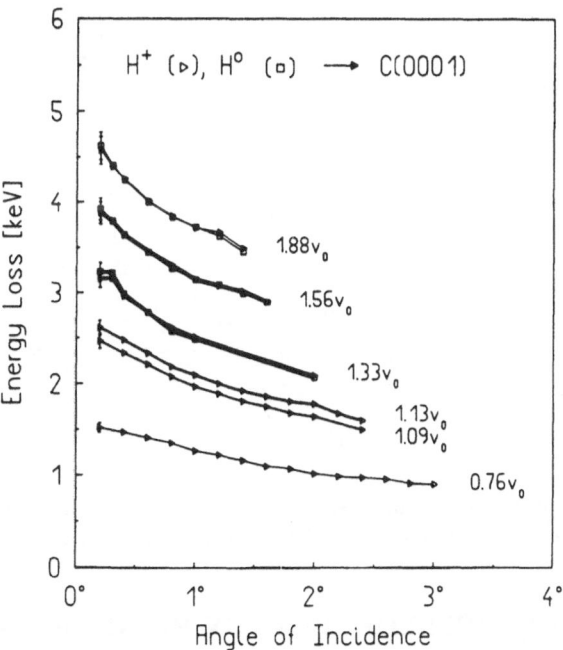

Figure 4. Measured energy loss of H^+ (triangles) and H^0 (squares) planar surface channeled at a graphite surface versus incidence angle Φ for different velocities as indicated. The symbols are connected by straight lines.

At $\Phi \approx 1.1°$ another multi-peaked structure appears. Now, the energy losses of the peaks are in a ratio of roughly 1:3:5. This points to a subsurface channeling[11]: the ions penetrate the surface layer and experience planar channeling below the surface. Subsurface channeling is most effective for incidence angles slightly larger than the critical angle for (ordinary) specular reflection Φ_s, which we define as the angle where the yield of specularly reflected H^+ has decreased by half. The critical angle corresponds to a critical transverse energy $E_{\perp s} \approx E_0 \Phi_s^2$, which is about 25 eV in the present case. For Φ not much larger than Φ_s, the transverse energy is high enough that the ions feel the individual atomic potentials, but there still remains some correlation in the scattering. In the bulk, such trajectories are refered to as quasi-channeling[12].

The small "bulge" between the main peak and the incident energy (fig.3) evolves from scattering at step- and hump-like structures at the surface. It decreases considerably upon annealing (cf. fig.1).

It should be noted that the energy loss is practically due only to inelastic energy transfer (electronic stopping). Nuclear stopping can be neglected: the elastic energy loss for a binary collision between H and C is $\Delta E/E_0 = 1 \cdot 10^{-4}$ for a typical scattering angle of $2°$. This is much smaller than the measured energy losses, which are about $\Delta E/E_0 = 5 \cdot 10^{-2}$ for the main peak. Add to this, in channeling the elastic stopping power is further reduced by orders of magnitude. This is supported by the absence of any beam dose effects on the measured energy spectra. Even after doses of $5 \cdot 10^{15}$ H^+/cm^2, no decrease of the yield of specularly reflected H^+ could be observed.

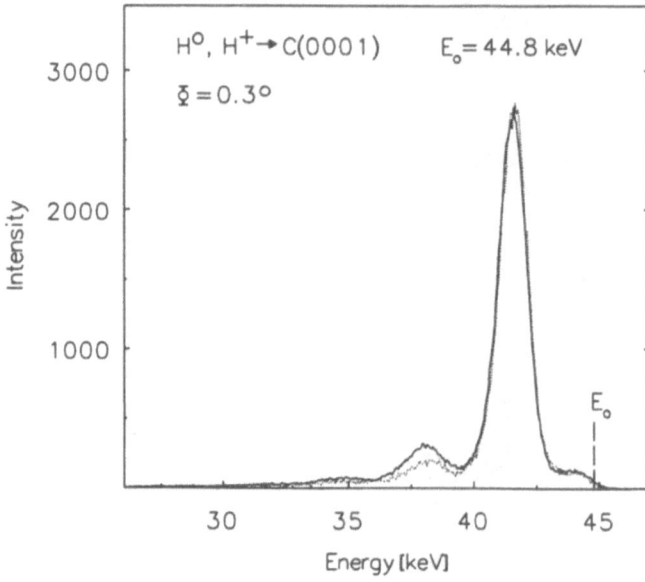

Figure 5. Energy spectra of H^+ specularly reflected from a graphite surface for an incident H^+ beam (dashed line) and H^0 beam (solid line) of energy $E_0 = 44.8$ keV ($v = 1.34v_0$).

DERIVATION OF THE STOPPING POWER NEAR THE SURFACE

The energy loss of the planar surface channeled projectiles is given by

$$\Delta E(v, E_\perp) = \int S(v, z)\, dx \qquad (1)$$

where $S(v, z)$ is the stopping power or energy loss per unit path length of travel near the surface and z and x are the position coordinates perpendicular and parallel to the surface, respectively. The integration is carried out along the trajectory. We can treat v as a constant parameter and, upon changing the variable of integration, we get

$$\Delta E(v, E_\perp) = 2 \int_{z_{min}(v, E_\perp)}^{\infty} S(v, z) \sqrt{\frac{E_0}{E_\perp - V(v, z)}} \, dz \qquad (2)$$

where $z_{min}(v, E_\perp)$ is the distance of closest approach and $V(v, z)$ the planar surface potential; it is the sum of the planar continuum potential and the dynamical image potential. With increasing projectile velocity v, the image potential decreases due to the non-adiabatic response of the target electrons. This should be roughly compensated by the lesser screening of the image attraction due to transient neutralisation. Thus, in a first approximation, the velocity-dependence of the planar surface potential may be neglected. Due to our limited range of velocities, it should be further possible to factorize $S(v, z)$ in $S(v) \cdot F(z)$. Eq.2 thus reduces to

$$\Delta E(v, E_\perp) = \tau(E_\perp) \cdot S(v) \cdot v \qquad (3)$$

with

$$\tau(E_\perp) = \sqrt{2m} \int_{z_{min}(E_\perp)}^{\infty} F(z) \sqrt{\frac{1}{E_\perp - V(z)}} \, dz \qquad (4)$$

being only dependent on the transverse energy. $\tau(E_\perp)$ can be considered as effective interaction time of the projectile with the surface.

Theoretical studies of the energy loss of particles moving slowly ($v \ll v_0$) parallel with a surface show that the stopping power near the surface is proportional to v and, thus, equal to the velocity-dependence of the theoretical bulk stopping power[13]. It is tempting to check whether the same holds for the medium-velocity range. According to eq.3, we plot in fig.6 the measured energy losses $\Delta E(v, E_\perp)$ of H^+ (open symbols)

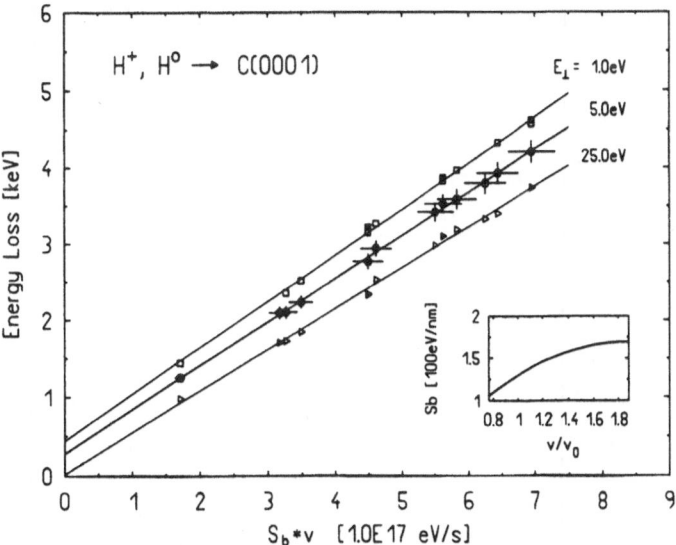

Figure 6. Measured energy loss ΔE of H^+ (open symbols) and H^0 (solid symbols) planar surface channeled at a graphite surface versus $S_b(v) \cdot v$. The straight lines are linear fits to the data corresponding to the transverse energy $E_\perp$ as indicated. The bulk stopping power $S_b(v)$ is shown in the inset[14].

and H^0 (solid symbols) versus $S_b(v) \cdot v$ ($S_b(v)$ is the stopping power in the bulk[14]) for some selected values of $E_\perp$.

Fig.6 shows that the measured energy losses ΔE actually lie on a family of straight lines with the transverse energy $E_\perp$ as parameter, which confirms our assumption that $S(v) \propto S_b(v)$. The slopes $\tau(E_\perp)$ of the straight lines hardly depend on $E_\perp$; they are about $5.5 \cdot 10^{-15}$ s on average. This, incidentally, corresponds to an effective interaction length of 12 nm at $v = v_0$.

A more detailed inspection of fig.6 reveals that the lines have small intercepts $\Delta E_2(E_\perp) > 0$. This indicates an additional contribution to the energy loss. ΔE_2 amounts to about a quarter of the total energy loss at the smallest transverse energies and velocities used; it is about zero at the critical transverse energy $E_{\perp s} \approx 25$ eV. According to eq.3, the corresponding stopping power $S_2(v)$ must be proportional to v^{-1}, since ΔE_2 does not depend on v.

In conclusion, the energy losses of the planar surface channeled H^+ or H^0 actually obey to $\Delta E(v, E_\perp) = \Delta E_1(v, E_\perp) + \Delta E_2(E_\perp)$; we have two contributions to the total electronic energy loss, i.e. two contributions S_1 and S_2 to the total electronic stopping power may be distinguished. The dependence of S_1 on the projectile velocity is the same as in the bulk (see inset of fig.6), whereas S_2 is proportional to the reciprocal of the velocity.

In order to get information on the position-dependences of the two contributions to the electronic stopping power, we numerically evaluated the integral of eq.1 using appropriate expressions for the planar continuum potential and the screened dynamical image potential[15]. We then fitted the experimentally derived slopes $\tau(E_\perp)$ and intercepts $\Delta E_2(E_\perp)$ to the numerical results for different reasonable functions F_1 and F_2. The result is shown in fig.7. The (predominant) contribution S_1 decreases exponentially with increasing distance z. The decay constant is about $3a_0$ (a_0 is the Bohr radius). The second contribution S_2 has a position-dependence which is markedly different: it is effective only within a small interval of distances some a_0 outside the surface plane.

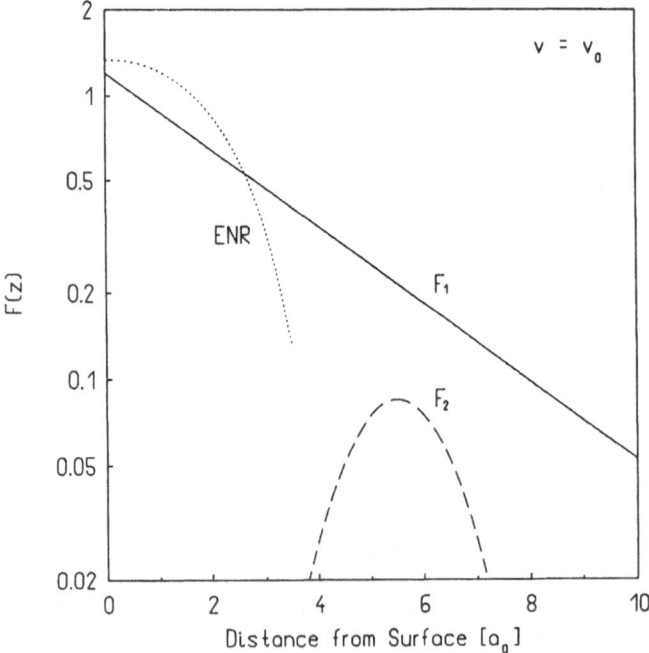

Figure 7. Dependence of the stopping power S_1 (solid line) and S_2 (dashed line) on the distance z from the surface plane as derived from the experimental data. Also plotted are the density-functional results from Echenique et al.[16] (dotted line).

In fig.7, we have plotted for comparison the theoretical stopping power obtained locally using the density-functional formalism. At larger distances from the surface, the experimentally derived stopping power decreases much more slowly than the theoretical one. This is attributed to nonlocal effects due to the large variation of the electron density at the surface. The physical origin of S_2 is unclear so far. Its effective location some a_0 outside the surface suggests surface plasmon excitation to be responsible.

ACKNOWLEDGEMENT

The authors would like to thank Dr. R. Möller for probing the sample with STM. This work was funded by the German Federal Minister for Research and Technology (BMFT) under the contract number 03SI2LMU1.

REFERENCES

1. C. Rau, Ion-induced electron emission from magnetic and non-magnetic surfaces, these Proceedings (NATO Advanced Research Workshop on Ionization of Solids by Heavy Particles), ed. R. A. Baragiola, Taormina (1992).

2. H. Winter, Electron emission in grazing ion surface collisions, these Proceedings (NATO Advanced Research Workshop on Ionization of Solids by Heavy Particles), ed. R. A. Baragiola, Taormina (1992).

3. Y. H. Ohtsuki, K. Koyama and Y. Yamamura, Skipping motion of the surface scattering of ion beams, Phys. Rev. B 20 (1979) 5044.

4. L. M. Siperko, Scanning tunneling microscopy of various graphitic surfaces, J. Vac. Sci. Technol. B 9 (1991) 1061.

5. R. Coratger, A. Claverie, F. Ajustron and J. Beauvillain, Scanning tunneling microscopy of defects induced by carbon bombardment on graphite surfaces, Surf. Sci. 227 (1990) 7.

6. R. Möller, Universität Konstanz, Konstanz (1992).

7. F. Stölzle and R. Pfandzelter, Observation of discrete energy loss peaks in the energy distribution of protons reflected from a graphite surface, Phys. Lett. A 150 (1990) 315.

8. R. Pfandzelter and F. Stölzle, Observation of skipping motion, Nucl. Instrum. Methods B 67 (1992) 355.

9. H. Winter, J. Leuker, M. Sommer and H. W. Ortjohann, Effect of image charge and charge exchange on the trajectoriy in grazing ion-surface collisions: "skipping motion" and acceleration of multi-charged ions, Proceedings "14th Werner Brandt Workshop", eds. R. N. Hamm et al., Oak Ridge (1992).

10. Y. H. Ohtsuki, R. Kawai and K. Tange, Skipping motion of ions at the surface, Nucl. Instrum. Methods B 13 (1986) 193.

11. K. Kimura, M. Hasegawa and M. Mannami, Energy loss of MeV light ions specularly reflected from a SnTe(001) surface, Phys. Rev. B 36 (1987) 7.

12. D. Van Vliet, The continuum model of directional effects, in: Channeling, ed. D. V. Morgan, Wiley and Sons, London (1973).

13. R. Nuñez, P. M. Echenique and R. H. Ritchie, The energy loss of energetic ions moving near a solid surface, J. Phys. C 13 (1980) 4229.

14. F. Schulz and J. Shchuchinsky, Proton stopping cross sections for carbon, aluminium and gold: new experimental data and critical analysis of the validity of empirical fit formulas, Nucl. Instrum. Methods B 12 (1985) 90; P. Bauer, Stopping power of light ions near the maximum, Nucl. Instrum. Methods B 45 (1990) 673.

15. R. Pfandzelter and F. Stölzle, Probing the stopping power near the surface by specular reflection of protons from graphite, Nucl. Instrum. Methods B (1992), in press.

16. P. M. Echenique, F. Flores and R. H. Ritchie, Dynamic screening of ions in condensed matter, in: Solid State Physics, eds. H. Ehrenreich and D. Turnbull, Academic Press, New York (1990).

HAS THE TRUE ION-INDUCED ELECTRON YIELD FROM COPPER
AND OTHER METALS BEEN MEASURED?

J. F. Kirchhoff[1], T. J. Gay, and E. B. Hale

Physics Department
University of Missouri-Rolla
Rolla, MO 65401

[1]Present address: Department of Physics
University of North Texas
Denton, TX 76203

INTRODUCTION

Ion-induced electron emission (IIEE) is a common phenomena which has been studied for many years. IIEE occurs when an energetic (about 10 keV/amu or greater) incident ion strikes a target material. This process is often called kinetic emission because kinetic energy is transferred from the incident ion to the constituents of the target via primary and secondary ionizing collisions. Experimental reviews of IIEE have been written by several authors in recent years (Baragiola, 1979a; Benazeth, 1982; Krebs, 1983; Hasselkamp, 1988; Brusilovsky, 1990; Hofer, 1990) and comprehensive theoretical articles also serve as good reviews (Sigmund and Tougaard, 1981; Bindi et al., 1987; Schou, 1988).

Recently, we have built an ultra-high vacuum apparatus to measure IIEE from targets whose temperature can be well controlled. In this paper, measurements of the room temperature yield, γ (electrons emitted per incident ion), during 100 keV H^+ bombardment are reported for copper targets after annealing at various temperatures. These measurement gave unexpected results. The data indicate that the commonly measured and accepted yield of 1.77 electrons/ion for copper at room temperature (Hasselkamp et al., 1981; Hippler, 1988; this work) is more that 10% higher than the yield, also measured at room temperature, in samples annealed above 300°C. The yield from these annealed samples was $\gamma = 1.59$ electrons/ion.

In the past, yield values for pure metals have been reported from samples cleaned by heavy ion sputtering, but not annealed. Hence, these surfaces have been severely damaged without removal of damage or strain. Our results suggest that it seems likely many of the currently accepted, "standard" yields from metals are not representative of the true, relaxed surface of the metal.

The next section describes the apparatus used to take our data. Then, the results are reported and an analysis and discussion of the implications of the data are presented.

Ionization of Solids by Heavy Particles, Edited by
R.A. Baragiola, Plenum Press, New York, 1993

EXPERIMENTAL DETAILS

Accelerator

The ion accelerator used in this experiment was the 200 keV ion implantation accelerator which recently had a second beam line added for IIEE studies. The beamline section of the apparatus is shown in Fig. 1. A microwave-discharge ion source was used to create the proton beam from a mixture of argon and methane gas. The beam was sent to the IIEE beam line by the mass selection magnet, B1. This line was differentially pumped to separate the ultra-high vacuum (UHV) in the target/detection chamber from the more modest vacuum in the ion source and accelerator sections. The differential pumping chamber was isolated from the UHV target chamber by a long, thin beamline (10 mm dia.) which reduced contamination. The target chamber, pumped by ion pumps, attained a typical base pressure of 2×10^{-10} Torr. With the beam on during data taking, the target chamber pressure was never greater than 4×10^{-9} Torr and usually was a factor of two lower. To enter the UHV chamber, the proton beam axis was changed from horizontal to vertical via electromagnet, B2, which was situated above the detection chamber. This arrangement removed neutrals from the beam.

Yield Measurements

Yields were computed from currents measured in the detection apparatus sketched in Fig. 1 and shown in greater detail in Fig. 2. This part of the apparatus was fashioned after that used by Baragiola et al. (1979b). The major difference in the present experiment was the change in dimensions and voltages of the electrode configuration. When the proton beam entered the target chamber, it was collimated to its smallest diameter (6.4 mm) by two grounded molybdenum collimating plates. The beam caused some electrons to be emitted from these plates which had to be prevented from reaching the detector. This was done using an electron stripper plate positively biased at 300 volts which collected all spurious electrons produced in the upper portion of the target chamber. An additional plate, called the electron suppressor, was used to create a barrier which would isolate electrons in the upper section of the chamber from those (released by the target) in the lower section. The -175 volts on the suppressor repels electrons to the stripper electrode or the detector. The stainless steel detector was a 100 mm diameter hemispherical shell. It surrounded the target holder and was biased at +90 volts to collect all the electrons emitted from the target.

Two electrometers were used to simultaneously measure the currents on the target and detector and convert them to an analog voltage signal. Both electrometer output voltages were measured and recorded using separate analog-to-digital channels of an IBM Data Acquisition and Control Board (DAAC) installed in a PC compatible computer. Unity-gain isolation amplifiers were used to isolate the DAAC from electrode bias voltages.

The yield is defined as the average number of electrons emitted from the target per incident ion. Under steady-state conditions, the yield can then be computed from the target current, I_t, and detector current, I_d, as

$$\gamma = \frac{|I_d|}{I_t - |I_d|} \tag{1}$$

since the only current on the detector is the emitted electron current, but the target current includes both the ions entering and electrons leaving the target.

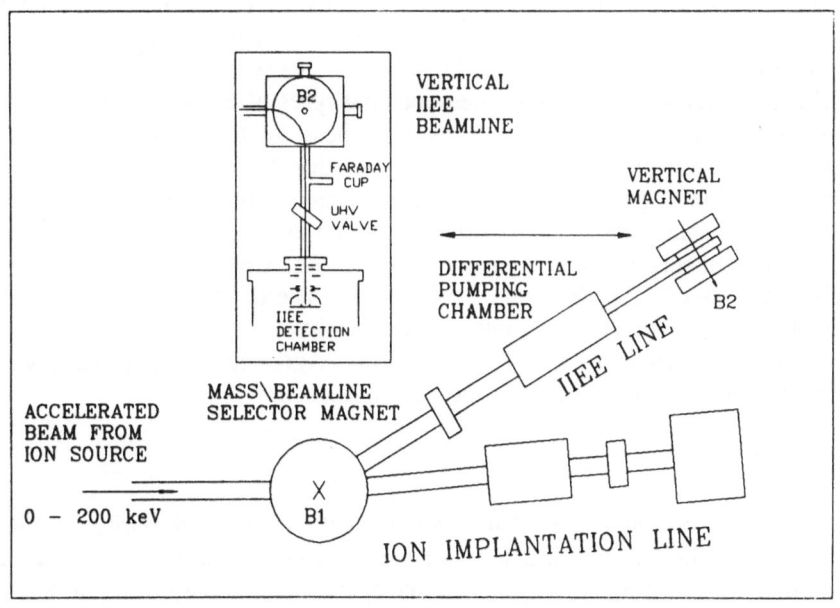

Figure 1. Schematic of IIEE beam line and target chamber.

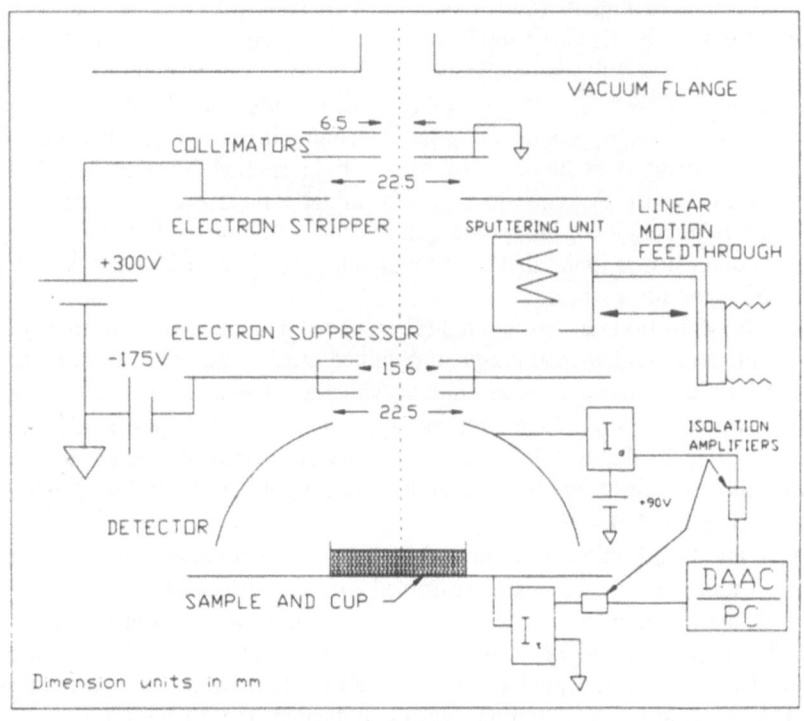

Figure 2. Electrode configuration in the target chamber.

The yield reported for a given set of conditions was obtained by reading I_t and I_d 25 times, obtaining γ using Eq. (1) for each of these pairs of currents, and then averaging these 25 γ values. Often, multiple readings were taken over a period of several minutes, on different days, and on different samples. The reported yields are those obtained by averaging yields on what was believed to be identical ion energy and annealing conditions.

Careful consideration was given to three potential types of errors and uncertainties: those that might be called configurational errors, i.e., those due to stray currents, neutrals in the beam, etc.; instrumental errors, e.g., those due to improper calibration of electrometers, amplifiers, and/or the DAAC board; and uncertainties due to temporal fluctuations in the current readings. The latter was found to cause the largest uncertainty in the yield measurements and was typically ± 0.01 e/ion, while the sample-to-sample reproducibility was typically ± 0.02 e/ion. The absolute uncertainty can only be estimated and is thought to be ± 0.04 e/ion. Of interest in this regard is how our results compare with yields previously reported by other researchers. Agreement was typically within ± 0.03 e/ion, as can be seen in Fig. 4 of the Results section.

Sample Preparation

The copper and gold samples used in this investigation were high purity (99.999%) foils obtained from Aldrich Chemical Co. *In situ* cleaning of the target surfaces was performed with a commercial 2 keV argon-ion sputter unit. This unit, a Physical Electronics, Model 04-161, was mounted on a linear motion feedthrough and could be positioned above the sample and the electron suppression plate as shown in Fig. 2. The sputter unit was operated in unfocussed mode in order that the sputter area was larger than the area hit by the proton beam.

Samples were sputtered with a range of 2 keV argon ion doses with currents on the target from 3 to 20 microamperes (corresponding to Ar back-fill pressures between 1 to 40×10^{-6} Torr) for periods from 10 to 45 minutes. The heaviest doses [500 to 1000 $\mu A \cdot mins$ (1 $\mu A \cdot min \approx 4 \times 10^{14}$ ions/cm^2)] were used on fresh samples in order to remove the surface oxide layer as well as carbon or other contaminants. Within about ten minutes of an initial yield measurement, accumulating surface contaminants caused subsequent yield measurements to increase with time. However, a light sputter "dusting" of 20 to 50 $\mu A \cdot mins$ was able to return the yield to the value initially measured. An estimate was made of the depth removal rate of the sputter gun. Using a 100 nm calibrated oxide grown on a tantalum sample, it was found that the sputter rate was about 0.8 to 1.0 Å/min at 10 microamperes of argon ion current.

Typically, the yield from an unsputtered (i.e., oxide covered) and unannealed copper sample was significantly higher than the yield obtained after sputtering. With increasing sputter time, the yield approached an asymptotic value as shown in Fig. 3. The lowest measured yield, which did not reduce after further sputtering, was reported as the yield. For the sample of Fig. 3, this was 1.78 e/ion. Such yield versus sputtering dose curves agree well with those obtained by other researchers (Baragiola *et al.*, 1979a; Hasselkamp *et al.*, 1980; Svensson and Holmén, 1981).

The sample and target holder were heated from below by a variable intensity quartz-halogen lamp. Light from the lamp was restricted to the bottom of the target holder by a stainless steel mask which minimized the effects of heating and out-gassing of the vacuum chamber walls at the higher anneal temperatures. Temperatures were measured with an iron-constantan thermocouple attached to the sample holding cup and read by the computer's DAAC board. Calibration checks indicated the uncertainty in sample temperature to be less than $\pm 5 \degree C$.

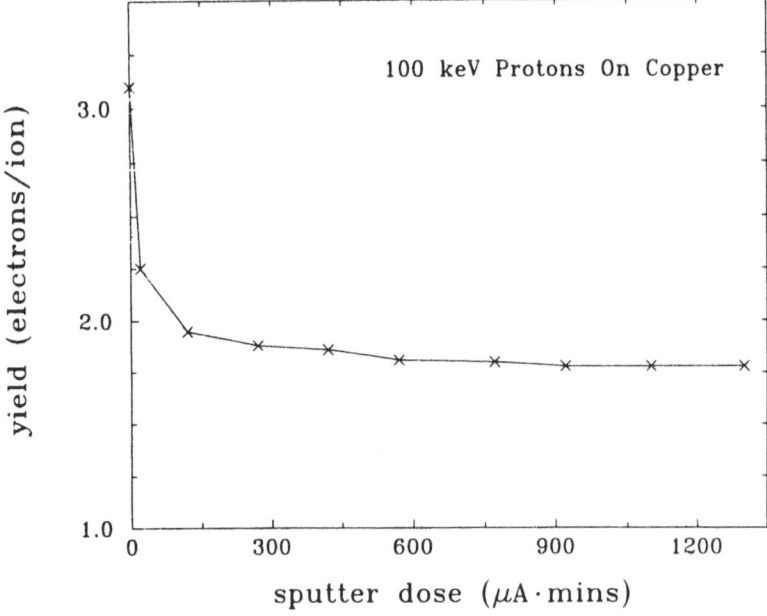

Figure 3. Change of yield with argon ion sputter dose after initial pump-down. Sample had not previously been sputtered, hence a zero-dose reading is indicative of an oxide coated sample.

RESULTS

Standard (Pre-Anneal) Yields

To verify if our apparatus was working as designed, yields for gold samples under proton bombardment in the energy range 50 to 180 keV were measured. The results obtained after 2 keV Ar^+ sputter cleaning are shown in Figs. 4(a). Also shown are other UHV results. Our data agreed reasonably well (within a few percent) with previous data.

Then, measurements were made on copper in the same energy range. Figure 4(b) shows these results, as well as those of others. The copper measurements of Holmén *et al.* (1981) and Svensson and Holmén (1982) are about eight to ten percent or more below the yields reported by other researchers. In their study, their copper sample was maintained at 400°C during measurements and sputter doses were kept low to minimize sputtering effects. However, the differences between their yields and those of others was not analyzed by them. As a result of our temperature dependent studies, we now believe there is good evidence (see Discussion section) that their yields (at their elevated temperature) were a consequence of both annealing and the elevated temperature.

Effect of Annealing On Yield

Each of the several polycrystalline copper foils studied was first sputter cleaned until the standard, room temperature yield was obtained. Then, each sample was held at an anneal temperature until the target chamber pressure returned to below 7×10^{-9} Torr. Heated samples were then typically allowed to cool to room temperature overnight and the base pressure returned to below 1×10^{-9} Torr. Each sample was then recleaned with a

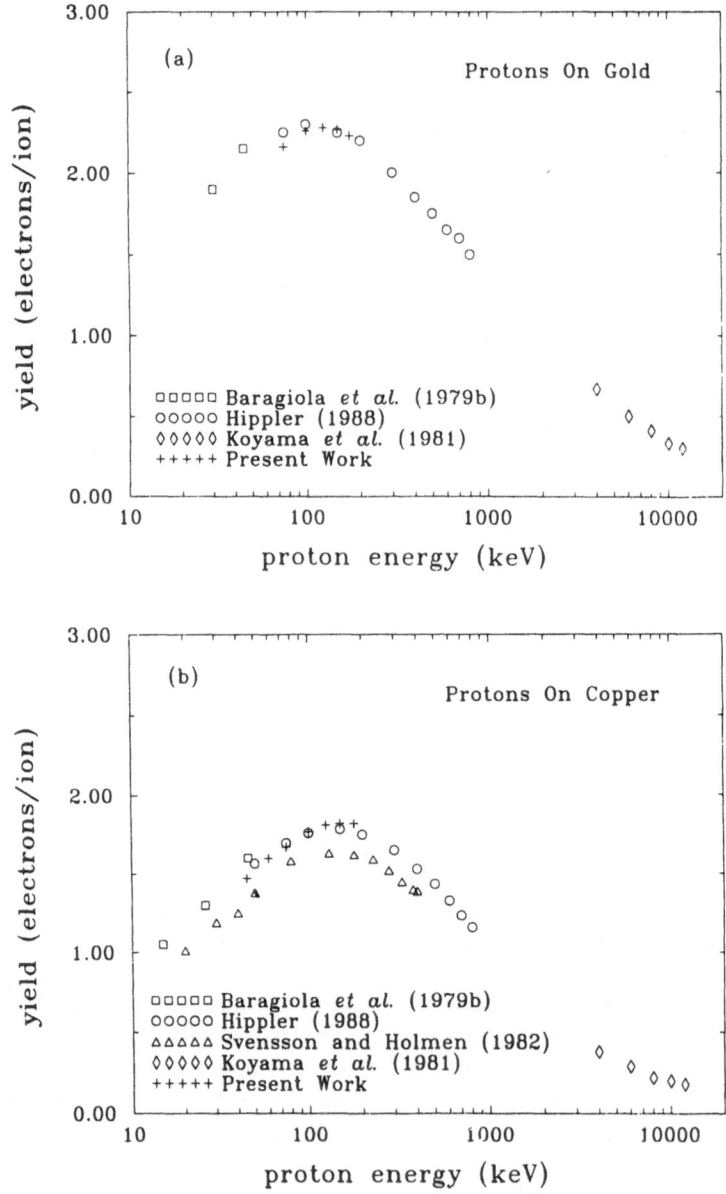

Figure 4. Comparison of IEEE yields from unannealed samples. (a) Yields from gold. (b) Yields from copper.

relatively light sputter dose to remove any surface impurities which had accumulated during annealing. The yield was then measured and recorded.

The measured room temperature yields after various annealing temperatures for several different copper samples are shown in Fig. 5. The yield is plotted versus the highest annealing temperature attained by the sample at the time of the yield measurement. The data suggests that at the higher anneal temperatures a transition to a lower yield value occurred. The high yield plateau value, near 1.77 e⁻/ion, was consistently obtained for anneals up to 200°C. The low plateau yield, near 1.62 e⁻/ion, was found for anneals above 300°C. Some data was also obtained in the transition region where full annealing of the entire surface was apparently not complete. This data indicates heat treatment had a major effect on the yield.

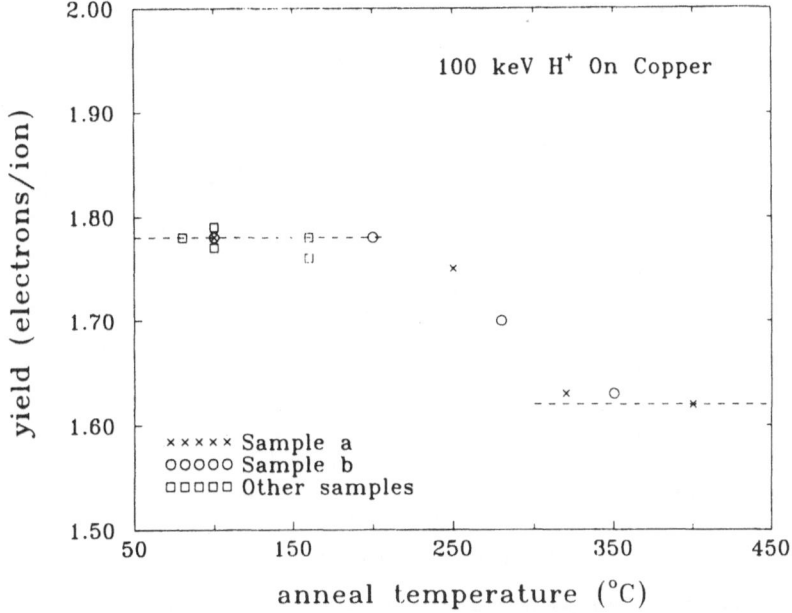

Figure 5. Room temperature yield as a function of annealing temperature.

To better understand the effects of annealing, a series of experiments was performed involving anneal and sputter cycling. In these experiments, care was taken to use only a very light, "dusting" sputter dose after cooling. After a yield of 1.77 e⁻/ion was obtained on a new sample, the sample was annealed near 210°C first for 25 and then for 12 hours. After each annealing, a yield below 1.76 e⁻/ion could not be obtained. However, when the same sample was subsequently annealed at 390°C for 11 hours, the measured yield was found to be 1.59 e⁻/ion. The yield after this and three more 390°C cycles averaged 1.59±0.02 e⁻/ion. The results from these experiments are summarized in Fig. 6. As in Fig. 5, this figure also clearly shows two distinct yield plateaus. In these experiments the low yield plateau value was 1.59±0.01 e⁻/ion. This value was definitely lower than the value found after the simple annealing experiments (see Fig. 5). We attribute this lowering to the smaller sputter dose used to just "dust" clean the sample. In all our experiments, cleaner samples have always produced lower yields and, as expected, the yields in the anneal/sputter cycled sample increased with time after cleaning, attributable to accumulation of surface contaminants.

DISCUSSION

Interpretation of Results

The data of both Figs. 5 and 6 clearly indicate that two distinctly different room-temperature yields can be measured in copper depending on the heat treatment of the sample. The room-temperature yield measured after low temperature annealing has been measured by several groups [see Fig. 4(b)]. However, the room temperature yield measured after higher temperature anneals has not been reported before. Interestingly,

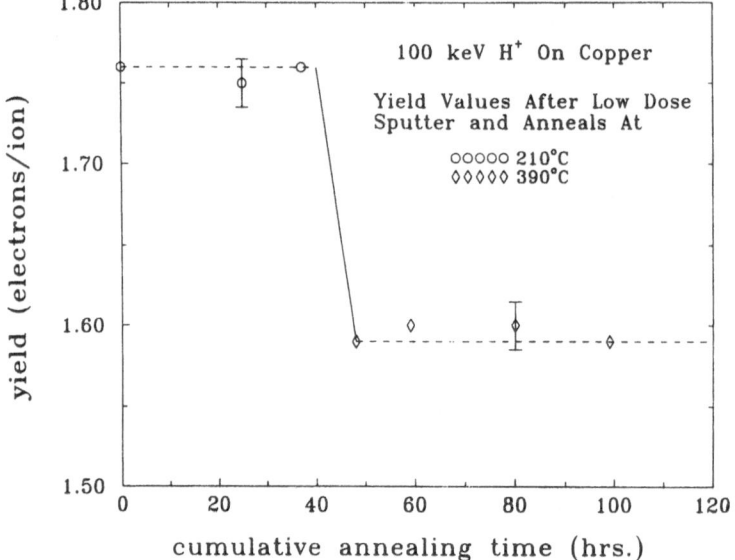

Figure 6. Room temperature yields measured following anneals and low dose sputter cleaning.

Svensson and Holmén (1982) measured the yield at 400°C and reported somewhat lower yields [see Fig. 4(b)]. Our analysis of their yield data indicate that their reduced yields were due mainly to annealing. However, their yield values were slightly higher than ours because the yield increases with increasing temperature (Kirchhoff, 1992). Hence, we believe their different results are due mainly to annealing, but also reflect a true temperature dependence in the yield.

We have considered what we believe to be the two possibilities for the lower yield measured at the higher anneal temperatures. One possibility is that heat treatment introduces surface impurities which influence the yield. A second possibility is that the treatment truly anneals and relaxes the surface to its natural, ordered state.

Surface impurities certainly accumulate during heat treatment. To investigate surface impurity effects we have carried out a series of Auger spectroscopy studies on our copper foils annealed at various temperatures. These studies have been reported in detail by Kirchhoff (1992), but, due to space limitations, we simply summarize our findings here.

The surfaces of samples originally placed in the UHV Auger spectrometer showed mainly copper, carbon, and oxygen, with much smaller amounts of chlorine, nitrogen, and sulfur. Argon-ion sputter cleaning of these samples with doses similar to those used in the IIEE apparatus removed the oxide layer and reduced all impurities below the detection limits, except for carbon and oxygen, which typically remained around 5 atomic percent. This agrees with the findings of Hippler *et al.* (1988) and Hasselkamp *et al.* (1990). Numerous annealing, cooling, and sputter cycles on these samples were done to study the surface segregation kinetics of these impurities. In all samples annealed in the temperature range from 200 to 400°C, copious quantities of carbon, oxygen, and sulfur accumulated on the surface (the latter only at the higher temperatures). These results were in quantitative agreement with the findings of Jenkins and Chung (1971) and McDonnell and Woodruff (1974). After cooling and low dose sputtering as done in our IIEE experiments, there was less carbon and oxygen than before annealing and no other impurities were found (except for about 2 atomic percent occluded argon). Of particular importance to our IIEE study was the finding that *no major or abnormal change in surface impurities was found in the 200 to 300°C temperature range*, i.e., in the range where the IIEE yield transition occurred as shown in Figs. 5 and 6. Thus, we believe that impurities are not responsible for the transition observed in the IIEE yields, and that the observed change with temperature is actually due to repair and relaxing of the surface damage.

The Auger studies clearly showed that, after initial heavy sputtering to remove the surface oxide, very light sputtering was able to remove other surface impurities which accumulated with time or as a result of heat treatment. This result is consistent with our observation that light sputtering after heating also reduced our measured yields slightly. However, additional sputtering increased the yield (due to induced surface damage). It is not clear how such an initial decrease followed by an increase could be due to impurities alone. The observed increase is consistent with sputter induced damage inflicted on a well annealed surface.

A study of the literature concerned with annealed and well-ordered surfaces provides strong support for our sputter damage hypothesis. Low-energy electron diffraction (LEED) studies have been done on metals for many years, and LEED spot diffraction patterns are only observed from clean, well-ordered surfaces [for a review of LEED see Jona (1977)]. Farnsworth *et al.* (1955) were apparently the first to report on the enhanced LEED imaging obtained from metal samples prepared by the dual process of annealing and argon ion bombardment. Since the first reports of improved surface conditions attained by cycling, the procedure itself has been extensively studied to understand the processes involved and its effect on surface properties, especially structure (Boggio and Farnsworth, 1964; Lee and Farnsworth, 1965; Farnsworth and Hayek, 1967; Farnsworth and Bellina, 1968; Reid, 1972; Kobayashi and Kato, 1969; Sickafus, 1970; Jenkins and Chung, 1971; McDonnell and Woodruff, 1974; Noonan *et al.*, 1978; Haas and Thomas, 1977; Gartland *et al.*, 1972). Because of the resulting well-ordered and clean surfaces, cycling became the standard preparation technique for most LEED metal surface studies (*e.g.*, see the references from 1960 on in the bibliography of Haas *et al.*, 1972).

It has always been clear that IIEE yields depended critically on sample preparation. It is now well recognized that older yield data measured without UHV conditions are of limited value (see any of the previously mentioned experimental review articles). Even in UHV systems, it is also well known that surface cleaning is always necessary to obtain the yield which is indicative of a clean target surface (Holmén *et al.*, 1981; Hasselkamp, 1988). The present results indicate that even further preparation is needed to obtain results indicative of pure, undamaged surfaces.

It is also well known that IIEE yields are very sensitive to surface effects. For example, Hasselkamp *et al.* (1980) observed effects due to a monolayer or so of foreign adatoms. Holmén *et al.* (1981) worried about sputter damaged effects on their yield measurements and thus preformed their experiments at high temperature. Numerous other investigators have observed yield changes with time due to surface changes. Thus, in retrospect, it is really not too surprising that the measurement of true yields might require the best surface preparation technique, which has been well established by LEED studies to be anneal/sputter cycling. We have shown with this work that such techniques may be required for IIEE studies as well.

It also seems quite likely to us that our results are not just restricted to copper, but are a more general finding which applies to other metals as well. For example, Svensson and Holmén (1981 and 1982) also measured the yields from heated (250°C) aluminum for various incident energies. At most energies, their yields were consistently low and, if their results are plotted against other UHV yield measurements, as in Fig. 4(b) (in particular, see Fig. 1 of Hasselkamp, 1988), the plot is very similar to that of Fig. 4(b). This suggest that annealing also reduces the yield in aluminum. There is also evidence that heat treatment can reduce the yield in indium (Kirchhoff, 1992).

Our results are likely to have application to electron-induced electron emission. In emission the escape process, rather than the creation process, is the one most sensitive to surface conditions (Schou, 1988; Hale *et al.*, 1993) and is the process common to both ion and electron-induced emission.

CONCLUSIONS

The room temperature IIEE yield in copper bombarded with 100 keV protons has been performed after various annealing and sputtering procedures. Moderate heat treatments $(300°C > T_a > 400°C)$ were found to reduce the yield from the often measured value of 1.77 e$^-$/ion to 1.59 e$^-$/ion. This reduction due to annealing is attributed to thermal removal of sputter damage and surface strain and the measured yield is characteristic of the ordered copper surface. Thus, the commonly measured yield is about 10% above the ideal surface yield. Results of previous LEED investigations suggests that surface ordering due to annealing is a common phenomenon observed in many metals. Thus, annealing, as well as light, dusting sputter cleaning, must now be considered as an integral part of surface preparation in fundamental electron emission studies.

ACKNOWLEDGMENT

We would like to thank the Research Corporation which generously supported part of this work.

REFERENCES

Baragiola, R.A., E.V. Alonso, J. Ferron, and A. Oliva-Florio, Ion-induced electron emission from clean metals, *Surf. Sci.* **90**, 240 (1979a).

Baragiola, R.A., E.V. Alonso, and A. Oliva Florio, Electron emission from clean metal surfaces induced by low-energy light ions, *Phys. Rev.* **B 19**, 121 (1979b).

Benazeth, N., Review on kinetic ion-electron emission from solids metallic targets, *Nucl. Instrum. Methods* **194**, 405 (1982).

Bindi R., H. Lanteri and P. Rostaing, Secondary electron emission induced by electron bombardment of polycrystalline metallic targets, *Scanning Microsc.* **1**, 1475 (1987).

Boggio, J.E., and H.E. Farnsworth, Low-energy electron diffraction and photoelectric study of (110) tantalum as a function of ion bombardment and heat treatment, *Surf. Sci.* **1**, 399 (1964).

Brusilovsky, B.A., Kinetic ion-induced electron emission from the surface of random solids, *Appl. Phys.* **A50**, 111 (1990).

Farnsworth, H.E., R.E. Schlier, T.H. George and R.M. Burger, Ion bombardment-cleaning of germanium and titanium as determined by low-energy electron diffraction, *J. Appl. Phys.* **26**, 252 (1955).

Farnsworth, H.E., and K. Hayek, Investigation of surface bombardment damage by LEED, *Surf. Sci.* **8**, 35 (1967).

Farnsworth, H.E., and J. Bellina, Jr., The application of LEED to the study of bombardment damage at crystal surfaces, *Trans. Am. Cryst. Assoc.* **4**, 45 (1968).

Gartland, P.O., S. Berge, and B.J. Slagsvold, Photoelectric work function of a copper single crystal for the (100), (110), (111), and (112) faces, *Phys. Rev. Lett.* **28**, 738 (1972).

Haas, G.A., and R.E. Thomas, Work function and secondary emission studies of various Cu crystal faces, *J. Appl. Phys.* **48**, 86 (1977).

Haas, T.W., G.J. Dooley, J.T. Grant, A.G. Jackson, and M.P. Hooker, 1972, A bibliography of low energy electron diffraction and Auger electron spectroscopy, p.155, *in:* "Progress in Surface Science", Vol. 1, S.G. Davison, ed., Pergamon Press Ltd., Oxford.

Hale, E.B., T.J. Gay, and J.F. Kirchhoff, 1993, Model for ion-induced electron emission from metals, *in:* "Ionization of Solids by Heavy Particles", Raul A. Baragiola, ed., Plenum, New York.

Hasselkamp, D., A. Scharmann, and N. Stiller, Ion induced secondary electron emission as a probe for adsorbed oxygen on tungsten, *Nucl. Instrum. Methods* **168**, 579 (1980).

Hasselkamp, D., K.G. Lang, A. Scharmann, and N. Stiller, Ion induced electron emission from metal surfaces, *Nucl. Instrum. Methods* **180**, 349 (1981).

Hasselkamp, D., Secondary emission of electrons by ion impact on surfaces, *Comments At. Mol. Phys.* **21**, 241 (1988).

Hasselkamp, D., S. Hippler, A. Scharmann, and T. Schmehl, Electron emission from clean solid surfaces by fast ions, *Annalen der Physik (Leipzig)* **47**, 555 (1990).

Hippler, S., Thesis, Justus-Liebig-Universität, Giessen, FRG, 1988.

Hippler, S., D. Hasselkamp, and A. Scharmann, The ion-induced electron yield as a function of the target material, *Nucl. Instrum. Methods* **B 34**, 518 (1988).

Hofer, W.O., Ion-induced electron emission from solids, *Scanning Microsc. Suppl.* **4**, 265 (1990).

Holmén G., B. Svensson, and A. Burén, Ion induced electron emission from poly-crystalline copper, *Nucl. Instrum. Methods* **185**, 523 (1981).

Jona, F., Past and future surface crystallography by LEED, *Surf. Sci.* **68**, 204 (1977).

Jenkins, L.H., and M.F. Chung, LEED and Auger investigations of Cu (111) surface, *Surf. Sci.* **24**, 125 (1971).

Kirchhoff, J.F., Ion-induced electron emission from metals, Ph.D. thesis, University of Missouri-Rolla, Rolla, MO, USA, 1992.

Kobayashi, H. and S. Kato, Observations on the photoelectric work function and LEED patterns from the (111) surface of an iron single crystal, *Surf. Sci.* **18**, 341 (1969).

Koyama, A., T. Shikata, and H. Sakairi, Secondary electron emission from Al, Cu, Ag and Au metal targets under proton bombardment, *Jpn. J. Appl. Phys.* **20**, 65 (1981).

Krebs, K.H., Recent advances in the field of ion-induced kinetic electron emission from solids, *Vacuum* **33**, 555 (1983).

Lee, R.N., and H.E. Farnsworth, LEED studies of adsorption of clean (100) copper surfaces, *Surf. Sci.* **3**, 461 (1965).

McDonnell, L., and D.P. Woodruff, A LEED study of oxygen adsorption on copper (100) and (111) surfaces, *Surf. Sci.* **46**, 505 (1974).

Noonan, J.R., H.L. Davis, and L.H. Jenkins, LEED analysis of a Cu (110) surface, *J. Vac. Sci. Technol.* **15**, 619 (1978).

Reid, R.J., LEED studies on the low index face of copper. I. The room temperature intensity-energy spectra, *Surf. Sci.* **29**, 603 (1972).

Schou, J., Secondary electron emission from solids by electron and proton bombardment, *Scanning Microsc.* **2**, 607 (1988).

Sickafus, E.N., Sulfur and carbon on the (110) surface of nickel, *Surf. Sci.* **19**, 181 (1970).

Sigmund, P., and S. Tougaard, 1981, Electron emission from solids during ion bombardment. Theoretical aspects, *in:* "Inelastic Particle-Surface Collisions", E. Taglauer and W. Heiland, eds., Springer-Verlag, Berlin-Heidelberg.

Svensson, B. and G. Holmén, Electron emission from aluminum ion bombardment aluminum, *J. Appl. Phys.* **52**, 6928 (1981).

Svensson, B. and G. Holmén, Electron emission from aluminum and copper under molecular-hydrogen-ion bombardment, *Phys. Rev.* **B 25**, 3056 (1982).

SPIN-POLARIZED ELECTRONS: SOURCES, TIME-RESOLVED PHOTOEMISSION, THERMOEMISSION

F. Meier[1], A. Vaterlaus[1], J.C. Gröbli[1],
D. Guarisco[1], H. Hepp[1], Yu. Mamaev[2], Yu. Yashin[2],
B. Yavich[3], and I. Kochnev[3]

[1]Laboratorium für Festkörperphysik
ETH Hönggerberg
CH-8093 Zürich
Switzerland

[2]Division of Experimental Physics
St. Petersburg Technical State University
St. Petersburg
195251 Russia

[3]A.F.Ioffé Physicotechnical Institute
Russian Academy of Sciences
St. Petersburg
194021 Russia

INTRODUCTION

The physics of polarized electrons has a long history and many aspects. A number of reviews deals with recent applications of spin-polarized electrons in atomic physics [1], solid state physics [2,3] and high energy physics [4]. The spin-polarization is defined as $P=(N\uparrow-N\downarrow)/(N\uparrow+N\downarrow)$ where $N\uparrow$ $(N\downarrow)$ is the number of electrons with spin direction parallel (antiparallel) to a quantization axis determined by the geometry of the experiment. In the spin-polarized photoemission experiments described below this direction is given by the surface normal of the sample.

The first topic covered in this paper concerns the development of efficient sources of highly polarized electrons. Until 1990 the standard source was based on optical spin orientation in GaAs. With this material the photoemitted

Ionization of Solids by Heavy Particles, Edited by
R.A. Baragiola, Plenum Press, New York, 1993

electrons have a maximum polarization of 50 %. Due to progress in materials technology substantial improvements have been made resulting in polarization values up to 90 %. - The subsequent section illustrates the role of spin-polarized electrons in time-resolved photoemission experiments. In time-resolved photoemission the light sources are pulsed high-intensity lasers. The resulting high photocurrent densities give rise to space charge which has the effect of randomizing the energies and momenta of the individual photoelectrons. Therefore, it becomes impossible to recover information from the photoelectrons on the electronic states occupied in the solid before excitation. However, there is one experimentally accessible property of the photoelectrons which is insensitive to space charge: the spin-polarization of the photoyield, i.e., of the total number of emitted photoelectrons. For the measurement of this quantity the use of pulsed high intensity lasers as light sources becomes admissible: therefore time-resolved photoemission is naturally linked to the spin-polarization. In this paper the characteristic time for establishing thermal equilibrium between the spin system and the lattice in ferromagnets is discussed. The technique has also been applied to study the magnetization reversal in magneto-optic storage materials [5], and the dynamics of the surface photovoltage in semiconductor/metal contacts [6]. - Finally, in the last section measurements on the spin-polarization of thermoemitted electrons from cesiated surfaces of Fe, Co and Ni are presented. Having a work function close to 1.4 eV these surfaces thermoemit electrons in measurable quantities already at 150 $^{\circ}$C, i.e., far below the Curie temperature. The thermoelectrons emitted from all three ferromagnetic transition metals were found to be unpolarized. A theoretical argument has been put forward showing this result to be of general validity.

SOURCES OF SPIN-POLARIZED ELECTRONS

In solids the optical excitation of electrons from an unpolarized ground state to a spin-polarized final state - optical spin orientation - is a process which has been studied in great detail [7]: As an application, the development of a polarized electron source combining high polarization with a high quantum yield has been a constant challenge.

Generally, optical spin orientation involves transitions with circularly polarized light. Exactly as in the case of atoms [8] the final state polarization is determined by the symmetry of the wave functions involved in the transition. For GaAs, the transition scheme for excitation from the valence band maximum (VBM) to the conduction band minimum (CBM) is shown in Fig.1a. Two transitions are excited simultaneously. They lead to oppositely polarized final states and have an intensity ratio of 3:1. Therefore the maximum polarization which can be achieved is P = 50 %.

A natural way for getting a higher polarization is to remove the orbital degeneracy of the VBM in order to eliminate one of the two competing transitions. This is accomplished by reducing the symmetry of the lattice from cubic to tetragonal. The natural tetragonal analogue of the GaAs-structure is the

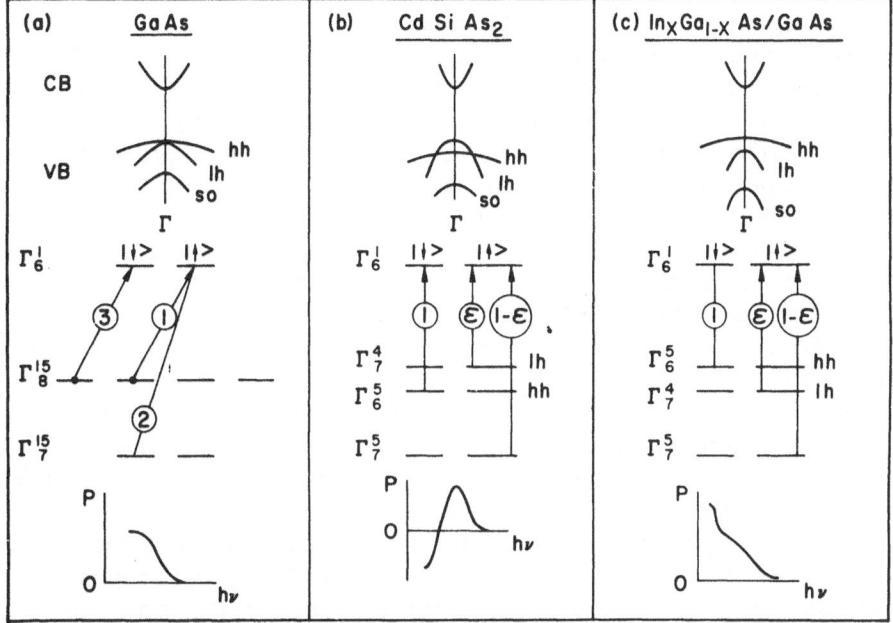

Fig.1. Optical spin orientation in (a) cubic GaAs, (b) tetragonal CdSiAs$_2$ with chalcopyrite structure and (c) tetragonal In$_x$Ga$_{1-x}$As epitaxially grown on GaAs(100). The transitions indicated are induced by circularly polarized light at Γ. The upper part of the figure shows the schematic band structure near Γ. hh: heavy hole valence band (VB),lh: light hole VB, so: spin-orbit split VB. The middle part shows the transition scheme at Γ. Note that for CdSiAs2 the lh-VB lies above the hh-VB whereas for the strained In$_x$Ga$_{1-x}$As the sequence of the hh-VB and lh-VB is opposite. The bottom part of the figure shows schematically the polarization of the photoyield as function of photon energy.

chalcopyrite structure. However, in the few experiments which have been reported no P-enhancement was observed [9]. The reason is that in the chalcopyrite structure crystals – at least the ones studied so far [10] – the c/a-ratio is smaller than 2. As a consequence, the light hole valence band (lh-VB) is lifted above the heavy hole valence band (hh-VB), see Fig. 1b. In the absence of hybridization, the transition from the lh-VB to the CBM is forbidden, whereas the transition from the hh-VB to the CBM gives P = +1 and the transition from the spin-orbit split valence band (so-VB) to the CBM gives P = −1.

It turns out, however, that the lh-VB is strongly hybridized with the so-VB which makes the transition (lh-VB)-CBM allowed, see Fig. 1b. This transition is oppositely polarized with respect to the (hh-VB)-CBM transition and has a linewidth comparable to the splitting between the lh-VB and the hh-VB: therefore, the two transitions cannot be energetically separated and, as a result, an even smaller threshold polarization was observed than for GaAs. Furthermore, the preparation of high quality, p-doped chalcopyrite crystals - in particular of a (100)-surface - is a difficult problem.

The situation is much more favorable when the hh-VB lies above the lh-VB, see Fig. 1c. Then, even for small splittings, the polarization should become enhanced with respect to GaAs at photothreshold. Such a system is realized by applying a tensile strain perpendicular to the surface of a fcc III-V semiconductor. A highly successful method is to induce the strain via the lattice mismatch between adjacent epitaxial layers like $In_xGa_{1-x}As$ on GaAs. $In_xGa_{1-x}As$ has a larger lattice constant than the substrate. Therefore, the overlayer lattice becomes compressed in the plane of the interface and elongated along the direction of the surface normal. Following this method a breakthrough in the development of polarized electron sources has been achieved: With the systems $In_xGa_{1-x}As$/GaAs [11] and GaAs/GaP_xAs_{1-x} [12] threshold polarizations of 70 % and 86 %, respectively, have been obtained.

In this paper, experiments on optically pumped strained layers of $In_xGa_{1-x}As$ (x=0.15) grown on GaAs are reported. The thickness of the overlayer ranged from 0.05 to 1.0 μ. The threshold polarization was largest for the thinnest layer: at T=150 K it was 80 %, see Fig.2.

The samples were grown at the Ioffé-Institute using the MOCVD-technique. On a GaAs(100) substrate a buffer GaAs-layer of 0.5 μ thickness was grown first in order to obtain a high-quality surface on which the $In_xGa_{1-x}As$-layer was deposited. The structure was uniformly p-doped with $(1 - 3) \times 10^{18}$ cm^{-3} magnesium. From luminescence spectroscopy a hh-lh-valence band splitting of about 50 meV was derived.

Since the optically polarized electrons are close to the CBM they can escape into the vacuum only if the vacuum level is lowered below the CBM. Then the surface is in a state of negative electron affinity (NEA). For this purpose a surface activation procedure must be followed involving the alternating deposition of cesium and oxygen until the photoyield reaches a maximum.   In this way an almost zero electron affinity could be achieved. The circularly polarized light was incident along the surface normal. The sample temperature could be controlled between 10 K and 400 K.

Fig.3 shows the polarization spectra P(hv) of the total photoyield for samples with an overlayer thickness of 0.05, 0.1, 0.2, and 1.0 μ. For the thinnest layer the polarization is

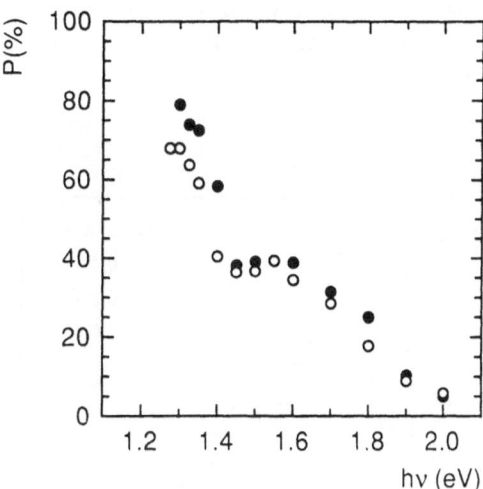

Fig.2. Polarization of the total yield of optically spin-oriented photoelectrons emitted from $In_xGa_{1-x}As/GaAs$ (x=0.15) as function of photon energy. The sample temperature was 220 K (open circles) and 150 K (full circles), respectively.

largest. This is attributed to the fact that the strain in these layers is rather uniform. For thicker layers the strain relaxes which leads to a reduction of the polarization. The threshold polarization of the 1μ overlayer is identical to the one of unstrained bulk GaAs.

Fig. 4 shows the quantum yield of the same samples from which the polarization data were obtained. The shape of the curves near photothreshold and the low saturation value of 10^{-2} are untypical for NEA. We conclude that the samples described in this paper had a small positive electron affinity.

For the interpretation of Fig.3 and 4 the question arises whether the GaAs substrate contributes significantly to the photocurrent [11]. The optical absorption coeffcient of $In_xGa_{1-x}As$ is of the order 10^4 cm^{-1} at a photon energy 0.2 eV above the absorption threshold [13]. Accordingly, for the 1 μ overlayer, the light intensity at this photon energy has dropped to 1/e when it reaches the GaAs substrate. Therefore, the number of excitations produced in the substrate is about equal to the one

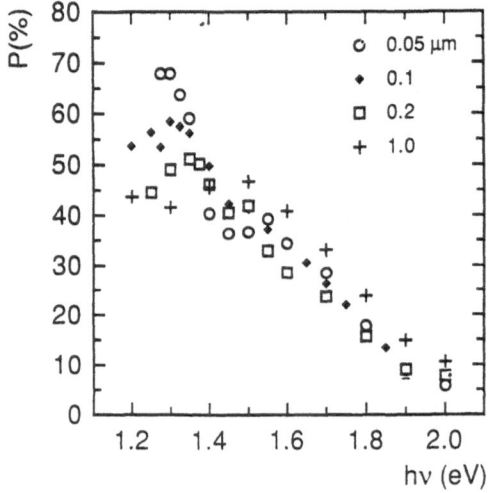

Fig.3. Polarization of the total yield of optically spin-oriented photoelectrons emitted from $In_xGa_{1-x}As/GaAs$ (x=0.15). The sample temperature was 220K.

produced in the overlayer. Furthermore, data on heterostructures consisting of a 1000 Å overlayer of GaAs on top of AlGaAs show that the ballistic mean free path in GaAs is below 1000 Å [14].Related work is in qualitative agreement with this value [15].Therefore, in the present experiment, it is not only the small number of excitations in the substrate but also the attenuation of the transmitted current in the overlayer which makes strong emission from GaAs questionable. A more thorough verification is needed regarding the interpretation of the hν-dependence of polarization and yield in terms of a superposition of overlayer and substrate emission. This is possible, e.g., by examining extremely thin (< 500 Å) or extremely thick (> 1 μ) overlayers.

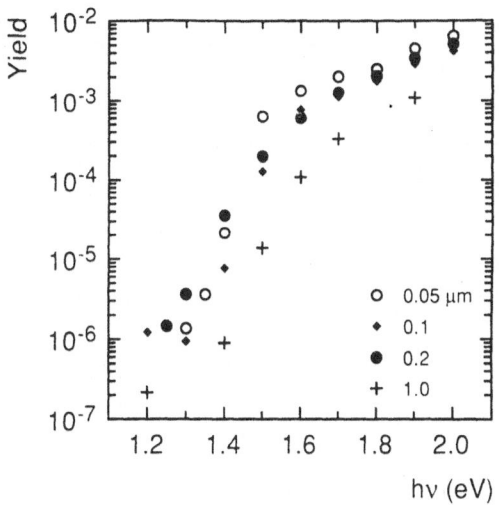

Fig.4. Quantum yield of the same samples as used for the P(hν)-measurements shown in Fig.3.

TIME-RESOLVED PHOTOEMISSION

The study of fast dynamic processes in " real time" - not, e.g., indirectly via line widths - has become one of the great challenges in recent years [16]. Progress is tightly bound to advances in pulsed laser technology: today, pulse durations as short as 6 femtoseconds have been achieved.

Photoemission is one of the most powerful tools to probe the electronic properties of solids. However, in spite of the large diversity of applications, photoelectron spectroscopy is almost exclusively used to investigate systems in thermal equilibrium. In such experiments the parameter "time" does not appear. The conspicuous absence of photoemission in studies of time-dependent, dynamic phenomena has its origin in a single cause: space charge.

The time-resolution in photoemission is given by the exposure time of the system to the light source: therefore high time resolution requires short light pulses. Still, during this short time enough photoelectrons must be emitted to give a measureable signal. Therefore, the pulses must be sufficiently intense. In time-resolved photoemission with picosecond resolution the required light intensity is so high that space

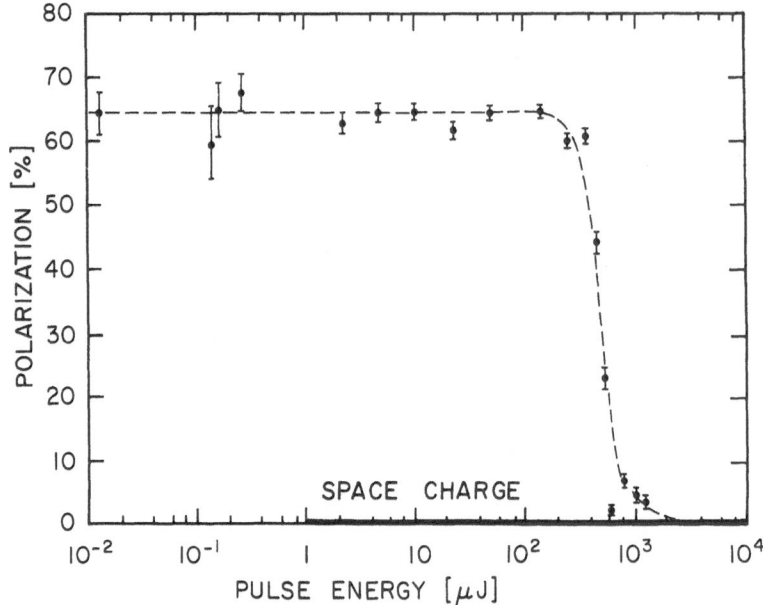

Fig.5. Polarization of the photoyield emitted from polycrystalline Fe as function of pulse energy using a KrF-excimer laser (hν = 5 eV). The space charge region as inferred from the nonlinearity of the quantum yield is indicated by the thick solid line on the energy scale.

charge is always present. It randomizes the momenta and energies of the photoelectrons and therefore it destroys all information the photoelectrons carry on the properties of the solid from which they are emitted. Therefore, an angle- and energy-resolved photoemission experiment is not suitable for work under space charge conditions. Experimentally, space charge is easily recognized by the non-linearity of the quantum yield [17].

There is however one property of the photoelectrons which is insensitive to space charge: the spin-polarization of the total number of emitted electrons, i.e., of the total photoyield. The reason is that the electromagnetic interactions of the electrons within the space charge cloud are internal forces: therefore the total spin angular momentum is conserved during emission. The experimental verification is given in Fig.5 which shows the polarization of the photoelectrons emitted from Fe as function of pulse energy. The measurements were done using a KrF-excimer laser ($h\nu$ = 5 eV) with a pulse duration of 16 ns. Over many decades of the pulse energy the polarization is constant and has the same value as the one found with a low-intensity continous lamp. Only above 1 mJ the polarization drops: Then, the pulse energy is sufficiently high that the Curie temperature is reached or exceeded in the laser focus. Measurements of the quantum yield show that space charge effects are already present at pulse energies as low as 1 μJ. The figure shows that space charge does not influence the polarization: the expected insensitivity of P with respect to space charge is confirmed.

From single pulse experiments of the kind described above information is obtained on the spin-lattice relaxation time in ferromagnets [18]. The pulse energy of the laser is transformed into lattice heat very quickly, in a characteristic time which is of the order of 1 ps [19]. In order to learn more about the characteristic time to establish thermal equilibrium between the lattice and the spin-system in a ferromagnet the polarization of the photoelectrons emitted from polycrystalline Fe was measured as function of pulse energy, see Fig. 6. Two lasers were used with pulse durations of 30 ps and 20 ns. The energy scale is calibrated in units of E_{melt}, the critical energy where the surface starts melting in the laser focus. For the ns-pulses the P(E)-dependence is identical to the one found in Fig. 5. However, for the ps-pulses the result is totally different: even for pulse energies up to 4xE_{melt} no significant change of the polarization is observed. Evidently, during a laser pulse of 30 ps thermal equilibrium is not established between the lattice and the spin system. This experiment, therefore, gives a lower limit for the spin-lattice relaxation time.

A more accurate determination of the spin-lattice relaxation time has been made with ferromagnetic gadolinium using a pump-probe technique. Gadolinium was chosen because of its favourable material properties: It has a low Curie temperature and a low work function. In the pump-probe-experiment the beam of an excimer laser is split into two pulses. One of them is directed over a variable beam delay and thereafter pumps a 10 ns dye laser with a photon energy of 2.15 eV. No measurable electron emission is induced by this pulse, it just heats the sample. The absence of multiphoton excitations is due to the moderate laser pulse intensities required to heat Gd above the Curie point. The other part of the excimer beam pumps a 60 ps dye laser having a photon energy of 3.2 eV. The spin polarization of the electrons emitted by this second pulse is measured. By varying the delay distance the probing pulse can be moved with picosecond accuracy from 2.5 ns before to 7 ns after the onset of the heating pulse. The energy of the probing pulse is low enough to cause negligible heating of the sample.

The Gd-film was evaporated onto a conically shaped polycrystalline Fe substrate and magnetically saturated in an external field of 0.38 Tesla applied perpendicular to the

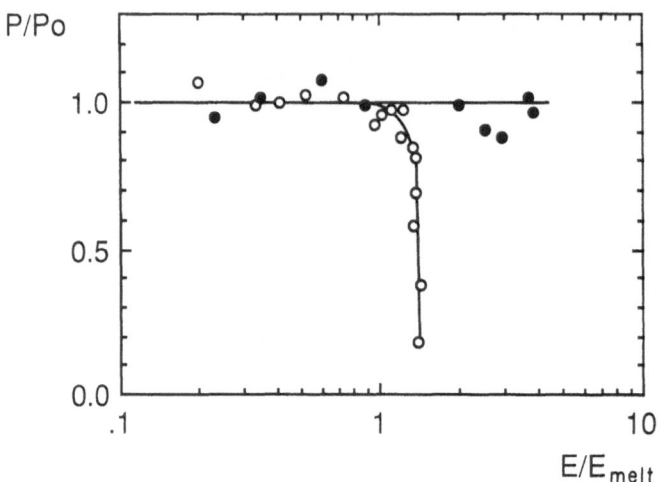

Fig.6. Polarization of the photoyield emitted from polycrystalline Fe using 20 ns (open circles) and 30 ps (full circles) laser pulses. The polarization is normalized to the spin polarization P_O measured with a low intensity continuous lamp. The energy is normalized in units of E_{melt}, the threshold energy for melting the surface.

surface. The photoelectrons are emitted from a surface area which is much smaller than the diameter of the heating pulse. Accordingly, an almost homogeneously heated surface area contributes to the photoemission signal.

The effect of the heating pulse becomes apparent in Fig. 7a. For $t < 0$ the probing pulse arrives before heating has started, therefore P_O (= 38 %) measured at a negative time corresponds to the equilibrium magnetization at the initial temperature $T_O = 45$ K. For $t > 0$ the polarization decreases. Due to the modest energy of the heating pulse the sample temperature remains below the Curie temperature. With the help of the measured equilibrium polarization as function of temperature - $P(T)$ - it becomes possible to replace each value of the polarization by the corresponding spin temperature. Therefore, to each change of the polarization ΔP a corresponding change of the spin temperature $\Delta T_{spin} = T_{spin} - T_0$ can be assigned. The result is plotted in Fig. 7b. Note the unique feature of the spin-polarized pump-probe experiment, namely that the magnetization acts as a thermometer indicating at each instant of time

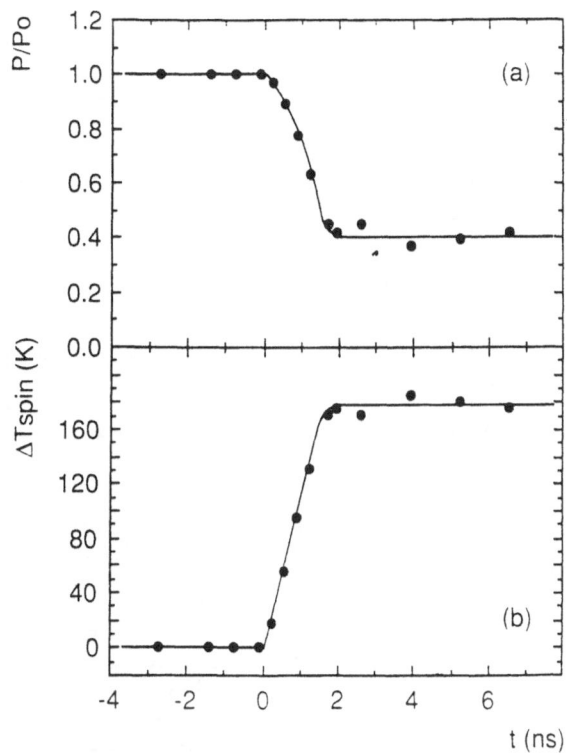

Fig.7.a: Pump-probe experiment using a 60 ps (hν=3.2 eV) laser pulse as probing pulse and a 10 ns (hν=2.15 eV) laser pulse as heating pulse. The sample is held at an initial temperature of 45 K. The reduced spin polarization of the photoyield emitted by the probing pulse is plotted as function of time-delay between probing and heating pulse. Zero time-delay corresponds to the onset of lattice heating.b: Rise of the spin temperature determined from measurement (a) using the equilibrium P(T)-curve.

the spin temperature. For the measurement of the relaxation times of the lattice and the electron gas similar thermometers do not exist. The form of $\Delta T_{spin}(t)$ as evident from Fig.7 is caused by the particular intensity profile of the heating pulse. This intensity profile is not of any basic significance, however when it is simple it facilitates the interpretation of the data [20].

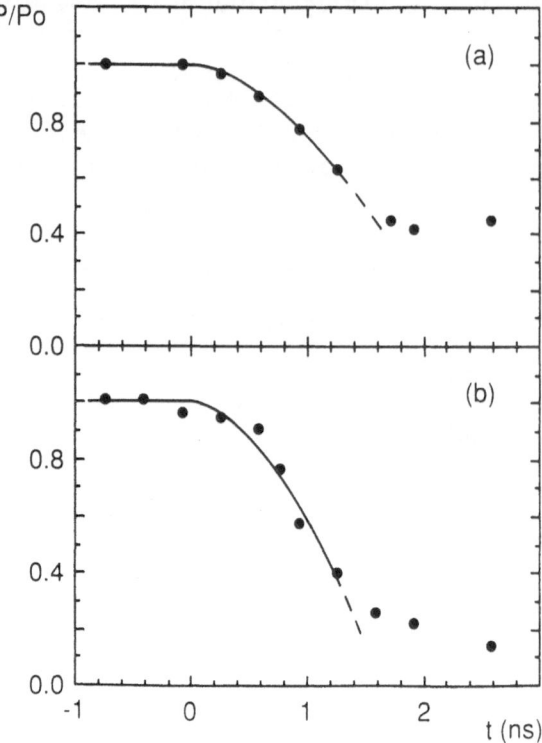

Fig.8. Pump-probe experiment using a 60 ps (hν=3.2 eV) laser pulse as probing pulse and a 10 ns (hν=2.15 eV) laser pulse as heating pulse. For measurement (b) the heating pulse energy is twice as large as for measurement (a). The reduced spin-polarization of the photoelectrons emitted by the probing pulse is plotted as function of the time-delay between probing and heating pulse. Zero time-delay corresponds to the onset of lattice heating. Solid lines: calculated P(t)-curves involving no adjustable parameters, see text.

Time-resolved measurements performed at two different energies of the heating pulse are plotted in Fig. 8 on an extended t-scale. In Fig. 8b the heating pulse energy is twice as large as for Fig. 8a.

Using these two measurements we next derive a numerical value for the spin-lattice relaxation time. The rate equation for the temperature transfer between the spin system and the lattice is [21]

$$C_{spin} \frac{dT_{spin}}{dt} = G \ (T_{lattice} - T_{spin}) \qquad (1)$$

C_{spin} is the specific heat of the spin system and G is the phonon-magnon coupling constant. The characteristic time for the equilibration of the temperature is then given by

$$\tau_{sl} = \frac{C_{spin}}{G} \qquad (2)$$

C_{spin} (not G !) depends on temperature. However, in the following τ_{sl} is understood to be a T-independent, averaged (45 < T < 225 K) quantity which makes it possible to solve Eq.(1) analytically. Next, we make use of the special temporal intensity profile of the heating pulse, namely that it increases in very good approximation linearly in time. Then it can be

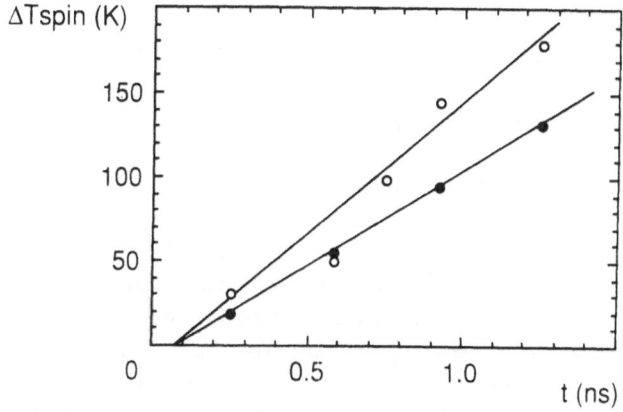

Fig.9. $\Delta T_{spin}(t)$ derived from the data of Fig. 8a (full circles) and Fig.8b (open circles). The straight lines correspond to the fit $\Delta T_{spin}(t) = q(t-\tau_{sl})$. The open circle data point at 0.6 ns deviates from the straight line as does the corresponding measured polarization value in Fig. 8b from the P(t)-curve.

shown [20] that also the lattice temperature increases linearly in time, $T_{lattice} = T_0 + qt$ where q is the rate of the temperature increase. With this form of $T_{lattice}$ the solution of Eq.(1) becomes

$$T_{spin}(t) = T_0 + q\left[t - \tau_{sl}(1 - e^{-\frac{t}{\tau_{sl}}})\right] \qquad (3)$$

Again, as for Fig. 7b, $\Delta T_{spin}(t)$ is derived exclusively from experimental data. Eq.(3) shows that $\Delta T_{spin}(t) = q(t-\tau_{sl})$ for $t > \tau_{sl}$ Indeed, this linear relationship is found in Fig. 9. Extrapolation of the straight lines to $\Delta T_{spin} = 0$ gives – independent of q – an intersection with the time axis at $t=\tau_{sl}$.

In this way $\tau_{s\ell}$ is found: it amounts to 100 ± 80 ps for both measurements shown in Fig. 8. Note that the assumption $t > \tau_{s\ell}$ is satisfied.

The zero of the time scale is obtained from a $P(t)$ measurement at high heating pulse energy. It must lie between the last point where $P/P_O = 1$ and the first point where $P/P_O < 1$. The error in $\tau_{s\ell}$ of ± 80 ps is mainly due to the uncertainty of the zero point of the time scale.

The rate of the temperature increase is given by the slope of the straight lines in Fig. 9; it is $q = 115$ K/ns and $q = 158$ K/ns for the measurements of Fig. 7a and 7b, respectively.

Using these experimentally determined values of $\tau_{s\ell}$ and q together with the $P(T)$-relation from Fig.7, the $P(t)$-curves are obtained immediately using Eq.(3). The result is shown as the solid lines in Fig. 8. Evidently the fit - without any adjustable parameters - is perfect.

SPIN-POLARIZED THERMOEMISSION

Spin-polarized electrons are observed in photoemission [22] and field emission [23] from ferromagnets. In the following, measurements on the spin-polarization of thermoemitted electrons are reported. It is shown that electrons thermoemitted from Fe [24], Co, and Ni[24] are unpolarized even if thermoemission occurs at temperatures far below the magnetic ordering temperature. In particular, the polarization of the thermoelectrons is by no means accounted for by a single-particle band structure which would predict $P = -75$ % for Fe [25], $P = -4$ % for Ni [26] and $P = -29$ % for Co [27].

The samples used were cylinders (5 mm diameter, 6 mm long) of single-crystalline Ni(111), Co(1000) and high-purity polycrystalline iron. Standard surface preparation techniques were applied: Repeated cycles of Ar^+-sputtering (1000 V) and annealing at 800 K. In order to lower the work function Φ a submonolayer of cesium was deposited onto the sample surface. In this way, work functions in the range of $1.40 < \Phi < 1.50$ eV were obtained for all 3 materials. Adsorbed submonolayers of alkali metals, in particular cesium, are known not to affect the polarization of electrons excited in the bulk material [28]. It was found that the cesium stuck better to surfaces which were mildly sputtered before deposition. Then, even at the highest measuring temperatures (450 K) no desorption of Cs took place ensuring a constant photothreshold.

The sample is located in the center of a superconducting coil which generates a magnetic field perpendicular to the emitting surface. The magnetic field has two purposes: it aligns the magnetization along the surface normal and, in addition, limits the diameter of the spot from which the electrons are collected by the electron optics to a few tenths of a mm, i.e., a small fraction of the total sample surface. This diameter has been determined by monitoring the counting rate of the photo-electrons emitted by a well focused laser beam (focus diameter 155 μm) when it is scanned over the sample surface. Obviously, no electrons from outside the sample surface reach the polarization detector.

In the present experiment the spin-polarization of all the emitted electrons is measured without energy discrimination. The thermoemission current from metals is given by the Richardson equation [29]

$$j(T) = A_O \, \delta \, T^2 \, \exp(-\Phi/k_B T) \qquad (4)$$

where $A_O = 120$ Amp/cm^2 degree^2 is a universal constant, T is the absolute temperature and Φ is the work function. δ is the average transmission coefficient for thermoelectrons: it is assumed to be 1 (0) for electrons inside (outside) the escape cone. The measured thermocurrent emitted from the cesiated metal surfaces was found to agree well with Eq.(5).

The results of the spin-polarized thermoemission experiments on Fe, Co, and Ni are shown in Fig. 10. The open circles represent the spin-polarization of the thermocurrent vs. applied magnetic field: within the experimental error it is always zero.

The data on Fe, Fig. 10a, were all obtained at the constant temperature T = 423 K. The work function (measured at 368 K) was 1.42 eV. The magnetic order of the Fe sample was confirmed by spin-polarized photoemission using light of photon energy hν = 2.15 eV. The full circles indicate the spin-polarization of the total emitted current consisting of thermoemission and photoemission. The light intensity was chosen such that both emission processes contributed about equally to the total current. Due to the large demagnetizing field of the iron sample magnetic saturation is not reached within the field range of 12 kOe. If the spin-polarized photocurrent is suppressed by switching off the light source only the thermocurrent remains: its polarization is zero as shown by the open circles. The measurement done with Co(1000), Fig. 10b, is completely analogous: The work function was 1.51 eV, the measuring temperature was 412 K.

The thermoemission data on Ni(111), Fig. 10c, were taken at 449 K. In this case the magnetization of the sample was checked at 406 K where the thermocurrent is negligible, .again using a photon energy of 2.15 eV. As shown by the full circles the photoelectrons are strongly polarized and P(H) saturates at an external field H = 5 kOe.

Fig. 10 gives conclusive evidence that electrons thermoemitted from Fe, Co, and Ni are unpolarized. This result cannot be accounted for by the single-electron band structure of these materials.

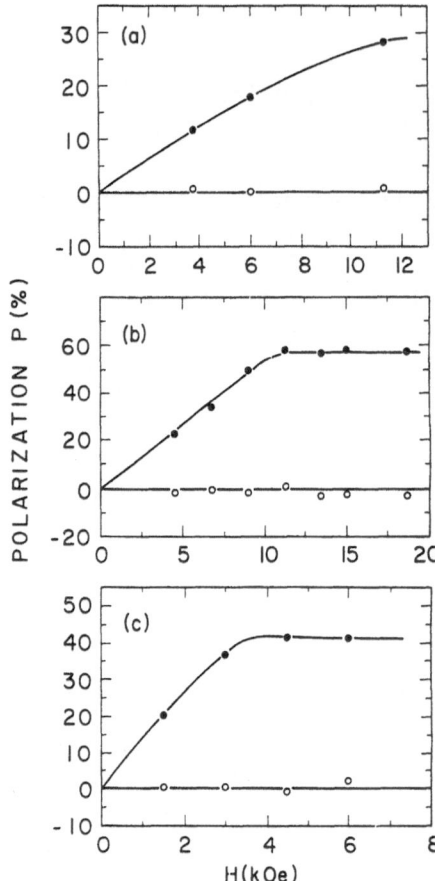

Fig.10. Open cirlces: polarization of thermoemitted electrons as function of the applied magnetic field. a: polycrystalline Fe T=423 K, b: Co(1000) T=412 K, c: Ni(111) T=449 K. Solid circles: a,b: polarization of thermoelectrons and photoelectrons, c: polarization of photoelectrons alone.

From band structure calculations the polarization values P = -75, -29 and -4 % are expected for the thermoemission current from Fe, Co, and Ni, respectively. However, in all three cases the measured value is P = 0 suggesting that the zero polarization is a general material-independent property of the thermoelectrons. This is indeed confirmed in a general argument put forward by Baltensperger and Helman [30]: In thermoemission the measured polarization is the equilibrium polarization of the electrons far - > 10 Å - from the surface: but there it is always zero.

We thank H.C.Siegmann for his support and K. Brunner for his expert technical assistance. The financial support by the Schweizerische Nationalfonds is gratefully acknowledged.

REFERENCES

1) J. Kessler, Polarized Electrons, Springer-Series on Atoms and Plasmas Vol 1, Springer-Verlag Berlin, 1985
2) J. Kirschner, Polarized Electrons at Surfaces, Springer-Tracts in Modern Physics Vol. 106, Springer-Verlag Berlin, 1985
3) Polarized Electrons in Surface Physics, Advanced Series in Surface Science (Editor: R. Feder), World Scientific Publishing Co., Singapore, 1985
4) High Energy Spin Physics Vol.2: Workshops: Proc. of the 9th Int. Symp., Bonn FRG, 6-15 Sept. 1990 (Editors: W. Meyer, E.Steffens, and W.Thiel), Springer-Verlag 1991
5) M. Aeschlimann, A.Vaterlaus, M.Lutz, M.Stampanoni, F.Meier, H.C.Siegmann, S.Klahn, and P. Hansen, Appl. Phys. Lett.59, 2189 (1991)
6) A. Vaterlaus, J.C.Gröbli, D. Guarisco, F.Meier, and M.Hecht, submitted for publication.
7) F.Meier and D.Pescia in "Optical Orientation", (Editors: F.Meier and B. Zakharchenya), Modern Problems in Condensed Matter Sciences 8, 295 (1984)
8) A. Kastler, J.Phys.Radium 11, 255 (1950)
9) see Ref.[4], F.Meier et al., p.11 ff.
10) The existence of chalcopyrites with c/a > 2 is not established: P. Villars, Intermetallic Phases Data Bank, Postfach 15, CH-6354 Vitznau, Switzerland (private communication); P.Villars and L.D.Calvert, Pearson's Handbook of Crystallographic Phases, Vol. 1- 4, ASM International, Materials Park, Ohio 44073, U.S.A.
11) T.Maruyama, E.L.Garwin, R.Prepost, G.H.Zapalac, J.S.Smith, and J.D.Walker, Phys. Rev.Lett. 66, 2376 (1991)
12) T.Nakanishi, H.Aoyagi, H.Horinaka, Y.Kamiya, T.Kato, S.Nakamura,T.Saka, and M.Tsubata, Phys.Lett. A158, 345 (1991)
13) Li chen, K.C.Rajkumar, and A.Madhukar, Appl.Phys.Lett 57, 2478 (1990)
14) F.Ciccacci, H.-J. Drouhin, C.Hermann, R.Houdré, G.Lampel, and F.Alexandre, Solid State Electron. 31, 489 (1988)

15) C.L.Petersen, M.R.Frei, and S.A.Lyon, Phys.Rev.Lett. <u>63</u>, 2849 (1989)

16) see, e.g., Ultrafast Phenomena Vol. I - VII, Springer Series in Chemical Physics, Springer-Verlag Berlin

17) A.M.Malvezzi, H.Kurz, and N.Bloembergen, Appl.Phys.<u>A36</u>, 143 (1985); see also Ref.[18]

18) A.Vaterlaus, D.Guarisco, M.Lutz, M.Aeschlimann, M.Stampanoni, and F.Meier, J.Appl. Phys. <u>67</u>, 5661 (1990)

19) J.G.Fujimoto, J.M.Liu, E.P.Ippen, and N.Bloembergen, Phys.Rev.Lett. <u>53</u>, 1837 (1984)

20) A.Vaterlaus, T.Beutler, and F.Meier, Phys.Rev.Lett. <u>67</u>, 3314 (1991); A.Vaterlaus, T. Beutler, D.Guarisco, M. Lutz, and F.Meier, Phys.Rev.<u>B46</u> (issue: 1 September 1992)

21) S.I.Anisimov, B.L.Kapelovich, and T.L.Perel'man, Zh. Eksp. Teor. Fiz. <u>66</u>,776 (1974) [Sov.Phys.JETP <u>39</u>, 375 (1974)]

22) G.Busch, M. Campagna, P.Cotti, and H.C.Siegmann, Phys.Rev.Lett. <u>22</u>, 597 (1969)

23) N.Müller, W.Eckstein, W.Heiland, and W.Zinn, Phys.Rev.Lett. <u>29</u>, 1651 (1972); M.Landolt and M.Campagna, Phys.Rev.Lett. <u>38</u>, 663 (1977)

24) A.Vaterlaus, F.Milani, and F.Meier, Phys.Rev.Lett. <u>65</u>, 3041 (1990)

25) L.Fritsche, J.Noffke, and H.Eckardt, J.Phys.<u>F17</u>, 943 (1987)

26) H.Eckardt and L.Fritsche, J.Phys.<u>F17</u>, 925 (1987)

27) M.M.Steiner, Dept. of Physics, University of California, San Diego, La Jolla (private communication); see also: C.M.Singal and T.P.Das, Phys. Rev. <u>B16</u>, 5068 (1977)

28) G.Lampel and M.Eminyan, J.Phys.Soc.Japan <u>49</u>, 627 (1980)

29) G.A.Haas and R.E.Thomas, in Techniques of Metals Research (Editor: G. Passaglia), Interscience Publ., New York, Vol. VI, Pt.1, p.91 ff.

30) J.S.Helman and W. Baltensperger, Mod.Phys.Lett. <u>B5</u>, 1769 (1991)

ION-INDUCED ELECTRON EMISSION FROM

MAGNETIC AND NONMAGNETIC SURFACES

C. Rau,[1] N. J. Zheng,[1] M. Rösler[2] and M. Lu[1]

[1]Department of Physics and Rice Quantum Institute
Rice University, Houston TX 77251-1892, USA
[2]permanent address: KAI, Berlin, Germany

INTRODUCTION

Ion-surface scattering experiments enable us to study the electronic and magnetic properties of surfaces of thin films and of bulk materials which, at present, receive great attention. This intense and broad scientific interest arises from the fact that the development of new electronic and magnetic devices of dimensions in the submicron region requires a fundamental understanding of topmost surface and interface layer electronic and magnetic properties of materials such as Ni, Fe, Cu, Al, etc. to be used in such new devices.

Contrary to spectroscopies such as UV- or X-ray-photoelectron spectroscopy, ion-based techniques provide an extreme real-space sensitivity.[1] Photoelectrons or electron-induced secondary electrons emitted from the topmost surface layer are often difficult to distinguish from those that originate in subsurface and deeper layers. Often the emitted electrons originate up to 5 or more layers beneath the surface. Although this probing depth is relatively small, mainly bulk properties will be detected because the electronic structure becomes that characteristic of the bulk within a few atomic layers of the surface.[1-5].

Among several extremely surface-sensitive spectroscopies, electron capture spectroscopy (ECS), which utilizes grazing-angle surface reflection of fast ions at magnetic surfaces, has been shown to provide a powerful means to study the two-dimensional (2D) magnetic properties (critical behavior, ferromagnetic order and magnetic anisotropies) of surfaces of ultra-thin films and of bulk materials with unprecedented sensitivity.[5,6]

Quite recently, in a methodological advancement of ECS, spin-polarized electron emission spectroscopy (SPEES) was implemented and has already provided a series of remarkable results on surface electronic and magnetic structures and on ion-surface interaction mechanisms.[7,8]

Despite many breakthroughs in the interpretation of surface-related experimental data obtained using various surface-sensitive spectroscopies (electron capture spectroscopy (ECS),[5,6] spin-polarized electron emission spectroscopy (SPEES),[7,8] spin-polarized metastable deexcitation spectroscopy (SPMDS)[9] and ion neutralization spectroscopy (INS)[10,11]) at surfaces of magnetic and nonmagnetic materials, there is still no complete and fundamental understanding of how these data are linked to the local surface electronic band structure of these materials.

The unraveling of the physical processes inherently involved in the above mentioned spectroscopies is not only of broad theoretical interest, it can also greatly enhance the power and applicability of these techniques.

The aim of the present paper is to provide new and challenging information about SPEES experiments performed at Al, Cu and Fe surfaces which can further help to advance our understanding of the physics of SPEES.

A very promising approach to elucidate details of the physics of electronic processes involved in ion-surface interaction includes not only the determination of the angle-resolved energy distribution (ARED) of the emitted electrons but also the detection of the sign and magnitude of the spin polarization (ESP) of the emitted electrons which can be used as an additional "label" to identify various processes occurring in ion-surface interaction processes.[8]

Recently, we reported on experimental data using SPEES during reflection (angle of incidence $\alpha = 1°$) of hydrogen and helium ions at magnetic surfaces.[7,8] From these experiments, there is evidence that the ARED and the ESP of electrons emitted during ion-surface reflection contains most valuable information on the spin-polarized surface electronic structure of Ni(110) and on the physics of various electron excitation processes occurring during ion-surface interaction.

We find that the grazing-angle, ion-induced ARED of the emitted electrons is <u>significantly</u> different from that of electron- or ion-induced secondary electron spectra and shows pronounced energy-dependent features. It is especially the nearly complete suppression of secondary electron cascading processes that makes SPEES very attractive for exploring details of a series of electron excitation processes (single electron excitation, Auger electron emission, plasmon excitation·and decay, etc.) occurring during grazing-angle particle surface interaction.

In this paper, we present information about SPEES experiments performed at Al, Cu, Fe and oxidized Al surfaces. Changing α from $0.2°$ up to $45°$ allows us to systematically vary the <u>probing depth</u> of the incident ions from the topmost surface layer to interface and deeper located layers. The use of Ne^+ ions instead of H^+ ions allows us to connect our data to published data obtained using INS.[10] Using angles of incidence of $\alpha = 45°$ enables us to link our SPEES data to well-known electron- or ion-induced secondary electron spectra (cascade maximum at low electron energies).[12-15]

At grazing angles of incidence, the ion-induced emitted electrons originate from the <u>topmost</u> surface layer. Electron transport and scattering processes occurring in subsurface and deeper layers are predominantly negligible. Therefore, using SPEES at clean or oxygen-covered metal surfaces, we obtain new and more direct information on the various ongoing physical processes inherently involved in particle-surface interaction.

In the following, we discuss a series of new and unique SPEES experiments which clearly and unambiguously show that the presence of secondary electron cascades not only can prevent the direct experimental identification and verification of different electron excitation processes occurring at surfaces during particle surface interaction, but also limit the applicability of ion-induced electron emission experiments to answer many important questions in modern-day surface science and technology.

EXPERIMENTAL

In our SPEES experiments, we use small angle ion-surface scattering of energetic (5-150 keV) ions (H^+, He^+ or Ne^+) to study the emission of ion-induced electrons occurring during particle surface interaction (see Fig.1). Varying the scattering angle α from 0.2^0 up to 45^0 allows us to vary the <u>probing depth</u> of the incident ions from the topmost surface layer to interface and deeper layers.

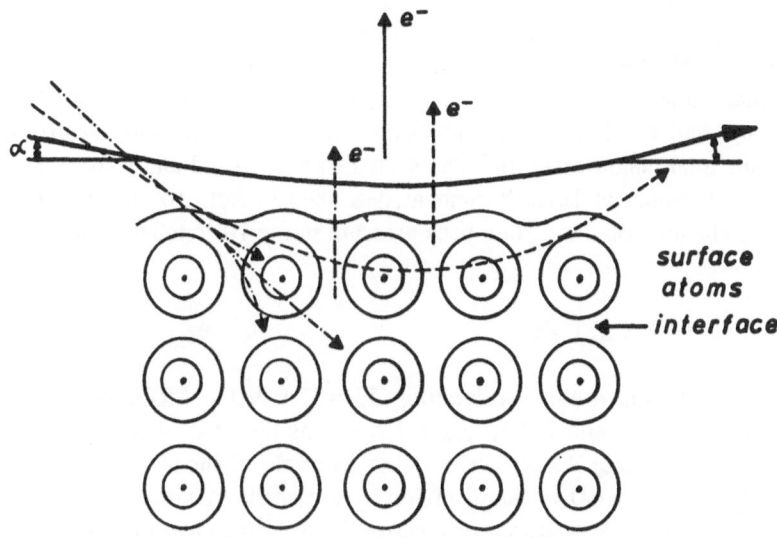

Figure 1. Scheme illustrating ion trajectories for various scattering angles, the emission of ion-induced electrons and the surface potential plotted on a plane perpendicular to the reflecting surface.

The distance d_{min} of closest approach of the ions towards a surface is well characterized by the energy component $E_\perp$ of the ions normal to the surface, $E_\perp = E_o \sin^2\alpha \approx E_o\alpha^2$. For $E_o = 25$ keV, $E_\perp$ amounts to 7.6 eV for a reflection angle $\alpha = 1^0$. Using appropriate planar surface potentials (see Fig.9 in Ref.10), d_{min} amounts to approximately 0.1 nm for $\alpha = 1^0$ which reveals that under these experimental conditions the ions cannot penetrate surfaces such as Ni, Fe, Cu or Al and are specularly reflected thereby causing the excitation of electrons solely to occur at the topmost surface layer. We note, however, that, for $E_\perp \geq 10\text{-}20$ eV - the exact value for $E_\perp$ depending on the specific (hkl) surface orientation and on the selected material - the ions are able to penetrate the surfaces and induce the emission of electrons from interface and deeper layers.

Using an einzellens system, we detect electrons emitted along the surface normal (emission cone angle 8°) of a nonmagnetic or remanently magnetized ferromagnetic target. For energy analysis an electrostatic energy analyzer (energy resolution: 300 meV) is used. For spin analysis, using a second einzellens system, the energy-analyzed electrons are accelerated into a electron spin polarization (ESP) detector. In this ESP detector, the electron beam is precisely focused on a Au target. The count ratio N_A/N_B of electrons elastically backscattered in two channeltrons A and B, positioned at $\Theta = 135^o$ to the incoming beam direction, provides a direct measure of the ESP whereas the sum of the count rates $N_A + N_B$ provides a measure of the total intensity N(E) of the emitted electrons. We note that N(E) can also be directly measured by replacing the ESP detector by a single channeltron and recording directly the intensity N(E). For zero ESP calibration, the Au target is replaced by an Al target. Further zero ESP calibration can be performed by using nonmagnetic samples for the ion-surface reflection or by heating ferromagnetic targets above the Curie temperature where only non-polarized electrons are emitted. The component of the ESP, P, along the direction of magnetization of the ferromagnetic target is given by P $=S(\Theta)^{-1}(N_A\text{-}N_B)/N_A+N_B)$, where $S(\Theta)$ is the Sherman function. We note that P > 0 is related to a predominance of so-called majority-spin electrons (ESP parallel to the total magnetization), and P < 0 refers to a predominance of minority-spin electrons (ESP antiparallel to the total magnetization).[5] The samples are prepared and characterized in a target preparation chamber at 1×10^{-10} mbar as described in Ref. 10 and then transferred to the measuring chamber operating at a base pressure in the 10^{-11} mbar region.

RESULTS AND DISCUSSION

In Fig.2 we present for H^+ with energy $E_0 = 27$ keV the ARED of electrons emitted from rough (unpolished) and flat (average flatness as detected using STM: 2 nm within 200 nm along the surface) Al surfaces. Using an angle of incidence $\alpha = 45^o$ for the fast ions, we find that the ARED is characterized by a cascade maximum at 2 eV and closely resembles well-known, classical ion- or electron-induced secondary electron spectra which are material-independent.[16,12-15]. For $\alpha = 45^o$, the ions with $E_\perp = 13.5$ keV penetrate the Al surface and excite electrons in deep lying layers which causes secondary electron cascades in the bulk which can be observed experimentally after escaping through the Al surface. Nearly exactly the same cascade maximum in the ARED is observed when the ions are directed with $\alpha = 1^o$ towards a rough Al surface which, on an atomic scale, due to the surface roughness provides incidence angles α ranging from 0° to 90°.

Using $\alpha = 1^o$ at flat Al surfaces results in a ARED which is significantly different from the previously observed ARED's. The ARED lacks completely the low energy shoulder with a maximum at around 2 eV and peaks now at around 3 eV. This implies that, due to preferential electron excitation in opmost surface or subsurface layers, the excited electrons can directly escape the surface without being multiply scattered thus preventing the built-up of a low-energy secondary electron cascade maximum.

We note that we already observed the suppression of secondary electron cascades in the ARED using SPEES at atomically flat surfaces of Ni(110) single crystals.[7,8] For Ni(110), the peak in the ARED is located at around 5 eV, a fact which could point towards a material dependence in the ARED of our SPEES spectra. In order to investigate such a possible material dependence, we used SPEES at Cu and Fe surfaces.

Fig. 3 shows the ARED of electrons using SPEES at flat Cu, Fe and Al (see Fig.2) surfaces. For these experiments we used again 27 keV H$^+$ ions and $\alpha=1°$. For Cu and Fe surfaces, we find ARED's that are similar in shape compared to the ARED of Al. For Cu, the maximum is located at around 5 eV and for Fe at around 4 eV. In addition, the ARED's for Cu and Fe show between 0 and 3 eV a low-intensity shoulder similar to that found in our SPEES experiments at Ni(110) surfaces where the electron spin was used as an additional label to identify various electron excitation processes.[8]

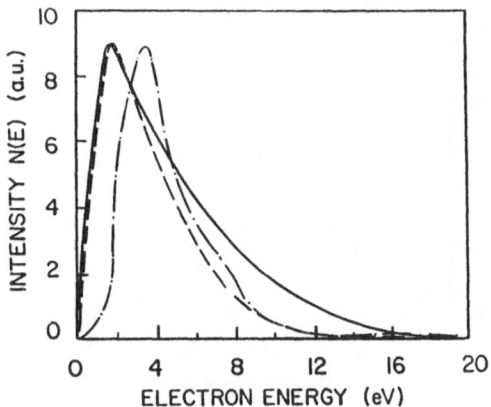

Figure 2. Normalized, angle- and energy resolved intensity distribution N(E) of electrons emitted during surface scattering of 27 keV H$^+$ ions at rough (___ , - - -) and flat (-.-.-) Al surfaces for angles of incidence α of 1° (- - -, -.-.-) and 45° (___).

This low-energy shoulder can be explained, in agreement with our Ni(110) experiments, to be due XVV Auger electron transitions where X denotes the 1s unfilled state of the incoming H$^+$ ion ($E_X=13.58$ eV) and V denotes valence band electronic levels of the investigated material. For the work function ϕ of Cu and Fe, we use $\phi_{Cu}=4.65$ eV[17] and $\phi_{Fe}=4.75$ eV.[17]

The maximum kinetic energy of the emitted Auger electrons amounts then to $E_X-2\phi=$ 4.28 eV for Cu and 4.08 eV for Fe which is in agreement with the experimental data. Due to the overlap of such a low-energy shoulder XVV Auger process with the main peak in the ARED for Al, we cannot, at present, indicate evidence for the existence of these XVV processes until a complete and detailed theory of SPEES is available.

We remark that the distinct location of the main peak in our ARED spectra for Fe, Cu, Al and Ni[7,8] found so far could point towards a material dependence of grazing-angle ion-excited electron emission spectra which could link these data to the local surface densities of filled and unfilled electronic states of these materials. Note that, in addition, details of the shape of these ARED's could be influenced by deviations of the surfaces from perfect atomic flatness.

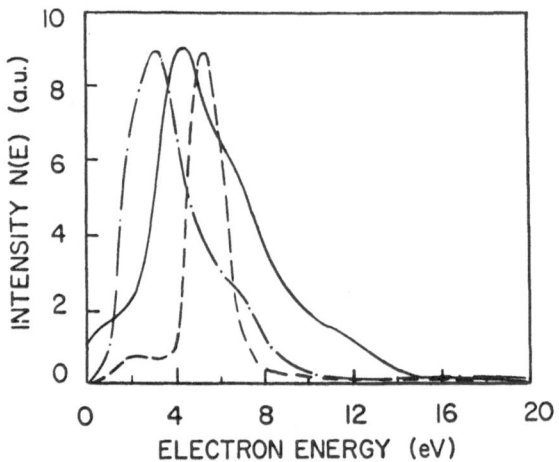

Figure 3. Normalized ARED of electrons emitted during surface scattering of 27 keV H^+ ions at Al (-.-.-), Cu (- - - -) and Fe (_____) surfaces using an angle of incidence $\alpha = 1°$.

In further SPEES experiments, we studied the ion-induced angle- and energy-resolved electron emission at Al surfaces using 150 keV instead of 27 keV H^+ ions. This allows us to investigate the ion-induced excitation and decay of surface and bulk plasmons which should not occur for H^+ ions with energies less than 40 keV.[16]

Fig. 4 gives the ARED of electrons emitted from Al surfaces using 150 keV H^+ ions and $\alpha = 0.5°$ which gives for $E_\perp = 11.4$ eV indicating that the ions are reflected at the topmost surface layer. We note that, as detected using a cylindrical mirror analyzer for Auger electron analysis, the Al surfaces were covered with approximately 0.1 monolayer of oxygen after initial sputter cleaning. The ARED is characterized by two peaks located at around 5 and 11 eV and a shoulder peak located at around 2-3 eV.

For the determination of the energetic location of plasmon peaks for pure and oxidized Al, we refer to electron energy loss data[18,19] where the bulk plasmon energy is 15.4 eV, the surface plasmon energy is 10.3 eV, and the energy of the surface plasmon of the oxide layer (or interface plasmon of the metal-metaloxide interface) is 7.1 eV. From these energies, we have to subtract the work function of Al which amounts to ϕ_{Al}=4.2 eV. This yields the following values for the excitation and emission of electrons due to plasmon decay: 11.2 eV for the bulk plasmon, 6.1 eV for the surface plasmon and 2.9 eV for the surface plasmon of the oxide layer.

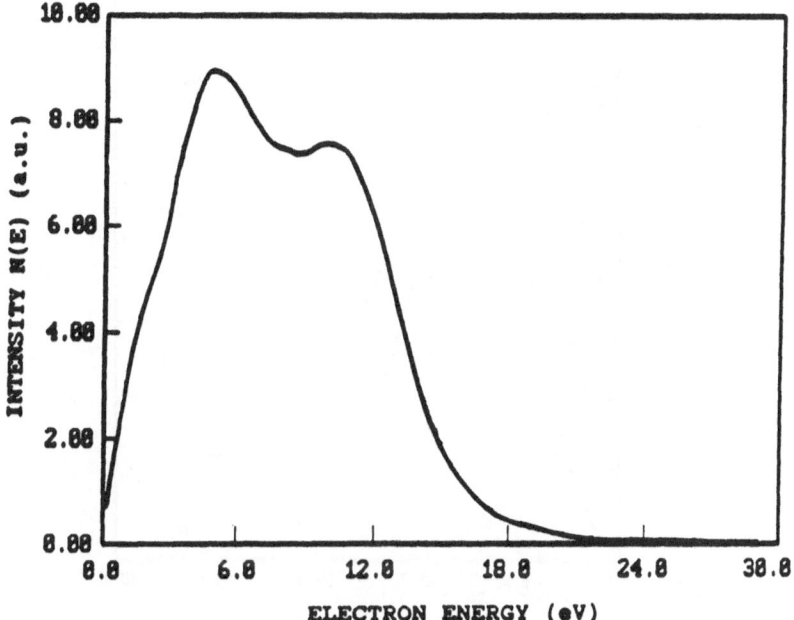

Figure 4. ARED of electrons emitted during surface scattering of 150 keV H$^+$ ions at Al surfaces using an angle of incidence α=0.5°

At first sight one is tempted to correlate these three features found in our ARED spectra at 11 eV, 5 eV and 2-3 eV with these three plasmon peaks. We note, however, that for oxygen at Al surfaces a (2s2p2p) Auger electron transition cannot be neglected. For oxygen on Al[18] the 2p level is found to be located 7.1 eV below the Fermi level and the 2s level 27.8 eV below the vacuum level which yields for the emitted Auger electrons an energy of (27.8-2(7.1+4.2)) eV=5.2 eV. The energetic position of this peak is close to the peak at 5 eV found in our ARED spectra.

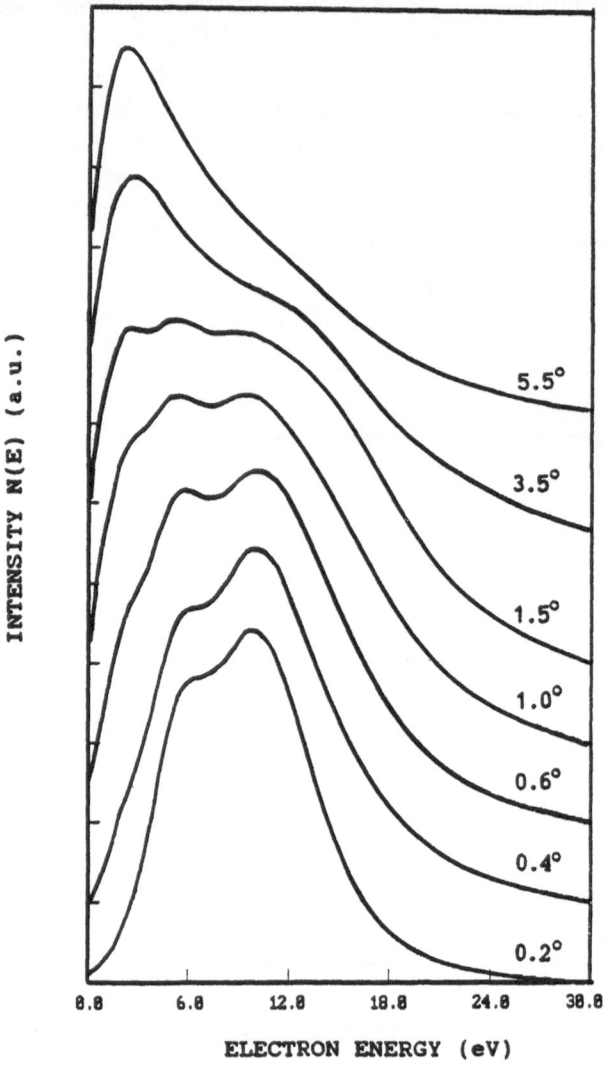

Figure 5. ARED of electrons emitted during surface scattering of 150 keV H$^+$ ions at Al surfaces using angles of incidence α ranging from $=0.2°$ to $5.5°$

In order to find out whether the peak located at 5 eV could be also due to this (2s2p2p) Auger electron decay, we have further sputter cleaned the Al surface to reduce the amount of oxygen to less than 0.05 monolayer.

In Fig. 5 we present ARED data for electrons emitted during surface scattering of 150 keV H$^+$ ions at Al surfaces with a 0.05 monolayer coverage of oxygen for α ranging from 0.2° (E$_\perp$=1.8 eV) up to 5.5° (E$_\perp$=1.378 keV). The ARED for α=5.5°, where the ions probe the bulk of Al, resembles closely the before-mentioned classical secondary electron spectra with a peak located at around 2.5 eV. Reducing α from 5.5° to 0.2°, results in the appearance of two peaks located at 11 eV and 6 eV. In addition, at the low energy part of the ARED we again observe a peak at around 2-3 eV. We note that for

$\alpha \leq 0.6^\circ$ ($E_\perp = 16.5$ eV), the ions do not penetrate the surface and are specularly reflected. The presence of atomic steps at the Al surface, however, could cause penetration of the ions through interface and subsurface layers during the surface reflection.

The peaks at 11 eV and 2-3 eV are in agreement with the assumption that they are induced by electron emission due to the decay of ion-excited Al bulk plasmons and Al-oxide surface plasmons (or Al-Al-oxide interface plasmons). As regards the peak in the ARED which has slightly shifted from 5 eV (see Fig.4) to 6 eV (see Fig. 5) by reducing the oxygen surface coverage, we remark that this could be an indication that in the first case (see Fig. 4) the predominant contribution to this peak could be due to (2s2p2p) Auger electron decay from oxygen, whereas in the latter case (see Fig. 5), the predominant contribution to this peak could be due to electron emission originating from Al surface plasmon decay. Experimental and theoretical work is in progress to perform a more quantitative study of these phenomena.

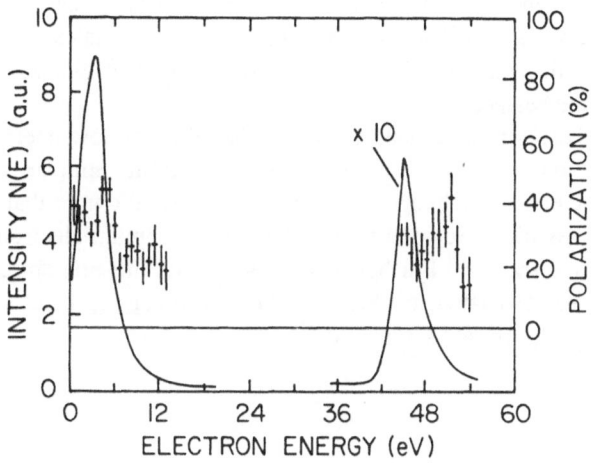

Figure 6. ARED and ESP as function of the energy E of electrons emitted from clean Fe surfaces for 25 keV Ne[+] ions and $\alpha = 45^\circ$.

In further SPEES studies, we investigated the angle- and energy-resolved electron emission from magnetized Fe surfaces. Fig.6 gives for 25 keV Ne[+] ions and $\alpha = 45^\circ$ the ARED and the ESP as function of the energy E of electrons emitted from clean Fe surfaces. The ARED is similar to that obtained in secondary electron emission experiments.[13,15] For electrons with energies around 10 eV, we observe an ESP of P=25% which increases up to P=45% with decreasing electron energy. From Fig.6 we further observe a pronounced peak in the ARED at around 45 eV which can be attributed to the emission of spin-polarized MVV Auger electrons[20,8] with an average ESP of 30% which is close to the bulk magnetization of Fe (28%).

At present, it is a point of considerable interest of whether the measured ESP of the emitted electrons reflects the net magnetization of a material. From ion-induced[15,21,22,8] and electron-induced electron spectra[14], there is evidence that the ESP of electrons emitted at high energies (≈ 10 eV above the vacuum level) scales roughly with the average net magnetization. From our experiments, we find an average value for the ESP around 25% for E around 10 eV. This value is in good agreement with the average magnetization of Fe which amounts to approximately 28%. These findings imply that SPEES could be profitably used as a simple magnetometer for the investigation of the average magnetization of new magnetic materials.

We note that the enhancement of the ESP from 25% to 45% for electron energies decreasing from 10 eV to lower energies is a very interesting effect. In order to find out whether this effect can be caused by a spin-dependence of the electron mean free path, we performed SPEES experiments for $\alpha = 1°$. From the results of these experiments, where the electrons are excited at the topmost surface layer, we observe a similar enhancement of the ESP for electron energies decreasing from 10 eV to lower energies.

These findings exclude the possibility that these strong enhancements of the ESP are predominantly due to the spin dependence of the electron mean free path[23] which is caused by an excess of unfilled minority-spin electronic states in Fe over unfilled majority-spin electronic states in which excited spin-polarized electrons can be scattered during transport to the surface.

We correlate these strong enhancements of the ESP for low-energy electrons to so-called Stoner excitations (electron hole pairs with antiparallel spin orientation) which occur during inelastic electron exchange-scattering processes. We note that the role of Stoner excitations was successfully discussed in the literature to explain electron- and ion-induced secondary electron spectra.[14,15] Further details as regards the fine structure of these ESP spectra, which can be associated to details of the spin-polarized band structure above the vacuum level[24], will be reported elsewhere.[25]

CONCLUSIONS

In conclusion, we remark that the absence of secondary electron cascade processes, which is an essential feature of SPEES, allows us a refined and detailed analysis of ARED and ESP spectra from angle- and energy-resolved SPEES experiments performed at non-magnetic and magnetic surfaces. The experiments presented here, not only provide important and fundamental information about a series of most interesting physical effect (single electron excitation, spin-polarized Auger electron emission, plasmon emission and decay, etc.) inherently involved in particle surface interaction processes but also about electronic and magnetic properties of a material.

ACKNOWLEDGMENTS

This work was supported by the National Science Foundation, by the Welch Foundation and by the Texas Higher Education Coordinating Board.

REFERENCES

1. F.B. Dunning, C. Rau and G.K. Walters, Comments Solid State Physics, 12, 17 (1985).

2. C.L. Fu, A.J. Freeman, and T. Oguchi, Phys. Rev. Lett. 54, 2700 (1985).

3. C. Rau and G. Eckl, in: Nucl. Phys. Meth. in Mat. Res, eds. K. Bethge et. al. (Vieweg, Braunschweig, 1080) p. 360.

4. C. Rau, Comments on Solid State Physics, 9, 177 (1980).

5. C. Rau, J. Magn. Magn. Mater 30, 141 (1982).

6. C. Rau, Appl. Phys. A49, 579 (1989).

7. C. Rau and K. Waters, Nucl. Instr. Meth. B33, 378 (1988); C. Rau and K. Waters, Nucl. Instr. Meth. in Phys. Res. B40, 127 (1989).

8. C. Rau, K. Waters and N. Chen, Phys. Rev. Lett. 64, 1441 (1990).

9. M.S. Hammond, F.B. Dunning, and G.K. Walters, Phys. Rev. B45, 3674 (1992).

10. H.D. Hagstrum, Phys. Rev.150, 495 (1966).

11. P.A. Zeijlmans Van Emmichoven, P.A.A.F. Wouters and A. Niehaus, Surf. Sci. 195, 115 (1988).

12. D.E. Harrison, Jr., C.E. Carlston, and G.D. Magnuson Phys. Rev. 139, 737 (1965).

13. R.A. Baragiola, E.V. Alonso, and A. Oliva-Florio Phys. Rev. B61, 121 (1979).

14. M. Landolt, in "Polarized Electrons in Surface Physics", ed. by R. Feder (World Sci. Publ. Co., 1985), Chap.9; and Refs. cited therein.

15. J. Kirschner, in "Surface and Interface Characterization by Electron Optical Methods" ed. by A. Howie and U. Valdre (Plenum Publ. Co., 1988), p. 297; and Refs. cited therein.

16. M. Rösler and W. Brauer, "Particle Induced Electron Emission I," Springer, Berlin (1991) (Springer Tracts in Modern Physics, Vol.122).

17. J. Hölzl and F.K. Schulte, "Work Function of Metals,"Springer, Berlin (1979) (Springer Tracts in Modern Physics, Vol. 85).

18. C. Bendorf, G. Keller, W. Seidel, and F.F. Thieme, Surf. Sci. 67, 589 (1979).

19. C.J. Powell and J.B. Swan, Phys. Rev. 115, 869 (1959); 640 (1960).

20. M. Landolt, Appl. Phys. a41, 83 (1986); and M. Landolt, R. Allensbach, and M. Taborelli, Surface Science 178 311 (1986).

21. J. Kirschner, K. Koike, and H.P. Oepen, Phys. Rev. Lett. 59, 2099 (1987).

22. J. Kirschner, K. Koike, and H.P. Oepen, Vacuum 41, 818 (1990).

23. D. Penn and P. Apell, and S.M. Girvin, Phys. Rev. Lett. 55, 518 (1985).

24. E. Tamura and R. Feder, Phys. Rev. Lett. 57, 759 (1986).

25. C. Rau, N.J. Zheng, and M. Rösler, to be published.

SOME PHYSICAL DESCRIPTIONS OF THE CHARGING EFFECTS IN INSULATORS UNDER IRRADIATION

Jacques Cazaux

LASSI/GRSM
Faculté des Sciences, BP 347
51062 Reims, France

INTRODUCTION

Charge phenomena occur in a wide variety of insulator materials when they are subject to various types of radiation. The goal of the present paper is to analyze these effects in order to try to elucidate them. For this goal, first the physical properties that insulators have in common are indicated. Next, based on the electrostatic, electrokinetic and secondary emission properties of insulators under irradiation, some explanations of the macroscopic effects of charging are given. The full description of these effects, using the electrostatic approach, requires the knowledge of the microscopic trapping mechanisms of charges; some of these mechanisms are suggested. This approach invalidates the former one based on the total electron yield curve. At last, various practical consequences are indicated or suggested.

PHYSICAL PROPERTIES OF INSULATORS

What is the common point between a polymer and a crystal of magnesium oxide, a biological specimen and a piece of glass? The unique common point is that all have a very poor DC conductivity, γ. To be more precise, insulators are, by definition, the materials having the lowest values for γ. The range of the conductivity is from around 10^{-8} Ω^{-1} m^{-1} (at the blurred frontier with semiconductors) up to 10^{-16} Ω^{-1} m^{-1} for excellent insulators (paraffin, TiO, sulfur).[1] Neglecting the specific case of ionic conductivity, it can be deduced

Ionization of Solids by Heavy Particles, Edited by
R.A. Baragiola, Plenum Press, New York, 1993

(from the values of γ and using reasonable order of magnitude for the mobilities) that the density of the quite free carriers is less than 10^8 cm^{-3} for insulators while it is in the 10^{12} cm^{-3} - 10^{20} cm^{-3} range for semiconductors. It can also be deduced that the electronic conductivity of monocrystalline insulators is extrinsic by evaluating the expected density of carriers resulting from the thermal excitation of electrons from the valence band to the conduction band, (i.e., through a band gap, E_g, greater than 3 eV — up to 12 eV for LiF). This conductivity is due to crystal imperfections and residual impurities, and not due to their intrinsic composition. The above analysis explains the fact that the exact value of the electrical conductivity of insulators is very sensitive to their elaboration conditions.

From the point of view of the chemical composition, one may observe that insulating materials offer a diversity far greater than that of metals or semiconductors: if the family of insulators contains a restricted number of pure elements (diamond, S and P), it however includes most compounds (oxides, sulfides, polymers, glasses, ceramics, biological and mineral materials), very different from each other from the point of view of their other physical and chemical properties.

If almost all the valence electrons of an insulator are bound, they participate in the polarization of the insulator subject to the influence of an electric field E. For most materials, the polarization is approximately proportional to the applied electric field and the dielectric properties of these materials are described by their dielectric constant ε.

When irradiated by incident particles (ions, electrons, X-ray photons, etc), the atomic mechanisms involved in insulators are qualitatively the same as those in metals. They include the elastic scattering of the incident particles, the possible emission of atoms from the specimen, the excitation of the valence electrons (plasmon excitation) as well as the ionization of the electronic core levels of the atoms composing the insulator, and the de-excitation processes (photon and electron emissions, etc) that follow these ionization processes. In particular, secondary electron emission is a general effect taking place in insulators placed in vacuum and irradiated by any kind of incident particles, even neutral particles such as X-ray photons. Consequently, the corresponding loss of negative charges leads to charging effects. The main aspect of the investigation of insulators is that, contrary to the case of a metal, there are insufficient conduction electrons (or mobile holes) to restore quickly the neutrality of the specimen. The consequences of this non-neutralization of charging are:

i) The possible modifications of the local composition of the specimen, such as oxygen desorption in oxides, and the migration of the mobile ions (driven by the electric field built up). Examples of such effects are given on Figs. 1 and 2.

ii) The modifications of the physical parameters of the interactions due to the acceleration or the slowing down of the incident particles (when they are charged) such as the slowing down of incident electrons and their deflection by the electric field built up in the specimen.

iii) The change of spectral distribution and total yield of the charged species emitted by the specimen: secondary ion emission in SIMS and secondary electron emission in almost all techniques. These effects concern, for instance, the shift of the Auger and XPS lines in AES and XPS (Figs. 3 and 4).

iv) In addition, the secondary electron emission δ in insulators generally reaches larger values than in metals when irradiated under the same experimental conditions (same incident particles and primary beam energy). Fig. 5, 6 and 7 illustrate this effect. For instance, the maximum value of the secondary electron emission coefficient, δ_{max}, of monocrystalline insulators subject to electron bombardment may reach values greater than 10 (up to 20 or more) while that of metals is close to unity. This is the case for alkali halides[12] and for MgO (Fig. 5a)

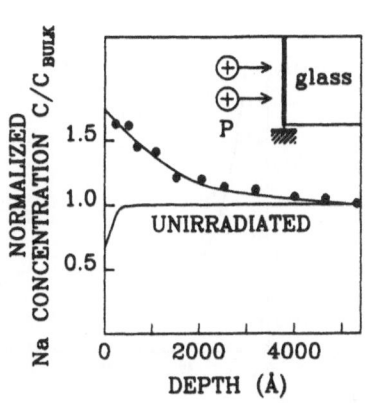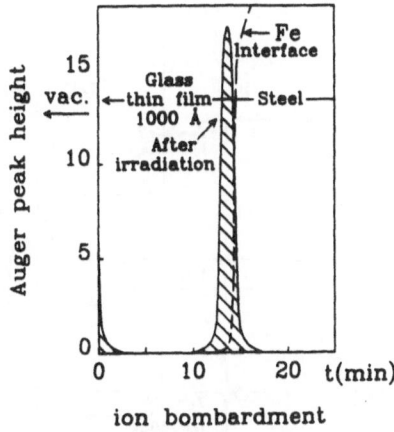

Figure 1 (left): Normalized sodium concentration profile vs. depth for unirradiated and 600 keV proton irradiated ($\bullet$) soda lime silica glasses covered by a thin Al layer; experimental results reported by Battaglin et al.[2]

Figure 2 (right): Na concentration depth profile in a vacuum/100 nm thick soda glass/metal system irradiated with 3 keV electrons.[3]

The situation is similar for X-ray irradiation where the total yield ratios between alkali halides (CuI, CsI) and metals (Al, Au) have been found[8] to be situated between 10 and 80. It is also the case for heavier projectiles such as Ar ions (fig.7) where the electron yield, y, increases with the thickness of the oxide layer while the full width at half maximum of the spectral distribution decreases (like for electron irradiation: Fig. 6, right).

In a simple three step model of secondary electron emission —generation, transport, escape into vacuum— the first step is obviously a function of the nature of the projectile

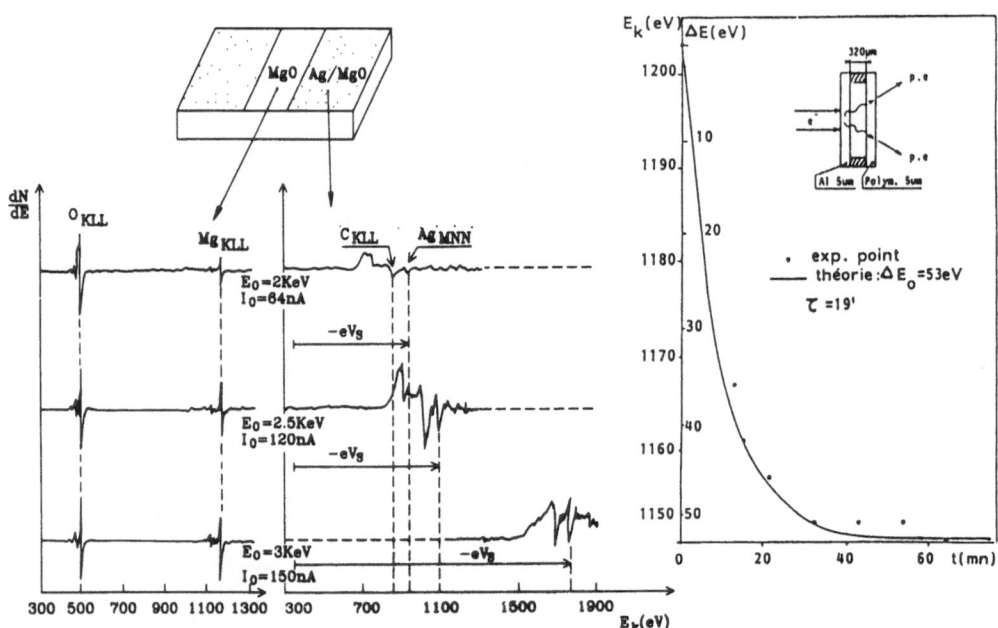

Figure 3 (left): Auger analysis of a MgO single crystal. On the left, there is no shift in the Auger spectra obtained on the clean part of the specimen when the primary beam energy is changed. On the right, there is a negative shift in energy in the same experimental conditions when the probe is set on the coated part of the specimen (the coating is a thin layer of silver, 3nm thick). The specimen and the investigated zones are shown in the inset (from ref. 4).

Figure 4 (right): Positive shift, as a function of irradiation time, of the C_{1s} photolines emitted by a polycarbonate film irradiated by Al Kα X-ray photons. The experimental arrangement is shown in the inset.[4]

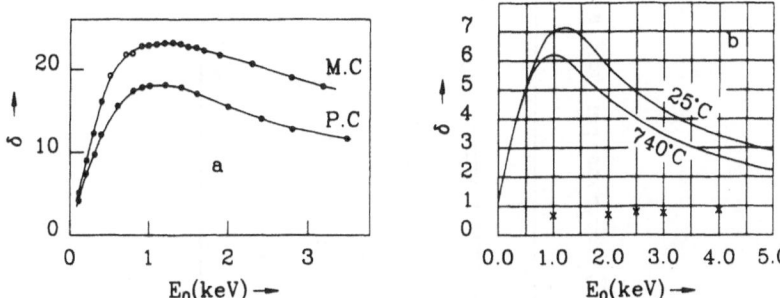

Figure 5: a) (left) Secondary electron emission of monocrystalline (M.C.) and polycristalline MgO samples as a function of the incident electron beam energy.[5] b) (right) Change of the secondary electron emission of MgO with temperature. Figure taken from ref. 6 on which crosses have been added to represent the total yield deduced from specimen current measurements.[7]

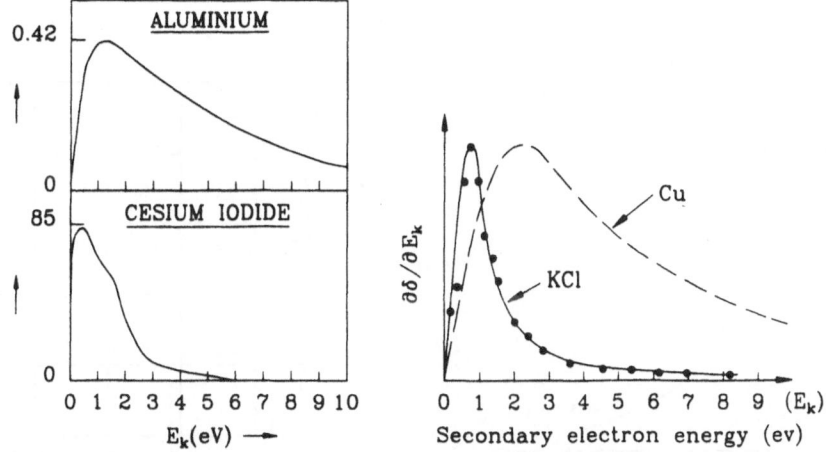

Fig. 6: Left: Spectral distribution of the secondary electron emission of a metal (Al) and an insulator (CsI) under X-ray irradiation (Cu-Kα).[8] Right: Spectral distribution of the secondary electron emission of an insulator (KCl)[9] and of a metal (Cu)[10] under electron irradiation at $E_0 \simeq 1$ keV. The two curves have been normalized to have the same maximum value.

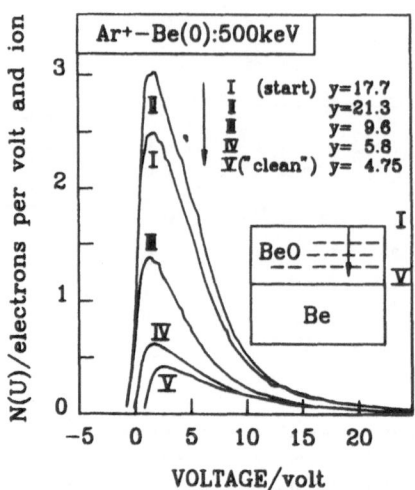

Fig. 7: Spectral distribution of secondary electrons from a beryllium target covered by a thin oxide film during high energy Ar+ ion sputtering.[11]

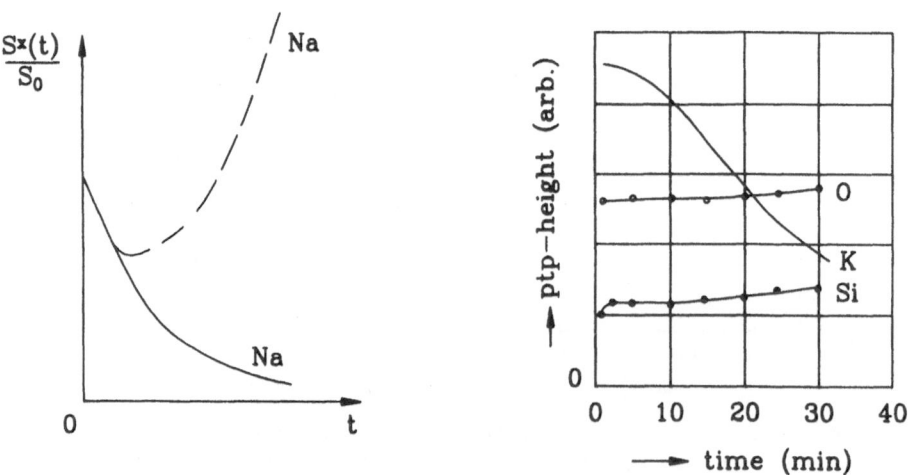

Fig. 8 (left) Experimental change of the Na-Kα signal in electron probe microanalysis of a glass specimen.[13] The full line corresponds to the specimen coated by a conducting layer while the evolution indicated by the dashed line appears when this coating is destroyed.[14]

Fig.9 (right) Experimental time dependence of Auger signals during electron irradiation of a uncoated soda lime glass specimen.[15]

and its energy while the two last steps are only a function of the nature of the specimen. To explain the main experimental features observed in the secondary electron emission of insulators (with respected to the metallic case) a general analysis of these two last steps will be performed below. One difficulty in the experimental investigation of secondary emission from insulators is that the yield may change as a function of time. To compare to theoretical predictions, it is necessary to collect the results at the early beginning of the irradiation when no electric field is established inside the insulator.

When the first secondary electrons have left it, the specimen becomes charged and an electric field begins to be established. This electric field modifies the trajectories of emitted charged species (electrons and ions) as well as the trajectories and interactions of the incident particles (if they are charged) inside (and sometimes outside) the specimen. In turn, the field first established is modified, and the main consequence is that charging effects of insulators are complex phenomena because they are time dependent: using standard methods, the trajectories (and interactions) of incident particles can only be calculated at the beginning of the irradiation.

ELECTROSTATIC AND ELECTROKINETIC STUDY OF INSULATORS

To establish a coherent and more quantitative explanation of the observations described above, it is convenient to start from the charge conservation law illustrated on Fig.10 for the signs:

$$I_o = I_r + \frac{dQ}{dt} + I_s \qquad (1)$$

I , I_s: incident and specimen beam currents; I_r is the current associated with the particles going out of the specimen into vacuum; it includes the charged particles issued from the specimen (such as the secondary electrons) as well as the primary particles being backscattered. In the case of electron irradiation $I_r = I_o y = I_o (\delta + \eta)$ where y is the total yield, δ and η being the secondary and backscattered fractions, respectively. dQ/dt describes the algebraic rate of increase of the charge trapped in the specimen. Eq. 1 applies at any time t_o for any kind of irradiation ($I_o = 0$ for irradiation by neutral particles such as X-ray photons) and any kind of specimen. In the case of ion bombardment, the quotient method widely used to determine the total yield[11] is also derived from Eq. 1 where the charge state of the incident ion, z, is taken into account using also an experimental arrangement similar to that shown in Fig. 10. In the case of a metallic sample, for instance, dQ/dt=0 because the neutrality of the specimen is quickly restored via the specimen current, I_s; thus, $I_o = I_r + I_s$ .This steady state is not spontaneously attained in the case of an insulating sample (see below). In the case of a permanent irradiation it is necessary to consider the time evolution of each term involved in Eq. 1.

SECONDARY ELECTRON EMISSION

At the early beginning of the irradiation, $t_0 = 0$, $I_s = 0$ and the specimen is not charged, $Q(t = 0) = 0$. Assuming that dQ/dt can be neglected, Eq. 1 is reduced to $I_0 = I_r$ $(= I_0 (\eta + \delta)$ for electron irradiation) and corresponds to the experimental situation used for the determination of δ by the pulse method.[6] Keeping outside the detailed theories of secondary electron emission from solids,[12,16-18] we wish only to focus the attention on two points concerning this emission from insulators (with respect to the metal case). In a three step description, these points concern the two last steps: transport and emission into vacuum.

The transport of secondary electrons towards the surface is described by the attenuation length of the secondaries, λ, which is expected to be approximately one order of magnitude larger in insulators than in metals because of the lack of electron-electron

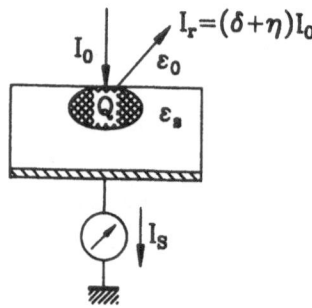

Fig.10: Conventional positive signs are used here for the charge conservation law (Eq. 1).

interaction between the generated secondaries and the (absent) conduction electrons in insulators. This explanation has been given for a long time (see Schou's paper in this volume and refs. 8, 19-20) and is illustrated in Fig.11 where the corresponding trend of λ as a function of the kinetic energy E_k is suggested.

In this figure, it is assumed that the interactions of electrons having a kinetic energy at around or greater than few tens of eV are approximately the same for metals and insulators; the minimum corresponds to the strongest probability for collective excitation of the valence electrons (plasmon excitation). The evolution of λ in the kinetic energy range of the secondaries ($E_k < 5eV$) explains easily the more pronounced peak shape (and the apparent narrower energy width) in the spectral distribution of the secondaries observed for insulators with respect to metals (cf. Figs. 6 and 7). In insulators, when E_k is less than the band gap energy, the inelastic interaction of conduction band electrons is restricted to the electron-phonon interaction which increases with temperature and may explain the decrease of δ as a function of T (Fig. 5b). The decrease of δ for polycrystalline specimens (Fig. 5a) may be associated to the limitation of λ by the grain size.

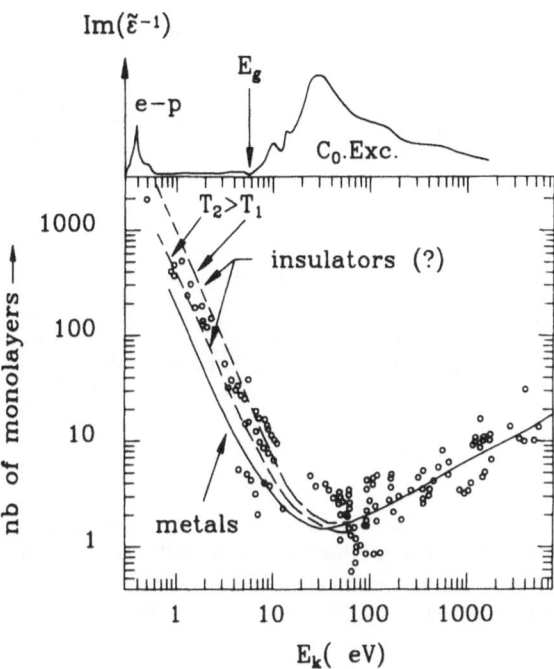

Fig.11: Attenuation length of electrons as a function of their kinetic energy. The open symbols are from the compilation of Seah and Dench.[21] Above 10-30 eV, λ is correlated with the inelastic mean free path of X-ray induced photoelectrons, below this limit the secondaries are counted when emitted into vacuum even when they have suffered inelastic events (as mentioned by Sigmund, this volume). Lines have been added in the low kinetic energy range to suggest the different behavior in metals (full line) and in insulators (dashed lines). The loss function of electrons in insulators is indicated on the top inset (e.p. : electron phonon interaction ; c.e: collective excitation; Eg: band gap).

Concerning the escape of the secondaries into vacuum, here again the case of insulators differs from that of metals. As shown in Fig. 12 (left), the work function of metals is lowered by the electric image effect. As first suggested by Schottky, this lowering is due to the fact that a charge q in the vacuum at a distance z from a metal/vacuum interface is subject to the repulsive potential of its image (at a distance -z into the metal), V

$$= \frac{-q}{4\pi\varepsilon_0} \frac{1}{2z} .$$

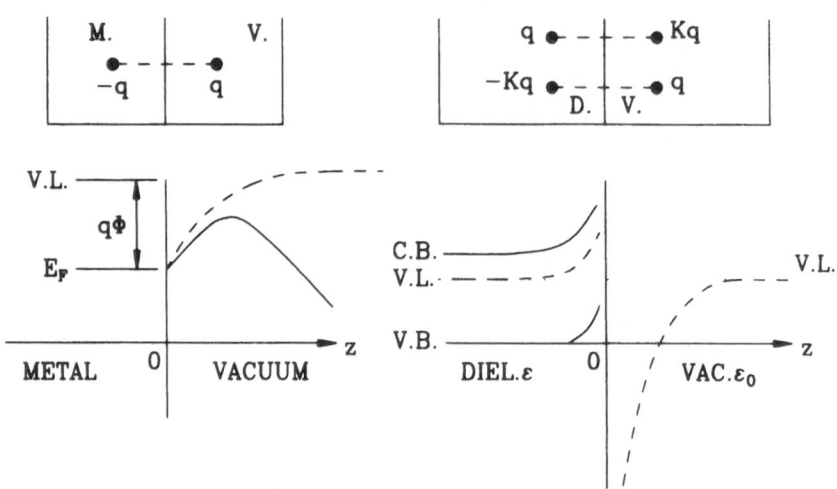

Fig.12. Left: classical energy diagram for a free electron near a plane metal surface and influenced by its electric image (see inset), without (dashed line) and with (full line) an applied electric field. Right: Suggested energy diagram for an electron near a plane dielectric/vacuum interface. This diagram takes into account the influence of the image charge Kq (see inset) on this electron.

In the case of a dielectric ε bounded by another dielectric ε_0 (plane interface), the field created by a point charge q in the medium ε is that created by the charge itself and by its image charge of weight Kq, where $K = (\varepsilon-\varepsilon_0)/(\varepsilon+\varepsilon_0)$. In the case of a vacuum/dielectric interface, $0 < K < 1$ since ε is greater than ε_0 (while K is -1 for the metal/vacuum system, a result obtained by setting $\varepsilon_0 = \infty$).[22,23]

The obvious consequence of this classical analysis is that a real charge q is subject to the potential of its image. It is a repulsive potential, $V = Kq/8\pi\varepsilon z$, when the charge is in the dielectric; it is an attractive potential $V = -Kq/8\pi\varepsilon_0 z$ when the charge is in vacuum (with $\varepsilon > \varepsilon_0$). The potential function experienced by a point charge going from the dielectric towards vacuum is shown in Fig. 12 (right). It is an asymmetric function, with a rather low barrier (when ε is large) followed by a deep well. The combination of the two (barrier and well) may reflect electrons and prevent some of them from escaping into vacuum. This combination may explain why the electrons injected in the conduction band in electron irradiated insulators having a negative electron affinity (i.e. the vacuum level in the band gap as it is the case at least for MgO[19] and for solid argon (Grosjean and Baragiola, this book) are not all reemitted into vacuum. This internal reflection effect is an increasing function of K so that the secondary emission yield is expected to be a function of K^{-1}. To verify this hypothesis, a compilation[12] of the maximum values of δ has been plotted in Fig. 13 as a function of K^{-1} for various alkali halides and other insulating materials.

The fact that the yields are loosely a linear function of K^{-1} is satisfying after considering, on one hand, the experimental difficulties to measure δ and, on the other hand, the influence of many other physical factors (generation, transport) involved in the secondary electron emission mechanisms.

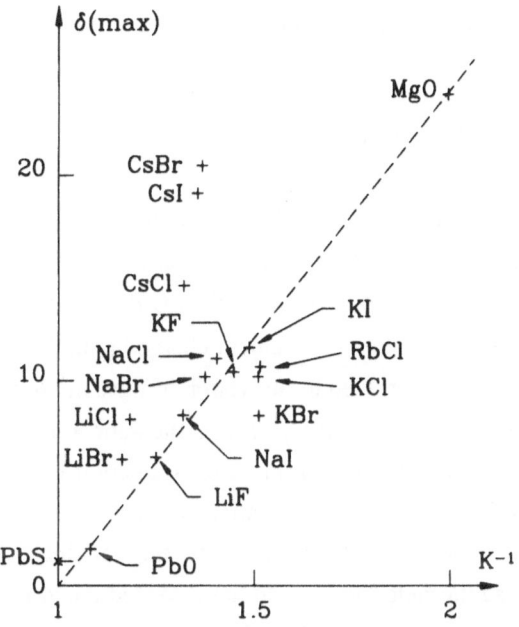

Fig.13: Plot of the experimental values of δ_{max} for various insulating materials[24] as a function of the inverse of the factor K associated to the weight of the images, obtained from tabulated values of dielectric constants.[1,25]

To the author's knowledge, the relation between the emission yield and the potential function at the dielectric/vacuum (or dielectric/dielectric) interface seems to be new and full of promising explanations concerning not only all the emission mechanisms from insulators into vacuum but also some self-trapping effects at these interfaces (see below the section: Additional questions and comments). Work to explore these various aspects is in progress.

Coming back to the time dependence of the secondary electron emission from insulators, the difficulties for a quantitative analysis can be summarized as follows. After the early beginning of the irradiation (t > 0), the first emitted secondaries have left positive ions in the specimen and electron emission is modified by the potential thus created but also by all the other electrostatic sources, including the electric potential created by the incident particles first trapped in the bulk. In the case of positive (or neutral) incident particles, the surface potential function tends to be positive leading to a decrease in the secondary electron emission (the slowest electrons being reattracted by the surface). In the case of negative incident particles (such as electrons), the surface potential at any point of space, and particularly at the surface, can only be deduced from electrostatic calculations taking into account the positively charged layer left by the emitted secondary electron but also the negatively charged region where the incident particles may be trapped. The results depend obviously upon the trapping mechanisms inside the investigated insulator and not only from the balance between the incoming and the outgoing particles at the early beginning of the irradiation (as is commonly done by using the total yield curve: see below).

ELECTROSTATIC APPROACH

To evaluate the electric field, E, and potential, V, at any time and point in space (inside the insulator and outside it), one needs to know the charge distribution $\rho(\vec{r})$ and its time evolution, $\partial \rho / \partial t$ (which allows also to evaluate the $\partial Q / \partial t$ term in Eq. 1 by an integral over the volume of the specimen). When the incident charged particles stay fixed (implanted) in the specimen, the problem is restricted to the search of the solutions of the Poisson equation satisfying the boundary conditions, starting from a theoretical or experimental implantation profile. More difficult is to solve the problem of irradiation by light and mobile particles such as electrons. From elementary considerations in that case, it can be established that all the incident electrons are not trapped, or if they are, the trapping time is very short because if this were not the case the field they would have created could reach the disruptive value (>10^7 V/cm) in less than one second (for $J_0 = 1\mu A/mm^2$) and the specimen would be destroyed.[14] Consequently, for electron irradiations the solution of the electrostatic problem requires the knowledge of the microscopic mechanisms of trapping, mechanisms depending upon the composition and the crystalline structure of the specific specimen being irradiated (as well as the specific irradiation conditions). To keep the general aspect of the description of charging effects it is possible to delay these difficulties by using reasonable models for the charge distribution inside the irradiated insulators. Using next the classical laws of electrostatics it will be easy to deduce the general form of the electric field and of the electric potential inside and outside the specimen in order to explain the main aspects observed during the irradiation of insulators. For uncoated insulators, the model distribution (Fig.14a) can be assumed to be composed of two cylinders having the same diameter, d_0

336

(the spot size of the incident beam). The height of the first one (positively charged) corresponds to the attenuation length of the secondaries. The height of the second one corresponds to the range R of the incident particles in the material and the sign of the charged distribution is the same as that of the primaries. This cylinder only remains when the insulator is coated by a thin conducting layer (at earth potential) while only the first cylinder remains in the case of irradiation by neutral particles (such as X-ray photons).

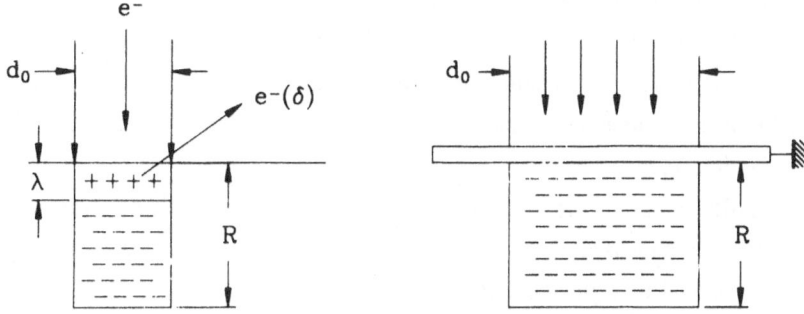

Fig.14: Examples of models for the charge distribution of irradiated insulators. Left: uncoated insulator. Right: coated insulator.

Of course, such naive models may be improved (by using more realistic concentration profiles of implanted iona) but even in their simplified forms they lead to many fruitful informations in the case of electron irradiation[14,22,23] (which can be easily transposed to other irradiations). Among others, they are:

- The electric field inside the dielectric is not uniform, it diverges (or converges).

- For a given charge distribution, the field characteristics change when the insulator is coated.

- The potential on an uncoated surface is a function of position (and not a constant as assumed in the total yield approach); its shape depends upon the incident spot size diameter while its sign depends upon the sign of the algebraic sum of the positive and the negative charges trapped at the surface and in the bulk.

This type of analysis allows to explain the direction of migration of mobile ions in glasses (Figs. 1 and 2) and the corresponding change of the emitted X-ray signal (Fig.8). It also explains the shift and distortion of the energy spectra the emitted particles; it is clear, for instance, that the trajectories of the secondary electrons in vacuum are influenced by the electrostatic lens produced by the electric field built-up at the surface and around it. Moreover, since the electrons are emitted at various places of the specimen where the potential function is not the same, their spectral distribution is distorted, and in the case of a positive charging, there is no reason why this distribution should present a sharp cut-off.

Thus, it is possible to explain the change of the observed spectra by the change of the experimental conditions (incident probe size defocused or not, change of the width of the field aperture of the analyzer) or the change of the specimen environment or thickness (the boundary conditions lead to different fields if the specimen is in the form of an unsupported foil, or if it is backed by the metal of the specimen holder, or if it is coated by a thin conducting layer at earth potential). All these remarks seem obvious (they are derived from classical electrostatics) but they may explain many conflicting results obtained on samples of the same nature but investigated using different equipment. In particular, to prevent the deflection of the incident charged beam into vacuum, the insulator to be investigated is often coated by a thin conducting layer but it is also often forgotten that the electric field inside such a specimen may take very large values mainly at the metal/dielectric interface.

The subtle change of the interface situation from a clean dielectric/vacuum interface to a dielectric/conducting layer interface may lead to drastic changes in the trapping effects at these interfaces and to significant changes in the direction of the electric field build up. During the analysis of a dielectric bounded by vacuum, a contamination layer may grow at its surface and this conducting layer (even not connected to earth potential) will attract electrons from the bulk towards the interface (attraction by the electric image of opposite sign) and this new interface may lead to a negative surface potential. The same mechanism may also explain the negative charging observed in alkali halides under electron irradiation at low temperatures when the metal atoms at the surface produce a continuous conducting layer[26] (attracting again the bulk electrons).

It must also be kept in mind that the measured parameter is often the potential (through the shift of the electron spectra) while the physically important parameter is the electric field (and its associated Coulomb force). In the irradiation of a thin oxide on a conducting material, the change of the surface potential when going from the surface to the interface may be too small to measure while the local field may be sufficient to induce effects such as the migration of the species (for a thickness t = 100 Å; a voltage drop of 0.1 Volt corresponds to a field of 10^5 V/cm).

ELECTRON TRAPPING MECHANISMS

To go deeper into the electrostatic approach, and for the case of electron irradiations, it is necessary to obtain more information on the electron trapping mechanisms. By analogy with the semiconductor situation, it is believed[27] that incident electrons can be trapped on structural defects (either induced by the irradiation itself or pre-existing in the material) or on impurities. It is thus believed that, for perfectly monocrystalline insulators and in the absence of radiation damage, the trapping effects on incident electrons may only occur on surface defects but not in the bulk (after the usual interactions with matter that lower their initial kinetic energy down to the bottom of the conduction band, what kind of additional interaction may trap them?). On the contrary, in polycrystalline or amorphous materials, it can be assumed that these trapping centers are more or less uniformly distributed inside the solid. Within this hypothesis, the schemes shown on Fig.14 are only valid for this last class of materials.

In the case of a monocrystalline specimen coated by a metallic film (not connected to earth potential), the negative cylinder (Fig.14) is reduced to a disc, the electrons being trapped only at the metal/dielectric interface. The resulting potential of the film is thus expected to be negative and this prediction has been experimentally verified —see Fig.3 and reference 7— while the total yield approach predicts a positive charging in such a situation (see comments below).

The hypothesis of electron trapping at defects is also supported by the hole drilling experiments in metallic oxides. In electron beam nanolithography, Humphreys et al[28] observe that hole drilling begins at the surfaces in the monocrystalline form of Al_2O_3 while it begins at the mid plane of the foil and along a circle having the radius of the incident electron beam, in the amorphous form of Al_2O_3.

A possible explanation for such a phenomenon is that in monocrystalline Al_2O_3 the trapping defects are only located on the two surfaces of the foil leading to an electric field component mainly normal to the foil (field created by two discs) and located at the center of the irradiated surfaces (entry and exit). In the opposite case of the amorphous material, the trapping centers (and the charge distribution) is uniformly distributed inside the foil and the main electric field component is the radial one in the mid-plane of the foil (field created by a uniformly charged cylinder) (Fig. 15). In others words, hole drilling occurs exactly in the regions where the electric field generated by the trapped charges is expected to be maximum.

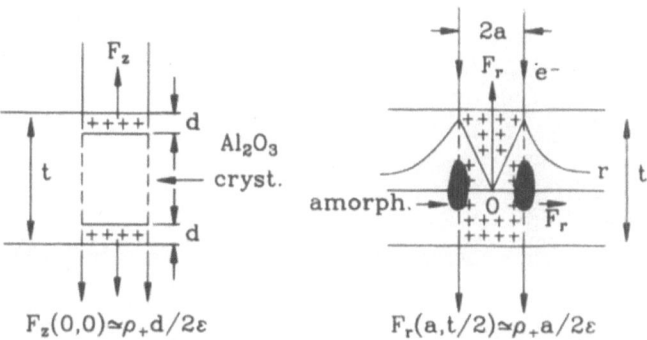

Fig.15: Direction of the electric field, F, generated by trapped charges (during hole drilling). Left: located at the surface for an irradiated monocrystalline thin insulating specimen. Right: uniformly distributed inside the foil for an amorphous thin specimen.

If it is assumed that electrons are trapped at defects, one needs to consider also some detrapping processes to limit the increase of the resulting electric field. Based on an approach similar to that used for semiconductors, the study can be summarized as follows:[27] during irradiation some incident charged particles are trapped with a probability σ (in fact an effective capture cross-section) on some of N (per unit volume) trapping centers distributed in the insulator. The corresponding increase in density of the trapped charge ρ_c per unit time is:

$$\partial\rho_c/\partial t = J_oN\sigma \tag{2}$$

where J_o is the incident beam density. During the same time interval, recombination mechanisms have canceled out (detrapping process with a time constant τ):

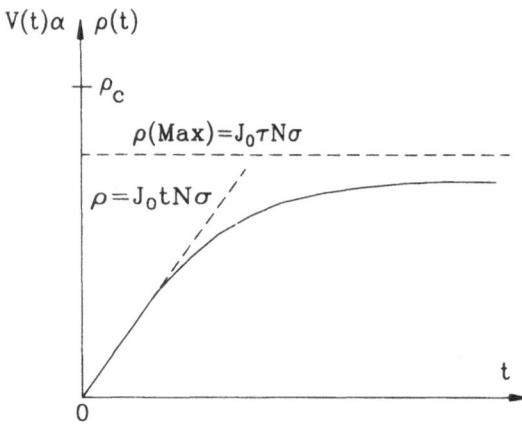

Fig. 16: Time dependence of the trapped charge density (and electric potential it creates) as deduced from Eq. 5.

$$\partial\rho_r/\partial\tau = -\rho/\tau \tag{3}$$

The local equivalent of charge conservation (previously expressed by Eq.1) is now $\partial\rho/\partial t = \partial\rho_c/\partial t + \partial\rho_r/\partial t$ hence,

$$\partial\rho/\partial t + \rho/\tau = J_oN\sigma \tag{4}$$

which, when solved, gives:

$$\rho(t) = J_oN\sigma\tau \,(1-e^{-t/\tau}) \tag{5}$$

The final result resembles the charging of a capacitor through a resistor (even if this analogy is not very satisfying) and the electric field and potential at a given point of the space are expected to follow the same time dependence (Fig. 16).

When applied to a perfect single crystal specimen, Eq. 5 leads to non charging of the bulk ($\rho = 0$ when $N = 0$) and to a surface charge of density $Q_s(t)$, associated to the corresponding surface density of trapping defects N_s (N becomes N_s). Eq. 5 also predicts, for instance, the increase of the surface potential V_s as a function of the beam intensity (J_0) and it shows the reduced role of the increase of the primary beam energy E_0 (at constant current density) because, in Eq. 5, E_0 may only act on the effective capture cross section, σ. Moreover, some charging effects, such as hole drilling, seem to start only when a critical field has been reached (i.e., a critical charge density ρ_c). From the time evolution of the trapped density —see fig.16— it can be seen that the critical charge density ρ_c is only obtained above a critical value of the beam density, $J_c = \rho_c/N\sigma\,\tau$, and that these effects are dose rate dependent ($\propto \partial D/dt$). When such a critical field is not required and when the charging effects are directly proportional to the electric field thus produced, the corresponding effects seem to be dose-dependent ($\propto D$) because the electric field at the beginning of the irradiation is proportional to $J_0 \cdot t$ (i.e., to the incident dose D) when t « τ.

In practice, the analysis of experimental results can be complicated by additional effects (contamination, surface neutralization by the secondaries, flashover when a critical value is attained), but some of them can also be taken into account[27] by additional terms in Eq. 5 to explain some observed time evolutions.[29] The same analysis may be applied to X-ray irradiation (where J_0 will represent the secondary emission current density) to explain, for example, the results shown on Fig.4. For all types of irradiation, a difficulty also results from the fact that the detrapping processes are certainly accelerated by the electric field being established and, thus, τ is no longer a constant.

At last it should be pointed out that the time constant for charging is different from the time constant for some consequences of the built-up electric field, such as ion migration in glasses (Figs. 8 and 9).

NEUTRALIZATION BY SECONDARY ELECTRONS AND BY THE SPECIMEN CURRENT I_S

As in metallic specimens, the physical reason of this specimen current in insulators is to restore the neutrality of the irradiated sample; its direction depends upon the sign of the majority charges being trapped into it. When this sign is positive the neutralization is due partly to the secondary electrons (a part of them in vacuum are reattracted and a part of them inside the solid cannot escape into vacuum: the two contributions lead to a decrease of δ) and partly to the incident electrons in the bulk (in the case of electron irradiation) which are also reattracted by the thin positively charged layer (associated to the main part of the secondary electron emission). These two electronic contributions acting like space charges may induce several changes in the electric potential as a function of depth.[7] In this situation the contribution of the specimen current is expected to be very low because of the poor conductivity of the irradiated specimen outside the irradiated region. In that case the exception is given by specific arrangements such as partial metallic coatings of the surface (or grids) for facilitating the neutralization by the surface conductivity.

Quite different is the situation where the sign of majority charges is negative. In electron irradiation, the incident electrons injected into the insulator contribute to dQ/dt when trapped, but the great majority of them are not and they differ from the internal electrons existing prior to irradiation of the specimen. The density of the latter is below 10^8 cm^{-3} (and this poor density explains the high electrical DC resistivity of insulators) while the local density associated to the former may be very large. For example, it has been shown that it is possible to inject current densities of around 1mA/mm^2 or more into insulators having an electrical resistivity of 10^{15} Ω cm. Being injected by the incident beam, these additional electrons play the role of some kind of dynamical doping. After having lost most of their initial kinetic energy through the usual interactions with the atoms of the solid, they are unable to interact with the valence electrons when their kinetic energy reaches values less than the band gap energy. In the conduction band their behavior is similar to that of the secondaries with the only possibility to interact with phonons and thus with a large inelastic mean free path and with a large scattering time.[7] Being in a negatively charged region, these excess electrons are pushed towards the earth potential where they produce the specimen current, I_s.

As first mentioned by Hobbs,[30] the corresponding current is non-ohmic in the sense that its value is independent of the (poor) conductivity of the material. A consequence of this analysis is to invalidate the popular use[29,31-35] of Ohm's law, $V = R\,I$, between the surface potential V_s and I_s via some leak resistance R. The same remark holds also when the charging of an insulator subject to irradiation is assimilated to the charge of a capacitor through the so-called leak resistance R.

The experimental proof of the non-validity of Ohm's law is illustrated on Fig. 3 where a clean and uncoated surface of a MgO crystal does not charge during electron irradiation ($V_s \simeq 0$) while I_s is a significant fraction of the incident beam intensity.[7] The application of Ohm's law to this case leads to the surprising result of a leak resistance, R, near zero for materials having an electrical resistivity in the 10^{15} Ω.cm range!.

For the future, we believe the possibility of injecting large current densities into insulators via the incident beam is full of promising applications. One interesting consequence of the above analysis is that the sign of the majority of trapped charges can be deduced from measurement of the sign of the specimen current (more easy than the measurement of the total yield); the experimental E_{C1} and E_{C2} values (see below) will be obtained when I_s is found to be zero.

ADDITIONAL QUESTIONS AND COMMENTS

On the Total Yield Approach

This approach is widely used in many publications to determine the sign of the surface potential V_s of uncoated insulators irradiated by incident electrons; we have mentioned above that this approach fails to explain some experimental results. The reasons are detailed in this subsection.

The total yield approach is based on a comparison between the number of incoming electrons (I_0 in Eq. 1) and the number of outgoing electrons (I_r in Eq. 1). When the total yield, y (with $y = \delta + \eta$), is greater than unity, V_s is expected to be positive. Consequently, from the shape of the curves shown on Fig.5, this positive interval is expected to be situated from E_{C1} (with $E_{C1} \simeq 0.1$-0.3 keV) up to $E_0 = E_{C2}$ (with $E_{C2} > 5$ keV), for monocrystalline MgO. In fact, this approach is only correct at the early beginning of the irradiation where the sign of Q (or dQ/dt) can be deduced from Eq. 1 with $I_s = 0$. It fails when this situation is assumed to remain during the irradiation where the reattraction of the secondary electrons by the positive surface potential is only considered but without any regards of the electric field established (which depends on the difference in the density of trapping centers, capture cross sections and time constants for the positive and the negative charges being trapped). This means that when the steady state is attained (dQ/dt = 0 but Q ≠ 0) the electric field built-up has modified the secondary electron emission in such a way that there is no reason for which the sign of the majority trapped charges would be the same as that of the early beginning of irradiation and no reason for which the E_{C2} point would be the same as that deduced from the ideal δ curves (cf. Fig. 5).

When $E_{C1} < E_0 < E_{C2}$, three different situations can be established, leading to three different results:

The first situation is illustrated on the left of Fig. 14 where it is assumed that the trapping centers are more or less uniformly distributed into the solid (amorphous materials with $N(+) \simeq N(-)$). If σ and τ are of the same order of magnitude for the two types of charges, the specimen is expected to be positively charged because the large secondary electron emission induces a high density of positive charges.

The second situation is obtained when there are almost no traps for the two types of charges as is the case for a quite perfect single crystal —Fig. 3 at left. In this situation, the surface potential is close to zero because of the rapid neutralization of the positively charged region (associated to the secondary electron emission by a fraction of the secondaries emitted into vacuum and also by a fraction of the mobile primaries inside the specimen). All these electrons produce negative space charge; the expected shape of the potential as a function of depth is shown on Fig. 17 (left).

The third situation is obtained when the trapping effects for electrons are very large; for instance at the metal/insulator interface of a thin metallic coating not connected to earth potential: Fig.3 left) . By the electric image effect, some incident electrons in the bulk are attracted to this interface where they are trapped while the secondary electron emission region is restricted to the thickness of the coating. Such a situation leads to negative charging (Fig. 17, right). In addition to the results thus obtained on MgO[7], such negative charging obtained in the 0-10 keV interval have also been reported by Hofmann[32] for a ZrO_2-stabilized polycrystalline Al_2O_3 showing surface contamination.

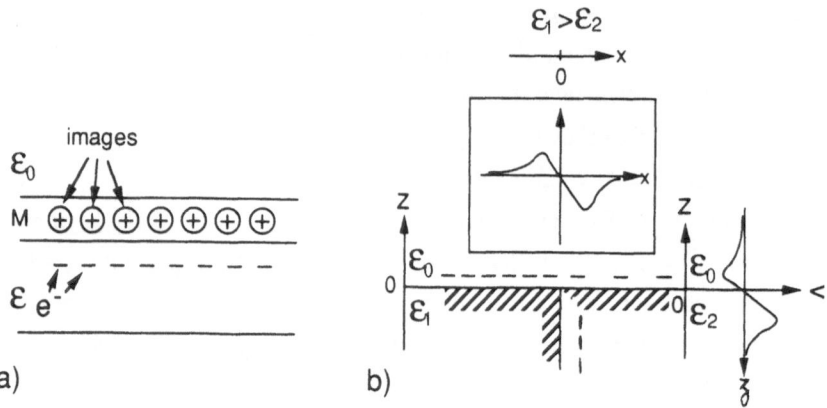

Fig.17: Sketch of the charge distribution (and of the potential function, V(z) for an uncoated single crystal (left) and a metal-coated crystal (right) subject to electron irradiations.

On Trapping Mechanisms

From the above analysis it is clear that the irradiation of insulators by electrons is a complex situation which depends on time. Little is known on the time constant and the trapping mechanisms. The time constant is certainly less than a few seconds and less than the time constant involved in ion migration (as shown in Fig.8). Among many others, the questions are: what type of impurity is the most efficient for trapping, or what is the order of magnitude of the capture cross-section (σ in Eq. 5)? Does the detrapping process (τ in Eq. 5) depend upon the local field?

We have suggested above that the electrostatic steady state for a given material (MgO) depends upon its crystalline state. It is thus clear that the behavior under irradiation of a glass will be different from that of a polymer which in turn is certainly different from that of a monocrystalline metallic oxide. It is outside the scope of the present paper to study each type of insulator in detail. Thus, in the following, some new possibilities are only considered to understand the possible role of impurities in the charging of a monocrystalline specimen and the possible role of grain boundaries in the charging of a polycrystalline insulator. For the role of impurities in a monocrystalline insulator, the question (derived from their role in the study of semiconductors) is to know if the electrons are trapped on the shallow levels situated (in energy) in the band gap of the insulator or if they are trapped in wells in its conduction band. The last hypothesis is based on the fact that in semiconductors the impurities are randomly distributed into the crystal and, at thermodynamic equilibrium, the Fermi level remains constant. The result is that there are local band bendings, with bumps at the places where clouds of acceptors are and wells at the places where clouds of donors are[36] (band bending similar to that of a pn junction).

Despite the difficulty of locating the Fermi level in the case of insulators (but we only need its energy to remain fixed), this analysis may be transposed to the case of monocrystalline insulators having residual impurities on which the electrons may be trapped during the irradiation (see Fig.18).

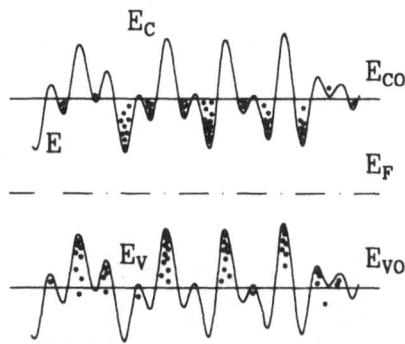

Fig. 18: Sketch of the covariant fluctuations of the band bendings associated to a statistical distribution of impurities in an insulator (inspired from ref.36).

When the electric field is established, a macroscopic potential is superimposed on the local band bending; detrapping effects may occur, leading also to the possibility of an avalanche breakdown. In order to elucidate the role of impurities when charging monocrystalline insulators, the need to characterize their concentration and distribution at the sub-ppm level is obvious.

In the case of surfaces, interfaces or polycrystalline materials, the possible influence of the image forces has to be outlined. Following the analysis illustrated previously in Fig.12, a metallic coating on an insulator leads to attraction of the majority charges in the dielectric by their images (in the metal) of opposite sign.

A vacuum/dielectric interface induces a depletion region of majority charges (because of the image effect of the same sign in vacuum) into the insulator (but close to the interface) while the possible excess charges in vacuum are expected to be attracted by this interface. The same idea may be applied to polycrystalline materials composed of grain boundaries (dielectric constant ε_2, with $\varepsilon_2 < \varepsilon_1$). In this last situation, the excess charges will be located mainly at the central regions of the grains and in the external regions of the grain boundaries with also a depletion layer more pronounced at the grain/vacuum interface.

The grain/grain boundary interface has been studied recently by Blaise and Le Gressus[37] with the use of a polaron model. The approach suggested here differs from it in various aspects; the results obtained in these three situations are illustrated on Fig. 19.

On Practical Solutions to Reduce Charging Effects

A practical consequence of the electrostatic approach is that the unique solution for canceling exactly the charging effects (i. e., the field and potential created by the trapped charges) is to neutralize exactly the charges creating these effects ($V = 0$ and $E = 0$ elsewhere when ρ and σ are null). Such a charge neutralization may occur spontaneously, for instance when the charges are distributed at the surface of the insulator and when this surface is surrounded by a poor vacuum (the molecules of the surrounding atmosphere neutralizing the excess charges), as in environmental Auger studies.[38]

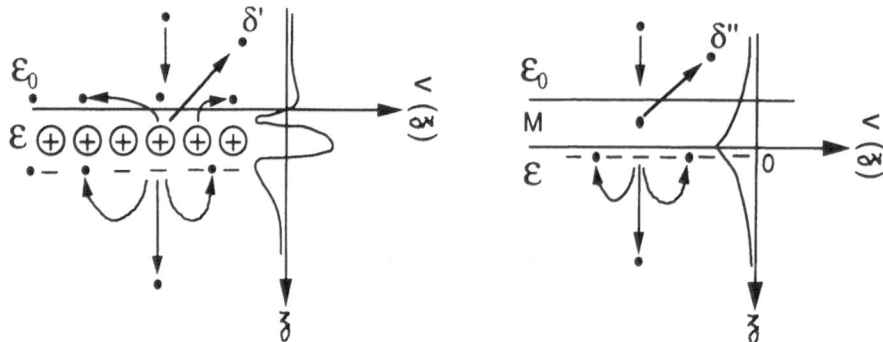

Fig.19: Influence of image forces on the excess charges (here electrons) in the case of a) a metal/dielectric interface and b) a dielectric/vacuum interface and a dielectric/dielectric interface ($\varepsilon_1 > \varepsilon_2$). The hatched areas represent depletion regions. Note the charge transfer between ε_1 and ε_2 (grain/grain boundary) and the accumulation of charges on the vacuum side.

In other cases, the use of additional irradiations by charged particles (i. e., use of flood guns to counterbalance the excess charges resulting from the primary irradiation) is very popular in surface analysis.[19,23,32,39] Using this solution, it must be kept in mind that the neutralization is exact only when the penetration depth of the neutralizing charges is very close to the depth distribution of the charges responsible for the charging effects. The fact that the surface potential is stabilized at a value close to zero as well as the fact that the surface potential is null because of the metallic coating at earth potential do not ensure that the electric field at the surface and inside the specimen is zero and do not prevent spurious effects to occur (such as the migration of the mobile ions).

Another elegant solution consists in illuminating widely the specimen and the sample holder with photons having an energy greater than the band gap energy of the investigated insulator (visible or UV radiation). The electron hole pairs generated by photo absorption induce a surface photoconductivity which allow the excess charge to be evacuated to the earth (see Fig. 20).

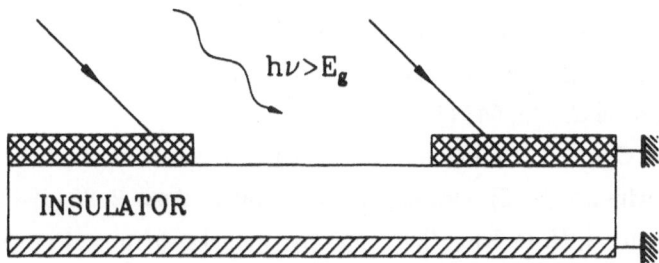

Fig.20: Schematic drawing of a practical solution to reduce charging effects by irradiation with ultraviolet light.

At last, when successive signals (or spectra) are acquired during the irradiation of an insulator, it is possible to extrapolate these results to the zero dose limit to deduce the expected signal (or spectrum) prior to the irradiation.[23]

On Electron Stimulated Desorption

The above electrostatic analysis strongly supports the fact that the ion migration in glasses is driven by the electric field built up in the specimen during irradiation. Also, it has been suggested that this electric field may explain the results obtained by Humphreys et al in electron beam nanolithography (see Fig.15). Owing to the fact[30] that the field strength may reach values as high as 10^5 V/cm, the open question is: why is the interpretation of results obtained in electron stimulated desorption based only on (sophisticated) microscopic mechanisms without any regard to the possible influence of this macroscopic field?[26,40-42]

CONCLUSION

Dealing with the wide class of materials named insulators, irradiated by a great variety of incident particles, this paper cannot pretend to be exhaustive. A large number of phenomena (such as the anomalous contrast observed on MgO crystals in Scanning Electron Microscopy[43,44]) or methods (i. e., for measuring the charge distribution[45]) have been omitted. Its modest goal was to remember that the macroscopic laws of electrostatics and electrokinetics (in principle well known but often forgotten) are able to explain several effects besides the obviously needed, valuable and more sophisticated microscopic interpretations.

REFERENCES

1. D. R. Lide. Ed., CRC Handbook of Chemistry and Physics. 72 th Ed. (CRC Boca Ann. Arbor, 1991-1992)

2. G. Battaglin, G. Della Mea, G. De Marchi, P. Mazzoldi, A. Miotello, M. Gulielmi. Alkali migration in glasses on electron, proton and heavier ion irradiations. Journal de Physique Colloque C2, 43, 645 (1982)

3. F. Ohuchi, M. Ogino, P. Holloway, C. G. Pantano. Electron beam effect during analysis of thin glass whith Auger Spectroscopy. Surface Interf. Anal. 2, 85 (1980)

4. J. Cazaux, O. Jbara, H. H. Kim. Some unconventional experiments cross correlating the techniques in surface analysis. Surf. Sci. 247, 360 (1991)

5. N. R. Whetten, A. B. Laponski. Secondary Electron emission from MgO thin films. J. Appl. Phys. 30, 432 (1959)

6. A. J. Dekker. Secondary electron emission, in: Solid State Physics, F. Seitz and D. Turnbull Editors (Academic Press NY) Vol. 6, 251 (1958)

7. J. Cazaux, K. H. Kim, O. Jbara, G. Salace, Charging effects of MgO under electron bombardment and non-ohmic behaviour of the induced specimen current. J. Appl. Phys. 70,960 (1991)

8. B. L. Henke, J. A. Smith, D. T. Attwood. 0.1-10 keV X-ray induced electron emission from solids: models and secondary electron measurements. J. Appl. Phys. 48, 1852 (1977)

9. N. R. Whetten. Cleavage in high vacuum of alkali halide single crystals: secondary electron emission. J. Appl. Phys. 35, 3279 (1964)

10. S. Luo, D. C. Joy, Monte Carlo calculation of secondary electron emission. Scanning Microscopy Suppl. 4, 127 (1990)

11. S. Hassemkamp, Secondary electron emission, in: Particle Induced Electron Emission II Springer Tracts in Modern Physics 123, 7 (1992)

12. S. A. Schwarz Application of a semi-empirical sputtering model to secondary electron emission. J. Appl. Phys. 68, 2382 (1990)

13. M. P. Borom, R. E. Hanneman: Local compositional changes in alkali silicate glasses during electron microprobe analysis. J. Appl. Phys. 38, 2406 (1967)

14. J. Cazaux, Electrostatics of insulators charged by incident electrons beams. J. Microsc. Spectrosc. Electron 11 293 (1986)

15. R. G. Gossink, H. Van Doveren, J. A. T. Verhoeven. Decrease of the alkali signal during Auger Analysis of glasses. J. of Non-Cryst. Sol. 37, 111, (1984)

16. A. Dubus, J. Devooght, J. C. Dehaes, Approximate solutions of the Boltzmann equations for secondary electron emission. Scanning Microscopy 4,1 (1990)

17. J. Schou. Secondary electron emission from solids by electron and proton bombardment Scanning Micros. 2. 2., 607 (1988)

18. M. Cailler, J. P. Ganachaud, Secondary electron emission from solids. Scanning Microscopy Suppl. 4, 57 (1990)

19. M. Liehr, PA. Thiry, J. J. Pireaux, R. Caudano. Characterisation of insulators by high resolution electron-energy-loss spectroscopy: Application of a surface-potential stabilization technique. Phys. Rev. B 33, 5682 (1986)

20. H. J. Fitting, D. Hecht, Secondary electron field emission. Phys. Stat. Solid. a 108,265 (1988)

21. M. P. Seah, W. A. Dench: Quantitative electron spectroscopy of surfaces; a data base for electron inelastic mean free paths in solids. Surf. Interf. Anal. 2,53, (1980)

22. J. Cazaux: Some consideration of the electric field induced in insulators by electron bombardment. J. Appl. Phys. 59 1418 (1986)

23. J. Cazaux and P. Lehuede. Some physical descriptions of the charging effects. J. Electr. Spectros. Rel. phenom. 59, 49 (1992)

24. R. S. Knox and K. J Teegarden in Physics of color centers, W. B. Fowler Ed. Academic Press NY p 625 (1968)

25. K. I. Grais and A. M. Bastawros. Study of secondary electron emission in insulators and semiconductors. J. Appl. Phys. 53 5239 (1982)

26. J. Fine and M. Szymonski, Electron Stimulated Desorption Mechanisms in Insulators, Le Vide, les Couches Minces 260 (Sup), 69 (1992)

27. J. Cazaux and C. Le Gressus, Phenomena relating to charge in insulators: Macroscopic effects and microscopic causes. Scanning Microscopy 5, 17 (1991)

28. C. J. Humphreys, T. J. Bullough, R. W Devenish, D. M. Maher, P. S. Turner. Electron beam nano-etching in oxides, fluorides, metals and semiconductors. Scanning Microscopy Supplement 4, 185 (1990)

29. T. Ichinokawa, M. Iiyama, A. Onoguchi, T. Kobayoshi,. Charging effect of specimen in scanning electron microscopy Jap. J. of Appl. Phys. 13, 1272 (1974)

30. L. W. Hobbs. Murphy's law and the uncertainty of electron probes. Scanning Microscopy Supplement 4, 171 (1990)

31. L. Reimer, Scanning Electron Microscopy (PW Hawkes Ed.). Springer, Berlin, Springer Series in Optical Sciences p 119 (1985)

32. S. Hofmann, Charging and charge compensation in AES analysis of insulators. J. Electr. Spectros. Relat. Phenom. 59, 15 (1992)

33. J. Cazaux, On the surface potential of insulators investigated by electron microscopies. Instit. Phys. Conf. Series 83-2, 375 (1988)

34. M. Brunner and R. Schmid, Charging effects in low-voltage scanning electron microscope metrology. Scanning Electron Microscopy II, 377 (1986)

35. D. Köhler, G. Koschnek, E. Kubalek. A contribution to the scanning electron microscope based microcharacterization of semi-insulating GaAs substrates, Scanning Microscopy 3. 3, 765 (1989)

36. B. Pistoulet, G. Hamamdjian. Mixed conductivity and potential fluctuations in semi-insulating GaAs, Phys. Rev. B 35, 6305 (1987)

37. G. Blaise, C. Le Gressus, Charging and flashover induced by surface polarization relaxation process. J. Appl. Phys. 69, 6334, (1991)

38. G. Ohlendorf, W. Koch, V. Kempter and G. Borchardt. Application of environmental electron spectroscopy to AES of bulk ceramics. Surf. Interf. Anal. 17, 947 (1991)

39. J. J. Pireaux, M. Vermeersch and R. Caudano. Study of insulators with high resolution electron energy loss spectroscopy J. Electr. Spectros. Rel. Phenom. 59, 33 (1992)

40. T. E Madey, D. E. Ramaker, R. Stockbauer, Characterization of surfaces through electron and photon stimulated desorptions. Ann. Rev. Phys. Chem. 35,215 (1984)

41. M. J. Dresser, Electron Stimulated Surface Chemistry, Scanning Microscopy, Sup. 4, 193 (1990)

42. R. A. Baragiola, T. E. Madey, A. M. Lanzillotto. Multiply charged ions from electron bombardment of SiO_2. Phys. Rev. B 41, 9541 (1990)

43. A. L. Bleloch, A. Howie, R. H. Milne, High resolution secondary electron imaging and spectroscopy, Ultramicroscopy 31, 99 (1989)

44. J. Liu, P. A. Crozier, C. G. Hembree, F. C. Luo, J. M. Cowley, J. A. Venables, Variation in the secondary electron emission from MgO. in proceedings of the XIIth Int. Cong. for Elect. Microscopy. San Francisco Press. Vol. 2, 336 (1990)

45. H. J. Fitting, P. Magdanz, W. Mehnert, D. Hecht, T. Hingst, Charge trap spectroscopy in single and multiple layer dielectric. Phys. Stat. Sol. (a) 122, 293 (1990)

SECONDARY ELECTRON EMISSION FROM INSULATORS

Jørgen Schou

Euratom Association-Risø National Laboratory
Physics Department
DK-4000 Roskilde, Denmark

INTRODUCTION

Electron emission from insulators induced by charged particles plays an important role for many applications. Generally, the electron yield from insulating materials is quite high compared with metals and semiconductors consisting of atoms of similar atomic numbers. This means that insulators are used frequently in sensitive amplification devices as ion-electron converters and electron multipliers. The combination of high electron yield and low charge mobility, which is a characteristic property of many insulators, leads to a rapid charge up of the bombarded surface. This charge up of surfaces is a serious problem for many experiments in surface physics as well as in plasma-surface interactions. Nevertheless, many aspects of electron emission from these materials have been studied comparatively little.

The experimental activity on insulating materials has been fairly low recently. Hasselkamp (1992) includes a brief discussion of results for insulators. Electron emission from insulators for primary electron impact has been discussed comprehensively by Hachenberg and Brauer (1959) and ion-induced emission by Krebs (1968,1976). In general, the measurements are difficult to carry out because of the charge up of the samples irradiated. In addition, a large number of measurements have been performed under vacuum conditions with poorly characterized surfaces. Results obtained under such conditions have to be critically evaluated (Schou (1988), Hasselkamp (1992)).

Most of the treatments of secondary electron emission divide the emission into three stages:
  i) production of internal secondaries including cascade electrons,
  ii) migration of some of the secondaries to the surface, and
  iii) ejection of the electrons from the surface.
This division is feasible for metals and insulators in most cases. Ion-induced electron emission deviates from electron-induced emission in the production stage alone, unless the recoiling target atoms produce a substantial fraction of the secondaries. Therefore, in the following both types of primary particles will be included.

Ionization of Solids by Heavy Particles, Edited by
R.A. Baragiola, Plenum Press, New York, 1993

FEATURES OF SECONDARY ELECTRON EMISSION FROM INSULATORS.

The experimental results from insulators induced by electrons or ions show many similarities with metals. The striking points, at which emission from insulators deviates from that of metals are:

A) The yield δ from insulators is much higher than that from metals (and semiconductors) of similar atomic numbers. Occasionally, the yield is up to an order of magnitude higher for insulators than for comparable metals (Hachenberg and Brauer (1959), Konig et al. (1975), Seiler (1983)).

B) The yield δ(d) for an insulating film of thickness d increases with thickness up to more than 500 Å. This variation is usually characterized by an escape depth, which may vary strongly from one material to another (Seiler (1967)).

C) The energy spectra from insulators show a narrow pronounced peak at very low energies (Konig et al. (1975)). This phenomenon is known from oxidized metal surfaces as well (Hofer (1990) and Hasselkamp (1992)).

D) The yield δ from insulators depends significantly on the target temperature (Hachenberg and Brauer (1959), Hasselkamp (1992)). Generally, the yield decreases with increasing temperature. Even though a few measurements on metals unexpectedly have shown a dependence on temperature (Rothard et al. (1992)), the overwhelming fraction of the existing measurements for metals does not indicate any systematic relationship with the target temperature.

The absolute yields depend on the specific projectile-target combinations as in the case of metal targets. Nevertheless, the variation from one insulator to another may be substantial. The experimentally observed energy dependence is roughly similar, even though cascade production of internal secondaries is likely to be much smaller in insulators without free electron than in metals (Sigmund and Tougaard (1981)).

The starting point for the analysis is the yield formula derived for metals on the basis of transport theory (Schou (1980,1988), Sigmund and Tougaard (1981)). Even though the treatment is incomplete for insulating materials with a large escape depth, it describes the trends in a qualitatively correct manner. The yield for electron- as well as ion-induced emission from a particle with primary energy E is

$$\delta \approx \frac{c}{4}(\frac{dE}{dx})\int_{U_o}^{\infty} \frac{dE_o}{E_o(dE_o/dx)}(1-\frac{U_o}{E_o}) , \tag{1}$$

where E_0 is the energy of the slow, internal electrons, dE_0/dx the stopping power for these electrons, U_0 the magnitude of the surface barrier, c a constant and $(dE/dx)_e$ the electronic stopping power for the primary particle.

Since all stages for the emission process from insulators deviate somewhat from those of metals we shall in the following discussion treat each of the three stages i-iii) separately. Many of the comparisons will be performed for solid nitrogen which is a simple insulator with well-known stopping powers and ionization cross sections for both protons and electrons.

PRODUCTION OF INTERNAL SECONDARIES

All liberated electrons in insulating materials originate from orbitals with a nonvanishing binding energy. Hence, the production from secondary or higher cascade generations becomes much smaller than that from metals. A consequence of the binding energy is that W, the energy required to produce one ion-electron pair, frequently is about 20-40 eV for insulators (ICRU (1979)). The ratio of W to the ionization energy is about a factor of two, but it varies considerably from one type of insulators to another (Sigmund (1993)).

Sigmund and Tougaard (1981) as well as Schou (1988) have pointed out that a large value of W leads to a yield dependence similar to that of the ionization cross section rather than that of the electronic stopping power. The analysis of the

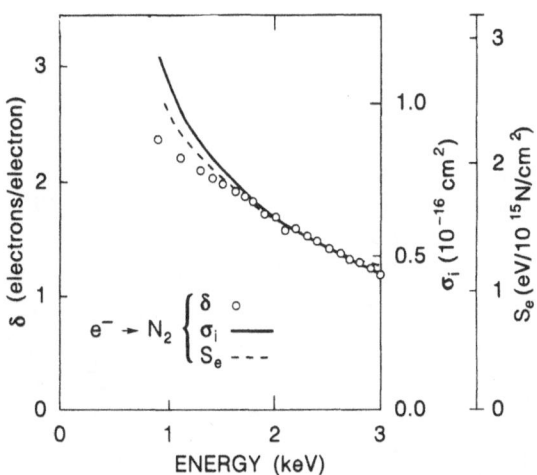

Figure 1. The secondary electron yield from electron-bombarded solid nitrogen (Sørensen and Schou(1978)). The stopping power and ionization cross section from Green and Sawada (1972) have been normalized to the experimental curve at 2 keV.

possible production channels is hampered by the similarity between the electronic stopping power and the ionization cross section over the entire accessible range of primary energies (Schou (1988)). In many cases both quantities are not known for the specific combinations of beam particles and insulators that have been studied experimentally.

For elemental gases the stopping power and the ionization cross section are fairly well known for electrons as well as for protons (Green and Sawada (1972), Kieffer and Dunn (1966), Andersen and Ziegler (1977), and Rudd et al. (1985)). Even though many of these quantities have been compiled or measured for gas phase targets, they will be used for solids as well. Solid nitrogen is one of the few solidified gases for which secondary electron yield measurements exist.

The secondary electron yield δ is shown as a function of primary energy from 0.9 to 3 keV in Figure 1. The stopping power and the ionization cross section from Green and Sawada (1972) are shown as well. The yield curve is below both curves at 2 keV, since the range of a 2-keV-electron practically corresponds to the migration (escape) depth of the slow electrons. Above 2 keV there is hardly any difference between both the electron stopping power and the ionization cross section and the experimental curve. The ion-induced curve for solid nitrogen (Børgesen et al. (1983)) is shown in Figure 2 together with the stopping power from Andersen and Ziegler (1977) and the ionization cross section from Rudd et al. (1985). For the ion case the agreement is significantly better for the ionization cross section and the experimental curve than for the experimental curve and the stopping power. This behaviour agrees with the expectation that the high value of W at these energies is accompanied by an energy dependence of the yield which is much closer to that of the ionization cross section than that for the stopping power. In formula (1) the corresponding change would be to replace $c(dE/dx)_e/E_0$ by an integral over the ionization cross section σ_i from E_0 to E (Schou (1988)).

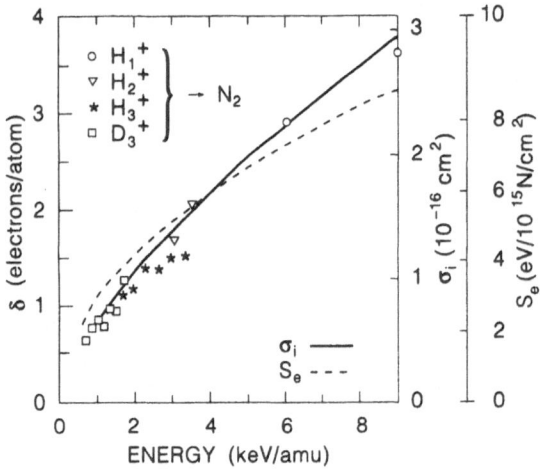

Figure 2. The ion-induced secondary electron yield from solid nitrogen (Børgesen et al. (1983)). The stopping power from Andersen and Ziegler (1975) and the ionization cross section from Rudd et al. (1985) have been normalized to the experimental curve at 4 keV.

MIGRATION OF THE INTERNAL SECONDARIES

The most important feature for insulators is the low stopping power for the slow internal secondaries. This is a consequence of the empty conduction band in these materials, so that internal secondaries are not exposed to any slowing down from electron-electron interaction below a certain energy. This threshold is determined by the lowest possible excitation, for example, the production of an exciton from the filled valence band. Since the electrons are slowed down very inefficiently from the interaction with atoms alone, the migration depth for these

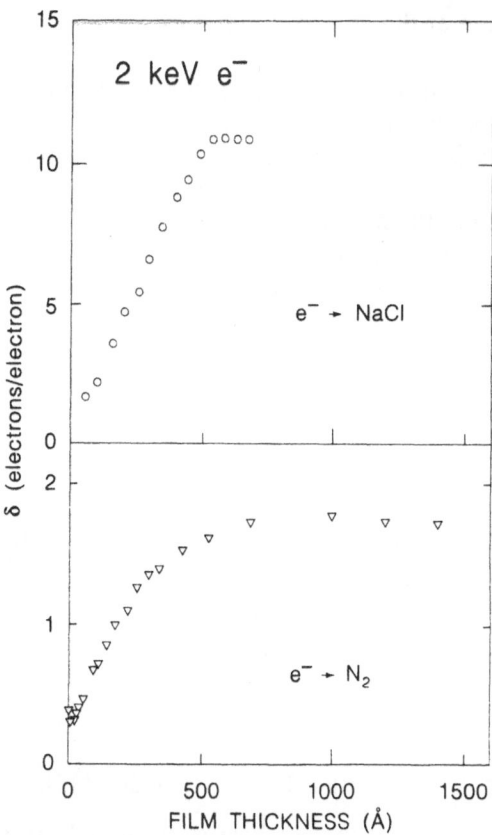

Figure 3. The secondary electron yield as a function of film thickness for solid nitrogen (Sørensen and Schou (1978)) and sodium chloride (Bronshteyn and Protsenko (1970)) at the primary energy 2 keV.

electrons is much larger in insulators than in metals. As a first approximation one might replace the total low-energy stopping power (dE_0/dx) in the denominator with the nuclear stopping power $(dE_0/dx)_n$, which is much smaller (Schou (1988)). For electrons which have suffered a non-negligible energy loss during migration, formula (1) has to be changed considerably (Sigmund (1993)).

Since the slow electrons interact with the atoms, the varying modes of density fluctuations may influence the stopping of the electrons in the absence of electron-electron interactions. The decrease of the secondary electron yield for high temperatures of insulators is caused by an enhanced electron-phonon coupling (Hachenberg and Brauer (1959)).

Usually, measurements of the escape depth are performed on a substrate of similar atomic number as the insulator. In this manner the effect of the primary backscattered electrons is minimized. Two of the relatively few examples of this type from the literature are shown in Figure 3. The electron yield from solid nitrogen was measured as a function of the film thickness on a carbon substrate by Sørensen and Schou (1978). The yield increases up to an effective escape depth of about 500 Å. Solid nitrogen is particularly appropriate for this type of measurements since it has a negative electron affinity (see below). This means that all electrons that arrive at the surface will be ejected.

The measurements of 2-keV-electrons incident on NaCl deposited on a graphite substrate covered with a thin nickel layer indicate an escape depth similar to that of solid nitrogen (Bronshteyn and Protsenko (1970)). The yield is larger than that for nitrogen by a factor of five, although many properties are similar for the two materials or even less favourable for emission from the alkali halide. This unexpected difference in the magnitude of the yield reflects the fact that very little is known about the low-energy stopping of electrons (Fano (1988), Nieminen (1988), Sanche (1991)) and about ionization cross section in solid ionic compounds.

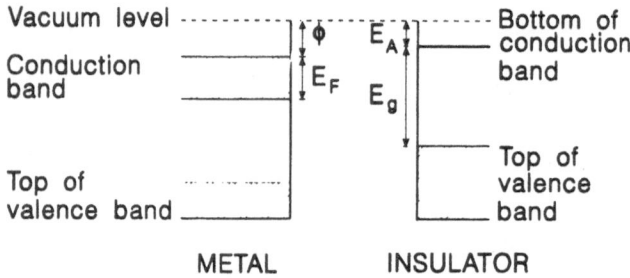

Figure 4. Schematic surface barrier for metals and insulators.

ELECTRON ESCAPE FROM THE SURFACE

The surface barrier for insulators is determined by the electron affinity E_A which is the energy difference between the bottom of the conduction band and the vacuum level. E_A is positive, if the bottom lies below the vacuum level, as shown in Figure 4. Actually most of the solid insulators studied have a positive electron affinity except for the solids neon, argon and nitrogen (Poole (1975), Jortner and Gaathon (1977) and Bader et al. (1984)).

The electron affinity plays a similar role for insulators as the ordinary surface barrier for metals, for which

$$U_o = \Phi + E_F \ .$$ (2)

Electrons for which the component of kinetic energy perpendicular to the surface is less than E_A will not escape, but be reflected into the bulk of the material. If E_A is positive one may replace U_0 with E_A in formula (1).

For most of the materials studied E_A, is slightly below 1 eV (Bader et. al. (1984), Poole et al. (1975)). The precise location of the band levels is determined by the fundamental electron-target particle interaction. For the solid rare gases this interaction has been studied to the extent that the general features may emerge (Jortner (1977)). The background polarization of the material tends to lower the bottom of the conduction band, whereas the scattering on the neighbor atoms has the opposite effect. Actually, this latter effect is a consequence of the uncertainty principle, since any localization of the electrons leads to an increase of the kinetic energy.

The combination of this small surface energy barrier for electrons and the absence of any slowing-down from conduction electrons at these low ejection energies leads to the narrow shape of the energy spectra that have been observed in a large number of experiments.

ACKNOWLEDGMENTS

The author thanks Peter Sigmund and Wolfgang O. Hofer for valuable discussions and for the necessary encouragement to write this contribution. This work was stimulated by the outstanding workshop organized by Raul Baragiola. Bjarne Stenum is acknowledged for having checked the manuscript.

REFERENCES

Andersen,H.H. and Ziegler,J.F. (1977) "Hydrogen Stopping Powers and Ranges in All Elements", Pergamon, New York.

Bader,G., Perluzzo,G., Caron,L.G. and Sanche,L. (1984) Phys. Rev. B 30, 78.

Bronshteyn,I.M. and Protsenko,A.N. (1970) Rad. Eng. Electr. Phys. 15, 677.

Børgesen,P., Sørensen,H. and Chen,H.-M. (1983) Nucl. Instr. Meth. 212, 517.

Fano,U. (1988) Rad. Phys. Chem. 32, 95.

Green,A.E.S and Sawada,T. (1972) J. Atmos. Terr. Phys. 34, 1719.

Hachenberg,O. and Brauer,W. (1959) Adv. Electr. Electron Phys. 11, 413.

Hasselkamp,D. (1992) in: "Particle induced electron emission II", Springer Tracts in Modern Physics 123, p. 1.

Hofer,W.O. (1990) Scan. Micr. Suppl. 4, 265.

ICRU (1979) "Average Energy Required to Produce an Ion Pair", Rep. 37 ICRU Public., Bethesda, MD 20814, USA.

Jortner,J. and Gaathon,A. (1977) Can. J. Phys. 55, 1801.

Kieffer,L.J. and Dunn,G.H. (1966) Rev. Mod. Phys. 38, 1.

Konig,W., Krebs,K.H. and Rogaschewski (1975) Int. J. Mass Spectr. Ion Phys. 16, 243.

Krebs,K.H. (1968) Fortschr. der Physik 16, 419.

Krebs,K.H. (1976) in: "Physics of ionized gases", B. Navinsek, ed., University of Ljubljana, p. 379.

Nieminen,R.M. (1988) Scan. Micr. 2, 1917.

Poole,R.T., Jenkin,J.G., Liesegang,J. and Leckey,R.C.G. (1975) Phys. Rev. B 11, 5179.

Rothard,H., Groeneveld,K.-O. and Kemmler,J. (1992) in: Particle induced electron emission II", Springer Tracts in Modern Physics 123, p. 97.

Rudd,M.E., Kim,Y.-K., Madison,D.H. and Gallagher,J.W. (1985) Rev. Mod. Phys. 57, 965.

Sanche,L. (1991) in: "Excess electrons in dielectric media", CRC Press, p. 1.

Schou,J. (1980) Phys. Rev. B $\underline{22}$, 2141.

Schou,J. (1988) Scan. Micr. $\underline{2}$, 607.

Seiler,H. (1967) Z. Angew. Phys. $\underline{22}$, 249.

Seiler,H. (1983) J. Appl. Phys. $\underline{54}$, R1.

Sigmund,P. and Tougaard,S. (1981) in: "Inelastic particle-surface collisions",
    E. Taglauer and W. Heiland, eds. Springer, Berlin, p. 2.

Sigmund,P. (1993) These Proceedings.

Sørensen,H. and Schou,J. (1978) J. Appl. Phys. $\underline{53}$, 5311.

SECONDARY ELECTRON EMISSION FROM ALKALI HALIDES INDUCED BY X-RAYS AND ELECTRONS

A. Akkerman*, A. Breskin, R. Chechik**
and A. Gibrekhterman

Department of Physics
Weizmann Institute of Science
Rehovot, 76100 Israel

INTRODUCTION

Secondary electron emission (SEE) from solids, induced by X-rays and charged particles (electrons, ions), is a rather complex process. In a simplified description it can be divided into three independent stages: 1) production of secondary electrons by the primary radiation; 2) transport of secondary electrons in the solid towards the exit surface; 3) escape through the surface. All three stages involve processes which occur inside the solid and thus require, for their exact mathematical representation, the solution of a many-body quantum mechanical problem. Consequently semi-empirical theories and models were developed, which in general explain the main features of the SEE. During the last few years several reviews of SEE have been published[1,2,3]. It was shown that simple models[1] may be used to reproduce the secondary electron yields and other integral characteristics of the process, for nearly-free-electron metals. For noble and transition metals, semiconductors and insulators the simple semi-empirical models are not as successful. It was also shown[2,3] that by using microscopic approaches for the basic interaction processes in the models of SEE they can be improved and may be used to calculate, for example, the energy spectra of the secondary electrons.

It is known from experiments that several semiconductors and dielectrics, and in particular alkali halides, have a very high secondary electron yield (SEY) which is by an order of magnitude larger than the SEY from metals. The experimentally derived dependence of the maximal SEY versus the optimal energy (to produce the maximal SEY) for these two groups of materials, were presented by Schwarz[4] (see fig. 5 in his article).

The high SEY from alkali halides is mainly due to their low work function, which is of the order of a few tenth's of an eV. Another important reason for their

---

* Also Soreq Nuclear Center, Yavne, Israel
** The Hettie H. Heineman Research Fellow

Ionization of Solids by Heavy Particles, Edited by
R.A. Baragiola, Plenum Press, New York, 1993

effective electron emission is the exceptionally low probability for energy losses, by inelastic collisions, of the secondary electrons in the low energy range ($E_e \leq E_G$; E_G is the gap energy). The most important process at these low energies is the electron-phonon interaction, with an average energy loss or gain per collision of ∼0.01 eV. Consequently the spectrum of secondary electrons is rather narrow, compared to similar spectra from metals. One should note that according to a somewhat arbitrary definition, "true" secondaries are considered as those with energies inferior to 50 eV, and "primaries" (that is, secondaries or slowed down initial electrons) are those having energies superior to 50 eV.

The general features of secondary emission from alkali halide layers, irradiated by soft X-rays and electrons, have been known for a long time (see for example ref. 5 and other references therin). Henke et al.[6] developed a model of SEE which includes a postulated treatment of secondary electron excitation function without details of the mechanisms of interaction. It uses Kane's one-dimensional random walk model[7] for electron transport, with a few fitting parameters. Secondary emission spectra predicted by this model agree satisfactorily with the experimental data, except for a region which is supposed, by the authors of ref. 6, to be responsible for plasmon decay. Some questions concerning the emission of primary electrons, like for example their absolute yield, are not solved by this model. A simpler and more straightforward, semi-analytical, model for SEE was proposed by Fraser[8]. He uses three parameters: the escape probability P(0) of secondary electrons, created at a distance x from the emitting surface, when $x \rightarrow 0$; the mean energy for secondary electron generation η; and the escape length L_s. These parameters cannot be predicted by the model, and are usually extracted from a fit to experimental data. Both models considered above are not based on "first principle" approach.

In order to have a coherent picture of the SEE from alkali halides it is necessary to understand the whole process of slowing down of primary and secondary electrons, generated by energetic electrons and X-rays. For that purpose it is desirable to include all the known processes of electron interaction in a single theoretical model. A step in this direction was made by McDonald et al.[9], using Monte Carlo simulations of the generation and the transport of secondary electrons. Llacer and Garwin[10] used the same method for low energy electron transport calculations, influenced only by electron-phonon interactions. But the over simplification of the interaction process, used in these models, does not help to elucidate the whole mechanism of creation, multiplication, slowing down and emission of secondary electrons in alkali halides. The effectiveness of the Monte Carlo method for SEE simulations was widely demonstrated in ref. 2.

In the present article we report on the development of a new physical model using recent, reliable, theoretical microscopic cross-sections for electron interactions in alkali halides[11]. Monte Carlo calculations based on this model were performed. We have chosen CsI and KCl as representatives to verify the model, and calculated the most important integral characteristics of SEE induced by X-rays and electrons over an energy range of 1-30 keV. The aim of this work is to improve the understanding of the SEE process and to present new data, which can be employed in the development of new radiation detection systems based on alkali halide convertors[12,13].

THE SIMULATION MODEL

The first stage in the simulation of the SEE induced by X-ray absorption, is the calculation of the spatial distribution of the primary photoelectrons in the solid. For high energy X-rays, also Compton electrons should be considered. For thin layers the simulation of the spatial distribution of photoelectrons is done using the forced Monte Carlo scheme[14]. The total X-ray attenuation coefficients are taken from the tables of Saloman et al.[15] and the shell partial photoelectron cross-sections from Scofield's data[16]. The angular distribution of the ejected photoelectrons is used in Fisher's form[17]. Auger electrons following the photo- and electron ionization processes are also taken into account as primaries. The problem is thus reduced to secondary emission following primary electron transport in the layer.

Many of the previous simulation works of SEE from metals, induced by electrons (see for example ref. 2,18), were performed using the direct Monte Carlo scheme (for details see 19,20). The basic assumptions used in this scheme are: 1) the electrons interact at random points in the bulk of the target; 2) the type of interaction (e.g. elastic or inelastic) is selected randomly, according to the relative cross-sections of the processes; 3) the inelastic scattering includes several mechanisms of interaction: ionization, excitation, collective interaction, etc; the particular mechanism is selected according to its relative cross-section; 4) the energy and the angle of deflection of each scattered primary electron (i.e. the most energetic electron) and of the ejected secondary electron (low energy electron) are sampled from the cumulative probability distributions, obtained by integration of the double differential cross-section of the corresponding interaction. Hence, the results of the simulation calculations depend on the adequacy of the cross-sections used to describe the interaction process.

As already pointed out we cannot use the exact theoretical cross-sections, because this implies solving the many-body interaction problem in the solid. On the other hand, systematic experimental data for electron interaction cross-sections in solids is rather scarce. Only a few indirect data is available and can be used: the stopping power for electrons transmitted through a layer, the mean free path for inelastic scattering extracted from the optical properties, and electron energy loss spectra in thin films . The last type of data show the existence of a collective excitation process, namely plasmon creation, as well as exciton creation and ionization of the inner atomic shells. It seems, therefore, reasonable to use in the simulation model two types of interactions: individual interaction with electrons in the unperturbed inner shells of free atoms and excitation of electrons from the outer shells, which contribute to the solid effects (valence band), in a collective manner. These include plasmon and electron-hole excitations, and interactions of slow electrons with the lattice vibrations. As was mentioned above, models based on these approaches for metals show results which are in satisfactory agreement with experimental data.

Another model, which treats the valence electrons as tightly-bound, was proposed by Szajman et al.[21] In this model there are no collective excitations and the only inelastic process is the ionizations of bound shells of free atoms. The agreement of the Inelastic Mean Free Paths (IMFP) calculated in the framework of this model with experimental measurements is good but it does not necessarily mean that this model will correctly predict the SEE characteristics. Calculations based on Szajman's et al.[21] model are presently in progress and will be compared with the calculations based on our model. The most crucial and important characteristic for verification of the model is the energy spectrum of secondary electrons.

A summary of the main processes of electron interactions included in our model and affecting the SEE are presented in table 1. The table also lists the main theoretical approaches and approximations used to describe these processes.

Table 1. Summary of all the processes included in our simulations, the methods of cross-section calculation and the relevant references.

Elastic Scattering.	Inelastic Interactions			
	Ionization	Plasmon excitation	Electron-hole excitation	Electron-phonon interaction
Partial wave expansion[22]	Binary encounter approx.[23]	Quinn's theory[24]	Dielectric function theory[25,26]	Time-dependent perturbation theory[10]

Elastic Scattering

Elastic scattering plays an important role in the transport of electrons in the solid, in particular for low energy electrons. This was pointed out in ref. 2, is evident from our calculations and was recently convincingly argued by Sigmund[27]. We do not use elastic scattering cross-sections which are affected by the solid, like those obtained[2] using the "muffin-tin" potential in Al, mainly due to the lack of such potentials for the relevant atoms. Rather, we followed the standard way[22], calculating the differential and total cross-sections by the "partial wave expansion" method. Solving the Dirac equation we used the screened potential proposed by Green et al.[28], without inclusion of polarization and exchange effects. The calculations were done for Ar and Xe atoms, since their atomic numbers, Z=18 and 54, correspond exactly to the mean values of that of KCl and CsI. We gradually reduced the potential to 0 at a distance r_0, which is equal to half the interatomic separation in CsI and KCl, thus including in a very approximate way the solid structure. The results of our calculations of the differential cross-sections for CsI, compared to Fink's et al[29], are shown in fig. 1. The total macroscopic elastic cross-section, or the IMFP, were calculated using Ar and Xe atoms with the bulk density of KCl (ρ=1.984 g/cm^3) and of CsI (ρ=4.51 g/cm^3) respectively.

Figs. 2 and 3 compare the calculated cross-sections with the experimental data[30]. For Ar our results agree quite well with the experimental ones, with maximal deviation of about 30%. For Xe the agreement of the calculated IMFP with the experimental data is poorer, with deviations of up to 100% for electron energies $E_e < 100$ eV. A somewhat better agreement is obtained with calculations which included polarization and exchange effects, but a marked difference still remains. For energies lower than 20 eV there exist no experimental data of the IMFP, and therefore we assume that they are constant and equal to the value at 20 eV. The most relevant data in this range exist for solid Xe, where it was shown experimentally[31] that the cross-section decreases drastically for energies below 2 eV. If this is also true for more complicated systems, such as alkali halides, it may explain the high transparency of the material to low energy electrons, as reflected in the secondary emission electron spectra. But as also follows from ref. 31 there is a large discrepancy between the various experimental data sets, which is a result of trapping of low energy electrons by impurities and other defects in the solid. Since a real layer of alkali halide is certain to contain many impurities and lattice defects, the material transparency at the very low energy range may not be very large. Our calculations in fact do show great sensitivity of the results to the elastic cross-section at the very low energy range. In the absence of good relevant experimental data we have to use our estimated values of the cross section.

Inelastic Scattering

Valence band excitations. In our calculations we assumed that the eight outer electrons in KCl and in CsI are the valence electrons. This follows from the estimated[6] plasmon energies $E_{p\ell}$ of these alkali halides. We assume the following

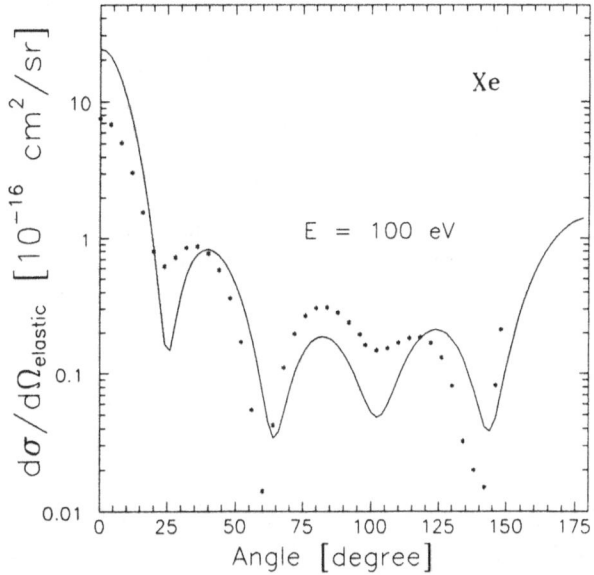

Figure 1. Differential elastic cross-section for electrons of 100 eV on free Xe atoms. Solid line - our calculations, dots - calculations of Fink et al.[29]

configuration for these electrons:

KCl: $3p^5$ from Cl and $(4s^1+3p^2)$ from K

CsI: $5p^5$ from I and $(6s^1+5p^2)$ from Cs.

The incident electrons interact with these valence electrons only by the plasmon and the electron hole $(e-h)$ excitation processes. For simplicity, the IMFP for plasmon excitation, $\lambda_{p\ell}^{-1}$, was calculated using Quinn's theory[24]. The $e-h$ cross-section was derived from Lindhard and Ritchie's theory[25] of the complex dielectric constant $\epsilon(\omega, k)$, and we have taken into account the gap energy in alkali halides, using the approach developed by Tossati and Parravicini[26] for semiconductors. In this case the IMFP for the $e-h$ excitation by an electron of energy E_e is obtained from the imaginary part of the inverse of a model dielectric function $\epsilon(\omega, k)$, by calculating

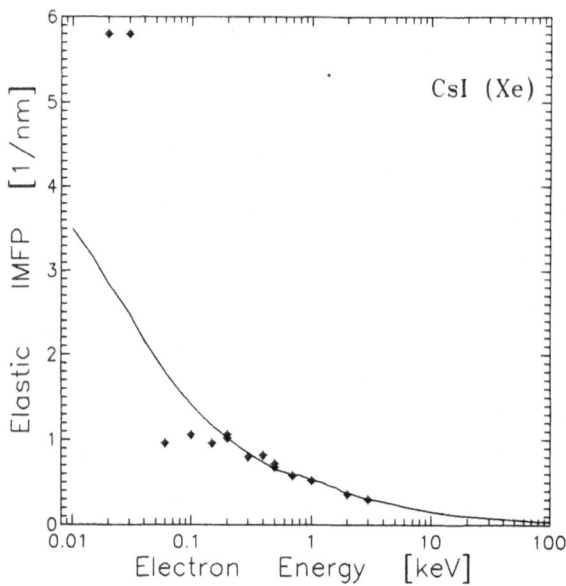

Figure 2. Elastic IMFP in Xe assuming the density of CsI. Solid line - our calculations, dots - experimental data of de Hear et al.[30]

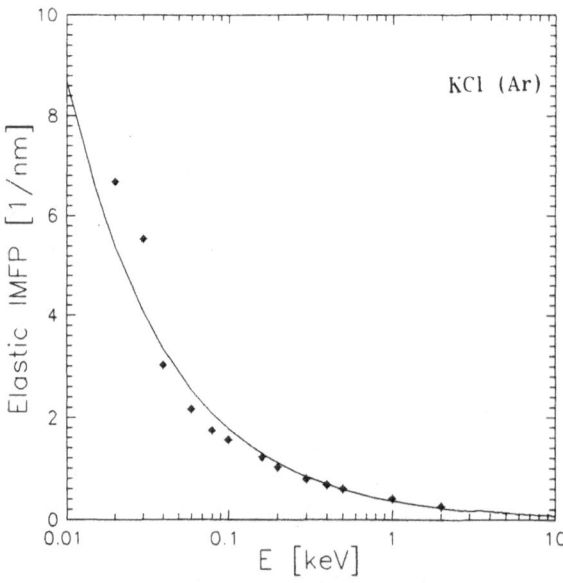

Figure 3. Elastic IMFP in Ar, assuming the density of KCl. Solid line - our calculations dots - experimental data of de Hear et al.[30]

the integral:

$$\text{IMFP}_{e-h}(E_e) = \lambda_{e-h}^{-1} = \frac{1}{a_0 \pi (E+1)} \int_0^{E-g} dx \int_{z_{min}}^{z_{max}} Im\left(-\frac{1}{\epsilon(x,z)}\right) dz \qquad (1)$$

where x is the energy transfer $\hbar\omega$ in units of Fermi energy E_F, z is the momentum transfer k in units of $2(2m_e E_F)^{1/2}$, m_e is the electron mass, z_{min} is defined as the smaller of the two quantities: $\frac{1}{2}[(1+x-g)^{1/2}-1]$ and $\frac{1}{2}[(1+E)^{1/2}-(1+E-x)^{1/2}]$, $z_{max} = \frac{1}{2}[(1+x-g)^{1/2}+1]$, E and g are the electron energy and the gap energy in E_F units, $a_0 = 5.29 \cdot 10^{-2}$ nm.

Table 2. Comparison of the calculated $\bar{E}$ with the estimated $\bar{E} \simeq E_{p\ell} + E_G$ and check of the sum rules. All energies are given in eV.

Material	$\int dE \cdot Im(-\frac{1}{\epsilon})$ [eV]		$\int dE \cdot E \cdot Im(-\frac{1}{\epsilon})$ [eV]^2		E_G	$E_{p\ell}$	$\bar{E}$	$E_{p\ell}+E_G$
	Calcul.	Sum rule	Calcul.	Sum rule				
NaCl	13	13.1	257	290	8.5	13.6	20	22.1
KCl	11	10.5	207	208	8.4	11.5	19	19.9
KBr	11	10.0	182	181	7.4	10.7	17	18.1
KI	9.1	9.4	146	148	6.0	9.7	16	15.7
CsI	12.2	10.6	182	178	6.2	10.7	15.2	`16.9

The total IMFP for valence band excitations, λ_v^{-1}, in our model is:

$$\lambda_v^{-1} = \lambda_{p\ell}^{-1} + \lambda_{e-h}^{-1}. \qquad (2)$$

In eq. (1) it would be better to use the experimentally derived $Im(-\frac{1}{\epsilon(\omega,k)})$. But presently only $Im(-\frac{1}{\epsilon(\omega,0)})$ can be extracted from optical measurements or from energy loss spectra of electrons transmitted through thin films. For alkali halides these show[32] a very complicated structure, and the methods (originaly proposed for metals) for reconstruction[33] of $Im(-\frac{1}{\epsilon(\omega,k)})$ from the optical data are inefficient in this case.

An approximate approach to the evaluation of the MFP for valence band excitations in solids, λ_v, using the centroid $\bar{E}$ of the optical loss function $Im(-\frac{1}{\epsilon(\omega,0)})$, was proposed by Szajman and Leckey[34].

$$\lambda_v [nm] = 0.212 E_e \bar{E} / \left\{ E_{p\ell}^2 ln[8E_e \bar{E} / \left(E_{p\ell}(E_{p\ell} \cdot \bar{E})^{1/2} + \frac{2}{3} E_F E_{p\ell} \right)] \right\} \qquad (3)$$

where $\bar{E} \simeq E_{p\ell} + E_G$, E_e and E_G are the electron energy and the gap energy respectively. Table 2 summarizes the parameters to be used in formula (3) for several alkali halides. (They are taken from ref. 34 and completed by us for CsI). It can be seen that the sum rules:

$$\int dE \cdot E \cdot Im\left(-\frac{1}{\epsilon}\right) = \frac{1}{2}\pi E_{p\ell}^2 \qquad (4)$$

$$\int dE \cdot Im\left(-\frac{1}{\epsilon}\right) = \frac{1}{2}\pi E_{p\ell}^2 / \bar{E} \qquad (5)$$

are fulfilled quite well. The authors[34] estimated the usefulness of formula (3) at electron energies $E_e >200$ eV. Experimental data of the total MFP related to valence band excitations exist for a few alkali halides[21] A comparison with these data may demonstrate the usefulness of the various methods for λ_v calculations discussed above. In fig. 4 we present the experimental MFPs versus electron energy data for NaCl together with MFPs calculated using eq. (3) (dashed curve) and our approach (full curve). For relatively high energies ($E_e \geq 200$ eV) all data agree quite well. Marked differences between the calculated and experimental MFPs can be seen for lower electron energies. But as was mentioned in ref. 21 the derived experimental λ_v data at the lower electron energies contain some indeterminate experimental errors. Hence, only indirect verification of our input data can be used: comparison of the calculated SEE characteristics with experimental ones.

In our calculations we assumed a Gaussian distribution of electronic states in the valence band, with a full width at half maximum ΔE_v. It is presented in table 3 together with other experimentally derived[35] parameters: $E_{p\ell}$, affinity χ, and gap energy E_G. Note the difference between plasmon energies given in table 2 and in table 3. We do not know the reason for these discrepancies. In our calculations we used data from table 3, where E_{min} is the minimal energy of an electron arriving to the vacuum level after the plasmon decay.

Table 3. Several alkali halides characterisitics (in eV)

Material	$E_{p\ell}$	E_G	χ	ΔE_v	$E_{min}=$ $E_{p\ell} - E_G - \Delta E_v$
LiF	25.3	13.6	1.0	3.7	8.0
NaCl	15.5	8.5	0.4	2.2	4.8
KCl	14.0	8.4	0.4	1.5	4.1
KBr	13.5	7.4	0.8	1.4	4.7
NaI	13.3	5.8	1.5	3.3	4.2
KI	11.4	6.3	1.2	2.4	2.7
RbI	11.1	6.1	1.1	2.1	2.9
CsI	10.7	6.2	0.1	2.3	2.2

Core electron excitation and ionizations In our model all inner electrons were treated as unperturbed shell electrons, for which the binding energy was taken from ref. 36. The formulae for the ionization cross-section, most frequently used in Monte Carlo calculations, are those deduced from the "binary-encounter approximation" theories of Gryzinski[23], and Vriens et al[37]. In fig.5 we compare the results of the calculations of the total ionization cross-section for Cs, using these two theoretical approaches and the compiled experimental data[38]. It can be seen that in general Gryzinski's approach applies better to the experimental cross-sections, although it overestimates them in the region of the maximum. In our calculations we prefer to use Gryzinski's formulae both for the differential (in energy) and the total cross-sections, mainly because of the their simplicity of sampling in Monte Carlo algorithms. The angles of scattering were obtained from the kinematics. Another process of energy loss is by inner shell excitations. At present this process cannot be treated precisely because of the lack of suitable cross-sections. But since these inner-shell excitations

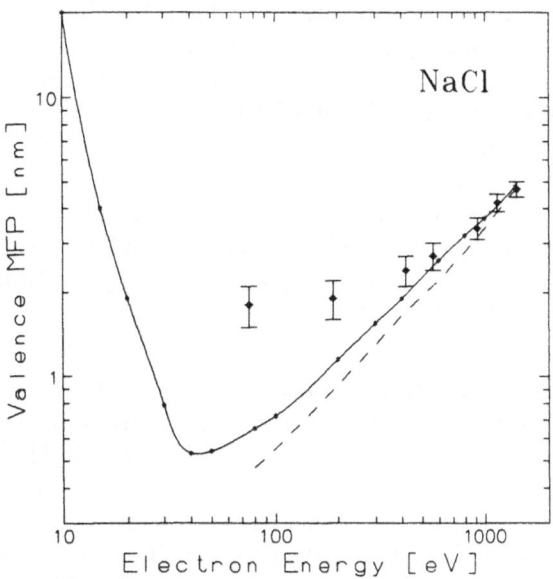

Figure 4. MFP for valence band excitation versus electron energy. Points - experimental data.[34] Full curve - our calculated data, dashed curve - results of calculations using eq (3).

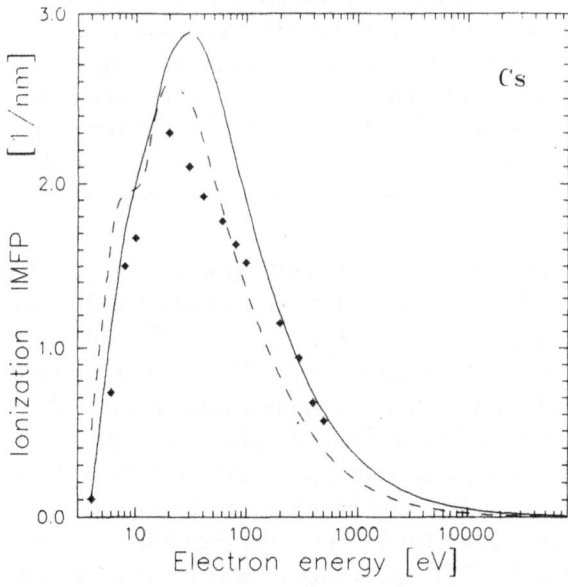

Figure 5. Ionization IMFP for Cs as function of electron energy. Full curve - Gryzinski's theory.[23] Dashed curve - Vriens theory,[37] dots - experimental data.[38]

do not affect the secondary electron emission, they were included as a mean energy loss ΔE along the free electron's path $t : \Delta E = t \cdot (-\frac{dE}{dx})_{exc}$. The values of $(-\frac{dE}{dx})_{exc}$ were obtained by supposing that Gryzinski's formulae are correct even for losses lower than the binding energy. From our estimation it follows that the uncertainties related to this process do not influence significantly the final results.

Figs. 6,7 show the energy dependence of the IMFP for various processes of inelastic interaction of electrons in CsI and KCl. To simplify the figure we present the total IMFP for ionization as the sum of K and Cl and of Cs and I atoms. It can be seen that a drastic drop in the cross sections occurs at energies below the binding energy of the last outer shell included in our calculations. Hence, the main mechanism responsible for the creation of cascades of secondary electrons is no longer effective at energies below E_G. Consequently, the secondary electrons with energies $E_e < E_G$ move in the solid without multiplication.

Electron-phonon interaction Secondary electrons with energies $E_e < E_G$ lose (or gain) energy by interacting with the lattice via electron-phonon scattering. In alkali halides only longitudinal optical phonons can be excited. The time dependent perturbation theory[10] was used to obtain the mean free paths for loss (gain) of a certain energy and the angle of scattering. The amount of energy loss (gain) is: 0.026 eV and 0.01 eV for KCl and CsI respectively. The energy dependence of the IMFP for loss (gain) due to phonon scattering in KCl and CsI is also shown in figs. 6,7.

Relaxation of the atomic shells A vacancy created in the inner shell as a result of an X-ray photon absorption or energetic electron ionization may be followed by radiative (fluorescence radiation) or nonradiative (Auger and Coster-Kroning) emission. There is a large branching of possible transitions, in particular for heavy atoms. Since most of the transition probabilities are unknown, and can not be estimated with sufficient accuracy by the existing theories, we include only averaged Auger energies in our calculations. They were calculated using binding energies of the corresponding shells. The probabilities of nonradiative transitions were calculated using the empirical formulae of Bambynek et al.[39]. For most applications of our model it is not necessary to take into account the radiative transitions because their role in secondary electron emission from thin layers is negligible.

Plasmon decay Only the decay of "bulk" plasmons was included in our simulations. We assumed that there is a 100% probability for this decay, followed by emission of electrons with energy $E_e = E_{pl} - E_G - X$. The parameter X takes into account the distribution of electronic states within the valence band and it's value was randomly selected from the above mentioned Gaussian distribution (truncated), with the parameters given in table 3. Since the plasmon excitation is one of the most important mechanisms for electron energy loss, any changes in the probability of its decay can influence the secondary electron emission. An important question remains: how far the plasmon propagates from its creation region until its decay. There is not much data available on the MFP for plasmon decay for the relevant materials. The correct way of treating this process is still not clear and further study should be carried out, theoretically and experimentally.

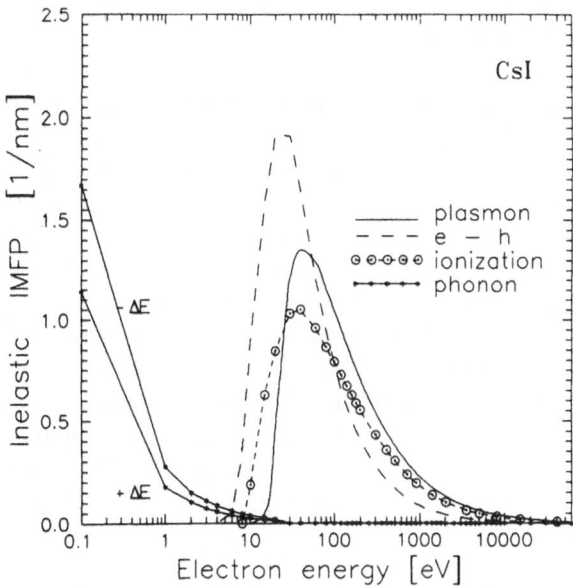

Figure 6. Inelastic IMFP in CsI for various interaction processes as function of electron energy. $-\Delta E$ and $+\Delta E$ curves correspond to electron-phonon interactions (loss and gain of energy respectively).

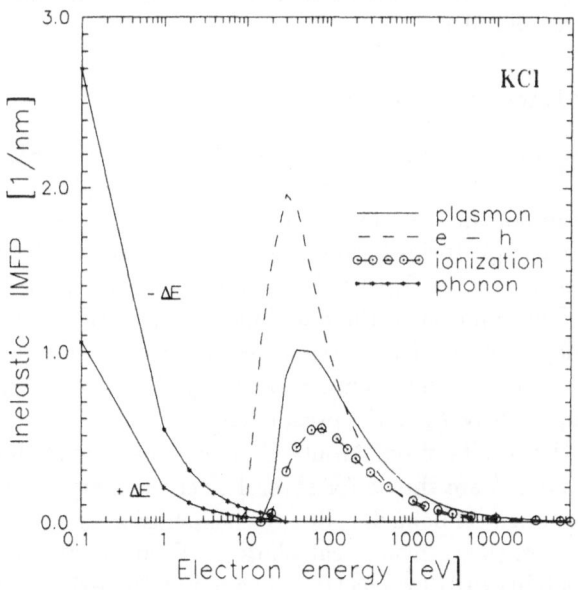

Figure 7. Inelastic IMFP in KCl. Notations as in Figure 6.

RESULTS AND CALCULATIONS

The results presented below were obtained using a PC-386 computer. The most time -consuming operation is the calculation of the low energy electrons transport. Therefore, we decided to divide the trajectory simulation process into two parts: a) production and transport of primaries and energetic secondaries, their multiplication and slowing down to the gap energy E_G; b) transport of low energy electrons ($\chi \leq E_e \leq E_G$). In the second part isotropically distributed electrons, of a given energy and set of distances from the exit surface, are followed by the Monte Carlo simulation. The results represent the electron escape probability versus distance. The transmission of low-energy electrons through the surface barrier was calculated, using the method proposed in ref. 2. Knowing the coordinates and energies of electrons created or slowed down to energies $E_e \leq E_G$ from the trajectory calculations in part a), we have convoluted them with the penetration probability from part b), to obtain the yield of secondary electrons. In addition to the time saved in computation , the division of the calculation into two steps allows us to compare our results with the diffusion model[40] for low energy electron transport and to evaluate the validity of this interpretation. However, using this approach we lose the possiblity of calculating the secondary electron spectra, because the escape probability is an average characteristic of the emerging electrons.

The minimum number of simulated electron trajectories in each of our calculations is of the order of 10^3. The transport of low energy secondaries ($E_e < E_G$) was simulated at each distance by $5 \cdot 10^3$ trajectories, providing an overall statistical uncertainty of 3% in the calculated yields.

For reasons of convenience, all results of SEE calculations presented here are for electrons emitted in the "forward" direction, namely, in the direction of the incident radiation. It should be noted that according to Henke et al.[40] and also according to our estimations and recent measurements there are no significant differences between "forward" and "backward" SEE.

Transport of Low Energy Electrons

As was explained above, all inelastic interactions followed by SEE are negligible for electron energies below E_G. In this range only electron-phonon scattering and elastic scattering are important and dominate the transport of the created secondary electrons. The electron trajectories become very long and involve very small energy transfers of the order of $\sim (1 - 3) \cdot 10^{-2}$ eV. Consequently the transport process resembles diffusion. The solution of the one-dimensional diffusion equation[40] has a nearly exponential dependence of the escape probability on the creation depth x. Also the often used formula for the escape probability $P(x)$ has an exponential form $P(x) = P(0) exp(-x/L_s)$, where L_s is the escape length.

In figs. 8 and 9 the results of our Monte Carlo calculations of the escape probability $P(x)$ for KCl and CsI are shown for several electron energies. These curves do not portray a pure exponential behavior, due to the inclusion of the elastic scattering of low energy electrons in the transport calculations. From the figures it can be seen that the escape probability $P(0)$ for electrons located at the exit surface is less than 1 (also suggested in ref. 10). Moreover, if we try to fit the curves by exponents, we obtain an energy dependence of the parameters L_s and $P(0)$. These are summarized in table 4, for energies up to 9 eV. Note that for KCl at $E_e = 9$ eV we obtain an L_s value lower than at $E_c = 7$ eV, because the gap energy E_G lies below 9 eV, thus enabling inelastic losses by the $\epsilon-h$ process.

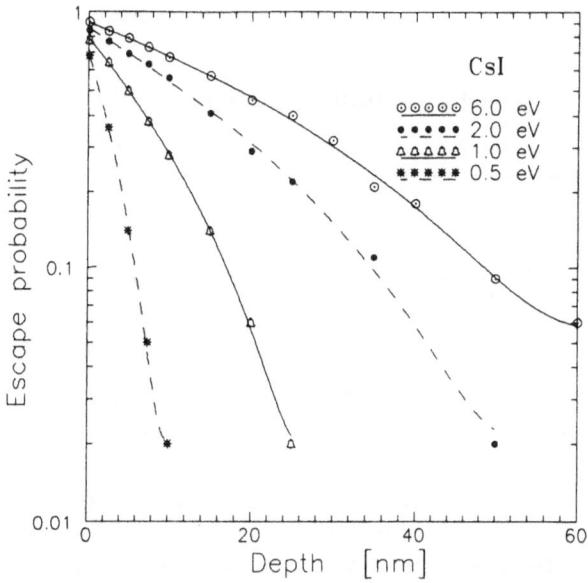

Figure 8. Escape probability from CsI for low energy electrons, as function of their creation depth. The curves are polinomial fits to our calculated points.

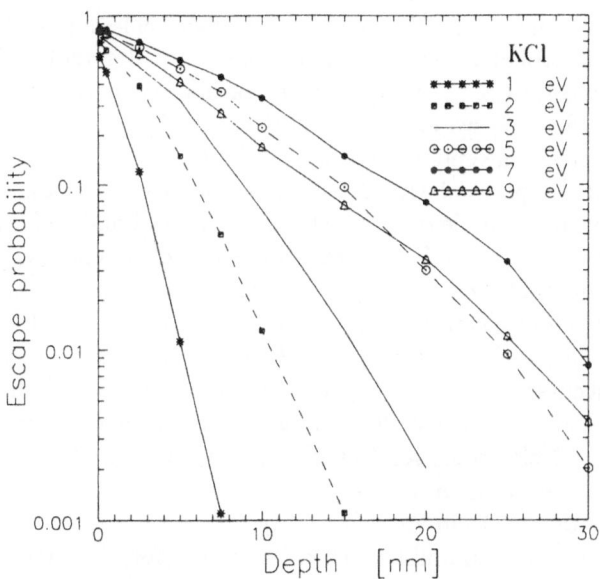

Figure 9. Same as Figure 8 for KCl.

Table 4. Parameters for the exponential fit to the calculated secondary electrons escape probability as function of depth.

CsI

E_e [eV]	6	5	3	2	1	0.5	0.15
P(0)	0.91	0.89	0.89	0.86	0.70	0.70	0.36
L_s [nm]	30.0	30.0	25.2	17.3	10.2	3.4	2.0

KCl

E_e [eV]	9	7	5	4	3	2	1
P(0)	0.80	0.85	0.80	0.78	0.75	0.7	0.6
L_s [nm]	6.3	9.00	7.5	5.9	4.4	2.9	1.4

As a matter of comparison Fraser[5], in his semi-empirical model, used for KCl the values P(0)=0.35, L_s=13.0 nm, η= 9 eV, and for CsI, the values P(0)=0.2, L_s=21.5 nm and η=7 eV (η is the mean energy for a secondary electron creation). These are values deduced from data of integral characteristics of SEE and they represent an averaging over the whole energy spectrum of secondary electrons.

The curves shown in figs. 8 and 9 were interpolated and the extracted values of $P(x)$ were used in the calculation of the transport of electrons of energies below E_G.

Secondary Electron Emission Characteristics

As a first test of the Monte-Carlo model we have calculated mono-energetic electron transmission probabilities through CsI layers of different thicknesses (see fig. 10). Only primaries ($E_e > 50$ eV) were considered in this case. From the transmission curves (fig. 10) we extracted the total projected electron range R for several electron energies. The energy dependence $R(E)$ (see fig. 11) agrees well with the experimental data[41], particularly at the higher electron energies. Separately we calculated secondary emission yields as functions of the KCl and CsI layer thickness. Typical examples of these curves, for several incident electron energies, are shown in figs. 12 and 13. From similar curves we calculated the maximal electron yields, δ_m, as function of the incident electron energy. Fig. 14 presents the relative secondary emission yield δ/δ_m versus the relative energy E/E_m (E_m is the energy at which the maximal yield δ_m occurs). The curve follows the "universal curve" shape suggested by Schwarz[4]. The absolute values of E_m and δ_m from our calculations are: for KCl E_m=1 keV and δ_m=12 electrons; for CsI E_m=2 keV and δ_m=23 electrons. These values are close to those quoted in ref. 4.

Experimental data of SEE induced by X-rays of energies up to 10 keV were obtained by Henke et al.[40] for CsI and by Eliseenko et al.[42] for KCl. We compare these results with our calculations. In fig. 15 the secondary yield dependence upon CsI thickness is presented for two X-ray energies E_γ=2.3 keV and E_γ=15 keV, together with the experimental data[40]. As expected, the yield at E_γ=2.3 keV saturates at thicknesses of the order of 100-200 nm. This is true for photon energies up to about 7 keV. The constancy of convertor thickness corresponding to the saturated yield

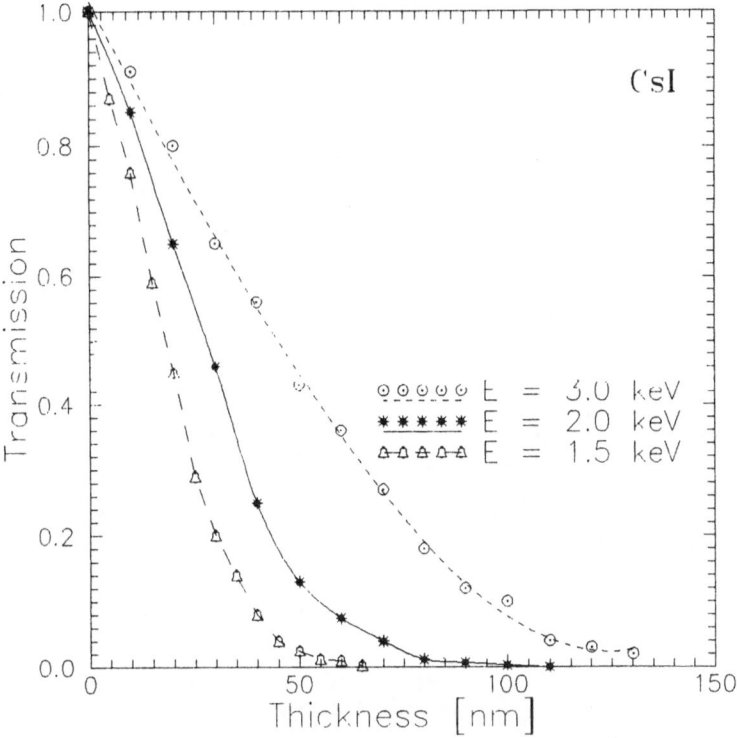

Figure 10. Electron transmission through CsI layers, for several energies indicated in the figure.

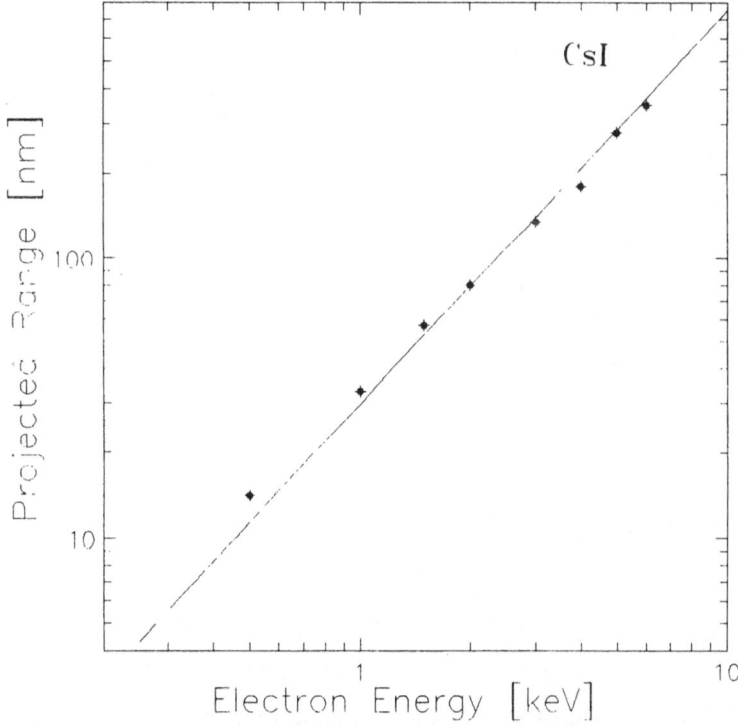

Figure 11. Energy dependence of the projected electron range for CsI. Full curve - experimental data.[41] The dashed curve fits our calculated data.

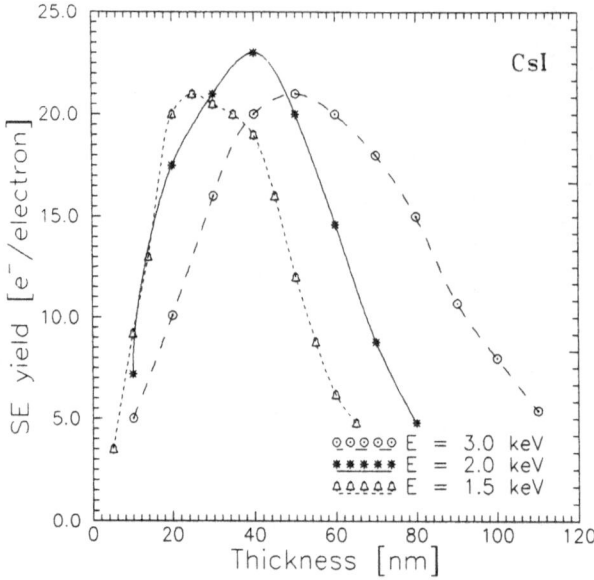

Figure 12. Forward secondary electron yield for several incident electron energies indicated in the figure, as function of the CsI layer thickness.

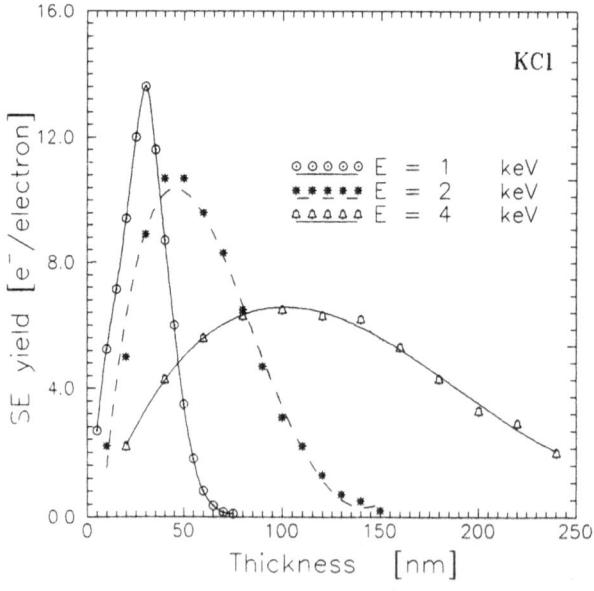

Figure 13. Same as Figure 12 for KCl.

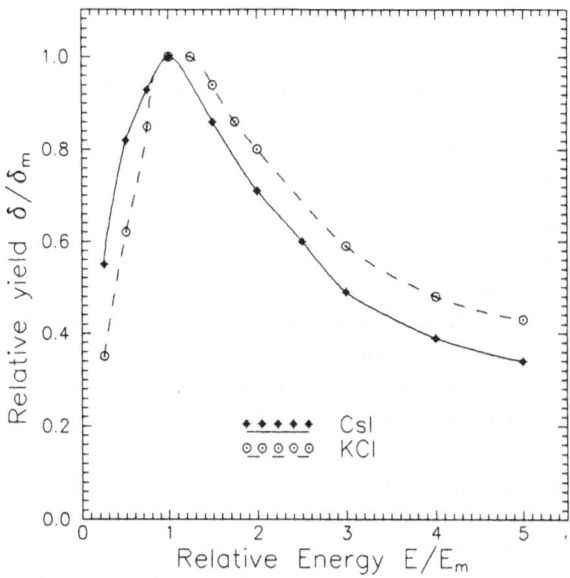

Figure 14. Calculated relative forward secondary yield δ/δ_m, versus relative electron energy E/E_m. Full curve - CsI, dashed - KCl.

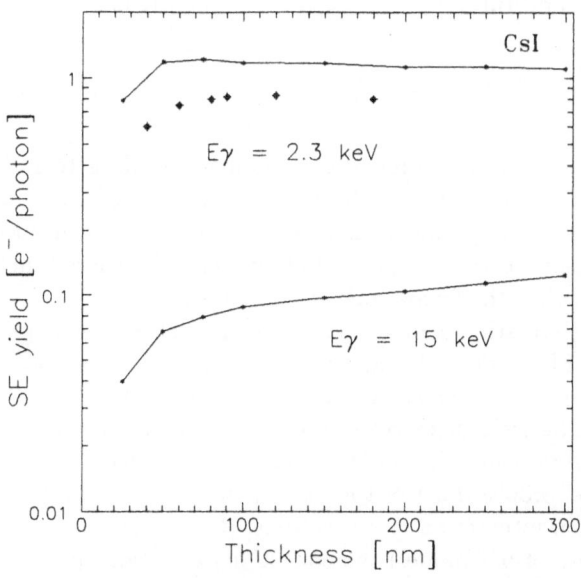

Figure 15. Forward secondary electron yield, per incident photon, as function of the CsI layer thickness, for two X-ray energies, indicated in the figure. Dots - experimental data[31] curves - our calculations.

over this energy range supports the interpretation that the secondary electrons are emitted from a finite thin layer, of a few escape lengths below the emission surface. At higher X-ray energies the "saturated yield" is obtained at larger CsI thicknesses, as shown in fig. 15. This is consistent with the picture that there is a small cascading process created by the primary electrons. The cascade of the secondaries consists of small "branches" of low energy electrons, which have a limited penetration depth as was shown in figs. 8 and 9.

The calculated energy dependence of the quantum yield Y_s (Y_s is the number of secondaries per normal incident photon) from a 100 nm thick CsI film irradiated by 1-30 keV photons, is shown in fig. 16, together with experimental data[40]. The overall trend of the $Y_s(E_\gamma)$ dependence follows the experimental data, with a systematic overestimation of the order of 20-30%. In the same figure the quantum yield Y_s versus photon energy for a 200 nm thick KCl film, is presented together with experimental data[42]. The experimental results in this reference were actually obtained at a grazing angle of 20°. To allow a comparison with our calculations they were converted to normal incidence. As can be seen from fig. 16 our calculations underestimated the experimental data by 20–30%.

Fig. 17 shows the calculated energy dependence of the primary-to-total yield ratio, Y_p/Y_t, for KCl and CsI films together with the existing experimental data[40] ($Y_t=Y_p + Y_s$, Y_t, Y_p and Y_s are the total, primary and secondary yields). There is a monotonic increase of the relative primary yield, and the K- and L-shell excitation effects are evident.

It should be noted that in the case of CsI the relative primary yield does not exceed a few percent as compared to a few tens of percent for KCl. This is an important feature of CsI, used as an X-ray convertor coupled to a gaseous electron multiplier; the number of emitted fast electrons, having a long ionization track in the gas, is very small and most of the yield is of slow electrons, thus preserving the excellent localization and timing properties of these X-ray detectors.

DISCUSSION AND CONCLUSIONS

A new physical model describing the phenomena leading to radiation-induced SEE in alkali halides is presented. The main advantage of this model is that it is free of preliminary assumptions such as secondary electron excitation function, semi-empirical formulae for the escape probability, etc. All these functions are automatically obtained within the framework of our model.

A Monte Carlo simulation code based on this model was developed. It was used for calculations of SEE from KCl and CsI, induced by X-rays and electrons. The results of absolute yields of SEE show a general agreement between calculated and experimental data. The yield dependence on the photon and electron energies is also well reproduced. The maximal discrepancy between the calculated and experimental data is of the order of ±30%. Part of the discrepancy may result from inaccuracy of the calculated electron interaction cross-sections, particularly at low energies, where (within the framework of our model) the collective excitations play a dominant role. Further improvements are needed in the calculational methods for cross-section estimations based on the use of the imaginary part of the dielectric function, $\text{Im}(-\frac{1}{\epsilon(\omega,k)})$, derived from experimental data. The role of Auger process should be estimated more accurately, including the Coster-Kroning transitions. Another source of discrepancy may be related to the difference between the macroscopic characteristics of the simulated films and those actually used in the experiments (often not specified by the

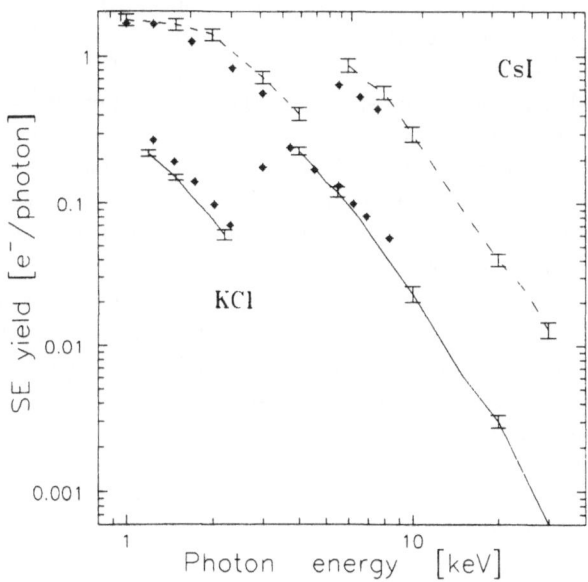

Figure 16. Forward secondary electron yield, for normal incident photon, as function of the photon energy, for 100 nm thick CsI and 200 nm thick KCl. Dots - experimental data, curves - fit to our calculated data, bars - calculational uncertainties. The experimental data for KCl were deduced from 20° grazing incidence - see text.

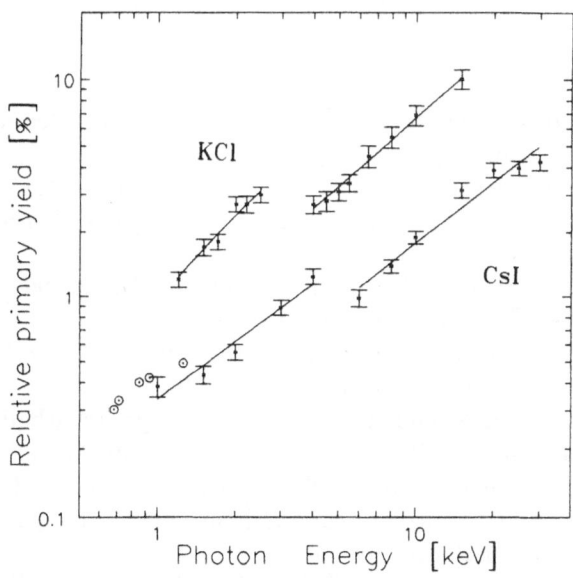

Figure 17. Ratio of primary-to-total yield, for KCl and CsI, as in Figure 16. Open circles - experimental data.[40] Dots - our calculations, lines - fit to the calculated data.

experimentalists). For example in our calculations we assumed the film density ρ_s to be equal to the bulk density ρ_v. There is, however, experimental evidence that this is not true for CsI and that ρ_s may be as low as 80% of ρ_v. We should also not forget the role of impurities and of the film structure. It is known from experiments that "spongy" CsI or KCl films present a very enhanced SEE.

Concluding the discussion we can state that the proposed model is able to predict the yields and other integral characteristics of SEE from alkali halides, induced by photons and electrons. The accuracy of the prediction is sufficient for most practical applications. In spite of some simplifications used in the model (interaction processes and solid state effects) the Monte Carlo calculations based on this model show a good agreement with existing experimental data.

The model presently covers a photon and electron energy range of 1-30 keV. It will be further expanded to higher incident energies. It will also enable the calculation of the energy spectra of the secondary electrons. Some calculations are in progress to predict SEE induced by relativistic particles crossing alkali halide convertors. This phenomenon is of importance for some applications of particle detection.

ACKNOWLEDGEMENTS

A.A. is grateful to Dr. J. Schou and Dr. M. Rösler for helpful discussions. He also expresses his gratitude to the Einstein Center for Theoretical Physics at the Weizmann Institute for its partial support. A.A. and A.G. are partially supported by the Center for Absorption in Science of the Israel Ministry of Absorption. This work was partially supported by the Minerva Foundation, Munich, Germany, The Israel Ministry of Science and Technology and the Foundation Mordoh Mijan de Salonique.

REFERENCES

1. J. Schou, Secondary electron emission from solids by electron and proton bombardment, Scanning Microscopy, 2:607 (1988).
2. M. Rösler, W.Brauer, J. Devooght, J.-C. Dehaes, A. Dubus, M. Cailler and J.-P. Ganachaud, Particle induced emission I, Springer-Verlag, Berlin, Heidelberg, New York, London (1991).
3. D. Hasselkamp, Particle induced electron emission II, Springer-Verlag, Berlin, Heidelberg, New York, London (1992).
4. S.A. Schwarz, Application of a semi-empirical sputtering model to secondary electron emission, J. Appl. Phys. 68:2382 (1990).
5. G.W. Fraser, The characterization of soft x-ray photocathodes in the wave length band 1-300 Å, II, Nucl. Instrum. Methods 206:265 (1983).
6. B.L. Henke, J. Liesegang and D.S. Smith, Soft-X-ray-induced secondary electron emission from semiconductors and insulators: models and measurements, Phys. Rev. B19:3004 (1979).
7. E.O. Kane, Simple model for collision effects in photoemission, Phys. Rev. 147:335 (1966).
8. G.W. Fraser, The characterization of soft X-ray photocathodes in the wavelength band 1-300 Å, I, Nucl. Instrum. Methods 206:251 (1983).
9. I.R. McDonald, A.M. Lamke and C.F.G. Delaney, Electron emission from alkali halides under soft X-ray bombardment, J. Phys. D6:87 (1973).

10. J. Llacer and E.L. Garwin, Electron-photon interaction in alkali halides. I. The transport of secondary electrons with energies between 0.25 and 7.5 eV, J. Appl. Phys. 40:2766 (1969).

11. A. Akkerman, A. Gibrekhterman, A. Breskin and R. Chechik, Monte Carlo simulations of secondary electron emission from CsI, induced by 1-10 keV X-rays and electrons. WIS-92/28/Mar-PH, J. Appl. Phys. in press.

12. A. Breskin, R. Chechik, V. Dangendorf, S. Majewski, G. Malamud, A. Pansky and D. Vartsky, New approaches to spectroscopy and imaging of ultrasoft to hard X-rays, Nucl. Instrum. Methods A310:57 (1990).

13. R. Chechik, A. Breskin, A. Akkerman, A. Gibrekhterman, I. Frumkin, H. Aclander and V. Elkind, Towards fast transition radiation imaging detectors with CsI convertors. Preprint WIS-91/78/Oct.PH. Presented at the IEEE Nuclear Science Symposium, Santa Fe, USA, November 1991; and A. Akkerman, A. Breskin, R. Chechik, I. Frumkin and I. Gibrekhterman, Ultrafast secondary emission 2D X-ray imaging detectors. Preprint WIS-91/74/Oct.-PH, presented at the European Workshop on X-ray Detectors for Synchrotron Radiation Sources, Aussois, France, 1991.

14. H.M. Colbert, SANDYL, A computer program for calculating combined photon-electron transport in comlex systems, SLL-74-0012 Sandia Lab. (1974).

15. E.B. Salamon, J.H. Hubbell and J.H. Scofield, X-ray attenuation cross section for energies 100 eV to 100 keV and elements Z=1 to Z=92. Atomic Data and Nuclear Data Tables 38:1 (1988).

16. J. Scofield, Theoretical photo-ionization cross-sections from 1 to 1500 KeV, UCRL-51326 LLNL (1973).

17. C.M. Davidson and R.D. Evans, Gamma-ray absorption coefficients, Rev. Mod. Phys. 24:79 (1952).

18. A.I. Ding and R. Shimizu, Monte Carlo study of backscattering and secondary electron generation, Surf. Sci. 197:539 (1988).

19. A.F. Akkerman and G.Ya. Chernov, Monte Carlo calculation of the electron transmission, reflection and absorption in solids in the energy range up to 10 keV, Phys. Stat. Sol. (b) 101:109 (1980).

20. A.F. Akkerman and A.L. Gibrekhterman, Comparison of various Monte Carlo schemes for simulation of low-energy electron transport in matter, Nucl. Instrum. Methods B6:496 (1985).

21. J. Szajman, J. Liesegang, R.C.G. Leckey and J.G. Jenkin, Photoelectron determination of Mean Free Paths of 200-1500 eV in potassium iodide, Phys. Rev. B18:4010 (1978).

22. A.F. Akkerman and G.Ya. Chernov, Elastic scattering of electrons by atoms in the keV range, Sov. Phys. Techn. Phys. 23:247 (1978).

23. M. Gryzinski, Classical theory of atomic collisions. I. Theory of inelastic collisions, Phys. Rev. 138A:336 (1965).

24. J.J. Quinn, Range of excited electrons in metals, Phys. Rev. 126:1453 (1962).

25. R.N. Ritchie, Plasma losses by electrons in thin films. Phys. Rev. 106:874 (1957).

26. E. Tossati and G.P. Parravicini, Elementary excitations in anisotropic semi-conductors, J. Phys. Chem. Solids 32:623 (1971).

27. P. Sigmund, Secondary electron emission: elementary questions and (few) attempts at answers, NATO Advanced Research Workshop "Ionization of solids by heavy particles", Giardini Naxos, Italy, 1-5 June (1992).

28. A.E. Green, I.L. Sellin and A.S. Zachor, Analytic independent-particle model for atoms, Phys. Rev. 184:1 (1976).

29. M. Fink and A.C. Yates, Theoretical electron scattering amplitudes and spin polarizations, Atomic Data 1:385 (1970).

30. F.J. de Heer, R.H.J. Jansen and W. van der Kaay, Total cross sections for electron scattering by Ne, Ar, Kr and Xe, J. Phys. B: Atom Molec. Phys. 12:979 (1979).

31. T. Goulet, J.P. Jay-Gerin and J.P. Patau, Monte Carlo simulations of low energy (< 10 ev) electron transmission and reflection experiments: application to solid xenon, J. Electr. Spectrosc. Relat. Phenomen. 43:17 (1987).

32. M. Creuzburg, Energieverlustspektren der alkalihalogenide und der metalle Cu, Ag und Au und Verglach mit optishen messungen, Z. Phys. 196:433 (1966)

33. S. Tougaard, Low energy inelastic electron scattering, Solid State Comm. 61:547 (1987).

34. J. Szajman and R.C.G. Leckey, An analytical expression for the calculation of electron mean free paths in solids, J. Electron. Spectrosc. Relat. Phenom. 23:83 (1981).

35. R.T. Poole, J.G. Jenkin, J. Liesegang and R.C.G. Leckey, Electron bond structure of the alkali halides. I. Experimental parameters, Phys. Rev. B11:5179 (1975).

36. K.D. Sevier, Atomic electron binding energies, Atomic Data and Nuclear Data Tables 24:373 (1979).

37. L. Vriens and T.F.M. Bonsen, Differential cross sections for ionization of the hydrogen atom by fast charged particles in the binary-encounter theory and Bethe theory, J. Phys. B1:1123 (1968).

38. H. Tawara and T. Kato, Total and partial ionization cross sections of atoms and ions by electron impact, Atomic Data and Nuclear Data Tables 36:167 (1987).

39. W. Bambynek, B. Crasemann, R.W. Fink, H.U. Freund, H. Mark, C.D. Swift, R.E. Price and P. Venugopala Rao, X-ray fluorescence yields, Auger, and Coster-Kroning transition probabilities, Rev. Mod. Phys. 44:716 (1972).

40. B.L. Henke, J.O. Knauer and P. Premaratne, The characterization of X-ray photocathodes in the 0.1-10 photon energy region, J. Appl. Phys. 52:1509 (1981).

41. I.M. Bronshtein and A.N. Protsenko, Transmission experiments in alkali halides, Sov. Phys. Sol. State 11:2113 (1970).

42. L.G. Eliseenko, V.N. Shchemelev and M.A. Rumsch, Quantum yields of the surface X-ray photoeffect at 1-10 Å, Sov. Phys. Tech. Phys. 13:122 (1968).

ELECTRON EMISSION FROM ION BOMBARDED SOLID ARGON

D. E. Grosjean and R. A. Baragiola

Engineering Physics
University of Virginia,
Charlottesville, VA 22901

INTRODUCTION

Rare gas solids (RGS) are good model systems to study ionization effects in insulators. Their electronic states are relatively well known[1]; they are monoatomic and bound by weak Van der Waals forces which allow them to be treated as dense gases with generally negligible chemical changes. The interaction of ionizing radiation with RGS produces lattice damage and desorption by the conversion of energy deposited as electronic excitation into kinetic energy. This conversion occurs by transitions involving repulsive potential energy curves and is generally known as desorption induced by electronic transition (DIET). We are studying the DIET mechanism in solid argon as described by Johnson and Inokuti[2], which also accounts for the bright luminescence observed during bombardment with ionizing particles. Previous studies have correlated sputter yields with luminescence yields in solid argon excited by MeV ions[3], low energy electrons[4], and photons[5, 6]. In order to determine the importance of this DIET mechanism in the decay pathways of electronic excitation, however, absolute correlation must be made. Furthermore, because this mechanism begins with an ionization event, the secondary electron emission processes during bombardment must also be understood and accounted for by any model for desorption and luminescence. In this paper we look at the beginning of this DIET mechanism by focusing on the secondary electron emission from bombarded solid argon.

The details of the energy pathways following ionization and excitation are not known with certainty; they are a subject of continuous research at the University of Virginia[3, 7, 8, 9, 10]. The DIET process for argon is shown schematically in figure 1, together with a simplified internuclear potential energy diagram for Ar pair potentials. The process begins with the formation of electron-hole pairs. The number of ionizations created by an energetic ion per unit path length, dI/dx, is given by nS_e/W, where S_e is the electronic stopping cross section, n is the number density, and W is the mean energy to create an electron-hole pair (W=23.6 eV for solid Ar)[12]. The number of ionizations per unit path length is larger than the number of ionizations produced directly by the projectile, since secondary electrons are also capable of producing ionizations[13].

Ionization of Solids by Heavy Particles, Edited by
R.A. Baragiola, Plenum Press, New York, 1993

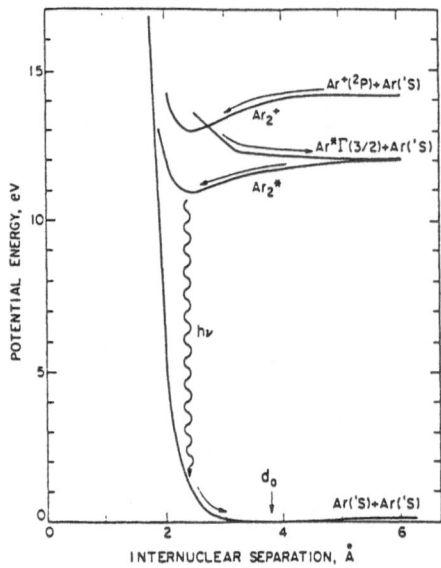

Figure 1: Solid state argon internuclear potential energy diagram. The arrows indicate the decay sequence as described in the text. From bottom to top, the curves are 1) Ar-Ar ground state, 2) Ar* exciton at ~12 eV, and 3) Ar$^+$ ion at ~14 eV. The cohesive energy is ~0.08 eV/atom for ground state Ar, and the ground state separation is $d_0 = 3.7$ Å (from reference [11]).

Electrons diffuse very rapidly, losing very little energy, especially if in states located below the exciton threshold at 12.2 eV when they can only cool very slowly by exciting lattice vibrations. Holes also diffuse very rapidly at the beginning, by resonant charge transfer between lattice sites. The hopping time for hole diffusion is $\tau_h = \hbar/\Gamma$, where Γ is the width of the Ar 3p-valence band; or $\tau_h \sim 10^{-15}$ s for $\Gamma \sim 0.6$ eV. Fluctuations in lattice vibrations can cause the formation of the more stable dimer ion Ar$_2^+$ after several lattice periods, or a time of $\sim 10^{-11}$ s. Unlike the simple hole, this dimer cannot diffuse very rapidly, since hopping requires the formation and breaking of bonds with atomic motion. Again, we can expect hopping times of the order of lattice vibration periods. After a time of the order of 10^{-11}-10^{-10} seconds from the primary ionization event, there can be a recombination between the dimer ion and a thermalized electron, and the system switches at the curve crossing to the Ar*-Ar potential curve. The state is more likely to be repulsive and will produce an energetic excited Ar* which can lead to sputtering if it is near the surface. Energetic Ar* have indeed been observed to be ejected from the solid[3, 14, 15]. Eventually, an excited Ar* moving in the lattice will lose too much energy to move as an atom inside the solid. The excitation can diffuse out, however more slowly than in the case of Ar$^+$, since in this case, two electrons are required for excitation transfer to a neighboring site. Through further multiphonon interactions, the more stable configuration of the Ar$_2^*$ excimer will form. The system will continue to relax to the bottom of the Ar$_2^*$ well; the only way out of the well is through the radiative transition shown in figure 1. The time for the excimer to relax and then emit a photon is about 1 ns, and the Ar lattice is transparent to the emitted photon. Transitions from the bottom of the Ar$_2^*$ well put the system on the highly repulsive part of the ground state potential, leading to sputtering, as well as radiation damage, and heat production.

The luminescence spectra and ejected atom energy spectra for ion bombarded solid Ar have been well studied[1, 4, 14, 15, 16, 17]. Figure 2 shows the luminescence spectrum of an Ar film during 1.5 MeV He+ bombardment. The peak labeled M at 9.8 eV corresponds to radiative decays of the relaxed Ar_2^* to the ground state. The peak labeled W corresponds to radiative decays from higher levels in the Ar_2^* well. Radiative decays from the electron-hole recombination directly to the ground state have not been observed. A distribution of internuclear distances in the Ar_2^* well leads to a distribution of repulsive energies upon photon emission, and thus a spectrum of ejected atom energies, which has also been measured[10].

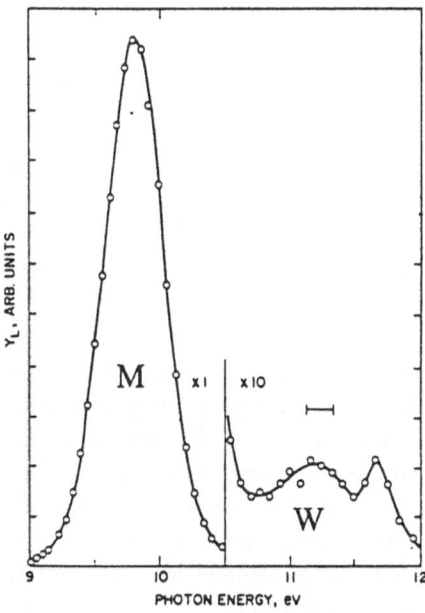

Figure 2: Ion induced luminescence spectrum of solid argon (from reference [3]).

To understand the energy pathways in solid Ar, it is important to determine whether the steps described above constitute the *primary* mechanism for sputtering. If they do, then luminescence and sputtering yields must at least be correlated. Reimann et al.[3] measured the relative luminescence yield of 9.8 eV photons and the absolute sputtering yield of MeV light ion bombarded Ar as a function of stopping power (figure 3). The yields correlated well with stopping power; correlation, however, is not enough to determine the importance of this decay pathway. *Absolute* luminescence yields are needed to determine the number of excimer decays that give rise to sputtered Ar atoms. There have been previous indications of high luminescence efficiencies under electron bombardment[19], but it was unknown whether much of the radiative decay path would be quenched at the higher excitation densities achieved under ion impact. Our measurements of the absolute luminescence yield are currently underway. Preliminary results indicate that ~50% of the electronic energy deposited by 1.5 Mev He ions is converted to 9.8 eV photons, agreeing with the electron impact studies and verifying that quenching has not occurred for our measurements. These measurements indicate that the excited Ar film is a very efficient light source and that the decay pathway being considered is indeed a major pathway for electronic energy[20].

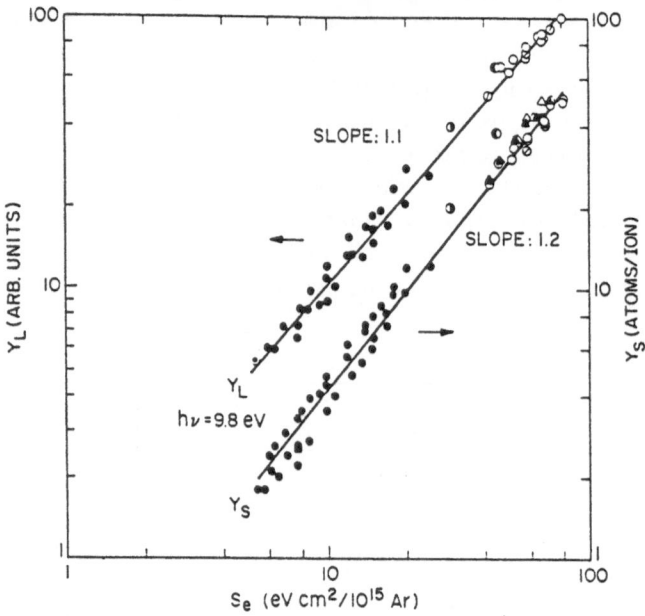

Figure 3: Stopping cross section dependence of sputtering and 9.8 eV luminescence yields (from reference [3]).

While much effort has been put into measuring the luminescence and sputtering of ion bombarded Ar, another aspect of the energy pathway to consider is the role of the electrons. We were guided by unpublished observations by W. L. Brown and C. T. Reimann that electron emission currents from thin Ar films under MeV ion impact were hundreds of times more intense than when bombarding metals or semiconductors. Such high yields can be understood by recalling two important properties of solid argon. One is the very large attenuation length of the electrons, of the order of thousands of Angstroms (large compared with 30 Å, a value typical for metals); this is caused by the large band gap. The other unusual property is that the bottom of the conduction band is above the vacuum level by about 0.3 eV (i.e., the electron affinity is negative)[21], which means that the solid will expel even thermalized electrons. We have recently undertaken the observation of ejected electrons to get a more complete picture of the energy pathways involved: we want to examine the electron that presumably begins the decay pathway described above.

EXPERIMENT AND RESULTS

The experiments are carried out in a UHV chamber with a base pressure of about 1×10^{-10} Torr connected to a 60 kV ion accelerator. Films are grown on a gold coated, quartz-crystal microbalance cooled to 16 K by vapor deposition of 99.999% pure Ar, further purified by a sorption getter pump. To minimize the effect of small temperature shifts on the frequency of the crystal, we use a heterodyne method with another crystal as a reference, on which no film is grown. A calibrated vacuum ultraviolet photodiode from VUV Associates is used for the luminescence efficiency measurements, and a vacuum ultraviolet spectrometer is also on the chamber for spectral measurements. To measure the ejected electron energy spectra, we use a retarding potential electrostatic analyzer surrounded by mu-metal to shield the earth's magnetic field. Figure 4 shows the experimental setup used to measure electron yields and energy spectra. Surrounding the target is a copper surface that is used as the anode to extract electrons from the bombarded Ar film, as well as for additional cryopumping. The ion beam plus electron current is measured by an electrometer connected between the target (cathode) and ground. Two experiments can thus be carried out: 1) scan the voltage on the anode while recording the current measured at the target, and 2) take ejected electron energy spectra for various anode voltages.

Figure 5 shows the results of the first experiment using 33 keV protons for several film thicknesses. In the 100 Å and 1000 Å cases, we note that the target current saturates above 150 and 250 volts anode voltage respectively. The electron yields γ in the saturation regions are very close to the expected number of ionizations produced by the incident proton. For example, in the case of a 1000 Å film being bombarded by 33 keV protons, we would calculate the electron yield to be

$$(nS_e d/W) = (5eV/\text{Å})(1000\text{Å})/(23eV) \sim 210$$

The escape depth of hot electrons in solid Ar has been measured to be $\sim 0.5\mu$ with no external electric field, therefore, the electron attenuation should be minimal[18], and we should be able to collect each electron formed if the substrate is biased to repel the electrons freed in the Ar film. In the saturation region, we observe that the luminescence intensity remains unchanged, thus, while we may be extracting all the electrons produced by the incident proton, the electrons are being replenished by the substrate allowing the luminescence to persist. The saturation indicates that there is an electric field set up within the film: if there were no field, the maximum number of isotropically emitted electrons that could be collected outside the film would be half of the total produced (the half emitted toward the front of the film –the rest would go to the substrate). The hysteresis effect seen in the figure (scan rate ~ 10 V/s) has yet to be fully characterized.

Also in figure 5 is the case for a 10,000 Å thick film. There is no saturation in this case, rather a dielectric breakdown with electron yields exceeding 1000. As the anode voltage increased, jumps in the target current, or discharges, were observed. Furthermore, in this discharge regime, the luminescence and target current persisted at the same levels for several seconds after the ion beam was removed, creating a state of sustained emission.

The effect of the depth of ionization path of the incident proton on the breakdown behavior was also studied, and the results are shown in figure 6. In this case, a 7500 Å Ar film was used and the proton beam energy was varied from 23 to 53 keV. The average penetration depth of 33 keV protons in Ar is 7500 Å, thus, the primary differences between the different beam energies are excitation densities and whether the incident beam traverses the film or not. It can clearly be seen that the breakdown voltage increases as beam energy increases.

The stability of electron emission from the film requires that the charge removed be compensated by the cathode. The mechanism for electron emission from the cathode (substrate) that explains these observations best is field emission, which can typically be observed with field strengths on the order of 10^7 V/cm. Field emission is described by the Fowler-Nordheim equation[22]:

$$j = 1.54 \times 10^{-6} \frac{\chi^2}{\phi} \exp(-\frac{6.83 \times 10^9 \phi^{\frac{3}{2}}}{\chi} f(y))$$

where j is the current density in A/m², ϕ is the work function of the substrate in volts (in this case barrier height, which is the work function of gold plus the electron

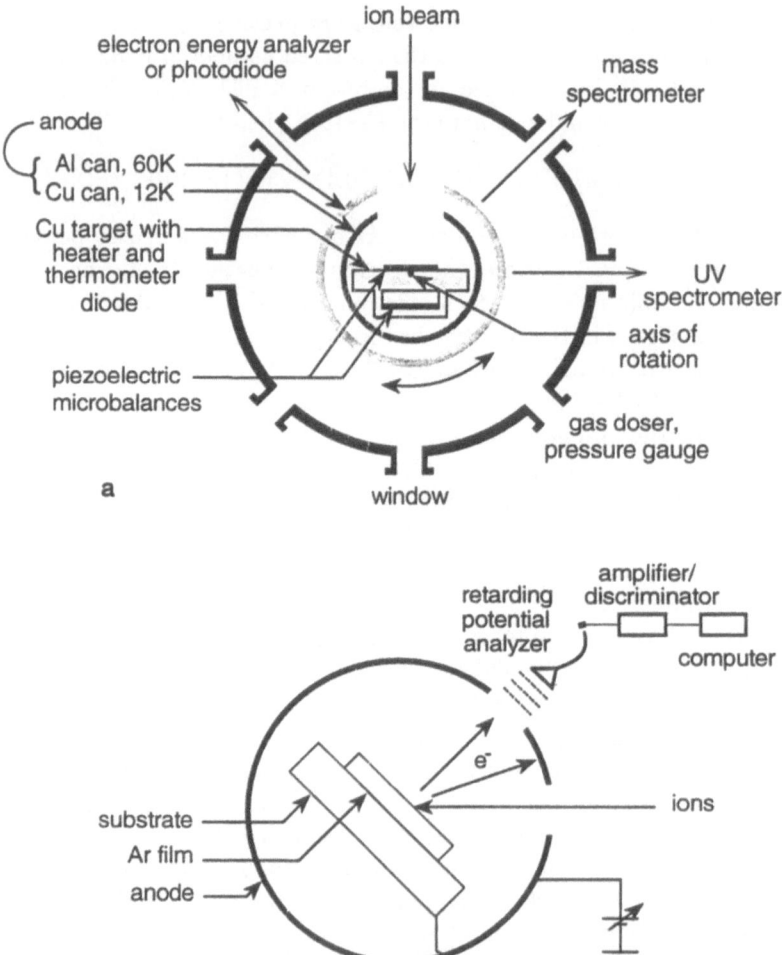

Figure 4: Schematic of experimental setup used to measure ejected electron yields and energy spectra: a) physical chamber and b) electrical setup.

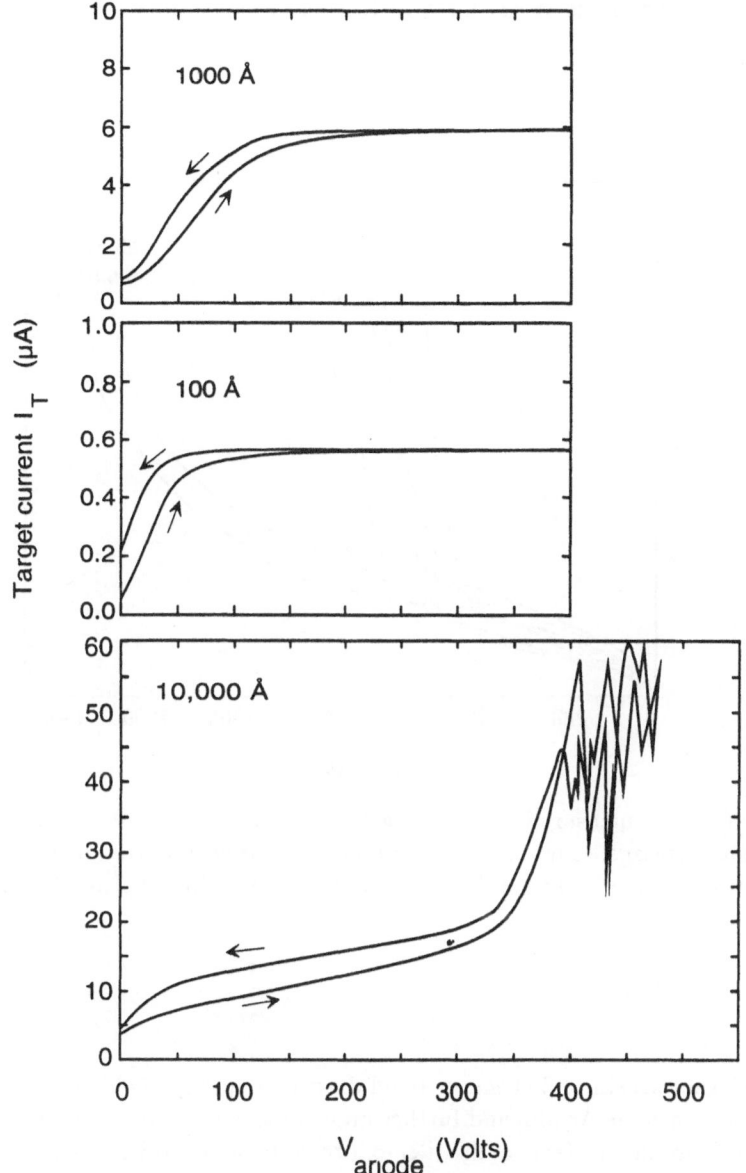

Figure 5: Current measured on the target for three Ar films of differing thicknesses as a function of applied anode voltage. The ion beam was 30 nA of 33 keV protons. The arrows indicate the direction the anode voltage was scanned.

affinity of Ar), $f(y)$ is the correction for the image force, and χ is the electric field in V/m. We applied this equation to the case of the 1000 Å Ar film for anode voltages in the saturation region (see figure 5). Substituting the values $j = 0.3$ A/m^2 (which was measured), $\phi = 5.4$ V, and $f(y) \approx 1 - 14 \times 10^{-10} (\chi/\phi^2)$, we numerically solved the Fowler-Nordheim equation for the electric field, yielding a value of $\sim 2.5 \times 10^7$ V/cm. This is strong enough for field emission to occur, and it corresponds with the electric field created by the anode, assuming that the field is constant throughout the film.

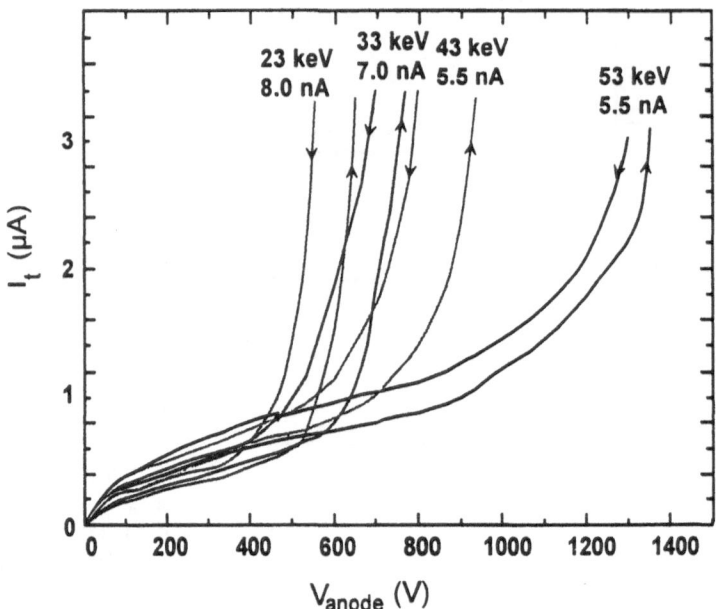

Figure 6: Current measured on the target for a 7500 Å Ar film as a function of applied anode voltage. The ion beam energy was varied but the beam current was kept constant at 6 nA. The arrows indicate the direction the anode voltage was scanned.

Figure 7 shows a simplified energy band diagram depicting field emission from the gold substrate into the Ar film and further impact ionization. In a thin film, electrons are emitted from the substrate and either exit the film or neutralize the surface charge, creating a saturation of electron emission current. In a thick film, however, electrons from the substrate can be accelerated enough to cause impact ionizations and thus an avalanche of electrons, producing the observed dielectric breakdown. The shape of the electric field inside the Ar film must be known to understand this mechanism better. An ejected electron energy spectra should help to map this electric field, because, in the absence of significant energy loss, electrons will acquire an energy equal to $q(V_{anode} - V_i)$ where V_{anode} is the anode voltage and V_i is the electrostatic potential at the point they are produced.

Energy distributions of electrons ejected normal to the Ar film were measured for a 1000 Å film at three anode voltages, and the results are in figure 8. For each case it

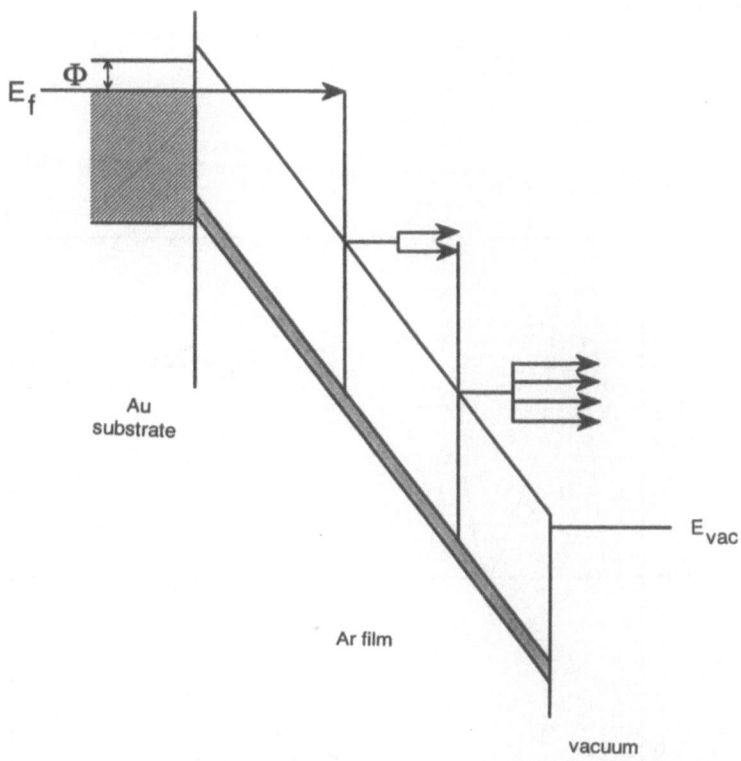

Figure 7: Simplified energy band diagram illustrating field emission from the gold substrate into the Ar film and impact ionization within the Ar film.

is seen that there are two groups of emitted electrons: one (I) emitted with energies close to qV_{anode} and one (II) emitted with very small energies. We attribute electrons in group I to those field emitted from the Au substrate and those from group II to those emitted from the surface region charged to a potential close to V_{anode}. The absence of a significant number of electrons with energies in between suggest that most of the film is at the same potential (close to V_{anode}) and that the high electric field (necessary to produce field emission from the Au substrate) occurs over a very thin region near the substrate. This is shown schematically in figure 9.

From this potential distribution we can get the electric field and, through Poisson's equation, the distribution of excess positive charges in the film. This is shown schematically in figure 9. The distribution of positive charge can be explained in the following way: electron-hole pairs are produced uniformly along the track of the incident ion. The electrons are extracted through the surface of the film by the positive anode potential, which also pushes the holes towards the substrate. Holes very close to the substrate (within a few Å) are removed by resonant or Auger neutralization[23]. This creates the hole distribution shown in figure 9 which exhibits low hole concentrations close to and far from the substrate and a maximum in between. The precise shape of the equilibrium distribution is determined by such competing processes as electron-hole generation, electron-hole recombination (substrate quenching as well as dimer formation and decay), hole drift and trapping, field emission, and impact ionization.

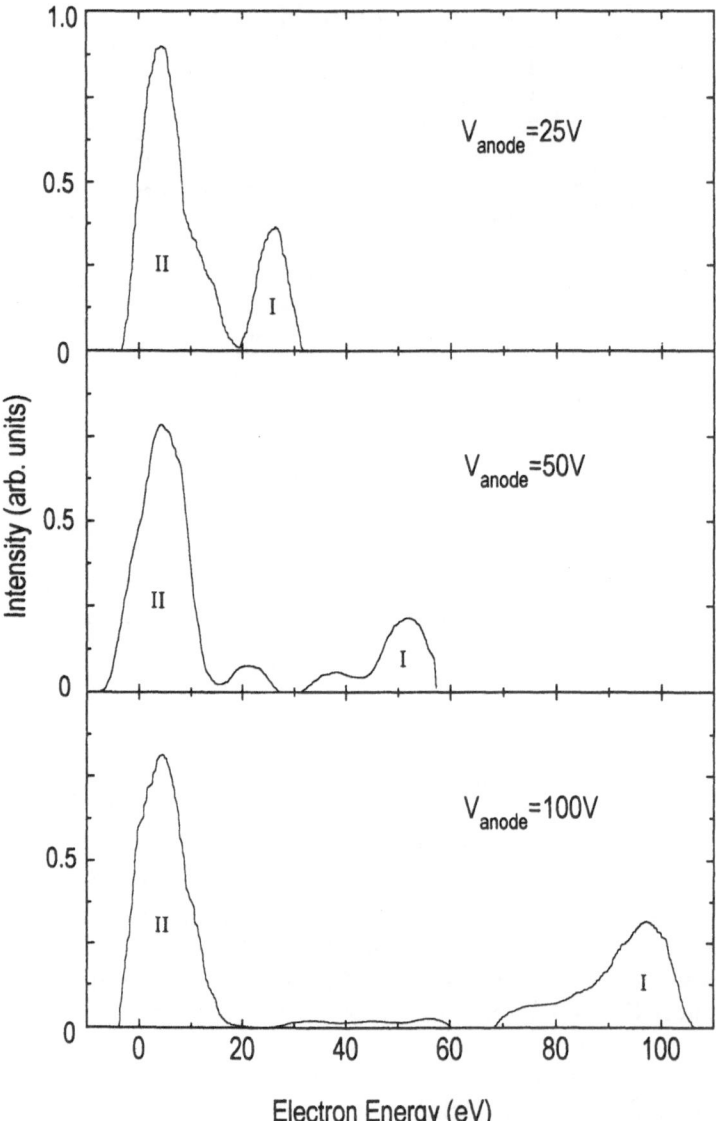

Figure 8: Ejected electron energy spectra for three anode voltages: 25, 50, and 100 volts. The two features referred to in the text are significant and the rest are noise.

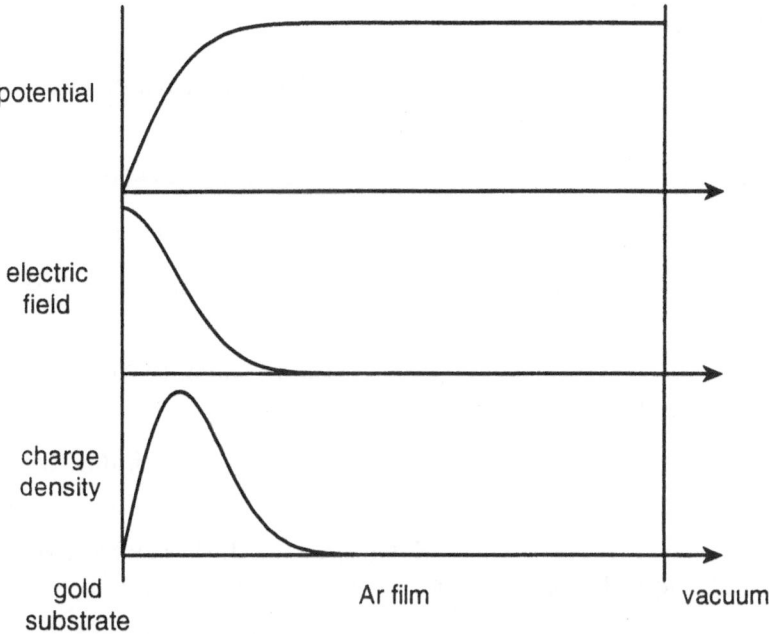

Figure 9: Schematic diagrams of electric potential, electric field, and hole concentration as indicated by the ejected electron energy spectrum using Poisson's equation.

CONCLUSIONS

When examining the electron emission from ion bombarded Ar, we found that field emission from the substrate stabilizes electron emission from thin insulators, and that field emission with impact ionizations causes dielectric breakdown in thick insulators. Thus, electrodeless breakdown can be triggered by ionizing radiation in the presence of an external electric field. Current work is focused on measuring the time evolution of luminescence and electron ejection when the Ar film is first exposed to the ion beam (before charging up) while the electric field is building. Measuring the time evolution of electron emission and luminescence is essential to clarify this picture. By determining the time constants involved in charging up the film and in luminescing, we hope to understand the process of electric field development and address the hysteresis effect. The question of luminescence efficiency could be better addressed by separating the electric field effects from the initial ionizations of the incident ion.

Acknowledgements

This work is supported by the National Science Foundation through grant DMR-9121272. We acknowledge stimulating discussions with R. E. Johnson, W. L. Brown, M. Shur, B. Gelmont, N. J. Sack, and C. T. Reimann. We also thank N. J. Sack for his help in setting up the experiment and M. Westley for his help in data collection.

References

[1] N. Schwentner, E. E. Koch, and J. Jortner, *Electronic Excitations in Condensed Rare Gases* (Springer-Verlag, Berlin, 1985).

[2] R. E. Johnson, M. Inokuti, Nucl. Instr. and Meth. **206** 289 (1983).

[3] C. T. Riemann, W. L. Brown, and R. E. Johnson, Phys. Rev. B, **37**, 1455 (1988).

[4] F. Coletti, J. M. Debever, and G. Zimmerer, J. Phys. (Paris) Lett. **45** L467 (1984).

[5] T. Kloiber, W. Laasch, G. Zimmerer, F. Coletti, and J. M. Debever, Europhys. Lett. **7** 77 (1988).

[6] P. Feulner, T. Muller, A. Poschmann, and D. Menzel, Phys. Rev. Lett. **59** 791 (1987).

[7] C. T. Reimann, W. L. Brown, D. E. Grosjean, M. J. Nowadowski, W. T. Buller, S. T. Cui, and R. E. Johnson, Nucl. Instr. Methods Phys. Res. B **58** 404 (1991).

[8] W. T. Buller and R. E. Johnson, Phys. Rev. B, **43** 6118 (1991).

[9] C. T. Reimann, W. L. Brown, M. J. Nowakowski, S. T. Cui, and R. E. Johnson in *Desorption Induced by Electronic Transitions DIET IV*, edited by G. Betz and P. Varga (Springer, Berlin, 1990), p. 226.

[10] D. J. O'Shaughnessy, J. W. Boring, S. Cui, and R. E. Johnson, Phys. Rev. Lett. **61** 1635 (1988).

[11] W. L. Brown, C. T. Reimann, and R. E. Johnson, Nucl. Instr. Methods Phys. Res. B **20** 9 (1987).

[12] T. Doke, A. Hitachi, S. Kubota, A. Nakamoto, and T. Takahashi, Nucl. Instr. and Meth. bf 134 353 (1976).

[13] R. A. Baragiola, Proceedings of the 9th International Workshop on Inelastic Ion-Surface Collisions, Nucl. Intr. Meth. (in press).

[14] I. Arakawa, M. Takahashi, and K. Takeuchi, J. Vac. Technol. A **7**, 2090 (1989).

[15] I. Arakawa, M. Sakurai, in *Desorption Induced by Electronic Transitions DIET IV*, edited by G. Betz and P. Varga (Springer, Berlin, 1990), p. 246.

[16] R. Pedrys, D. J. Oostra, A. Haring, A. E. deVries, and J. Schou, Nucl. Instrum. Methods Phys. Res. **33**, 840 (1988).

[17] M. K. Klein and J. A. Venables, *Rare Gas Solids* (Academic Press, London, 1977).

[18] E. Gullikson, Phys. Rev. B, **37**, 7904 (1988).

[19] E. E. Huber, Jr., D. A. Emmons, and Robert M. Lerner, Optics Comm. **11** 155 (1974).

[20] D. E. Grosjean, R. E. Johnson, R. A. Baragiola, and W. L. Brown, to be published.

[21] Z. Ophir, B. Raz, J. Jortner, V. Saile, N. Schwentner, E. Koch, M. Skibowski, and W. Steinmann, J. Chem. Phys. **62** 650 (1975).

[22] R. L. Ramey *Physical Electronics*, (Wadsworth Publishing Co., Belmont, California, 1961), p. 133.

[23] H. D. Hagstrum in *Inelastic Ion-Surface Collisions*, edited by N. H. Tolk, J. C. Tully, W. Heiland, and C. W. White (Academic Press, London, 1976), p. 1.

THE CONSEQUENCES OF ELECTRONIC EXCITATION

IN SEMICONDUCTORS AND FROZEN GASES

Walter L. Brown

AT&T Bell Laboratories
600 Mountain Avenue
Murray Hill, New Jersey 07974

INTRODUCTION

Ionization of semiconductors and insulators by heavy particles, electrons or photons always produces a transient in electrical conductivity as a result of the generation of electrons and holes in the solids. These transients are often of major importance. For example, they are the basis of silicon solid state particle detectors[1]. In some cases the transients may have very long time constants because of trapping of holes or electrons at defects in the solid which delays the process of electron-hole recombination. In many cases when the transient is complete the properties of the solid are exactly the same as they were before the ionization event. Silicon excited by individual low energy electrons or photons is such a case. However, in some cases there is a permanent change in the material as a consequence of the ionization. In these cases some of the energy of ionization has been transferred into atomic motion in the solid with a consequent change in its atomic structure.

Associated with the passage of a heavy ion through or into a solid, or even of a sufficiently energetic electron or photon, in addition to the generation of electron-hole pairs there may be direct momentum transferring collisions to the atoms of the solid which result in atomic displacements and permanent atomic rearrangements in the solid. Distinguishing the permanent effects of ionization from the permanent effects of momentum-transferring collisions can often be accomplished by changing the mass or velocity of the incident particle.

This paper will examine the effects of ionization in the semiconductors Si, GaAs and InGaAsP and the frozen gases Ar, N_2, and CO. In the case of Si, in particular, the effects of direct atomic displacements will be shown to be of dominant importance when the incident particles are MeV heavy ions even though these particles produce very high densities of ionization along their paths.

Ionization of Solids by Heavy Particles, Edited by
R.A. Baragiola, Plenum Press, New York, 1993

SILICON

Crystallization of amorphous silicon is a process that occurs thermally, but it may also be subject to "assistance" by bombardment with heavy particles, visible light and x-rays. This particular process has been selected as an example of an atomic transformation that may be influenced by ionization of the solid. Figure 1 illustrates the typical geometry in which experiments on this process have been carried out. An amorphous silicon layer a few thousand Angstroms to a few micrometers in thickness is prepared on a single crystal silicon substrate. This structure is conveniently produced by heavy ion bombardment, creating defects which accumulate with increasing ion fluence until the defective crystal becomes amorphous, but it also can be produced by deposition of silicon on a single crystal substrate under sufficiently clean conditions.

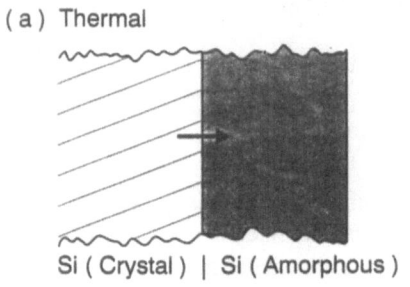

Si (Crystal) | Si (Amorphous)

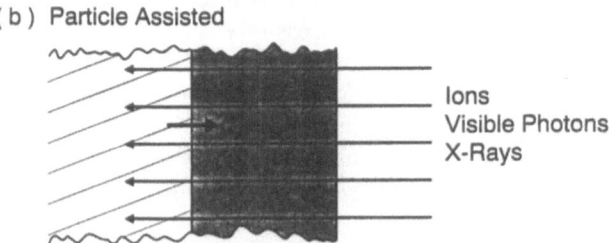

Figure 1. Schematic illustration of the solid phase epitaxial crystallization of a thin amorphous Si layer on a crystalline Si substrate: (a) Thermal only, (b) Particle assisted with ions, visible photons or x-rays.

The amorphous layer crystallizes epitaxially on the crystalline substrate in a solid phase transformation. The rate of movement of the interface is a strong function of temperature as shown in Figure 2[2]. Over the nearly ten orders of magnitude of the figure the data follow an activation energy of 2.7 eV. Note that the rate of transformation below 500 C is extremely slow. The inset in the figure shows the energetics of the transformation. The free energy of the crystalline state is, of course, lower than that of the amorphous, favoring the transformation, although the energy gained is only about 0.12 eV/atom. The 2.7 eV activation energy is a barrier to the transformation. Amorphous silicon is still four-fold coordinated (as in the crystal) and the nearest neighbor distances are the same, but the bonds are tangled and the bonding angles are

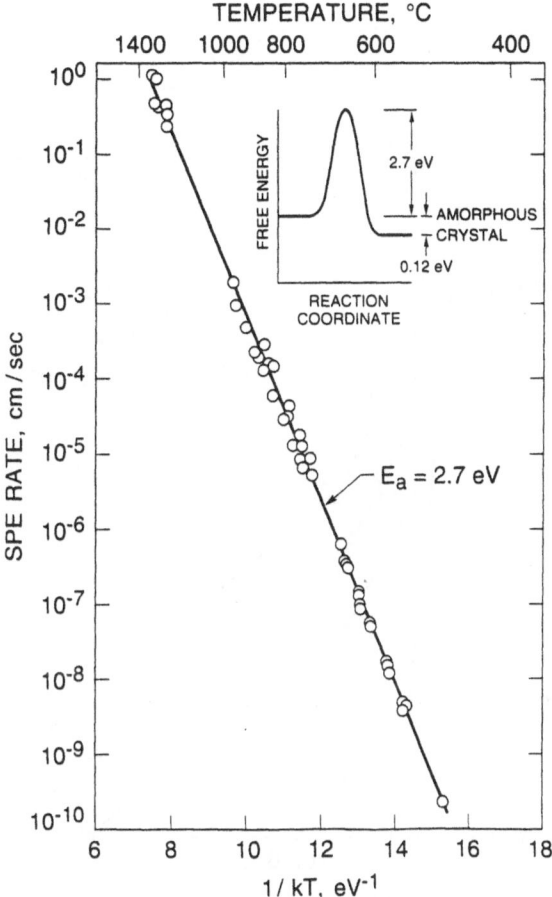

Figure 2. Rate of thermal solid phase epitaxial crystallization of amorphous Si (after Olson, et al, reference 2) as in Figure 1a. The inset shows the free energy of the amorphous and crystalline phases, which differ by 0.12 eV/atom, and are separated by a barrier of 2.7 eV, the activation energy of the straight line in the figure.

distorted. To "fix" the tangle it is necessary to break bonds and to reform them. That's what takes the energy. An ionization event in silicon removes a bonding electron, putting it into the non-bonding conduction band of the solid. Hence it might be anticipated that ionization would assist the crystallization process. Ionization can be produced by light (with above band-gap photon energy ~1.1 eV) or by x-rays or by ionizing particle bombardment.

Heavy Particles

Bombardment by heavy ions allows solid phase epitaxial growth to be carried out at temperatures far below 500 C. Results for bombardment with MeV neon ions are shown in Figure 3[3,4]. The amorphous layers in these experiments were much thinner than the range of the neon ions so that the ions all pass through the amorphous/crystalline interface, as indicated in Figure 1b, creating ionization and also displacing atoms in

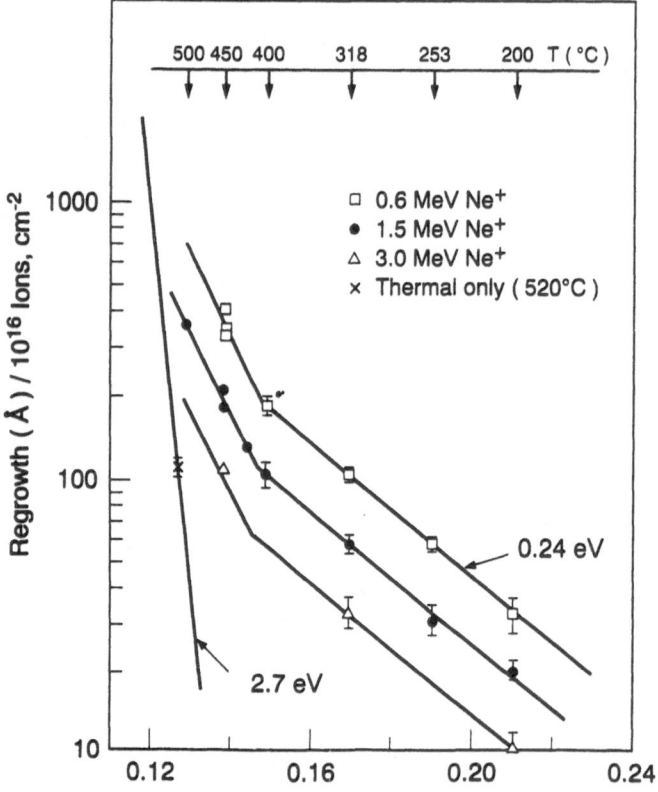

Figure 3. Solid phase epitaxial crystallization of amorphous Si as in Figure 1b, assisted by Ne ion bombardment at three different energies. The solid phase regrowth is proportional to the fluence of Ne ions but independent of ion flux below 400 C. A line for thermal-only solid phase epitaxy indicates what its rate would be for the typical flux used in the Ne-assisted experiments. The apparent activation energy of the ion-assisted growth is .24 eV rather than 2.7 eV for the thermal only case (after Williams, et al. reference 3).

direct momentum-transferring collisions. At temperatures of 400 C or below, the regrowth is found to depend linearly on the fluence of ions and not on their flux. The ordinate of the figure is thus given in regrowth thickness per unit fluence. The situation at temperatures above 400 C is more complicated and will not be discussed here. The thermal regrowth line (from Figure 2) is plotted on Figure 3 at a position appropriate to the particle flux used in obtaining the data of the figure. The apparent activation energy for the crystallization has been dramatically reduced from 2.7 to 0.24 eV. One is led to conclude that the major energy requirement for breaking bonds has been provided by the ion beam. This does not, however, allow one to conclude that the growth assistance is due to ionization.

Figure 3 shows results for three different neon ion energies. It is important to recognize how the ionization and direct momentum transferring energy transfers change with neon ion energy in this range. Figure 4 illustrates the situation schematically. All three of the neon energies are well above the maximum of the direct momentum-transferring energy loss processes. All three are well below the maximum in the ionization energy loss. The values for both processes for 0.6 and 3.0 MeV neon are indicated on the figure. By reference to Figure 3, it becomes clear that the decrease in regrowth rate with increasing neon energy is inconsistent with the increase in ionization with increasing energy, but is at least qualitatively consistent with the decrease in direct momentum-transferring energy loss with increasing energy. Quantitatively, Figure 5 shows that the data of Figure 3 are very well accounted for by a linear dependence on the direct momentum transferring processes, "nuclear stopping". Even though much more energy goes into ionization than into direct displacement of atoms, the latter is what makes the difference.

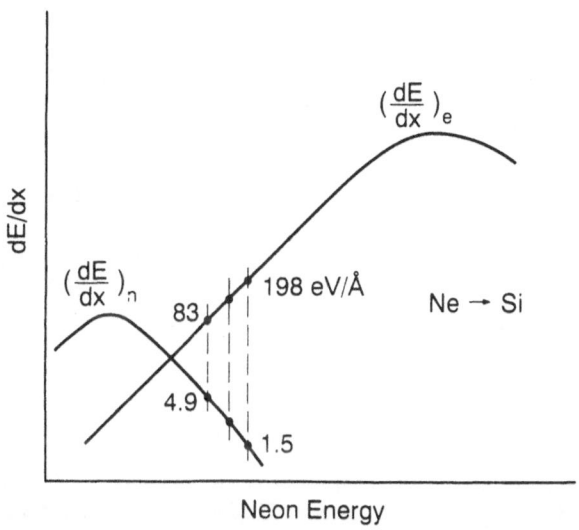

Figure 4. Energy dependence of the electronic $(dE/dx)_e$ and nuclear $(dE/dx)_n$ stopping powers of neon ions in the energy range of the experiments of Figure 3. The "nuclear" stopping power involves direct momentum-transferring collisions between the incident ions and the nuclei of the solid. The values on the curves are for 0.6 and 3.0 MeV and indicate that the electronic stopping power increases and the nuclear stopping power decreases with increasing neon energy in this energy interval.

Visible Light

Silicon can be ionized with photons of energy greater than about 1.1 eV, each photon giving rise to creation of a hole-electron pair. The consequences of intense ionization with pulses of light from lasers was the subject of wide-spread investigation in the late 1970's and early 1980's[5]. There was the possibility that with such excitation, atomic processes in the semiconductor could be modified. The issue of crystallization of amorphous silicon was a topic of particular interest. In considering this question, it is useful to compare the time scales of various electronic processes in silicon. If a laser pulse at 532 nm (a Nd:YAG laser at its doubled frequency) with an energy density of 0.2 J/cm^2 is incident on silicon, it creates approximately 6×10^{22} electron-hole pairs per cm^3. The inset of Figure 6 illustrates this initiating photon absorption event and electron-electron, electron-phonon and Auger processes which follow it. Figure 6 itself shows the characteristic times for these processes[6,7]. The Auger process, a three body process in which an electron and hole recombine, giving their energy to another electron, is a particularly strong function of the electron density. Among the processes shown,

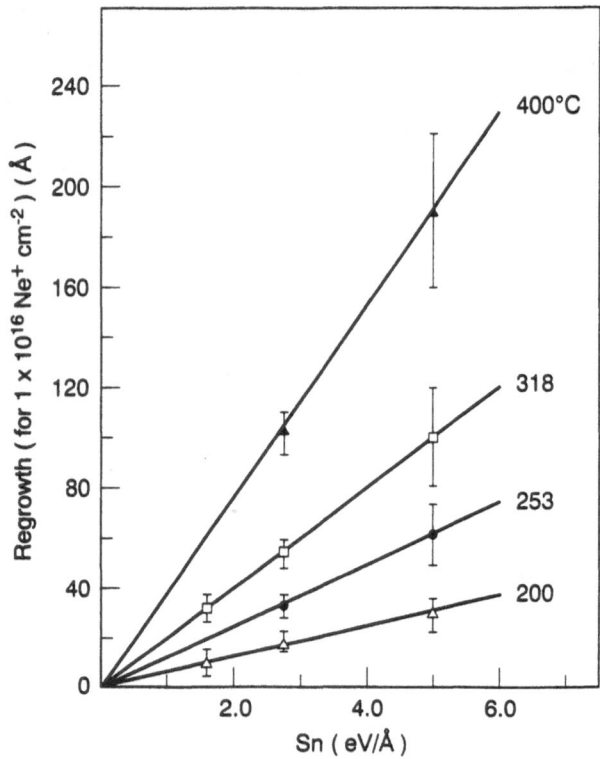

Figure 5. The variation of the neon ion-assisted regrowth (per 10^{16} ions cm^{-2}) at different temperatures plotted as a function of the nuclear stopping power. The data are from Figure 3 (after Williams, et al. reference 3).

only the electron-phonon process transfers energy to the lattice. The other processes are exchanging energy within the electronic system. However at high electron densities the Auger process is a very fast path for converting the band gap energy into heating electrons and keeping the electron-phonon process busy. The dashed portion of the phonon line in Figure 6 suggests that at very high electron densities there may even be a phonon bottleneck in the cooling process[7].

The time scales in Figure 6 lead to the conclusion that if the postulated 0.2 J/cm^2 initiating laser pulse were a picosecond in duration, most of the energy in the electronic system would have flowed to the lattice in 10^{-11} sec as heat. If the laser pulse were nanoseconds in length, although the same total number of electron-hole pairs would have been created, the density of electron-hole pairs present at any time would be much less because of Auger recombination and electron phonon relaxation. Based on the times in

Figure 6, in a nanosecond-long pulse the electron density cannot exceed 10^{20} cm^{-3}. The atomic density of silicon is 5×10^{22} atoms cm^{-3}. Considering each hole-electron pair as a broken bond, nanosecond laser pulses provide a maximum of 0.2% broken bonds that can possibly enhance recrystallization of amorphous silicon for nanoseconds of time. If the pulse length is extended to extend the time of potential enhancement the total energy density put into the system rises accordingly and melting will occur. The question thus comes down to whether there is an experimental regime in which ionization can make a difference that can be observed before melting.

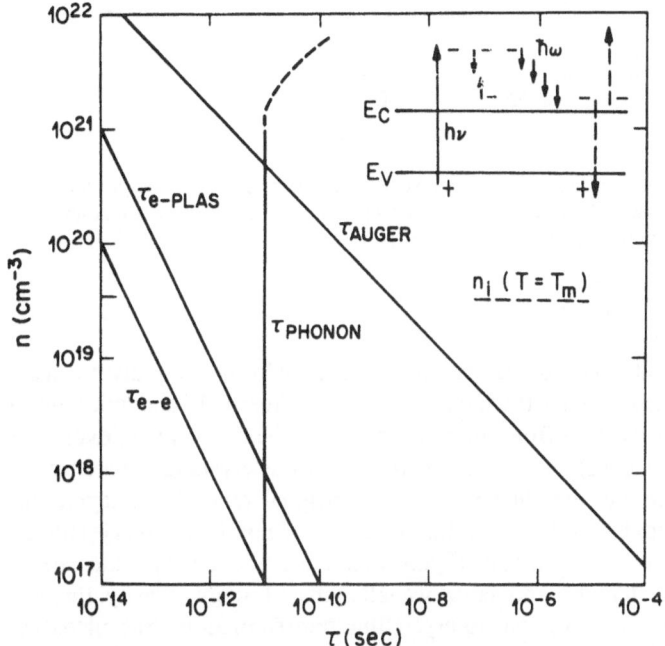

Figure 6. Characteristic times for electronic decay processes in Si as a function of electron density. The inset illustrates (from left to right) the creation of hole-electron pairs by photon absorption, electron-electron scattering, phonon emission, and Auger recombination. The dashed line marked $n_i(T=Tm)$ is the intrinsic equilibrium carrier concentration in Si at its melting point (after Brown, reference 6 and Yoffa, reference 7).

Figure 7 illustrates a typical experiment carried out to examine this question[8,9]. The inset shows the setup. High density electronic excitation is produced by the pulsed Nd:YAG laser. A cw helium neon laser measures the time dependent reflectivity as a monitor of the changes in the material. The figure shows the time dependent reflectivity for the excitation conditions of the inset. The reflectivity starts at the value of amorphous silicon and rises very rapidly, in a time shorter than the pulse length, to a value equal to that of molten silicon. The dip that follows the initial rise is due to the reduced reflectivity of silicon heated above its melting point while the laser pulse continues to supply energy. The dip is followed by a long plateau at the reflectivity of silicon at its melting point, and then by a fall to the reflectivity of crystalline silicon. The length of the plateau is a function of the energy density in the laser pulse and the explanation of the

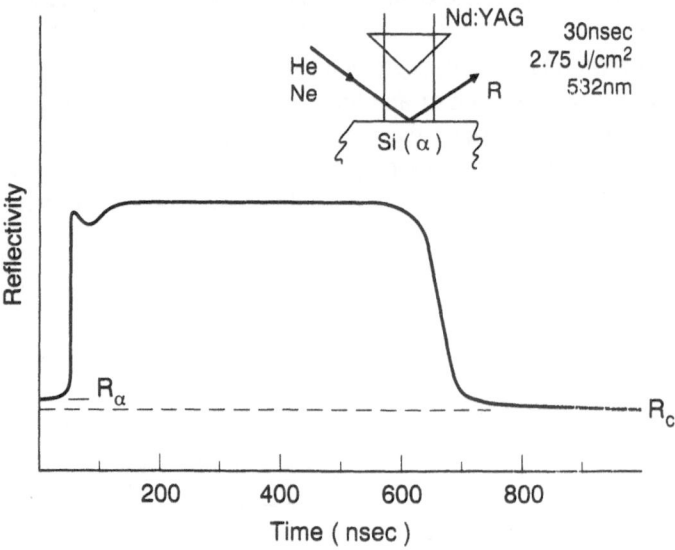

Figure 7. Time resolved reflectivity from Si before, during and following a Q-switched Nd:YAG glass laser pulse at 532 nm of 30 nsec duration and 2.75 J/cm2. The reflectivity is measured at 633 nm as shown in the inset (after Auston, et al., references 8 and 9).

result is straight forward. Laser energy is transformed rapidly to heat and melts the amorphous silicon. After the laser pulse is over, thermal conduction into the bulk of the silicon extracts the heat from the hot surface (in the case shown, over a period of several hundred nanoseconds) and the silicon resolidifies as crystalline material. If the maximum melt depth exceeds the thickness of the original amorphous layer, the recrystallized material is epitaxial on the crystalline substrate; if not, it is polycrystalline.

Experiments such as that of Figure 7 have been done for a wide variety of input pulse energies, pulse lengths and laser wavelengths. Energy density thresholds have been established for the amorphous to crystalline transformation. Sub-picosecond pulses have verified the picosecond time scale expected for energy to flow from the electronic to the lattice system. In all cases the phenomena seem to be completely explicable in terms of hot electron-hole plasmas and melting without evidence that the electrons and holes have a direct effect on the amorphous to crystalline transformation. Many other types of experiments have been done as well, all of them leading to the same conclusion. It is interesting to note that the highest solid phase amorphous/crystalline interface velocities in Figure 2 were produced using laser excitation below the melting point and there is no evidence for a new phenomenon in the consistent activation energy the data show. Nevertheless, the question of the influence of very high electron-hole densities continues to be explored, now with laser pulses and time resolved measurements in the femtosecond regime[10].

X-Rays

X-ray stimulation of solid phase epitaxy of amorphous silicon on crystalline silicon has recently been reported[11]. The x-rays were a broad energy spectrum from the synchroton at Tsukuba, Japan. The amorphous silicon layer in this case was approximately 3000 A thick, produced by plasma chemical vapor deposition on single crystal silicon. It contained about 10 atomic percent hydrogen. The initial synchrotron irradiation was for 72 hours in high vacuum at a power density of about 24 W/cm^2 and a temperature close to 20 C. A 5mmx5mm square was irradiated, defined by a mask. The authors estimate that about 10 photons were absorbed per atom during the 72 hour exposure. The irradiation produced no visible difference in the exposed region compared with the unexposed area around it. However, six months after the irradiation, the sample was annealed at 600 C for an hour and an obvious difference appeared as shown in Figure 8a. SEM's of the irradiated and unirradiated regions are shown in Figure 8b,c respectively. The much larger number of crystallites in part b suggests that the irradiation influenced the nucleation of crystal growth (which did not take place until the high temperature anneal). Since it is not reasonable to assume the "latent image" effect of the irradiation was retained electronically for six months, the authors conclude that the irradiation produced a structural change - not visible until after "development" by the high temperature anneal. The variety of orientations of the crystallites shows that the crystallized material is not epitaxial with the substrate, but is polycrystalline in both irradiated and unirradiated regions. This is probably a result of an insufficiently clean interface between the amorphous film and the underlying crystalline material and perhaps also to the hydrogen in the film which impedes crystallization.

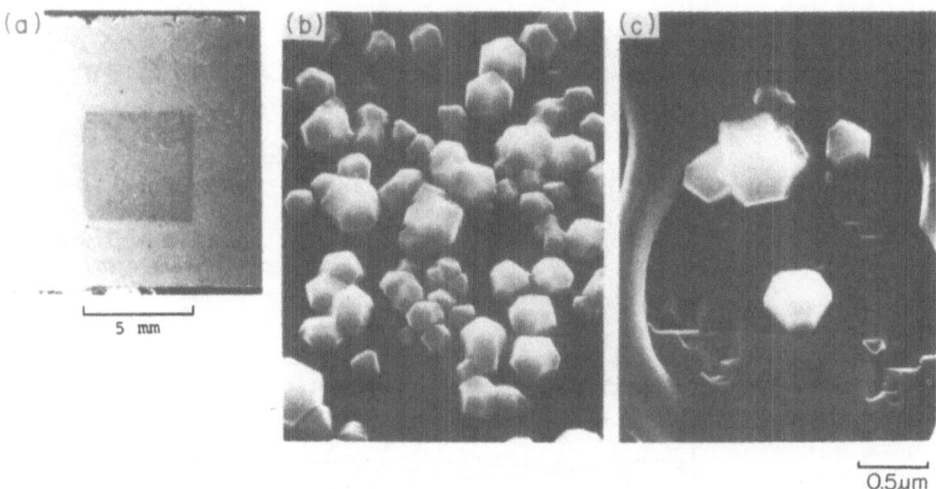

Figure 8. Effect or irradiation with synchrotron radiation on crystallization of plasma chemical vapor deposited amorphous silicon annealed at 600 C for one hour after the nominal room temperature irradiation. The sample was 300 nm thick and contained about 10 atomic percent hydrogen. In the optical micrograph (a), the irradiated 5mm x 5mm square at the center of the sample is clearly visible. Parts (b) and (c) are SEM's of the irradiated and non irradiated parts, respectively, as shown in (a). Irradiation has apparently increased the density of nucleation sites for silicon crystallite formation (after Sato, reference 11).

Raman measurements were made of the annealed material with the results shown in Figure 9[11]. The abscissa of the figure is the distance across the film, in a direction perpendicular to the orbital plane of the synchrotron, as illustrated in the inset to the figure. The ordinate is the ratio of the Raman intensities measured at a Raman wave number shift of 521 cm^{-1}, the peak of the Raman line for single crystal silicon. It is thus a measure of the "quality" of the material. The vertical intensity variation of x-rays in a synchrotron fan changes with x-ray energy. The dashed lines in Figure 9 show calculated variations at four different energies. Comparison of the variation in the Raman intensity across the 5mm of the exposed region with the calculated x-ray intensity variations leads the authors to conclude that x-rays of more than 3 keV must have been responsible for the material change observed.

Following these intriguing results, further experiments were carried out using higher intensity x-rays from a 54 pole wiggler at the synchrotron and inserting a beryllium absorber in the beam to cut out x-rays with energies less than about 3 keV. The setup is shown schematically in Figure 10. In this case the amorphous silicon that was irradiated was deposited in a different CVD process to avoid incorporation of hydrogen. The x-ray power was about 300 watts/cm^2 and although samples were mounted on a water cooled block their temperature rose to estimated temperatures of several hundred degrees C in about ten seconds. The actual temperature was difficult to determine.

Figure 11a shows Raman intensities for samples exposed to different fluences of photons and normalized to the intensity from dislocation free Si. The absorbed flux was about 3×10^{14}/cm^2 sec, so the exposures were much longer than the time for establishing

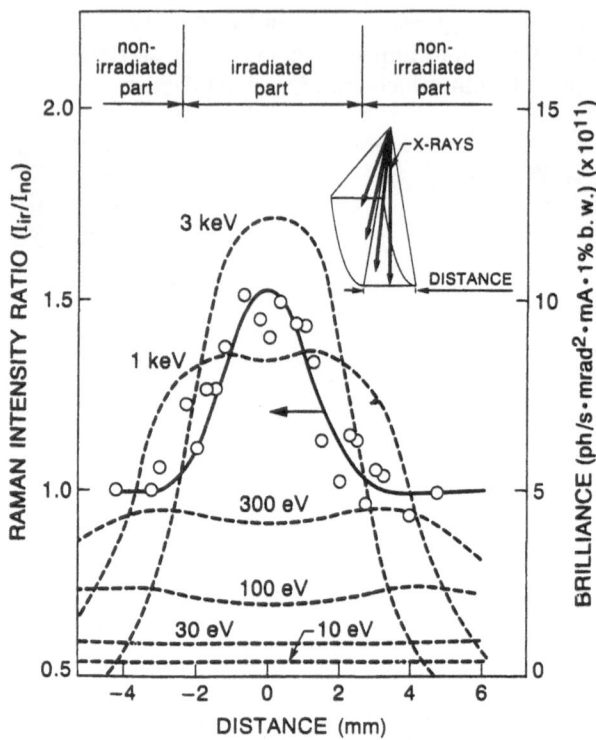

Figure 9. Comparison between crystallization of the sample of Figure 8 and the synchrotron intensity distribution in the direction perpendicular to the plane of the synchrotron. The inset shows this geometry. The zero on the abscissa scale of the figure is on the mid-plane of the fan of synchrotron radiation. The intensity distribution is more strongly peaked at the mid-plane for higher energy x-rays. The data points in the figure are the ratio of the Raman intensity measured at 521 cm-1 (the peak of the Raman line for single crystal silicon) for different positions across the square of Figure 8a in the distance direction (after Sato, reference 11).

a thermal steady state. The Raman intensities for similar amorphous layers crystallized thermally are shown in Figure 11b. Comparison of parts a and b of Figure 11 indicate that the x-ray crystallized material is of better quality than that thermally grown even at temperatures of 700 C. In contrast to the situation shown in Figure 8, the layers in both cases were epitaxial on the crystalline substrate, but TEM showed lower dislocation densities in the material crystallized with x-ray assistance[11].

The data of Figure 11 seem to rule out the conclusion that the x-rays are simply providing a very expensive heat source, although the lack of knowledge of the actual

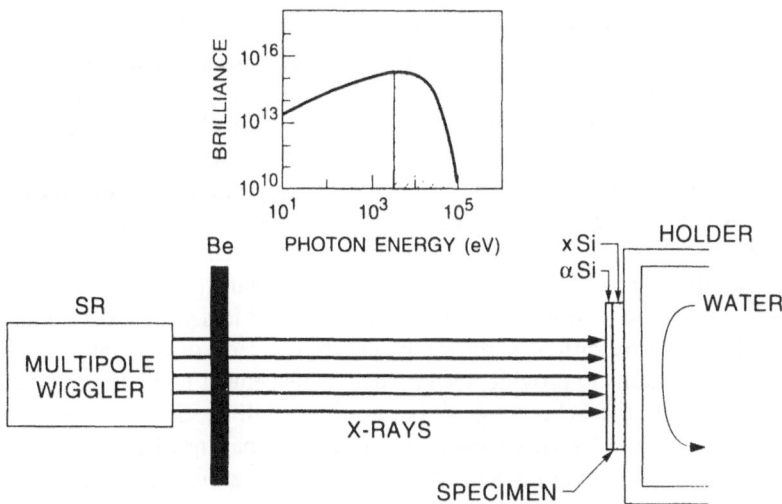

Figure 10. Setup using a multipole wiggler on the synchrotron for higher x-ray intensity and a beryllium absorber to remove x-rays with energies less than about 3 keV. The sample was a 300 nm film of pure amorphous Si (without hydrogen) deposited with low pressure CVD on a single crystal silicon substrate. The mounting block was water cooled. The x-ray intensity, about 300 W/cm^2, heated the sample to several hundred degrees C (after Sato, reference 11).

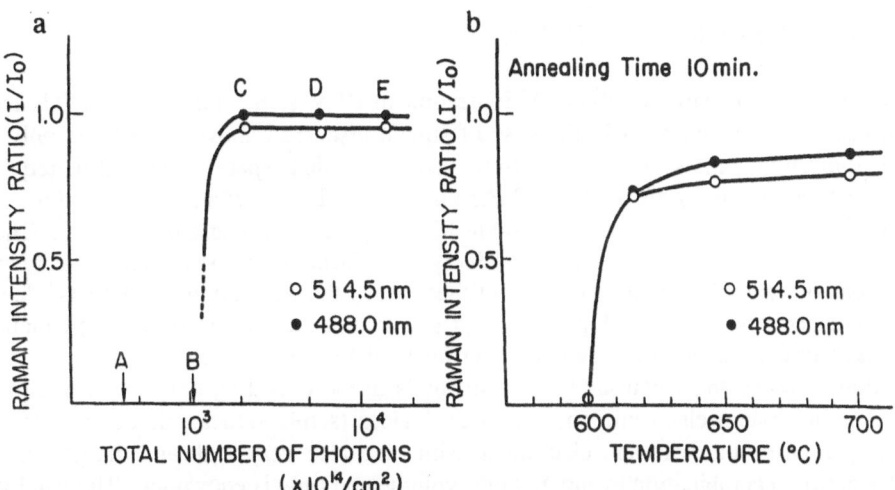

Figure 11. Comparison of the quality of crystalline Si grown by solid phase epitaxy with x-ray assistance with that grown in a thermal-only process. The crystalline quality is determined from the Raman intensity at 521 cm-1 of each sample, normalized to the intensity for dislocation-free single crystal Si. In (a) samples A, B, C, D and E were exposed (in the setup of Figure 10) for 2, 5, 10, 30 and 60 minutes respectively at approximately equal x-ray intensities. The abscissa is the total number of x-rays absorbed in the 300 nm layer of originally amorphous Si. Raman measurements were made at 514.5 and 488.0 nm wavelengths. In (b) similar measurements of crystalline quality are shown for samples annealed for 10 minutes in a furnace at different temperatures.

temperature in the experiments is a weakness in them. The authors propose an ingenious explanation for their results. The x-ray ionization energy for the K-shell of Si is 1.8 keV. Following the formation of a hole in the K-shell, a cascade of Auger processes will result in multiply charging Si atoms to charge states of 2, 3, 4 and even more[12]. The authors suggest that ionization is transferred to neighboring atoms and by the coulomb repulsion among these near neighbors an atom is displaced, forming a vacancy which assists the crystallization of the amorphous phase (analagous to the role of defects formed by momentum-transferring collions of heavy ions discussed above). Coulomb repulsion effects have never been observed in silicon, even in the tracks of very high energy heavy ions, presumably because of the rapid screening of any charge by the electrons of the material. It is not clear why the situation should be different in this case. It is interesting to note that a Si^{+4} ion has lost all of its bonding electrons and so has a chance to reform them in some different and less tangled configuration than it had originally. However, unless the atom moves before the bonds reform it is hard to see why the bonding should be different. It may be sufficient for the coulomb repulsion of the transiently localized charged cluster of atoms to provide a very small displacement to an atom or atoms of the cluster in order for the rebonding to be more crystalline than it was before. The formation of a vacancy by an energetic displacement may not be required. However, whether even a small atomic movement can take place before the repelling charges are screened by conduction electrons is uncertain.

It seems important to consider the relationship of these x-ray results to the effect of energetic electron bombardment (but with energies below the momentum-transferring displacement threshold energy of about 100 keV for silicon). Such electrons are also quite effective in producing K-shell holes. In any case, these results are the first which suggest that electronic processes in silicon can produce permanent material changes as neither MeV heavy ions nor visible light have done.

GROUP III-V SEMICONDUCTORS

The permanent atomic effects of ionization in III-V semiconductor materials have been much less elusive than in the case of silicon discussed above. In silicon only K-shell ionization seems to be effective, and even that special case still requires confirmation. In materials such as GaAs and InGaAsP the effects of ionization very clearly influence the reliability and usable life of devices formed from them[13]. These materials, of course, also experience "damage" effects from momentum transferring collisions of heavy ion bombardment and these mask the generation of "defects" due to electronic excitation from such particles, for example in the process of ion implantation. The most direct, unequivocal, relevant and unavoidable form of ionization is that from injection of holes and electrons by p-n junctions in the operating devices. The current densities in modern electronic and photonic devices (semiconductor diodes and lasers) are very high so the hole and electron densities are also very high and the number of hole-electron recombination events per unit volume and time is enormous. The band gap energy that is stored in a hole-electron pair is the order of 2 eV in these materials and in some fraction of recombination events a part of that energy is transferred to atomic motion. To illustrate these effects, three examples are discussed below.

GaAs IMPATT Diodes

Figure 12(a) shows the cross section of a GaAs IMPATT diode which has a platinum Schottky barrier contact to n-type GaAs. High temperature accelerated aging tests promote the diffusion and reaction of Pt with the GaAs to form PtAs and free Ga, which appears at the surface, and to produce defects which diffuse into the GaAs. The activation energy for the process is found to be about 1.6 eV[14]. Such an activation energy, extrapolated to an operating temperature of 225 C implies a mean device life of 10^6 hours. However, when the structure is tested under operating conditions of reverse bias across the Schottky barrier, the results are quite different. In operation there is a high current density flowing through the device as a result of avalanche multiplication of carriers. Under sufficient reverse bias, "hot" carriers are created in the high electric field of the Schottky junction with sufficient energy to generate additional carriers by impact ionization of GaAs.

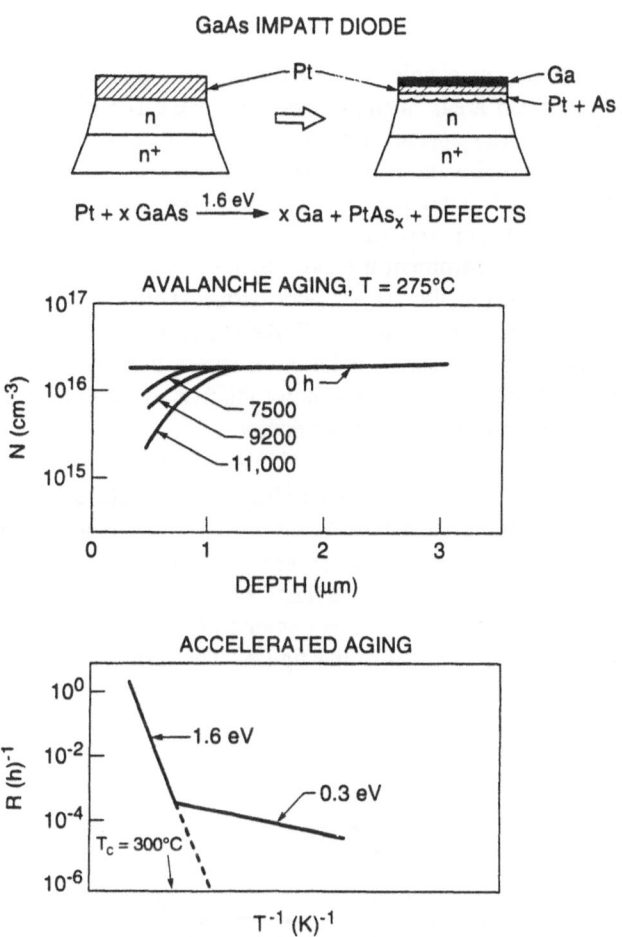

Figure 12. Schematic summary of the recombination enhanced failure mechanism in GaAs IMPATT diodes: thermally driven interface reaction between Pt and GaAs with 1.6 eV activation energy; compensation of donors by in-diffusing defects under avalanche aging conditions; failure rate vs 1/T under avalanche aging conditions illustrating a low activation energy failure mechanism at low temperatures (after Ballamy, et al reference 14).

Figure 12(b) shows the material changes that take place with avalanche aging at 275 C. The plot is of the effective donor concentration as a function of depth from the Schottky contact at different times. There is a progressive depletion of the original $2 \times 10^{16} \, cm^{-3}$ doping concentration as defects from the junction diffuse a micron or more into the GaAs. A plot of the accelerated aging (Figure 12c) against 1/temperature now shows two regions of very different activation energy. The 1.6 eV measured without avalanche operation is still dominant at high temperatures, but a much lower activation energy process has been added at temperatures below about 300 C. This new process has been identified as recombination enhanced diffusion of defects. The electronic energy that is available from electron-hole recombination is sometimes released in such a way as to contribute to the local vibrational amplitude of lattice atoms and hence to enhance diffusion. This electronic to vibrational conversion is particularly efficient when recombination takes place at defects as a result of the electronic localization they afford[13].

GaAs Lasers

A major degradation mechanism in GaAs-GaAlAs double heterostructure lasers appears as nonluminescent areas in the devices that have come to be called "dark line defects" or DLD[15,16]. DLD appear and move under operating conditions of the laser diodes. The structure of such a diode is shown in Figure 13. The active region is the 0.7 um thick GaAs layer which is bounded by p- and n-type GaAlAs layers to produce the double heterostructure. Defects introduced into the material on either side of the central 20 um width by proton bombardment increase the resistivity of those regions and restrict the current (and the lasing) to the 20 um wide stripe. Under device operation infrared electroluminescence is observed from the GaAs active region and as the operation is continued, DLD move into the active volume. Because they quench the luminescence, these DLD's are serving as localized non-radiative recombination centers which progressive degrade the optical output of the device.

By indexing the positions of the DLD's as they appear in the infrared electroluminescence observations and by careful TEM sample preparation, it has been possible to identify the defect structure of the DLD's. In all areas which exhibited a DLD, a network of dislocations was observed[16]. Similar networks are not present in other regions of the lasers. Figure 14 is a TEM of one of these dislocation networks[16]. An originating dislocation, marked D, is at one end of the network of dislocations which radiate down and to the right from it in the figure. The propagation of the networks has

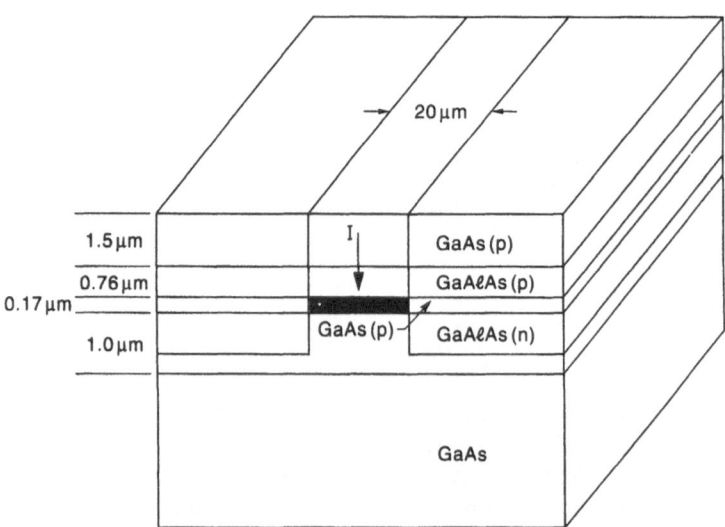

Figure 13. Schematic of a GaAs laser structure grown by liquid phase epitaxy. The 20 um width of the current carrying region is defined by proton bombardment of the adjacent regions to increase their resistivity. Light emission is from the GaAs region shown black (after Petroff and Hartman, reference 16).

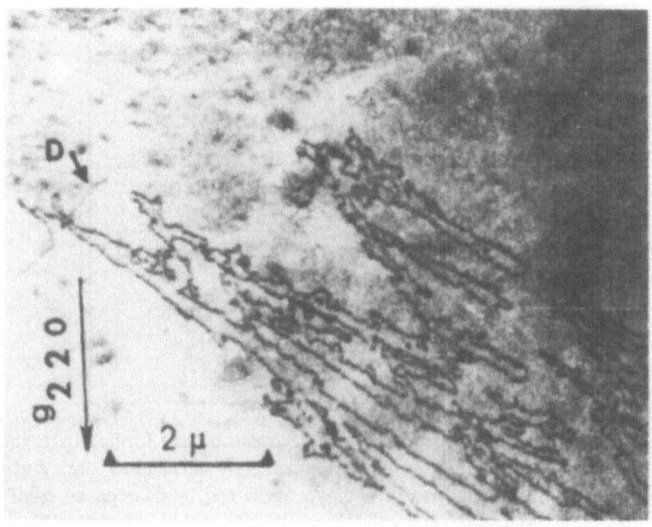

Figure 14. Bright field electron micrograph showing the dislocation network that forms a dark line defect in the laser stripe of Figure 13. The dislocation, marked D, crosses the two GaAlAs layers and the GaAs laser stripe (after Petroff and Hartman, reference 16).

been shown to be due to dislocation climb that takes place by the absorption and emission of point defects at the dislocation core. The climb, and hence the defects, are induced by the operation of the devices.

As in the case of the IMPATT diode discussed above, the ionization of the material is present as high densities of holes and electrons injected into the active region by the p-n junction operating a high current density. Many of these holes and electrons recombine radiatively, producing desired laser light, but some also recombine non-radiatively to provide a localized driving force for defect motion and possibly even for point defect creation at kinks in an existing dislocation[17]. The proton bombardment used in limiting the width of the active device stripe is not responsible for the DLD since DLD also form in non-proton irradiated structures.

InGaAsP Heterostructure Lasers

The structure of one type of InGaAsP/InP 1.3 um laser is shown schematically in cross section in Figure 15[18]. The active layer is the crescent of InGaAsP grown in a crystallographically etched V-groove in InP by liquid phase epitaxy. The crescent is about 2.2 um wide and 0.11 um thick in the center. The direction of current flow in the laser device is indicated. As in the IMPATT and GaAs lasers discussed in the preceeding sections, the device operates at a high current density. In accelerated studies of device degradation in this case it was approximately $4 \times 10^4 \text{A/cm}^2$. At 100 C and this current density there is a gradual loss of light output. Or for a constant light output there is a gradual increase in the operating current and in the threshold current for laser action.

Examination of the electroluminescent image of the active region again reveals dark line defects that encroach from the side walls of the V-groove into the active crescent. A TEM study of degraded devices provides the structural evidence for the degradation. Figure 16 is a micrograph together with its schematic representation[18]. An array of interstitial dislocation loops forms along the [111] V-groove sidewall interfaces with a density as high as 6×10^9 cm-2. The generation of the loops is believed to be a result of imperfections at the sidewall interfaces that provide interstitials that precipitate into

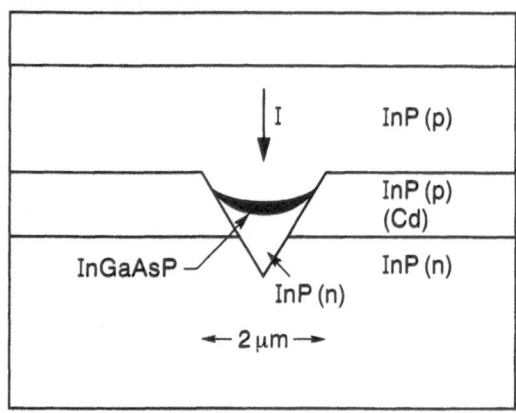

Figure 15. Schematic cross section of an InGaAsP heterostructure laser formed by liquid phase epitaxy. The active InGaAsP crescent layer and the InP layers immediately adjacent to it are grown in a V-groove crystallographically etched into a Cd doped p-InP layer previously grown on an n-InP substrate. The direction of current flow in the laser is shown by the arrow (after Chu et al. reference 18).

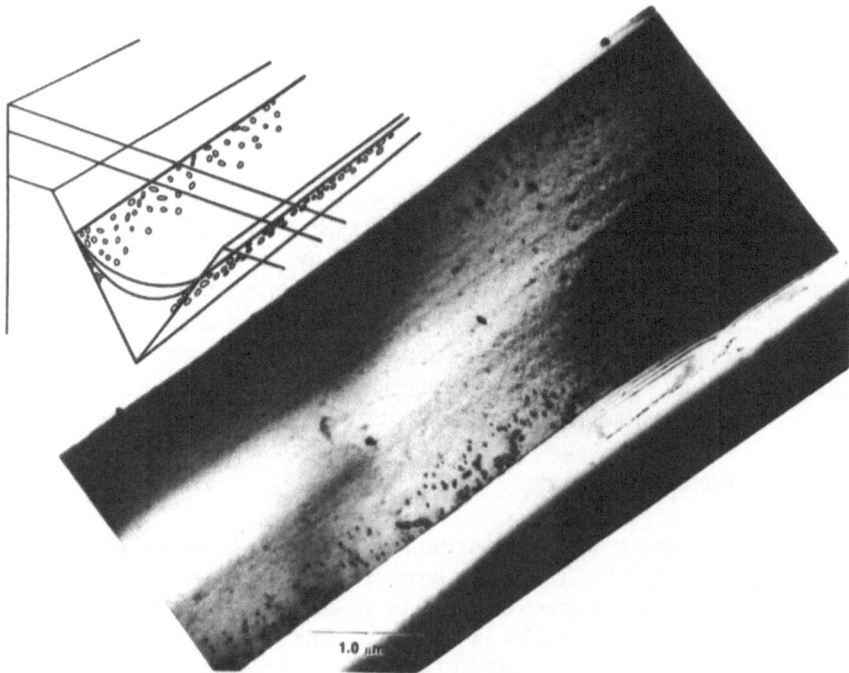

Figure 16. Top view TEM of the active InGaAsP crescent stripe of Figure 15 (shown in the inset). There is a uniform density of dislocation loops along the side walls of the V-groove adjacent to the stripe (after Chu et al. reference 18).

loops. If the loops remained at the sidewall interfaces they would not result in major laser performance degradation, but they don't. Under the influence of the current they migrate into the active crescent and grow rapidly in it. The nonradiative recombination of holes and electrons provides the stimulus for their enhanced diffusion. In a structure of this type and at the current densities at which it is operated, there are more than 10^{27}

recombination events/cm^3 sec taking place. With this enormous flow of energy density from the electronic system, even an inefficient process of conversion of that energy into atomic motion can have major consequences over time.

RARE GAS SOLIDS

Electronic excitation in solidified rare gases is transferred into atomic motion with much more dramatic consequences than in the semiconductors[19]. This is primarily due to the enormous contrast in the energy scales in these two classes of materials. In the rare

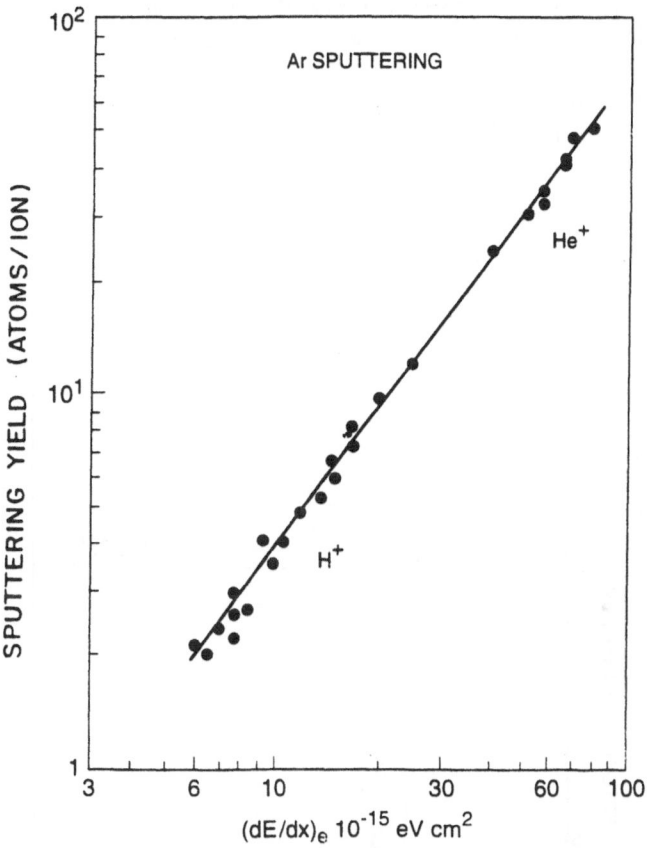

Figure 17. The sputtering yield of solid argon (argon atoms ejected per incident ion) at 12 K by MeV He and H ions as a function of the electronic stopping power of the incident ions. The dependence is nearly linear (after Reimann et al. reference 21).

gas solids the electronic energy of hole-electron (ion-electron) pairs is much higher than for any of the semiconductors discussed above (all of the rare gas solids are big band gap insulators) and in addition, the energy required for atomic motion is much lower. For solid argon the energy of a hole-electron pair is approximately 14 eV while the sublimation energy of the solid is only 0.06 eV[20]. Thus only a very small fraction of the total energy available through recombination needs to be transferred to atomic motion for major effects to occur. In contrast, for silicon the hole-electron pair energy is 1.1 eV and the sublimation energy is 3.8 eV and for the III-V semiconductors the band gap is 1.3 to 2.0 eV and the sublimation energy 4 to 3 eV. In these cases the electronic energy has to

411

be coupled much more subtly to atomic motion and, as we have seen above, the effects are observed only after prolonged periods of excitation.

One major effect of electronic to atomic energy transformation in the rare gas solids is sputtering - the ejection of atoms from the surface of a solid. Figure 17 shows the dependence of the sputtering yield (atoms ejected/incident ion) for MeV He and H ions at normal incidence on solid argon at approximately 12 K[21]. The abscissa is the electronic energy loss of the incident ion. Changes in the abscissa are made by changing the energy of the ion. The two groups of points are for H^+ and He^+, the highest stopping power in each group being for the lowest energy of that particle. The sputtering yield is nearly proportional to the electronic energy input to the solid near its surface. The yield is independent of the flux of ions, indicating that the excitation produced by each ion is acting independently. By examining the dependence of the sputtering yield on the argon film thickness it is possible to extract an effective depth (distance into the film) within which electronic excitation can lead to ejection from the surface. This distance is about 200 Å.

The energy conversion process that accounts very well for the observations is illustrated in Figure 18[21,22]. An incident MeV ion produces a track of individual ionization events:

$$Ar \rightarrow Ar^+ + e^-.$$

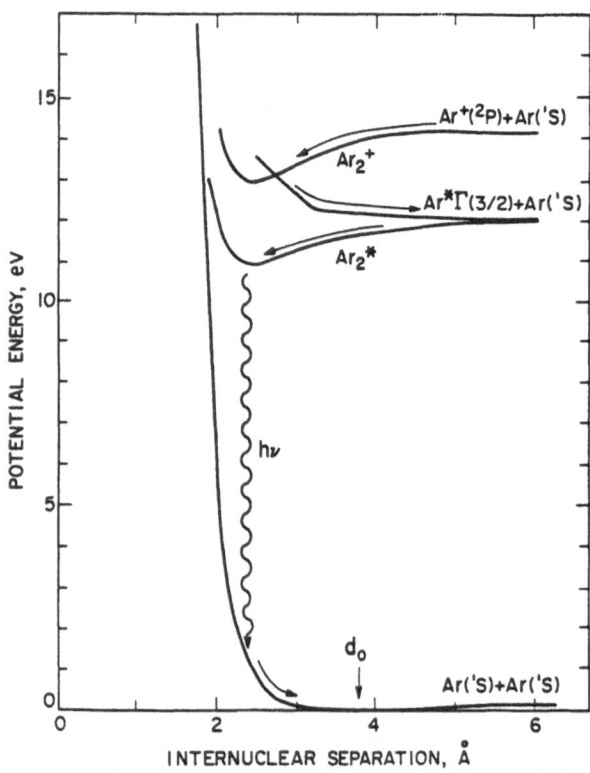

Figure 18. Solid-state internuclear potential energy diagram showing principal states of electronic excitation in solid Ar: Ar^+, the atomic ion, which combines with a ground state Ar atom and relaxes toward Ar_2^+; dissociative recombination of Ar_2^+ with an electron to produce Ar^*, an exciton, and a ground state Ar atom with kinetic energy; the relaxation of Ar^* in combination with a ground state Ar atom to form Ar_2^*, an excimer; radiative recombination of the relaxed excimer emitting a 9.8 eV photon and dissociating the two ground state Ar product atoms with kinetic energy (after Reimann et al. reference 21 and Johnson and Inokuti reference 22).

The Ar+'s are free to diffuse by charge exchange with their neighbors until they associate with a neutral Ar to form Ar_2^+. This dimer ion is no longer degenerate with its neighbors and is thus unable to diffuse. The 200 Å effective thickness of the ejection process is the diffusion length of an Ar^+. As the Ar_2^+ relaxes toward its equilibrium configuration by phonon emission it may capture an electron and dissociate:

$$Ar_2^+ + e^- \rightarrow Ar^* + Ar + KE$$

where Ar^* is an exciton. The Ar^* in turn associates with a neighboring Ar to form the excimer Ar_2^*. The equilibrium separation of this excimer is less than the equilibrium separation of ground state atoms. The excimer can only decay radiatively (with emission of a 9.8 eV photon) and when it does, the two atoms repel because of their too-close separation so that

$$Ar_2^* \rightarrow Ar + Ar + h\nu + KE.$$

The two little bursts of kinetic energy in this total decay sequence have energies ~ 1 eV so that if they occur in dimer ions or excimers close enough to the surface, atoms can be ejected from the surface, i.e. be sputtered.

There are a great many details of the electronically excited solid argon system that have been examined through optical emission as well as atomic ejection and the system has been excited by fast and slow electrons and by energy selected photons as well as by fast ions[21,23]. However, the most prominent electronic to atomic energy conversions seem to be quite well described by the sequence above. About 10% of the energy in the ion-electron pairs may appear in small bursts of kinetic energy of atomic motion and leads to easily measurable sputtering yields.

MOLECULAR ICES

Solid CO and N_2, have sublimation energies similar to that of argon and ionization energies the order of 5 to 10 eV. They also give evidence of conversion of electronic energy to kinetic energy of atomic motion through sputtering. The dependence of sputtering yield on the electronic energy loss rate of an incident ion is interestingly different from the case of argon. The results for N_2 are shown in Figure 19[24]. In the middle of the stopping power range of the figure the sputtering yield bends up in a transition from a linear to quadratic dependence. Note that the sputtering yield in the linear regime is a factor of about four lower than for argon (Figure 17), indicating a less efficient energy conversion process in N_2 than in Ar. The transition to quadratic behavior suggests the onset of a cooperative process. The sputtering yield is still a single ion phenomenon (as in argon) - it is not dependent on ion flux - but the excitations along the path of a single ion begin to influence each other. The fact that argon did not show this effect is due to the diffusivity of Ar^+. Around each ion track the electronic excitations are much more widely spread for argon than for N_2 so that the interactions between excitations would be expected to occur at much higher electronic stopping powers for Ar than for N_2. Further evidence for this difference in the diffusion length of electronic excitation in N_2 and argon is that the sputtering yield of N_2 is independent of film thickness to thicknesses less than the 200 Å diffusion length of Ar^+ in Ar.

The explanation for the linear to quadratic transition in N_2 is given schematically in Figure 20[24,25]. In Figure 20a individual excitations are well separated along the ion track. The circle around each excitation is intended to indicate that some part of the electronic energy in each excitation is converted to atomic motion. If the conversion takes place close enough to the surface, atomic ejection may occur. As the density of

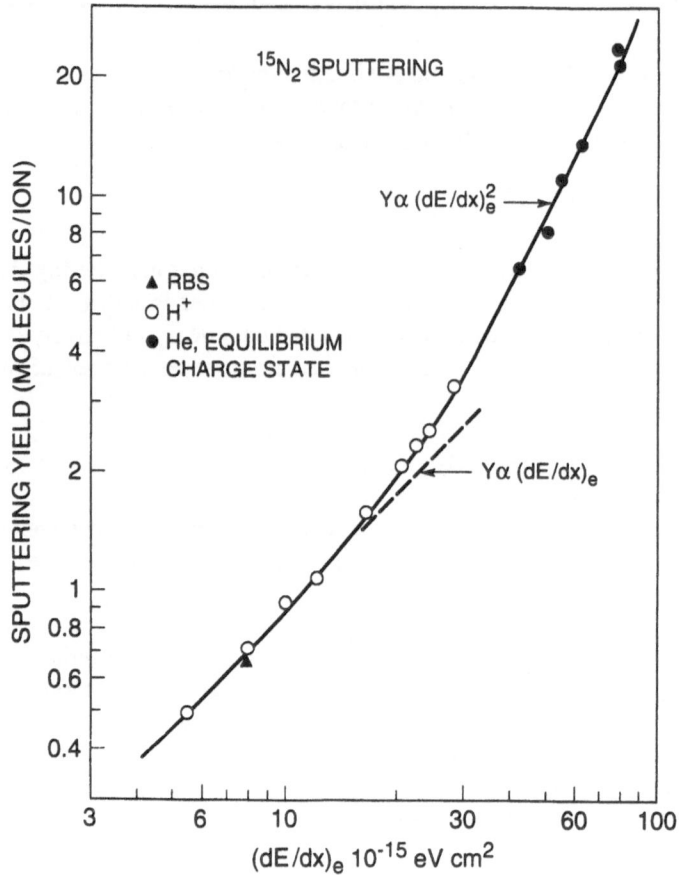

Figure 19. The sputtering yield of solid N_2 as a function of the electronic stopping power of incident MeV H and He ions (after Brown et al. reference 24)

electronic excitations along a track increases with increasing $(dE/dx)_e$ the situation approaches that of Figure 20b. Now the little energy conversions are no longer independent. They overlap, their individual energy inputs add up and because the probability of atomic ejection is nonlinearly related to the local energy density, the sputtering yield becomes quadratic.

The results for CO are different from those for both N_2 and Ar, as illustrated in Figure 21[24,26]. Over the entire range of electronic stopping powers measured the sputtering yield is quadratic in the stopping power. There is no evidence of a linear region at the low stopping power end of the data. It is possible that the localization of the electronic excitation around the track of an ion is even tighter for CO than it is for N_2. Measurements of sputtering yield vs film thickness have not provided a quantitative measure of the diffusion length of electronic excitation (ions) in the CO film. If the diffusion length is very short, or if the magnitude of the kinetic energy transfers from the electronic system are large, the overlap of neighboring kinetic energy bursts such as in Figure 20b will occur at lower stopping power in CO than it does in N_2. In this case the linear to quadratic transition in CO might be below the lowest stopping power examined. The fact that the absolute magnitude of the sputtering yield for CO is only slightly larger than that for N_2 at the lowest stopping powers does not support the idea that the energy transfers are larger, for if they were and they are in the quadratic regime, the sputtering

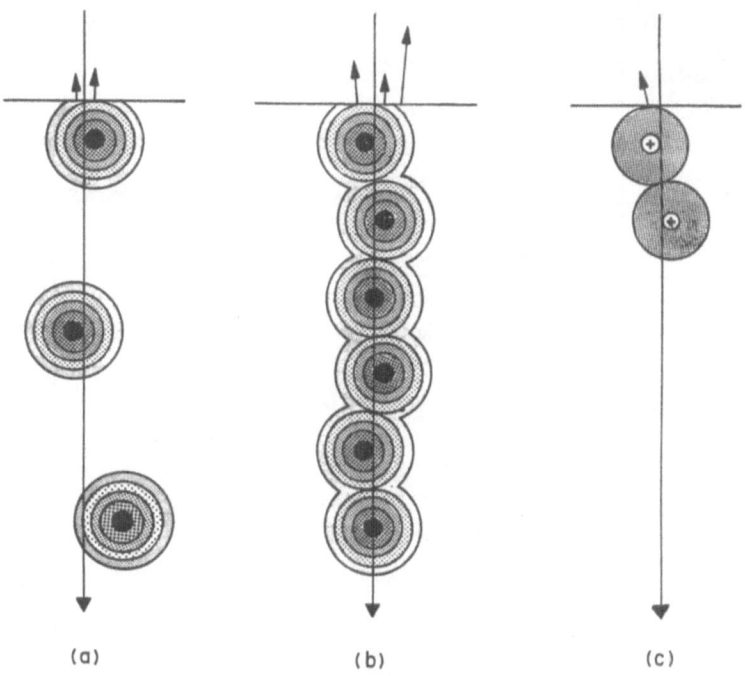

Figure 20. Pictorial representation of (a) the spherical spread of kinetic energy released at the site of individual electronic-to-atomic energy conversions along the path of an incident ion; (b) the overlap of the kinetic energy releases along the path of an ion as the electronic stopping power increases, increasing the density of excitation events; (c) a possible two-molecule electronic state involving the coupling of two independently ionized or excited molecules (after Brown et al. reference 24).

yield should be substantially larger for CO than for N_2 which is in the linear regime.

An alternative explanation for the observations depends on the assumption that there is an effective path for electronic to atomic energy conversion in CO which requires two independent excitation events on close neighbors[24]. Such a path leads to an intrinsically quadratic dependence of sputtering yield on stopping power. This explanation is, at best, quite ad hoc, since no electronic states have been identified which have this property. Figure 20c illustrates this possible mechanism. Even if such a two-excitation energy conversion route dominates the experimental results of Figure 21, it seems almost inconceivable that single excitation will not produce sputtering. Hence, at a low enough stopping power a linear dependence would still be expected. It would clearly be desirable to extend the CO measurements in a search for such a transition.

The molecular ices of O_2 and of H_2O fall one each into the classes of N_2 and CO presented above. In solid O_2 a transition from linear to quadratic behavior is observed[27]. It occurs a factor of about 2.5 lower in stopping power than the transition in N_2. These two cases have been discussed in the context of the model of Figure 20b which provides a satisfactory explanation for their behavior[25]. The case of H_2O, the first material in which electronic sputtering was studied in detail, remains a puzzle at least as perplexing as that of CO. The sputtering yield is quadratic to the lowest stopping powers at which it has been measured[28,29], and the yield is also much lower than for CO (or N_2). Of course, H_2O has a considerably higher sublimation energy than the N_2, O_2, CO and Ar group, and molecular fragmentation and new molecule formation may play an important

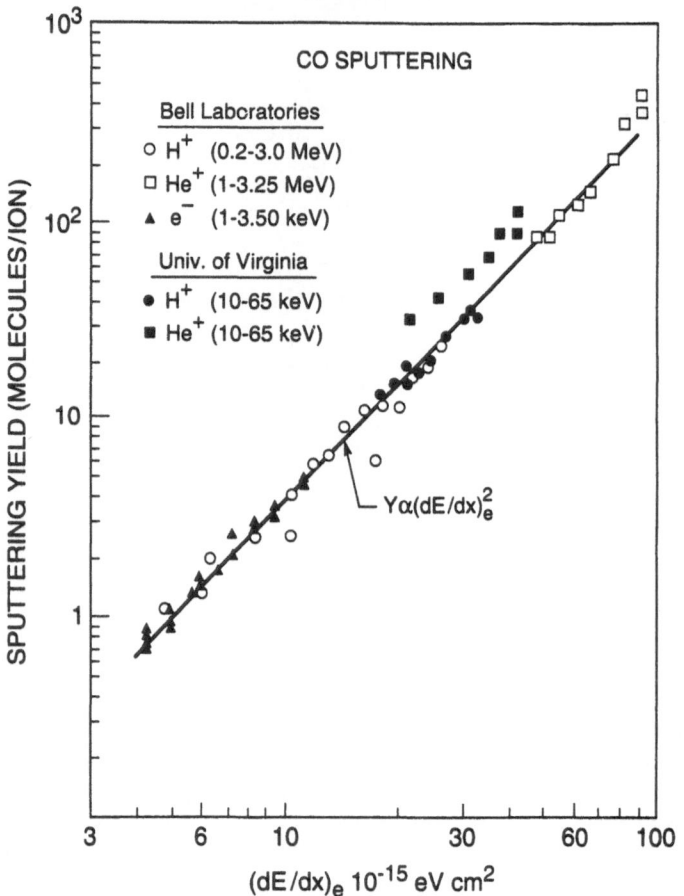

Figure 21. The sputtering yield of solid CO as a function of the electronic stopping power of MeV H and He ions and keV electrons (after Brown et al. reference 26).

part in the measured sputtering yield[30], although it is not thought to at very low temperatures. It is interesting to note that the persistent quadratic behavior is observed in two polyatomic molecules. The significance of this observation, if there is any, is not understood at this point.

SUMMARY

Electronic excitation in non-metallic solids can clearly be responsible for changes in the atomic structure of those solids. It is essential that the electronic excitation be localizable in significantly large units of energy - a property of all materials with an electronic band gap, but missing in metals. And, of course, it is essential that the transfer of energy from the electronic to the atomic system should be in units bigger than normal phonon energies. If (a) the energy available is large, (b) the energy transfer is efficient and (c) the energy required for atomic motion is small, the effects of electronic excitation are large. Such is the case of the rare gas and molecular gas solids. If the available energy is not large compared to the energy required for atomic motion, then special pathways for energy transfer and/or weak links in the atomic motion chain may still enable electronic excitation to alter the solid. Such is the case for the semiconductors. For them the fraction of electronic excitations that contribute to atomic motion is extremely small, but even that small fraction can be of tremendous importance in modern electronic and photonic devices.

REFERENCES

1. G. Dearnaley and D.C. Northrop. "Semiconductor Counters for Nuclear Radiation", E.&F.N. Spon Limited, London (1963).
2. G.L. Olson, S.A. Kokorowski, J.A. Roth, and L.D. Hess Mat. Res. Society Proceedings, 13:141 North-Holland, New York (1984).
3. J.S. Williams, R.G. Elliman, W.L. Brown, and T.E. Siedel, Phys. Rev. Lett. 55:1482 (1985).
4. W.L. Brown, R.G. Elliman, R.V. Knoell, A. Leiberich, J. Linnros, D.M. Maher, and J.S. Williams, p. 61 in "Microscopy of Semiconductor Materials", ed. A.G. Cullis, Institute of Physics, London (1987).
5. W.L. Brown, p. 337 in "Laser Materials Processing", M. Bass, ed., North-Holland, New York (1983).
6. W.L. Brown, p. 20 in "Laser and Electron Beam Processing of Materials", eds., C.W. White and P.S. Peercy, Academic Press, New York (1980).
7. E.J. Yoffa, p. 59 in "Laser and Electron Beam Processing of Materials", eds., C.W. White and P.S. Peercy, Academic Press, New York (1980).
8. D.H. Auston, C.M. Surko, T.N.C. Venkatesan, R.E. Slusher, and J.A. Golovchenko, Appl. Phys. Lett., 33:437 (1978).
9. D.H. Auston, J.A. Golovchenko, A.L. Simons, R.E. Slusher, P.R. Smith, C.M. Surko and T.N.C. Venkatesan, App. Phys. Lett., 34:635 (1979).
10. R.R. Freeman, private communications.
11. F. Sato, G. Katsuyuki, and J. Chikawa, Japanese J. of Appl. Phys., 30:205 (1991).
12. E. Shigemasa, K. Ueda and Y. Sato, Physica Scr., 41:67 (1990).
13. L.C. Kimerling, Rev. of Solid State Science, 4:335 (1990).
14. W.C. Ballamy and L.C. Kimerling, IEEE Trans. Electron Dev., 25:746 (1978).
15. D.V. Lang and L.C. Kimerling, Phys. Rev. Lett., 35:22 (1974).
16. P. Petroff and R.L. Hartman, Appl. Phys. Lett., 23:469 (1973).
17. L.C. Kimerling, Solid State Electronics, 21:1391 (1978).
18. S.N.G. Chu, S. Nakahara, M.E. Twigg, L.A. Koszi, E.J. Flynn, A.K. Chin, J. Appl. Phys, 63:611 (1988).
19. W.L. Brown and R.E. Johnson, Nucl. Inst. and Meth., 13:295 (1986).
20. M.L. Klein and J.A. Venables, "Rare Gas Solids", Academic, New York, (1977).
21. C.T. Reimann, W.L. Brown, and R.E. Johnson, Phys. Rev. B, 37:1455 (1988).
22. R.E. Johnson and M. Inokuti, Nucl. Instrum. Methods, 206:289 (1983).
23. C.T. Reimann, W.L. Brown, D.E. Grosjean, and M.J. Nowakowski, Phys. Rev. B, 45:43 (1992).
24. W.L. Brown, L.J. Lanzerotti, K.J. Marcantonio, R.E. Johnson, and C.T. Reimann, Nucl. Instrum. Methods B, 14:392 (1986).
25. R.E. Johnson, M. Pospieszalska, and W.L. Brown, Phys. Rev. B 44:7263 (1991).
26. W.L. Brown, W.M. Augustyniak, K.J. Marcantonio, E.H. Simmons, J.W. Boring, R.E. Johnson, and C.T. Reimann, Nucl. Instr. and Meth. B 5:304 (1984).
27. K. Gibbs, W.L. Brown, and R.E. Johnson, Phys. Rev. B 38:11 (1988).
28. W.L. Brown, W.M. Augustyniak, E. Brody, B. Cooper, L.J. Lanzerotti, A. Ramirez, R. Evatt, and R.E. Johnson, Nucl. Instr. and Meth., 170:321 (1980).
29. B.H. Cooper and T.A. Tombrello, Radiat. Eff., 80:203 (1984).
30. C.T. Reimann, J.W. Boring, R.E. Johnson, J.W. Garrett, K.R. Farmer, W.L. Brown, K.J. Marcantonio and W.M. Augustyniak, Surf. Sci., 147:227 (1984).

IONIZATION TRACKS

R.E. Johnson

Engineering Physics
Materials Science and Engineering
University of Virginia
Charlottesville, VA 22903

INTRODUCTION

'Tracks' produced in response to the ionization energy deposited by energetic particles penetrating matter have a long history in physics. Initially identified in bubble chambers and as etchable tracks in dielectrics,[1] altered regions called 'tracks' have more recently been seen in semiconductors and metals when very fast heavy ions are incident.[2,3] In addition to using etchable tracks to identify ions or to determine an ion's energy, tracks seen in materials collected from space have proven useful, for instance, for identifying their solar flare exposure ages.[4]

Here an overview is given of the nature of the processes leading to the formation of ionization tracks. By a 'track' we mean an altered (damaged), often cylindrical or extended, region in a material produced in response to the ionization energy deposited by fast ions. The initiating events and the resulting cascade of events, which complete the ionization stage, are briefly described. The relaxation of this excitation energy in the material is then discussed, noting differences between material types but placing emphasis on molecular insulators. The general nature of the equations for relaxation of and damage production by the 'hot plasma' associated with the high ionization density occurring in the core of the ionization 'track' is also discussed. The various models used in the literature to describe how electronic relaxation processes lead to track production are organized according to this description.

Examples of 'track' effects given in Table 1 are used to illustrate the physical processes discussed. Emphasis here is on the ejection of particles from the sample (electrons, ions, neutrals and photons) which is a surface manifestation of an ionization track.[5] In particular, a number of examples from the literature on electronically induced sputtering of condensed-gas and organic solids will be considered.[6,7]

PRIMARY IONIZATION

A fast ion incident on a solid produces electron-hole pairs (ionization), plasma excitations, and excitation of states within the band gap (when they exist).[8] These processes

Ionization of Solids by Heavy Particles, Edited by
R.A. Baragiola, Plenum Press, New York, 1993

are well studied except in the extreme case of highly charged, fast, heavy ions, which is discussed in other papers in this volume. Two often used quantities are obtained: (dJ/dx), the number of primary ionizations per unit path length, and $(dE/dx)_e$, the electronic energy loss per unit path length. The ratio of these quantities, $(dE/dx)_e/(dJ/dx)$, is the mean energy deposited per primary ionization produced; it amounts to $\approx$ 20-100 eV in typical insulators and increases slowly with increasing velocity.[9] Because this energy is above the electron-hole pair formation energy it is understandable that the kinetic energy of secondary electrons and, possibly, plasmons account for a significant fraction of the initial energy deposited.[8]

Table 1. Track Manifestations

Tracks in Dielectrics:	Etchable Tracks, Extended Defects
Biological Damage:	Inactivation of Molecular Species
Phase Changes	Amorphous - Crystalline, Crystalline -Amorphous, Volatile - Refractory, Insulator - Conductor, Adhesion, Solid-Gas.
Percolation 'Track':	Pathway for Irradiation-Produced Gases
Surface Ejection:	Secondary Electron,Secondary Ion, Electronic Sputtering, Luminescence.
Transients:	Track Surface Potential, Fast Ion Wakes

Aspects of the track dimensions are determined by the fact that the most energetic secondary electrons are produced in direct coulomb collisions (static screening radius ~ a_o) and that dipole excitations are the dominant ionization (electron-hole pair) formation process.[8,9] The latter determines a dynamic screening length, the so-called Bohr adiabatic radius, v/ω, where v is the incident ion velocity and $\hbar\omega$ a mean 'ionization' energy. The details of the radial track structure are determined by the kinetic energy spectrum of the secondary electrons, a topic of this conference. In addition, it is known that the ejection of secondary electrons from the surface region and the forward-directedness of the electrons produced by the primary ionization process leads to charge depletion where the track intersects the surface and, hence, to a transient out-of-the-surface electric field[10-12] which is another manifestation of the 'track' forming process (Table 1).

SECONDARY EVENTS

Whereas the primary excitations, occurring in times $< 10^{-16}$ sec, can, in principle be reasonably well described, the secondary process are less tractable, but are, of course, important in the so-called cascade contribution to secondary electron ejection.[8] Basically, the secondary electrons produced move through the material according to their initial energy distribution, the scattering (transport) properties of the material, and the density of electrons produced.[13] In addition, at high excitation densities, the positive charge (holes) can produce a space charge, limiting the motion of the electrons.[13] The energetic electrons ejected by

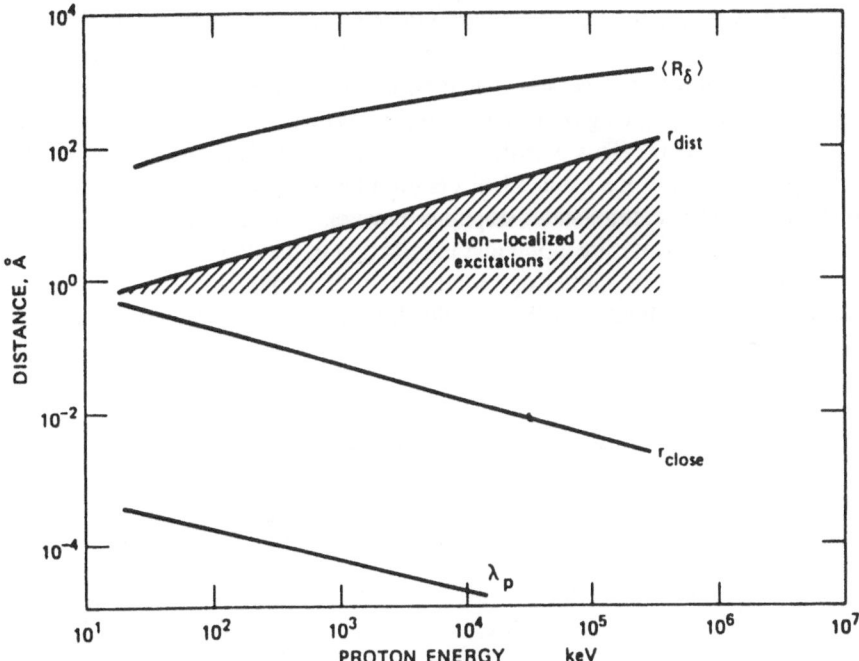

Figure 1. Radial Track structure vs ion energy measured from the ion's path: $<R_\delta>$ is the mean radial distance for penetration of delta rays (secondary electrons), approximate dimension of outer track (ultratrack); r_{dist} is the Bohr adiabatic radius, approximate maximum radius for primary excitations; r_{close} is the distance of closest approach of an ion to an electron in a lead-on collision; λ_p is the deBroglie wavelength of the proton (from reference 13).

the incident ion produce additional electron-hole pairs and excitations, with the fastest electrons determining the extent of the track as indicated in Figure 1.

There are a number of calculations[14] of the net spatial distribution of the excitation density deposited in the primary and secondary processes, and measurements in gases generally confirm the overall structure obtained. That is, the deposited energy distribution decays slowly out to about $r \approx v/\omega$, the inner tract (so-called infratrack), and then decays as $\sim r^{-2}$, due to the coulomb nature of the ejection of fast electrons and the cylindrical symmetry, forming the outer track (so-called ultra track).[13] A useful quantity is the W value, the energy deposited per electron hole pair (ionization) produced.[15] The ratio of W/I [where I is the ionization (band-gap) energy] exhibits differences between species according to whether or not there are many low lying states, as there are for molecular insulators and gases: e.g. W/I ~ 2.5 for molecular gases and ~ 1.7 for atomic gases. These differences mean that different fractions of energy are stored in electron-hole pair formation after the electrons cool, depending on the material.

In determining materials alterations leading to track production care needs to be taken as the hole transport properties differ significantly from material to material. That is, based on electron cooling rates in a number of materials and hole mobilities (in all cases the conduction electrons have high mobilities), it is found that significant hole transport can occur in some materials during the time over which the secondary electrons cool.[16] Therefore, although a calculation of the initial radial distribution correctly shows where the energy is first deposited, it does not necessarily represent an energy density profile at any later time. For molecular and ionic insulators, however, such profiles are representative, as hole diffusion is relatively slow[17] and electron cooling very fast due to vibrational excitation.

That the track structure changes with ion velocity, v, is indicated by the fraction of the energy deposited within v/ω or ℓ (the average lattice spacing), whichever is larger (Figure 1). For comparison we show in Figure 2 the fraction of the electronic energy deposited in the track <u>core</u> (region of dense excitation) which contributes to electronically induced sputtering of H_2O. This is extracted in reference 18 from the sputtering yield data for low temperature water ice by incident helium ions using a particular sputtering model and using equilibrium charge-state stopping powers. Although this result is model dependent, it clearly indicates the point that the <u>fraction</u> of the deposited energy contributing to material alteration (here sputtering) increases with decreasing ion velocity. As this sputtering process is non-linear in $(dE/dx)_e$, this fraction of energy is presumed to represent the energy in the track core.

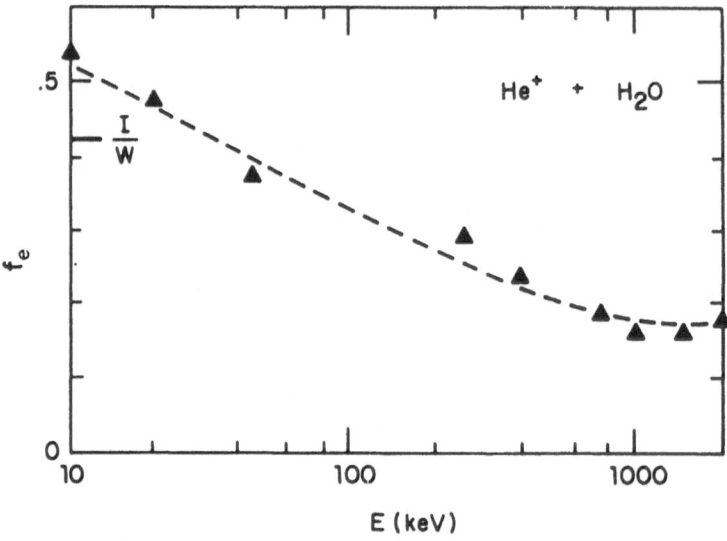

Figure 2. Fraction of the energy deposited in electronic excitations and ionizations in water ice at 77K by an incident He^+ which contributes to sputtering of water molecules, plotted vs. ion energy. Triangles data line drawn to guide eye. (I/W) is ratio of gas-phase ionization energy and W value (from reference 18).

OUTER TRACK: EXCITONIC REGION

Below some critical energy density an electron-hole pair 'decays' in the material as an individual event, as it does when it is excited by a single photon.[19] At such energy densities the behavior of individual excitons and holes determines the nature of the material alteration. That is, the holes (excitons) diffuse and eventually trap at defects or surfaces (e.g. vacuum, substrate) or, possibly, 'self-trap' if the material has a very low defect density. Once trapped the more mobile but cool electrons can recombine with the holes either via a potential curve crossing, leading to repulsive decay and kinetic energy production, or by an Auger-like process ($e + e + A^+ \rightarrow e + A^*$) producing a hot electron. The trapping process often involves dimerization $A^* + A + M \rightarrow A_2^* + M$, a ubiquitous process, which occurs in materials as different as rare-gas[20,21] and organic-molecular solids. In addition, holes/excitons may fuse at traps (e.g. $A^* + A^* \rightarrow A^+ + A + e$).

In the outer track region or at low excitation densities damage can be produced by an energetic repulsive decay process. In a molecular solid both the initial excitations and these decays can lead to bond breaking. This in turn can initiate a chemical reaction.[18] Clear examples of this are biological damage, where the individual decay/reaction events are described by Katz and co-workers[22] as 'hits'. These hits are described as being Poisson distributed, when calculating the 'damage' cross section generally observed in irradiation of organic and biomolecular solids.[23] For these materials, the primary form of damage at low excitation density occurs due to bond breaking followed by cross linking between neighboring molecules.[24]

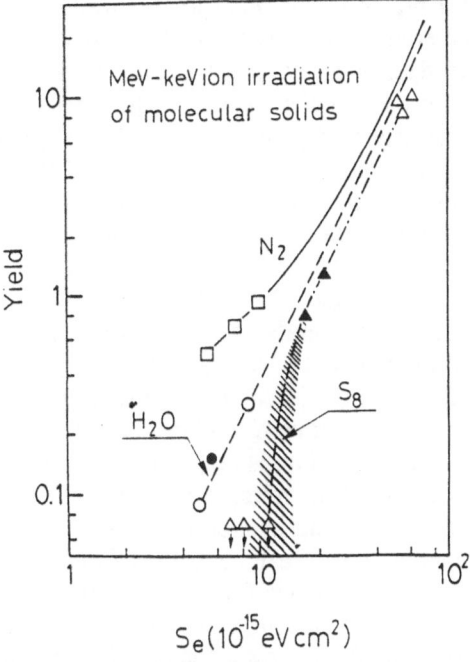

Figure 3. The erosion (sputtering) yield in <u>atoms</u> ejected (for water it is H_2O ejected) vs electronic stopping power per atom (H_2O). Different dependencies are seen at low values of S_e. The linear dependence for N_2 indicates ejection on a per excitation basis. The steeper dependence for H_2O and S_8 imply the inner track excitation density is important. At the largest S_e shown they have roughly the same dependence (for reference: open symbols are some data points above the maximum in S_e, dark symbols are below; from reference 35).

Aspects of the behavior of this region can be seen in the electronic sputtering of a different type of material, condensed Ar.[25] In this case the diffusion of holes leads to the redistribution of the initial excitation density, so that sputtering is dominated by the relaxation of individual excitations even at high excitation densities.[18,19,25] In addition, because an intermediate step in the sputter ejection process in these solids requires the emission of a photon,[19,25] the relaxation events are also significantly distributed in time, further limiting cooperative interactions between excitation events.

Although the inner track is often characterized by $r \sim v/\omega$, it is better defined as the region in which <u>interactions</u> between electron-hole pairs (the core plasma interactions) dominate the electronic relaxation processes and/or inner shell excitations occur with high probability. The outer edge of this region is where the excitation density is of the order of one trapped hole within a radius r_d, a 'damage effect' radius (distance over which an individual electronic relaxation process alters the solid). In the high excitation density region an added complication can occur: the radial motion of the electrons can be limited by the space charge. Modelling of the behavior of the cylindrical plasma in the high excitation density ('hot core plasma') region of the track has been considered by a number of researchers.[13,18,26-28]

That the inner track region is important for 'ionization track' effects is seen from the data for electronic sputtering of some condensed gas solids at 10K by fast protons and helium ions.[29] The yield data is displayed in Figure 3 vs electronic stopping power, $S_e[=(dE/dx)_e/n$, where n is the number density]. It is seen that at low $(dE/dx)_e$ the measured yields exhibit a variety of dependencies on $(dE/dx)_e$, but at the largest $(dE/dx)_e$ they all exhibit a similar dependence on $(dE/dx)_e$. For solid atomic Ar and molecular N_2 and O_2 the yields are found to be linear in $(dE/dx)_e$ at low $(dE/dx)_e$. However, for N_2 and O_2, the latter yields eventually increase more rapidly with increasing $(dE/dx)_e$, becoming quadratic at the higher excitation densities.[16,17,29] (At very high excitation densities steeper dependencies are obtained.[30]) In Figure 4 the measured yields for N_2 and O_2 are divided by the yield at low $(dE/dx)_e$ which depends linearly on $(dE/dx)_e$: i.e. $Y_e \propto (dE/dx)_e$. These scaled yields are plotted as a function of a 'damage effect' radius, $r_d = \Delta_o$, over the number of ionizations per unit path length produced, $\lambda_e = [(dE/dx)_e/W]^{-1}$. Here Δ_o is an average radial spread of the energy released during recombination and relaxation.[17] The consistency of these two sets of data using this scaling is remarkable. It suggests that a 'quasi-thermal track' is produced by the overlap of the released energy, so that 'plasma' processes in the core are, apparently, not critical for this sputtering process.

Two additional important details can be learned from the measurements on O_2 and N_2. First, it has been shown, using known primary ionization cross sections for both He^+ and He^{++} on gas phase N_2 and O_2 that <u>primary ionization alone is not</u> responsible for electronic sputtering.[17] This possibility was a subject of considerable debate in the early days of the study of electronic sputtering of condensed-gas solids,[18] stimulated by some of the much earlier work on track formation in dielectrics.[1] This means that the excitation density acting in this inner region contains a significant contribution from excitations by secondary electrons and, therefore, the radial extent of the region is <u>not</u> necessarily determined by the Bohr adiabatic radius, a common misconception.

Second, using He^+ and He^{++} as projectiles Brown et al[29] observed a difference of about 20% in the sputtering yield for low-temperature condensed N_2. This clearly showed that the holes did <u>not</u> diffuse significantly prior to relaxation in the vapor deposited samples of these simplest of molecular insulators.[17] This result also implies that the energy leading to sputtering <u>does not await the emission of a photon for these materials</u>. In the following we will ignore hole diffusion during the decay of the core plasma, a poor approximation in semiconductors,[16] but one probably valid in many track forming materials.

Holes in the core plasma can partially relax prior to recombination. That is, deep excitations can rapidly result in hole transfer to a neighbor, which we write in the notation of atomic physics as

$$A^{++} + A \rightarrow A^+ + A^+,$$
$$A^{+++} + A + A \rightarrow 3A^+.$$

Processes such as these can result in local coulomb repulsion between neighbors[31,32] to the extent that the 'free' and polarized-bound electrons do not shield the holes.

In this regime electrons cool and subsequently recombine with holes by both individual and interactive processes, also indicated in the notation of atomic physics in Table 2. Up to now we have ignored electron trapping, a process which may be important in certain materials. This has been studied recently by Sanche and coworkers,[33] using low energy incident electrons on low-temperature condensed-gas solids. Electron trapping is a critical aspect of the track formation model of Fleischer, Price and Walker,[1] leading to coulomb repulsion in the track core. Although this process clearly occurs in dielectrics, its importance in track formation has probably been overestimated. However, Sanche and co-

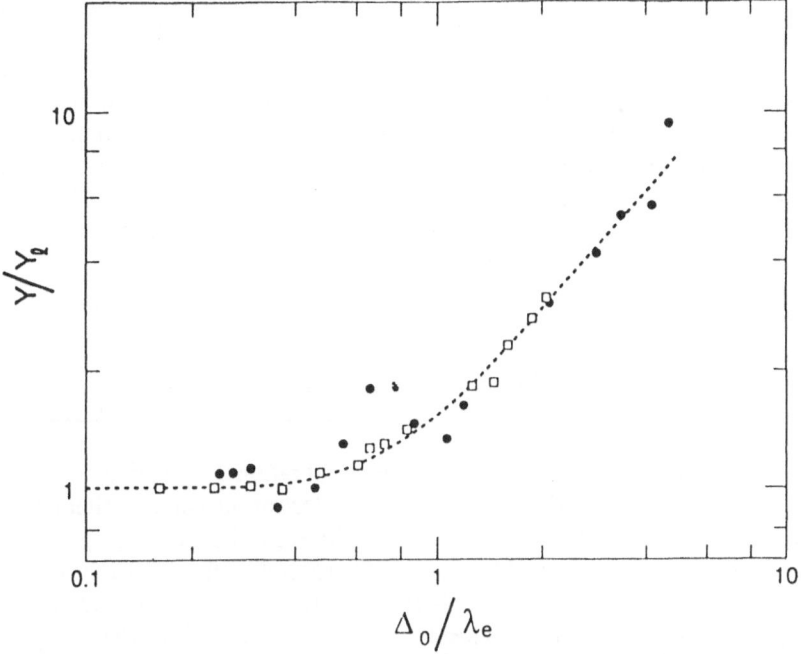

Figure 4. Yields of N_2 (open symbols) and O_2 (closed symbols) divided by the linear yield Y_ℓ and plotted vs the mean free path for excitation, $\lambda_e \equiv [W/(dE/dx)_e]$, scaled to the 'damage' effect radius, $r_d = \Delta_o$: 0.76 nm(N_2) and 1.02 nm (O_2). Note: yields in linear regime for O_2 are a factor of 2.3 larger than that for N_2 at same $(dE/dx)_e$. (from reference 17).

workers have shown that negative ion desorption is induced by electron trapping in condensed gas solids,[33] so that the importance of this process for track formation needs to be re-examined.

MODELS

The interactions in Table 2 all occur with varying importance. They can be used in a pair of coupled plasma diffusion equations, one for the ions (holes) and one for the excited, conduction band electrons.[26-28] In addition to a diffusion (mobility) term these equations contain the generation/loss terms and the electron-hole interaction terms (heating, cooling, and electric field). These terms are often not trivial to write down accurately and equations containing such terms rapidly become intractable. These equations are coupled to an equation for the excited species and the equations of motion of the lattice, the 'hydrodynamic' equations. For obvious reasons simplifications are important. Three principal classes of models for track formation are derived from these equations:

A.   Coulomb repulsion with time dependent screening
B.   Heating by electron-phonon interactions
C.   Deep excitations and repulsive decay

Such models are also discussed and evaluated in regards to electronically-stimulated sputtering.[5,18,30,34,35]

For models in class A (e.g. Fleischer, Price and Walker's coulomb repulsion model,[1] Watson and Tombrello's lattice potential model,[28] etc.[18]) the available 'heat' (kinetic energy) carried by the electrons is ignored. A lattice strain is produced by repulsive coulomb forces in the high density of closely spaced holes. The size of this effect depends on how screening by the electrons is treated.[18] Recently Dunlop et al[2] have invoked a model of this type to explain track formation in metals by fast, very heavy ions which deposit enormous energy densities. They showed that tracks were formed in those materials in which the soft phonon modes can be excited by the weak residual repulsion of the closely spaced but, in metals, rapidly screened holes.

Table 2.   'Track' Processes

1.	$e + AB\ (v=0) \rightarrow AB(v \neq o) + e$	vibrational excitation
2.	$e + A \rightarrow e + A^*$	electron-plasmon excitation
3.	$e + A^+ \rightarrow e + A^+$	electron-hole scattering
4.	$e(cool) + AB^+ \rightarrow (AB)^*$	recombination/lattice distortion
	$\rightarrow A + B^* + K.E.$	dissociative recombination
5.	$e + e + A^+ \rightarrow A^* + e(hot)$	Auger recombination
6.	$e + A^* \rightarrow A + e(hot)$	de-excitation
	$\rightarrow A^+ + e + e$	ionization
7.	$A^* + B \rightarrow (AB)^*$	dimerization
	$\rightarrow A + B + K.E.$	and repulsive decay
8.	$A^* + B^* \rightarrow (AB)^{**}$	exciton fusion

For models in class B the interactions between holes is ignored and the heat stored in the electrons, and then transferred to the lattice, is responsible for track formation. Ritchie and coworkers[13,26,27] have used this energy, fueled for instance by Auger recombination,[26] to cause melting of the lattice, which can then lead to track formation and sublimation or sputtering. This heat has also been assumed to induce a phase transition which causes a 'crystalline track' in amorphous metals and semiconductors.[3] Damage, rather than annealing, can also be produced and phase changes induced by fast ions have been widely reported for insulators.[36] In describing the ejection of biomolecules Williams and Sundqvist[37] noted that in a molecular solid the electrons efficiently excite vibrational modes producing a physical expansion. When the density of such excited (expanded) species is sufficiently high[6,30,34,38-41] permanent displacements can be produced in the solid.

The models in class C are a high excitation density extension of those processes occurring in the excitonic regime. Deep excitations may be required to produce defects in refractory solids[42] and can lead to local coulomb repulsion causing extended defects along the strain direction produced in the lattice.[31,32] Such models are inspired by the fact that at moderate excitations densities in refractory solids 'tracks' are seen to be disconnected, damaged regions.[43] Therefore, either statistical fluctuations in the total excitation density or special excitation processes are responsible.[44]

The above models are all, necessarily, over-simplified versions of the complete relaxation process occurring in the track core. That is, these processes all lead to material expansion[34] and, hence, one or the other can dominate in a particular material, causing both track formation and sputter ejection of surface material. Testing which processes dominate by measuring effects in the bulk is difficult, although recent work on crystalline to amorphous (or vice-versa) transitions[3,36] produced by fast ions in a number of materials and ion beam etching of polymers[45] are instructive. However, studying the surface ejecta can lead directly to quantitative information on track formation processes, as indicated above for low-temperature N_2 and O_2 solids.

SURFACE EJECTA

Table 3 gives some of the typically measured ejecta and rough times associated with both the ejection process and the corresponding track forming process. It is seen that measurements of the ejecta can be used to characterize the time evolution of the 'track'. It may be surprising that the most complete body of knowledge has been developed for organic and biomolecular solids. Mass spectrometric studies have lead to information on ejection of ions with masses ranging from 1 (protons) to ~40,000 amu.[6,30,34,38] This complements studies of ejection of electrons (the principal topic of this conference), luminescence studies, and studies of ejection of neutral species (sputtering).

Table 3. Times for 'track' Processes

Track Processes	Ejecta
$< 10^{-16}$ Primary Excitations	Primary Electrons
10^{-15} Plasma Oscillations	Cascade Electrons
10^{-14} δ-rays return	Auger Electrons
10^{-13} fast dissociation$^+$	Protons
10^{-12} electrons thermalize	Ions
↓	
10^{-11} chemistry$^+$	Neutrals
↓	
10^{-10} lattice relaxation$^+$	Heavy Molecules
10^{-9} or longer	Photons

+Processes can again be initiated following photo-emission.

427

From measurements of the ejection of ions of varying mass from organic solids, information on the radial and temporal characteristics of the track have been obtained.[6,38] That is, by measuring, in addition to the mass of the ion, the direction of ejection (or equivalently, the radial velocity distribution[46]) a rough idea of the radial origin of the prompt ejecta has been inferred for these materials. For instance, protons have been shown to be ejected primarily from the surface in the hot core region of the track at times for which there is still a significant deficit in the electron density at the surface due to the earlier secondary electron ejection.[10] Such a deficit was shown to affect the energy and angular distribution of the protons, and also accounts for the surface fields that affect electron emission.[11,12] In addition, it is found that the proton yield is determined only by the incident ion charge and velocity at high velocity and not by the ion type[47] (i.e. not by the z of the incident ion).

A further example from the studies of organic solids is that C_{2n} ($n > 10$) and, possibly, fullerenes (e.g. C_{60}) have been seen to be ejected from PVDF (poly-vinylidene diflouride) in response to a single fast, heavy ion impact.[48] These are ejected nearly back along the incident ion ('track') direction. This suggests that the ejected carbon molecule was formed in the extremely hot plasma in the track core, in which the total destruction and subsequent reformation of the bonds occurred, as well as the prior ejection of all of the light species (e.g. H, F as H_2, HF or F_2, probably HF is favored energetically). On the other hand, the large, intact biomolecular ions were found to be ejected with a radial velocity distribution which peaked at about 50 degrees from the normal[6,30,41,46] when the stimulating ion is incident normal to surface. Since this measurement is for intact molecular ions, and reformation of such complex species is unlikely, the ion must have been ejected from the edge of the track core. Additionally, its direction of ejection suggested that it came from the surface in response to the rapid material expansion caused by the transiently pressurized track.[6,30,41,48] Finally, studies of the threshold dependence of the ejection yield have lead to information on both the radial scale of this effect vs. $(dE/dx)_e$[50] and the ejection 'thresholds' for crater formation.[34,50,51] That is, the 'threshold' dependence was found to depend on the size of the cohesive energy of the material[34] and the physical size of the condensed gas molecule[35] or biomolecule ejected.[50] This allows a rough mapping of the increase with increasing $(dE/dx)_e$ of the size of the region affected by the cylindrical electronically excited 'track'.

CONCLUSIONS

From the above it is clear that it is very premature, if not impossible, to construct a complete model for track formation processes. First the nature of the 'tracks' differ from material to material and certainly the cause may differ according to what is measured [e.g. damage tracks, crystalline tracks, etchable tracks, etc. (Table 1)]. However, in addition to the recent observation of crater formation for a refractory solid,[51] a dynamic picture is emerging from the studies of molecular organic insulators[53] and condensed-gas solids. These van der Waals solids respond efficiently to the ionization energy deposited. That is, various species are ejected from the ion-excited material in response to the excited 'track' (hot core plasma) over distinctly different time scales (Table 3) and at different average radial distances from the ion's initial path. These ejecta, therefore, provide useful probes of the excited material, allowing the possibility of mapping the relaxation of the solid as a function of time and radial distance from the ion's path. Such measurements should be coordinated with more standard track measurements, as these are the only studies that are likely to be described with any certainty by models of the relaxation of the incident ion-excited plasma.

Acknowledgement

This work supported in part by a NATO grant (CRG-901013) between Alicante and Virginia and by the NSF Division of Materials (DMR-91-21272) and the NSF Division of Astronomy (AST-91-20078).

REFERENCES

1. R.L. Fleischer, P.B. Price, and R.M. Walker, "Nuclear Tracks in Solids", Univ of Calif. Press, Berkley (1975); G.E. Adams, D.K. Beailey, and J.W. Boag," Charged Particle Tracks in Solids and Liquids", Institute of Physics, London (1970).
2. A. Dunlop, P. Legrand, D. Lesueur, N. Lorenzelli, J. Morillo, A. Barbor and S. Boufford, Europhys. Lett 15:765 (1991).
3. S. Klaumünzer, Ming-dong Hou, and G. Schumacher, Phys. Rev. Lett. 57:850 (1986); C. Houpert, F. Stader, D. Groult, M. Toulemonde, Nucl. Instr. Meth B39:120 (1989).
4. M. Maurette and P.K. Price, Science 187:121 (1973); Bradley, J.P. D.E. Brownlee, P. Fraundouf, Science 226:1432 (1984).
5. P.K. Haff, Appl. Phys. Lett. 29:443 (1976); T.A. Tombrello, Nucl. Instr. and Meth. B2:555 (1984).
6. R.E. Johnson and B.U.R. Sundqvist, Physics Today, March 1992:28 (1992).
7. R.E. Johnson and J. Schou, (Review to appear in Mat. Fys. Medd. Dan. Vid. Selsk.)
8. M. Rösler, this proceedings.
9. M. Inokuti, Y. Itikawa, J.E. Turner, Rev. Mod. Phys. 43:297 (1971).
10. D. Fenyo, Ph.D. Thesis, Uppsala Univ.,Uppsala (1991); submitted to Phys. Rev. B.
11. G. Schiwietz P. Grande B. Skogoal, J.P. Biersack, R. Köhnbrück, K. Sommer, A. Schmold and P. Goppelt, I. Kadas, S. Ricz, V. Stettner, Phys. Rev. Lett 69: 628, (1992).
12. A. Akerman this proceedings.
13. W. Brandt and R.A. Ritchie, in "Physical Mechanics in Radiation Biology", eds. R.D. Cooper and R.W. Wood, U.S.A.E.C., Washington:20 (1974).
14. e.g. S.I. Khlupin and A. Akerman, Phys. Stat. St. 158:63 (1990).
15. H. Bischel et al., ICRU Report 31, U.S. Commerce Dept., Wash. D.C. (1979).
16. W.L. Brown this proceedings.
17. R.E. Johnson, M.K. Pospieszalska, and W.L. Brown, Phys. Rev. B.,44:7263 (1991).
18. R.E. Johnson and W.L. Brown, Nucl. Instru. Methods.,198:153 (1982).
19. W.L. Brown and R.E. Johnson, Nucl. Instru. Methods. B13:295 (1986); J. Schou Nucl.Instru.Methods. B27:188 (1987).
20. N. Schwentner, E.E. Koch, J. Jortner, Electronic Excitations in Condensed Rare Gases, Springer Verlag, Berlin, (1985).
21. D. Menzel, Appl. Phys. A51:163 (1990).
22. e.g. R. Katz and E.J. Kobetich, Radiat. Eff. 3:169 (1970).
23. H. Dertinger and H. Jung, Molecular Radiation Biology, Springer-Verlag, Berlin (1970).
24. O. Puglisi, G. Marletta, S. Pignatars, G. Foti, A. Trovato, A. Rimini, Chem. Phys. Lett. 78:207 (1981); G. Strazzulla and G.A. Baratta, Astron. Astrophys. 241:310 (1991). C. Calcagno and G. Foti Nucl. Inst. and Methods B 59/60; ll53 (1991).
25. C.T. Reimann, R.E. Johnson and W.L. Brown, Phys. Rev. Lett. 53:600 (1984).
26. R.H. Ritchie, and C. Claussen, Nucl. Instrum. Methods 198:133 (1982);R. Ritchie Proc. 8th Symp. Microdosimetry EUR8395:145 (1983).

27. R.H. Ritchie, A. Gras-Marti, and J.C. Ashley, in Proc. of 12th Werner Brandt Workshop:595 (1989). R.H. Ritchie, R.N. Hamm, J.E. Turner, H. A. Wright, J.C. Ashley and G.J. Basbos, Nucl. Tracks Radiat. Meas. 16:141 (1989).

28. C.C. Watson and T.A.Tombrello, Radiat Eff. 89:363 (1955).

29. W.L. Brown, L.J. Lanzerotti, K.J. Macantonio, R.E. Johnson, C.T. Reimann, Nucl. Instru. Methods B14:392 (1986).

30. K. Wien, Nucl. Inst. and Methods B65:149 (1992); Rad. Eff. Defects in Solids 109:137 (1989).

31. N. Itoh, Nucl. Instru. Methods.B 27:155 (1987);   N. Itoh, T. Nakayama, T.A. Tombrello, Phys. Lett. 108A:480 (1985).

32. T.A. Tombrello, Nucl. Instrum. and Methods B2:555 (1984).

33. L. Sanche, Comments Atom. Mol. Phys 26:6321 (1991).

34. R.E. Johnson, Int. J. Mass. Spec. and Ion Physics 78:357 (1987).

35. L. Torrisi, S. Coffa, G. Foti, R.E. Johnson, D.B. Chrisey, and J.W. Boring, Phys. Rev. B. 38:1516 (1988).

36. R. L. Hudson and M. Moore, J. Phys. Chem. 96:6500 (1992); A. Matsunagen, C. Kinoshita, K. Nakai and Y. Tomokyo J. Nucl. Mat. 179-181: 457 (1991) G. Strazzulla, G.A. Baratta, G. Leto, and G. Foti, Europhys. Lett. 18:517 (1991).

37. P. Williams and B.U.R. Sundqvist, Phys. Rev. Lett. 58:1031 (1987).

38. B.U.R. Sundqvist, in Sputtering by Particle Bombardment III, Springer-Verlag, Berlin, (1991). B.U.R. Sundqvist and R.D. Macfarlane Mass. Spec. Rev. 4:421 (1985).

39. D. Fenyo, B.U.R. Sundqvist, B.R. Karlsson, and R.E. Johnson, Phys. Rev. B 42:1895 (1990).

40. S. Banerjee, S.T. Cui, and R.E. Johnson, Phys. Rev. B 43:12707 (1991).

41. R.E. Johnson, B.U.R. Sundqvist, A. Hedin, and D. Fenyo, Phys. Rev. B 40:49 (1989).

42. M.L. Knotek and P.J. Feibelman, Phys. Rev. Lett. 110:964 (1978).

43. E. Dartyge and P. Sigmund, Phys. Rev. B32:5429 (1985); E. Dartyge, J.P. Duraud, Y. Langevin, and M. Maurette, Phys. Rev. B 23:5231 (1981).

44. T.A. Tombrello, Int. J. Mass Spec. Ion Physics, in press (1992); B.U.R. Sundqvist ibid.

45. L. Torrisi and G. Foti J. Mater. Res 5: 2723 (1990).

46. W. Ens., B.U.R. Sundqvist, P. Håkansson, D, Fenyö, A. Hedin, and G. Jonsson, J. Phys. (Paris) C2:9 (1989); Phys. Rev. B 39:763 (1989).

47. S. Della-Negra, K. Wein, Y. LeBeyec, B. Monart and K. Standing, Phys. Rev. Lett. 58: 17 (1987).

48. G. Brinkmalm, D. Barofsky, P. Demirev, D. Fenyö, P. Håkansson, R.E. Johnson, C.T. Reimann, B.U.R. Sundqvist, Chem. Phys. Lett. 191:345 (1992).

49. I. Bitensky and E. Parilis, Nucl. Instrum Meth. B 21:26 (1987).

50. A. Hedin, P. Håkansson, B. Sundqvist, and R.E. Johnson, Phys. Rev. B 31:1780 (1985).

51. F. Thibaudau, J. Cousty, E. Balanzat and S. Boufford Phys. Rev. Letts. 67, 1582 (1991)

52. R. Schmidt, Ch. Schoppmann, D. Brandl, A.Ostrowski, H.Voit, D.Johannsmann and W. Knoll, Phys.Rev.B44:560 (1991); D.Brandl, Ch. Schoppmann, R.Schmidt, B.Nees, A.Ostrowski, and H.Voit, Phys.Rev. B, 43:5253 (1991).

53. B.U.R. Sundqvist, Int. J. Mass Spectrom. Ion Physics, in press (1993)

TRACK EFFECTS AND THEIR INFLUENCE ON HEAVY ION ENERGY LOSSES IN SEMICONDUCTOR DEVICES

A. Akkerman, J. Levinson, D. Ilberg, and Y. Lifshitz

Applied Physics & Mathematics Dept.
Soreq NRC
Yavne 70600, Israel

INTRODUCTION

The Linear Energy Transfer (LET) concept is widely used in the context of Single Event Phenomena (SEP) in electronic devices induced by energetic heavy particles and protons in space missions. Very recently several experimental results have shown that the LET concept may not be adequate for a quantitative evaluation of a variety of SEP. Stapor et al.[1] showed that different charge collection values in specially designed electronic devices were obtained for ions having the same LET but different energies. In two recent conferences E. Stassinopoulos[2] gave a critical analysis for the validity of the LET concept in SEP while P. Sigmund[3] discussed the relevance of the LET concept for secondary electron emission. There are also some indications of the failure of using LET as a measure in radiobiological investigations.

All these facts encourage attempts to understand the observed failures of the LET concept.

RECOMBINATION EFFECTS

It is well established for a long time that there is a pulse height defect (PHD) in solid state silicon detectors (see for example[4,5]) irradiated by heavy ions. The major factors contributing to the PHD are: (i) loss of free electrons in recombination processes, (ii) loss of energy in the low-energy nonionizing collisions with the target atoms, (iii) loss of energy in surface dead layers, (iv) loss of free electrons trapped by lattice defects or impurities and others.

Ionization of Solids by Heavy Particles, Edited by
R.A. Baragiola, Plenum Press, New York, 1993

The PHD can be approximated[4] by an empirical relation

$$PHD = 10^{\beta} E_d^{\alpha},$$ (1)

where the parameters α and β are extracted from the experimental data obtained for PHD in a Si surface barrier detector[4]

$$\alpha(Z) = 0.0223(Z^2/10^3) + 0.5682$$

$$\beta(Z) = -0.1425\,(100/Z) + 0.0825.$$ (2)

In (1) and (2) E_d - is the deposited energy in MeV and Z is the ions atomic number. Measurements[1] of energy actually converted to collected charge in the gate and drain regions of a CMOS/SOS transistor were performed for energetic Ni and Cl ions. The sensitive volume from which the charge was collected was a Si layer with a thickness d=0.45 μm. Table 1 presents (i) the measured collected charge converted to energy losses - ΔE_c, for the gate region (a) and the drain region (b), (ii) the energy lost by the ion in the sensitive volume $\Delta E_t = (-\, dE/dx)_i \cdot d$ and (iii) the energy calculated after subtracting the PHD, $\Delta E_d = \Delta E_t$ - PHD.

Table 1. Comparison of various energies absorbed in a Si layer (d=0.45 μm) in the : a - gate region, b - drain region of a CMOS/SOS transistor.

Ion	Ni		Cl	
Energy MeV	389.5	19.7	46.4	18.6
ΔE_c, MeV	1.05[a]	0.67[a]	0.65[a]	0.47[a]
	1.73[b]	1.20[b]	1.18[b]	0.88[b]
ΔE_t, MeV	2.53	2.35	1.93	1.80
ΔE_d, MeV	1.90	1.70	1.64	1.52

It should be mentioned that the ion energies marked in Table 1 were taken so that they lay on both sides of the stopping power curve $(-dE/dx)_i = f(E_{ion})$, providing practically the same LET. It follows from the table that there is a large discrepancy between the predicted deposited energy ΔE_t, even including the PHD (ΔE_d), and the measured deposited energy ΔE_c. This suggests that PHD does not include all the mechanisms of charge recombination in electronic devices irradiated by fast ions.

It has been recently shown[6] that the Auger recombination is the dominant process when considering the fast ambipolar diffusion of carriers created in silicon by an energetic heavy ion. Semi-quantitative calculations[6], show that for Ni and Cl ions with the LET < 25 MeV.cm^2/mg the percent of charge losses (or the restricted energy losses $(-dE/dx)_r$) following Auger recombination does not exceed 20% of their initial values. Hence the Auger recombination cannot explain the lowering of the energy losses by more than a factor of two that is noted in Table 1.

TRACK ELECTRICAL FIELD

In this section an additional mechanism influencing the charge collection will be evaluated in an attempt to explain the discrepancies shown in Table 1. It is suggested that the electric field associated with the charge separation in the heavy ions track strongly influences the recombination in the near track region thus affecting both the secondary electron emission and the charge collection in the ion bombarded solid. Indeed this assumption was recently used[7,8] to explain the deviation of the functional dependence of secondary electron emission from metals and dielectrics from the expected z^2. The method used[8] seems to be applicable in our case since the secondary electron emission is related to the secondary electron creation and transport in the solid. Following Borovsky and Suszcynsky[8] the static electrical field in the track impedes secondary electron escape yielding a restricted charge collection. The effective LET value of $(-dE/dx)_w$ that corresponds to this restricted charge collection can be approximated by

$$\left(-\frac{dE}{dx}\right)_w = \left(-\frac{dE}{dx}\right)_i \frac{[T_m - (\bar{I} + e\Delta\phi_{trap})]\bar{I}}{(T_m - \bar{I})(\bar{I} + e\Delta\phi_{trap})} , \qquad (3)$$

where $\bar{I}$ is the mean ionization potential of the target atoms taken from ICRU report No. 37[9] (for Si $\bar{I} = 174$ eV), e - is the electron charge, $\Delta\phi_{tran}$ is the potential of the track field, $(-dE/dx)_i$ is the LET for the given ion energy and is evaluated from the data of Ziegler et al.[10], T_m - is the maximal energy, transferred by the ion to the δ-electron

$$T_m = \frac{2mc^2\beta^2}{1 - \beta^2} , \qquad (4)$$

and β is the ions velocity in c units. We have calculated $\Delta\phi_{tran}$ using the value of charge per unit length in the ion wake, as:

$$\Delta\phi_{trap} = \frac{2\pi e}{\varepsilon_0} \int_0^{r_0} D_0(r)rdr \qquad (5)$$

where ε_n is the energy necessary for a pair (electron-hole) creation in Si ($\varepsilon_n = 3.6$ eV). The right side of equation (5) was obtained performing the integration of the radial density distribution $D_0(r)$ of the energy transferred by an ion having the effective charge Z_{eff} and velocity β and measured from the ions track axis. The $D_0(r)$ distribution was calculated by Monte Carlo simulations and presented by an empirical formula in our article[11].

For calculations of the ions effective charge Z_{eff} in the expression for $D_0(r)$ we have used Ziegler's formula[10]. The estimated accuracy for Z_{eff} for high energies is in the order of 10 to 15%. Note that for smaller ion energies we do not know the accuracy of using the mentioned formula for Z_{eff}. The upper limit in the integral (5) was estimated to take into account ~80% of the total charge generated in the track. Usually r_n does not exceed 10 - 15nm.

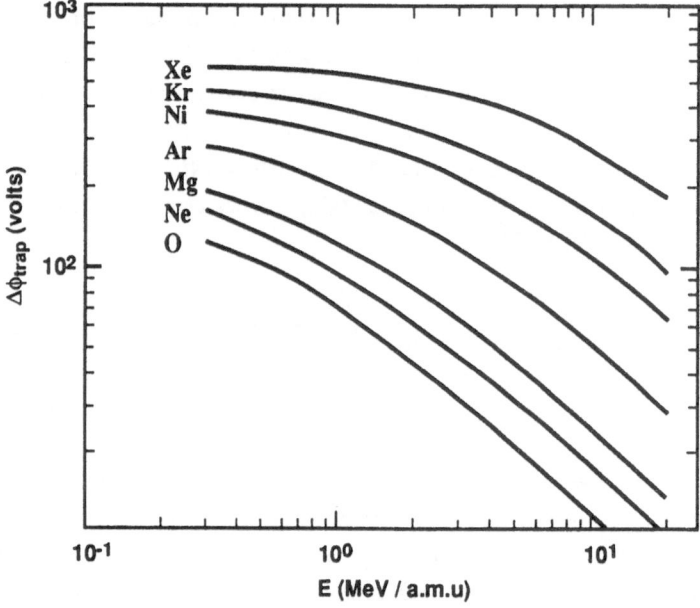

Figure 1. Track electrical field potential (in volts) versus energy for different ions.

RESULTS

The calculated dependence of $\Delta\phi_{tran}$ (retarding track potential) for various ions and energies is shown in fig. 1. It can be seen that for heavy ions and small energies there is a sufficiently high electrical field which, however, decreases when the ions energy increases. These trends are similar to the data[7] obtained for metals. The energy dependence of the ratio $(-dE/dx)_w/(-dE/dx)_i$ following from (3) is presented in fig. 2.

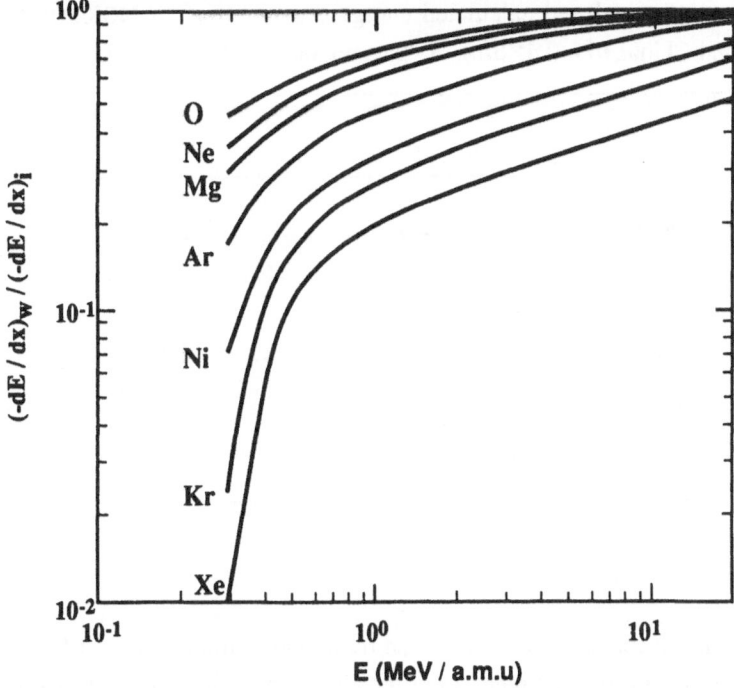

Figure 2. The energy dependence of the ratio of restricted stopping power$(-dE/dx)_w$ to LET (LET $= (- dE/dx)_i$) for different ions.

The carrier recombination in the track region, which actually introduces a time decreasing track potential was not included in the calculations performed above. This will consequently increase $(-dE/dx)_w$, deduced from formula (3). As mentioned above the main process which affects the recombination in the initial stages of track formation is the Auger effect. Hence one needs to solve a complicated diffusion equation extending Edmonds' equation[6] to take into account the track potential. Actually to this moment any exact solution of this problem does not exist. On the other hand the existence of a strong track potential will prevent diffusion of the low energy secondaries created by the ions, much more effectively for lower energy ions. Hence there is practically no diffusion in the track region and this simplifies the original equation[6]. The solution of the last one gives immediately the time dependent charge density $D(t)$ as:

$$D(t) = \frac{D_o}{(2CD_o^2 t + 1)^{1/2}} \ , \tag{6}$$

where C is a constant, estimated[6] as $C = 3.79 \cdot 10^{-31} cm^6/s$. The reduced LET $(-dE/dx)_w$ was calculated from (3), (5) and (6) to inlcude the recombination effects. Table 2 represents calculated values of $\Delta E = (-dE/dx)_w \cdot d$ for two cases; (i) ΔE_+ - prompt recombination included, (ii) ΔE_- - prompt recombination not included .

Table 2. Experimental and calculated energy transfer in the CMOS/SOS transistor for incident Ni and Cl ions (d = 0.45 μm). a - gate region, b - drain region.

Ion	Ni		Cl	
Energy MeV	389.5	19.7	46.4	18.6
ΔE_c, MeV	1.05[a]	0.67[a]	0.65[a]	0.47[a]
	1.73[b]	1.20[b]	1.18[b]	0.88[b]
ΔE_-, MeV	1.40	0.31	0.84	0.53
ΔE_+, MeV	1.50	0.77	1.00	0.84

The data presented for ΔE_+ were calculated for a border time of $t = 10^{-11}$s. This value is an overestimation of the real mean time for Auger recombination, hence the actual ΔE values should lie between the two calculated values.

It is evident from table 2 that the LET calculations using the track field approach yield a much better agreement between the calculated and experimental data than calculations including only the PHD effect or Auger recombination (see Table 1). Moreover, the marked energy dependence of the collected charge for ions with the same LET but different energies is qualitatively reproduced by the track field effect.

The track electrical field is actually dynamic rather than static and is governed by time dependent processes of secondary electron transport, diffusion of carriers and recombination in the wake. The present results should thus be further improved by solving this self consistent problem and by taking into account that the Auger recombinations are followed by the emission of secondary electrons which in several cases can penatrate the retarding track potential.

On the other hand for the SANDIA multi layer test structure investigated by Stapor et al.[1] with the CMOS/SOS transistor, the ratio of deposited charge to the collected one for the two energies mentioned in Table 1 and for both Ni and Cl ions never exceeds 1.15. This lies in the range of experimental uncertainties. Hence in this case one does not need to take into account the track field effects. At present it is not evident if this fact is an indication of the lack of generality of the discussed track field approach or it results from a specific combination of fields acting in this device. Additional experiments should be performed to clarify this issue.

In Fig. 3 the dependence $(-dE/dx)_w$ (without inclusion of recombinations) versus $(-dE/dx)_i$ for different ions is shown. The lower parts of the curves belong to the left side (from the maximum) of the stopping power curve and the upper ones to the right side of this curve. From the curves one can obtain the stopping powers (and simultaneously the energy of ions) for which the same reduced energy transfer (collected charge) can be achieved.

These predictions should be verified experimentally to confirm the usefullness of the track field approach. Secondary electron emission measurements from silicon accompanying swift ion penetration (reflection) similar to those performed by Borovsky and Suszcynsky[7,8] may be helpful. A study of the track field potential using Auger electron spectroscopy[12,13] (shift of the Auger emission lines) may also be relevant to the verification of the discussed approach.

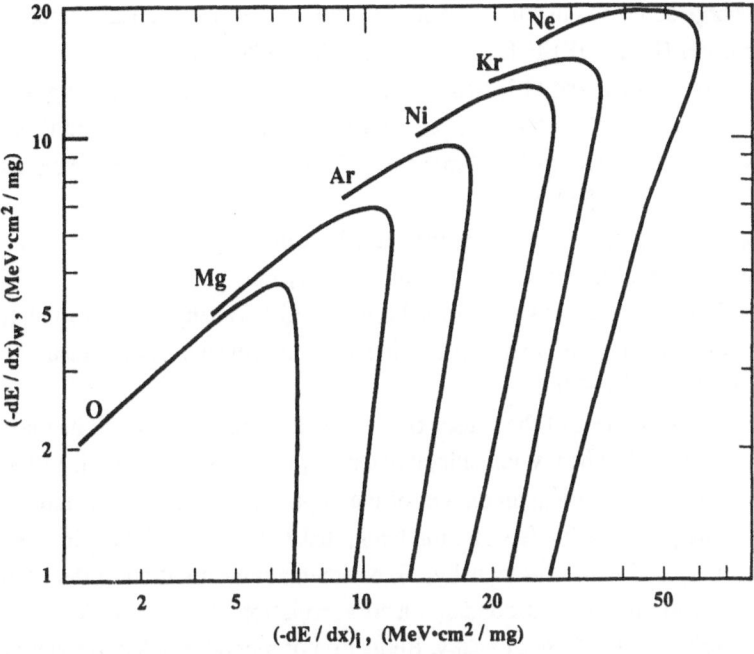

Figure 3. The restricted stopping power $(-dE/dx)_w$ dependence on LET for various ions. The upper parts of the curves belong to higher energies.

SUMMARY

The charge collected in devices was observed to be considerably lower than the predicted one using the LET concept. A possible explanation of this effect based on the track electrical field model was proposed and analysed. The main experimental trends including the energy dependence of the restricted energy transfer are qualitatively reproduced by the model. New experimental data of charge collection in devices are needed to verify the validity and generality of the discussed model for SEU simulations.

ACKNOWLEDGEMENT

The authors would like to express their gratitude to Z. Vager and W.L. Brown and G. Schiwietz for helpful discussions and to L.D. Edmons for providing detailed information of his calculational method.

This work is supported by the Israeli Ministry of Absorption.

REFERENCES

1. W.J. Stapor, P.T. Mac Donald, A.R. Knudson, A.B. Campbell and B.G. Glagola, Charge collection in silicon for ions of different energy but same linear energy transfer (LET). IEEE Trans. Nucl. Sci. 35: 1585 (1988).

2. E.G. Stassinopoulos and G.J. Brucker, Shortcomings in ground testing, environment simulations and performance predictions for space applications in: Proc. of RADECS 91, Montpelier, France Sept. 9-12. 1991, IEEE service center, Picataway, NJ: 3 (1991).

3. P. Sigmund, Secondary electron emission: Elementary questions and (few) attempts at answers, NATO workshop, Taormina, Italy, June 1-5, 1992.

4. J.B. Moulton, J.E. Stephenson, R.P. Schmitt and G.J. Woznjak, New method for calibrating the pulse height defect in solid state detectors, Nucl. Instrum. Meth. 157 A: A325 (1978).

5. E.C. Finch, An analysis of the causes of the pulse height deffect and its mass dependence for heavy-ion silicon detectors, Nucl. Instrum. Meth. 113:41 (1973).

6. L.D. Edmonds, Theoretical prediction of the impact of Auger recombination on a - charge collection from an ion track, IEEE Trans. Nucl. Sci. 36: 999 (1991).

7. J.E. Borovsky and D.M. Suszcynsky, Experimental investigation of the Z^2 scaling law of fast-ion-produced secondary electron emission. Phys. Rev. A43:1416 (1991).

8. J.E. Borovsky and D.M. Suszcynsky, Reduction of secondary-electron yields by collective electric fields within metals, ibid: 1433 (1991).

9. ICRU, "Stopping Powers for Electrons and Positrons", ICRU Report 37, ICRU, Bethesda, MD (1984).

10. J. Ziegler, J.P. Biersack and U. Littmark, Stopping and Range of Ions in Solids V.1 ed. J.F. Ziegler, Pergamon, NY (1985).

11 S.I. Khlupin and A.F. Akkerman, Radial energy transfer density distribution around the fast ion tracks in silicon and germanium, Phys. Stat. Sol. 158 (b):63 (1990).

12. G. Schiwietz, Electron emission from atoms and thin foil targets at incident energies of 300 keV/a.m.u. to 8.5 MeV/a.m.u., NATO Workshop, Taormina, Italy, June 1-5, (1992).

13. G. Schiwietz, P. Grande, B. Skogvall, J.P. Biersack, R. Kohrbruck, K. Sommer, A. Schmoldt, P. Goppelt, I. Kadar, S. Ricz and U. Stettner, The incluence of nuclear track potentials in insulators on the emission of target Auger electrons, submitted to Phys.Rev. (1992).

ENERGY TRANSFER, SCATTERING, AND ION EMISSION DUE TO HYPERTHERMAL NEUTRAL ATOM IMPACT AT SURFACES

Mark J. Cardillo

AT&T Bell Laboratories
600 Mountain Avenue
Murray Hill NJ 07974

I. INTRODUCTION

In this paper I present a brief summary of the principal observations of experiments carried out in my laboratory on the emission of ions associated with the scattering of hyperthermal (1-15 eV) rare gas atoms from single crystal (semiconductor) surfaces.[1] Over this energy range we have observed and characterized several phenomena: the excitation of substrate electron-hole pairs,[2] large fractional energy loss of the projectile which scatter into surprisingly sharply structured angular distributions,[3,4] and the above-mentioned collision induced emission of substrate ions (and neutrals) at projectile energies far below what is often considered the sputtering threshold.[5] These results have been discussed in detail elsewhere and a more substantial review has been recently published.[6] This paper is essentially an abbreviated version of these papers.

II. EXPERIMENTAL

The ultrahigh vacuum molecular-beam surface scattering apparatus used in these studies has been described elsewhere[6,7] as have the procedures used in the measurements. Briefly, a Xe, Kr, or Ar beam seeded in H_2 or He was expanded through a 100μ platinum nozzle that could be heated to 1300°C. The mole fraction of the heavy gas could be continuously varied from 0.1% to 2.0%. The energies obtained ranged up to 15.5 eV for Xe, 9.2 eV for Kr, and 5.5 eV for Ar. The speed ratios obtained in the raregas$-H_2$ mixtures were typically 13-20 ($=\upsilon/\Delta\upsilon$, full-width at half-maximum). The fluxes were calibrated against the signal from a flux gauge for pure rare gas beams expanded through hot nozzles (to prevent clustering). The beam was square-wave modulated at 80-1000 Hz for phase-sensitive detection, or chopped at the same frequencies to give pulses as short as -30μs for time-of-flight (TOF) analysis. After collimation the beam-spot size on the crystal was $1\times2mm^2$. The scattered beam was measured using a quadrupole mass spectrometer (QMS) which rotates about the axis of intersection of the molecular beam and the crystal surface.

Ionization of Solids by Heavy Particles, Edited by
R.A. Baragiola, Plenum Press, New York, 1993

The electron-hole pair creation efficiency was measured by applying a dc voltage bias (1-30 V, giving a current of $\sim 10^{-6}$ A) across two Ohmic contacts on the front surface of the crystal with the rare gas beam illuminating a fraction of the region in between as described in Ref. 2. The transient excitation currents observed due to the impinging rare gas atoms were 100 fA - 100 pA, measured on a lock-in amplifier after current-to-voltage conversion, amplification, and nullification of the dc currents (using a current amplifier). To understand the excited carrier recombination kinetics, and thus deduce the absolute efficiency of collisional $e^- h^+$ creation, carriers were also optically excited and their lifetimes monitored.

Relative ion ejection rates, when one species dominated the emission, were determined either with a channeltron mounted immediately above the crystal or by

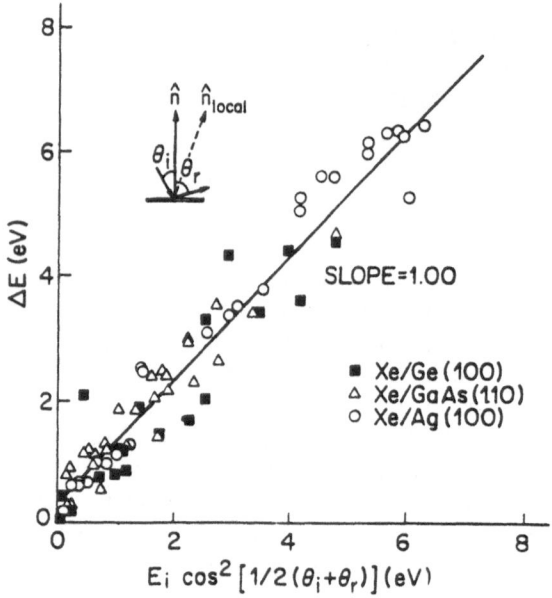

Figure 1. Energy loss in Xe surface collision vs the energy associated with local normal motion. A least-squares fit to the data gives a slope of 1.00.

measuring the crystal current to ground. The latter, when divided by a measured value of the incident flux, is taken to be the absolute ion emission efficiency. The ions extracted by the channeltron voltage are assisted by appropriate biasing of the crystal to typically + or -30 V. A second QMS at fixed angle (120° with respect to the incident Xe beam) was mounted on bellows in order to approach within ~2cm of the crystal and was used to identify the mass of the ejected ions. An additional extracting plate was mounted on the front of this QMS and served as a crude lens to help collect the positive ions. Negative ions were collected with the channeltron or the cylindrical mirror analyzer (CMA) of the Auger and identified by pulsed voltage ion time-of-flight. The output of the QMS was sent to a signal averager and recorded on a chart recorder. Mass calibration was provided by using the QMS as a residual gas analyzer.

III. RESULTS AND DISCUSSION

The scattering angle and energy distributions of Xe, Kr, and Ar atoms at hyperthermal energies have been measured from GaAs(110), InP(100), Ge(100), and Ag(100) surfaces[3]. Energy losses were measured to be large in these collisions, and a direct although not strict correlation between the mean energy loss and the energy associated with normal motion was found. A further correlation was obtained if all collisions were assumed to be *locally* specular. The local normal is defined as the bisector of the angle made by the incident and reflected atomic velocity vectors. The mean energy transfer was taken as the difference of the incident energy and the energy corresponding to the peak of the (board) time-of-flight spectra of the scattered atom, which is in all cases close to the mean energy transfer. The mean energy transfer so defined is found to follow the relation:

$$\Delta E = k E_i \cos^2 [(\theta_i + \theta_r)/2], \qquad (1)$$

where k is a constant for each rare gas scattered, θ_i and θ_r are the incident and reflected angles with respect to the surface normal (inset of Fig. 1), in the inset of Fig. 1, and E_i is the incident kinetic energy. This is illustrated by the data in Fig. 1 for Xe scattering from a variety of surfaces. Comparable correlations were found for Kr and Ar. The best-fit constant k from Eq. (1) are $k_{Xe} = 1.00, k_{Kr} = 0.92$, and $k_{Ar} = 0.63$.

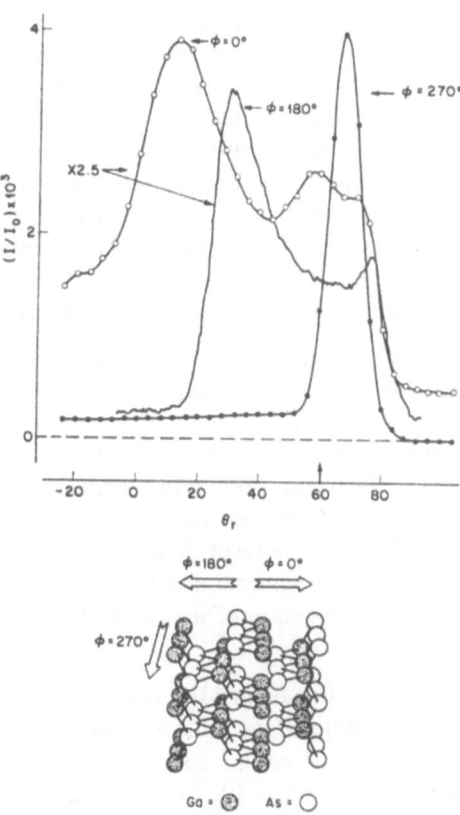

Figure 2. Scattered Xe atom angular distributions from a GaAs(110) surface along three different azimuthal directions, as shown, with the Xe atoms at $E_i = 3.3$eV and $\theta_i = 60°$. For scattering in the sagittal plane, the 270° azimuth appears smooth while the 0° and 180° azimuths appear corrugated. The tilt in the surface bond causes the scattering potential to be steeper for 0° than for 180°.

Despite the large fractional energy loss, the angular scattering distributions were found to be sensitive to the surface topography. In-plane scattering from a smooth surface, or a smooth direction of a corrugated surface, gave narrow, slightly supraspecular ($\theta_r > \theta_i$) angular distributions. Scattering from corrugated surfaces gave backscattered rainbowlike features as well as some scattering near specular in a fashion reminiscent of the envelopes of He diffraction peaks from such surfaces. This is illustrated in Fig. 2 in which the angular distributions for Xe scattered along the different azimuths of a GaAs(110) surface are shown under otherwise identical conditions. For the "smooth direction," the scattering is parallel to the rows and troughs of the surface and shows one peak in the angular distribution, similar to what is observed for Ag(100). For the "corrugated directions," the scattering is perpendicular to the rows and troughs and shows a backscattered peak as well as the nearly specular peak. As in He diffraction, if the crystal azimuth is rotated from 0° to 180°, the backscattered maximum is still observed, but is shifted ~10° closer to specular. The energy and angular scattering distributions for Xe on GaAs(110) were reproduced by the computer simulations reported in Ref. 4.

To summarize, the scattering observed in both the experiments and the simulations exhibit large fractional energy transfer which scales as the energy associated with local normal motion. In addition, the angular scattering distributions depend strongly on the surface topography (corrugation). Based on the success of the computer simulations, the microscopic collision mechanism could be inferred for the backscattered maxima and scaling of the energy loss with normal motion. For smooth surfaces or smooth directions of rough surfaces, the incident particle experiences what can be considered multiple hits of the substrate, each of which contributes to turning it around in a manner somewhat insensitive to impact parameter. *In coarse grain* this makes the substrate similar to a continuous body. This leads to near specular scattering for which the crystal normal is the appropriate reference. As the corrugation of the surface is increased, the center of the target atom is exposed. An increasing fraction of impacts then lead to rapid recoil of only one substrate atom. Some of these target atoms quickly rebound from the lattice, and rehit the incident projectile. This nearly binary collision mechanism leads to the observed backscattered maxima with the associated aspect that the energy transfer is along the line of centers of the projectile and target atoms and thus, yields the approximate correlation with a *local* normal in the scattering from strongly corrugated surfaces.

As noted above, for the higher range of E_i in these experiments the Kr and Xe scattering caused charged particle ejection from surfaces. Based on the known incident fluxes, and measured ion currents, the efficiencies of charged particle ejection were accurately determined for a range of incident energies and angles. In^+ and Ga^+ were the dominant species ejected from InP(100) and GaAs(100) surfaces, respectively. Alkali ions initially dominated from all surfaces including the nominally clean (500-eV Ar^+ sputtered and annealed) Si(100) and Ge(100) crystal surfaces. Ions due to contaminant species could also be observed, such as hydrocarbon fragments and In^+ from a contaminated Ge(100) surface. Occasionally and irreproducibly, some of the incident Xe atoms appeared as ions with low yield.

The dependences of the ion ejection on incident energy and angle are shown in Fig. 3 for Xe atoms incident on an InP(100) surface. The data were recorded with the crystal biased positively with respect to machine ground so as to record only cations. Note the absolute scale for the efficiency which reaches 10^{-3} charges ejected per 15.5 eV Xe atom at = 75°.

The dependences of ion ejection on incident projectile energy and angle are very different than for electron-hole pair excitation. The $e^- h^+$ excitation yield peaks at normal incidence[2], whereas the ion ejection peaks far from normal[1]. Furthermore, the ion ejection efficiency does not follow an exponential dependence on ΔE^{-1}. These observations are consistent with the nearly binary collisions observed for corrugated surfaces in the computer simulations in Ref. 4. Based on the large momentum transfer, particularly at grazing incidence, these nearly binary collisions might be expected to be responsible for removing particles from the surface. Although a collision at normal incidence collision can transfer a large amount of momentum to a single lattice atom, the projectile would "cap" the lattice atom (such as in the double-hit collision described in the previous section), preventing it from departing. The necessity for both large momentum transfer and high escape probability leads to a maximum near grazing incidence where the incoming atom does not interfere with the escape. For large momentum transfer at these angles, however, substantial corrugation in the lattice is required. One can also consider an adatom as an extreme case of high corrugation for which a large fraction of impact parameters lead to direct ejection with the consequent efficient of ejection of alkali ions as trace adatom impurities.

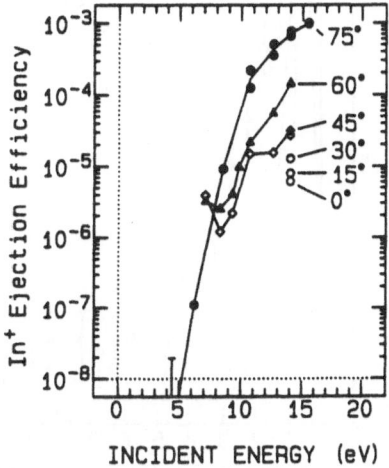

Figure 3. The absolute efficiencies for positively charged particle ejection as a function E_i for Xe atoms on InP(100) at the incident angles shown.

Ejected ion energies were measured in two ways. The CMA of the Auger electron spectrometer (AES) was used to analyze the energies of the ions coming off in the nominal acceptance angular range $42° \pm 6°$ from normal to the surface. Also, the energies of all the ejected ions were measured by retarding the ejected ions by biasing the crystal with respect to machine ground. The derivative of the ion current observed with respect to the voltage bias yields an approximate ion energy distribution. These two distributions are shown in Fig. 4. The solid line from the CMA is clearly different than the total ion distribution which has a very large high-energy tail. We conclude that the majority of the ions are ejected at higher than thermal energies, and that the energy distribution is strongly dependent on the angle of emission.

Using the QMS at close range (2cm.), neutral P_2 and P_4 and H^- were also observed as ejected particles from InP(100). However, no information was obtained to determine the relative numbers of ions and neutrals ejected. In general, the propensities for particles to be observed as ions appear to be about the same as those observed in secondary ion mass spectroscopy (SIMS).

Due to the relatively high ejection efficiencies obtained, impinging fast neutral Xe atoms were considered as a method of cleaning an InP (100) surface at the sub-monolayer level. Knuth and co-workers looked indirectly at particle ejection in this way,[8] observing

Auger peaks attributed to hydrocarbon contaminants on an epitaxially gown Ag(111) surface on a mica substrate decrease with exposure to a hyperthermal Ar atomic. With the incident fluxes used here, and the ion currents measured (not including neutral particles that may have been ejected), the surface removal rate was on the order of 1 monolayer/hr and was stable over several hours. Unlike sputtering with 500-eV Ar^+ ions, which degraded the resistivity of semi-insulating InP (100) orders of magnitude after a dose of 10^{18} ions/cm^2, the procedure of sputtering with 10^{20} Xe atoms/cm^2 at 14 eV did not measurably degrade the resistivity of the samples. Using Auger electron spectroscopy, contaminants could be seen to be quantitatively reduced, although strong chemisorbates never completely disappeared using this cleaning technique.

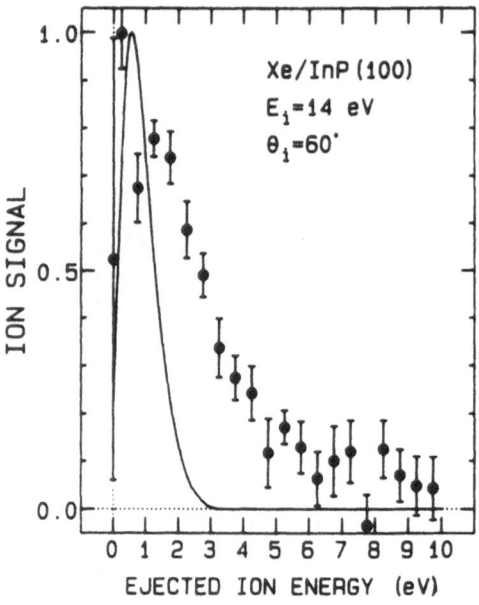

Figure 4. The energies of the ejected ions for 14-eV Xe incident on InP(100) at 60° measured in two ways as described in the text.

IV. SUMMARY

The scattering of hyperthermal rare gas atoms from semi-conductor surfaces has yielded a number of interesting phenomena. Electron-hole pair excitations have been observed above the band-gap threshold, the yields of which are consistent with the idea of a local hot spot with electronic equilibration. At the higher range of projectile, energies ion ejection becomes an important process. Both substrate and adsorbate ions of relatively low ionization potential are observed. The dynamics of ion ejection are in apparent contrast to $e^- h^+$ excitations. The ion yield increases with grazing incidence of the projectile and the onset is at energies considerably above the thermodynamic threshold. The scattered projectiles show backscattered (rainbow) maxima for corrugated surfaces which are sensitive to the surface topography despite the multiple-hit nature of the collision and massive energy transfer to the surface. Computer simulations indicate that as the corrugation of the surface increases, an increasing fraction of the collisions initially involve only one substrate atom-the collision is initially binary. The target atom rapidly recoils from the projectile with high energy, rebounds from the lattice, and returns to the region of impact. This yields a unified picture for the observed phenomena. For this mechanism most of the collision energy will reside transiently in the target atom and its nearest neighbors resulting in $e^- h^+$ excitations if electronic equilibration is fast. If the projectile has high parallel momentum it can clear the initial site such that the recoiling target atom can continue out and leave the surface. If the target atom is an adam, it is more exposed and a larger fraction of impact parameters can yield direct ejection.

REFERENCES

1. A. Amirav and M. J. Cardillo, Surface Sci **198**, 192 (1988).

2. A. Amirav and M. J. Cardillo, Phys. Rev. Lett. **57**, 2229 (1986).

3. A. Amirav and M. J. Cardillo, P. L. Trevor, C. Lim and J. C. Tully, J. Chem, Phys. **87**, 1796 (1987).

4. C. Lim, J.C. Tully, A. Amirav, P. L. Trevor, and M. J. Cardillo, J. Chem, Phys. **87**, 1808 (1987).

5. R. Behrisch, Ed., *Sputtering Threshold by Particle Bombardment I, Vol. 47 of Topics in Applied Physics* (Springer, Berlin, 1981).

6. P. S. Weiss, A. Amirav, P. L. Trevor, and M. J. Cardillo, J. Vac. Sci. Technol. **6**, 889 (1988).

7. M. J. Cardillo, C. C. Ching, E. F. Greene, and G. E. Becker, J. Vac. Sci. Technol. **15**, 423 (1978).

8. S. M. Liu, W. E. Rodgers, and E. L. Knuth, J. Chem. Phys. **61**, 902 (1974).

IONIZATION OF LiF BY HYPERTHERMAL MULTIPLY CHARGED IONS

T. Neidhart, M. Schmid and P. Varga

Institut für Allgemeine Physik, T.U.Wien
Wiedner Hauptstr. 8-10
A-1040 Vienna, Austria

ABSTRACT

Sputtering of a LiF thin film (0,25 µm) surface by singly- (He^+, Ne^+ and Ar^+) and doubly-charged (Ne^+ and Ar^+) ions with impact energies between 10 and 500 eV has been performed. The yield of sputtered Li^+ and F^- ions is only slightly higher for doubly-charged ions compared to singly-charged projectiles at impact energies below 100 eV, whereas the F^+ yield is substantially increased if doubly-charged projectiles are used. The experimental data are explained within the framework of a model [1,2,3] based on calculations by Walkup and Avouris [4]. A comparison is given with former measurements on a LiF single crystal[1,2,3].

1. INTRODUCTION

Most experimental information on sputtering concerns kinetic processes on metal surfaces [5]. Only few data are available for insulating targets, and information about the conversion of potential energy (ionization energy) of the primary projectile into kinetic energy of sputtered particles is still very scarce.

On the basis of experimental results [6,7], Bitensky and Parilis [8] have developed the model of "Coulomb explosion" for sputtering of insulators. Williams [9] reported on ion-stimulated desorption of F^+, where an interatomic Auger process, similar to the Knotek-Feibelman mechanism [10] in electron stimulated desorption, was proposed to explain the relatively high energy of emitted F^+. In ref. [11] a relationship between the potential energy of the primary ion and the F^+ yield has been demonstrated for LiF sputtering. With singly-charged He and Ne ions, a constant sputter yield down to zero impact energy was extrapolated from measurements with ion energies between 200 and 2000 eV and attributed to an inter-atomic Auger transition between the projectile and the F^- ion at the surface. In this process, the F^- is doubly ionized to F^+ and becomes desorbed because of the repulsive Madelung potential. Thompson and Taylor [12] investigated the F^+ emission on a LiF single crystal induced by Ar^+ and Ar^{++} ions with an energy between 1 and 5 keV. The F^+ yield was constantly larger for doubly charged Ar^{++} ions for the whole energy range.

We have set up a UHV experiment where singly- and doubly-charged ions with kinetic energies from 500 eV down to 10 eV can be used for ion-surface interaction studies

[13,14,15]. We use a quadrupole mass spectrometer (QMS) to detect particles sputtered at an angle of 90° by a mass-filtered ion beam. The target was a 0,25 µm thin film LiF evaporated on stainless steel. The surface was cleaned by sputtering with Ar^+ ions (500 eV, 40 µA/cm^2) and heated to 400°C. During the measurements the sample was heated to 400°C, where LiF is a good ionic conductor and a stoichiometric surface is restored. If the sample is sputtered at room temperature, it would be enriched with Li [16].

In the present work we seek further insight into the sputtering of insulators like LiF due to the conversion of the potential energy of the projectile into kinetic energy of sputtered particles. We show that the formerly [11] extrapolated constant sputtering yield for F^+ down to zero impact energy for ions with sufficiently large potential energy cannot be confirmed.

2. RESULTS AND DISCUSSION

For all projectiles, the dominantly emitted charged species were Li^+ and F^- ions. Furthermore, cluster ions like Li_2^+, LiF^+, Li_2F^+, F_2^-, LiF^- and LiF_2^-, have also be detected. To study the influence of the potential energy of the projectile, the emission of Li^+, F^+ and F^- has been studied systematically.

In fig. 1a-c the influence of the primary ion charge state on the Li^+ emission is presented for impact of singly- (He^+, Ne^+ and Ar^+) and doubly-charged (Ne^{++} and Ar^{++}) ions with different impact energies. Only below 100 eV a slightly higher Li^+ yield for all doubly-charged projectiles can be observed. The sputtering yields induced by the different ions can be extrapolated to threshold energies of 35-50 eV for Ar, 20-25 eV for Ne and 20 eV for He^+. For singly-charged projectiles, the threshold energy is slightly higher. The thresholds of the sputtering energies are in good accordance with measurements on a LiF single crystal[1,2,3]. The ratios of the different threshold energies can be estimated by the ratios of maximal energy transfer, which is possible in a head on collision. These data show that the emission of all Li^+ particles is a classical sputter process, dominated mainly by momentum transfer, and the potential energy of the projectile is not of great influence. The slightly higher Li^+ yield of $^4He^+$ with respect to $^3He^+$, which has been shown in ref. [3], is also a clear indication of the momentum transfer during sputtering.

For F^- emission (fig. 2 a-c) the thresholds of the sputtering energies are about 10-20 eV for Ar^+, 10 eV for Ne^+ and 20 eV for He^+ ions. The F^- yield is approximately equal to the Li^+ yield, which indicates a stoichiometric surface composition. Again, a slight difference between singly and doubly charged ions is observed at impact energies below 100 eV and the ratios of the different thresholds refer to the maximal energy transfer. In the case of the incident Ne ions, their masses are nearly equal to those of the F^- ions, the whole energy will be transferred and the threshold of the sputtering energy is less than that for He and Ar bombardment. According to Tosi [17] the threshold is as big as the 10 eV binding energy in LiF.

In fig. 3 a-c results for F^+ emission induced by impact of singly- (He^+, Ne^+ and Ar^+) and doubly-charged (Ne^{++} and Ar^{++}) ions are compared. From these results a clear influence of the potential energy on the emission of F^+ is evident. The thresholds of the sputtering energies show a strong depedence on the charge of the incident ions. For doubly charged Ne^{++} and Ar^{++} ions the thresholds are 10 and 20 eV. For singly charged He^+ and Ar^+ ions the thresholds are 50 and 100 eV. The results are in good accordance with the results of former measurements on the LiF single crystal. For the single crystal we measured a threshold of 25 eV for doubly charged He^{++} ions[3]. For Ne^+ induced F^+ emission it was not possible to determine the threshold, because at low energies the peak of scattered Ne^+ ions superposed the F^+ peak in the mass spectrum of the quadrupol mass spectrometer too much.

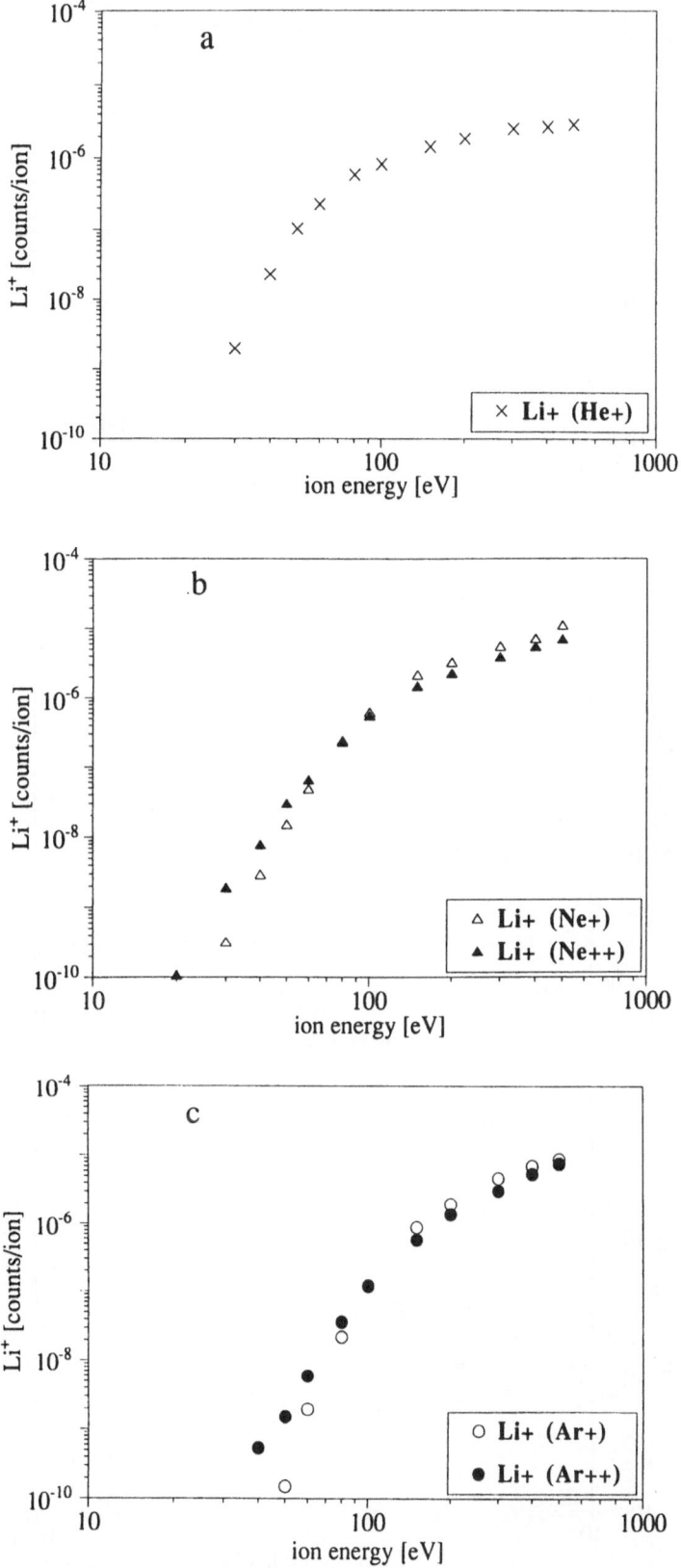

Fig. 1 a-c show the emission of Li$^+$ induced by singly and doubly charged He, Ne and Ar ions in dependence of the kinetic energy of the primary ions.

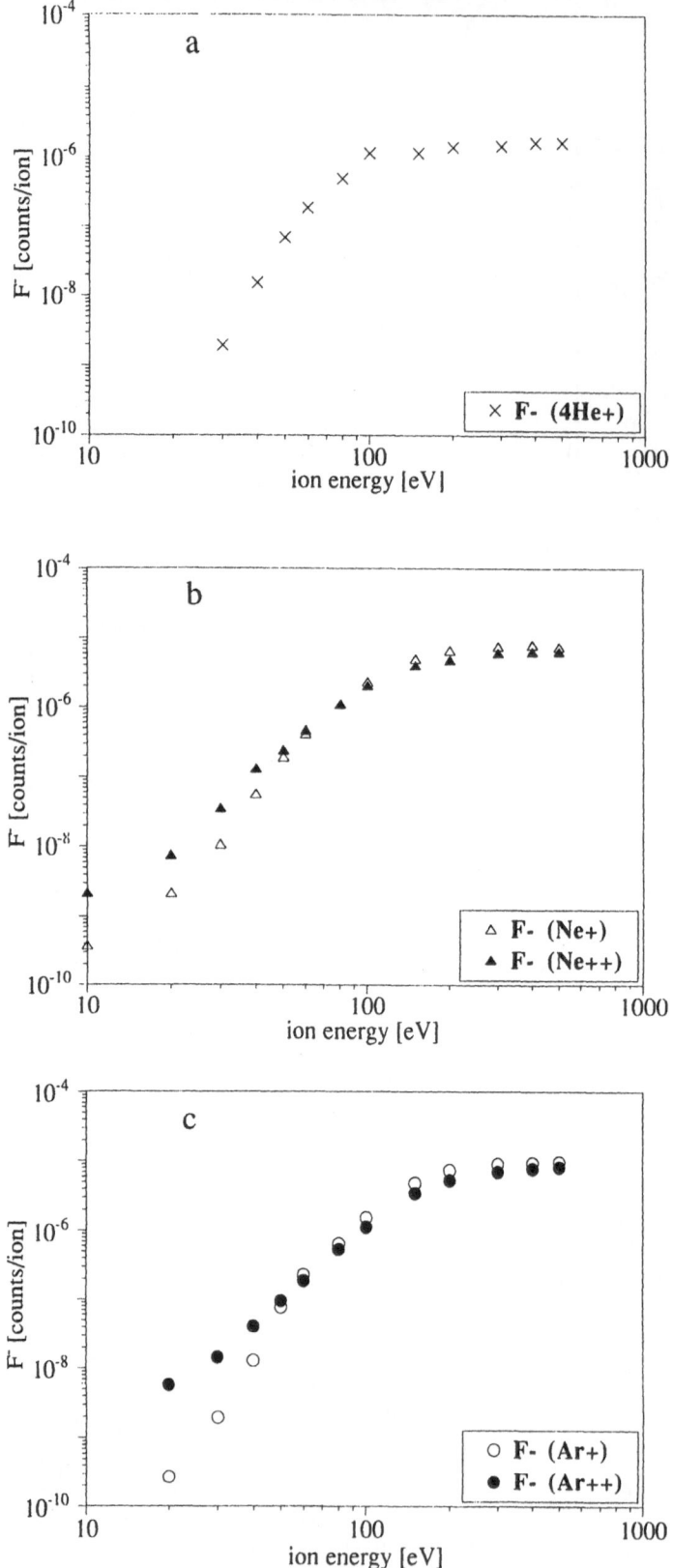

Fig. 2 a-c show the emission of F⁻ induced by singly and doubly charged He, Ne and Ar ions in dependence of the kinetic energy of the primary ions.

As it has been deduced from former measurements[1,2,3] on the LiF (100) single crystal the interaction between the potential energy of the incident ions and the surface of the insulator results in an interatomic Auger process. If the incident ion approaches the surface to a few Ångstrom, an electron of the valence band changes to an energetically more favorable free state of the primary ion and neutralizes it. The energy difference in the two states is transferred to a second electron of the valence band, which is emitted. An F⁻ anion on the surface changes its charge state to F⁺ and will be repelled by the surrounding positive Li⁺ ions in the LiF lattice. This interatomic Auger neutralization process (AN) is possible if the excess energy exceeds the binding energy of the fluorine 2p valence band. The binding energy of the fluorine 2p valence band is between 11 eV and 16 eV [18,19]. This condition is true for the primary ions He^+, Ne^+, Ne^{++} and Ar^{++}, but not for Ar^+. For Ar^+ only a resonance neutralization process (RN) is energetically possible. This explains the high threshold of the sputtering energy and the little yield for the F⁺ emission induced by Ar^+. If the F⁺ emission depends only on Coulomb repulsion, the thresholds of the sputtering energies should be equal for the different projectiles. The different thresholds point out a dependence on the kinetic energy too. As above, the threshold of the sputtering energy is for Ne^+ smaller than for Ar^+ or He^+.

From this data, it is evident that the potential energy of the projectile is of great importance for the production of F⁺, but, in addition, a momentum transfer seems to be necessary for the emission process as stated previously [1,2,3]. The repulsive energy between the F⁺ particle and the surface lattice is obviously not sufficient for the emission process.

Walkup and Avouris [4] have performed classical trajectory calculations on the behavior of an alkali halide lattice if one of the negative halide ions of the surface (eg. F⁻ in NaF) is suddenly changed into a positive one. In their calculations, it is clearly shown that the initial large repulsive energy of 11 eV between F⁺ and the NaF lattice becomes slightly attractive within 15-30 fs, due to a rearrangement of the lattice where the Na⁺ ions are pushed back and the F⁻ are attracted. This time is too short for desorption of F⁺, which therefore finally becomes trapped in a stable state on the surface, with a binding energy of 4 eV in a bridge position between two F⁻ surface atoms.

These results, which have been used to study electron induced desorption, may be applied to low energy ion sputtering as well. The electron transition processes (AN or doubly RN) take place when the projectile ion is still several Å away from the surface [20,21]. Therefore, the sudden change from F⁻ into F⁺ will be similar to the formation of F⁺ by electron impact, with a consequently similar reaction of the lattice. However, one has to consider the subsequent impact of the neutralized particle which can change the rearrangement of the lattice. The low current densities used in the present experiments exclude formation and emission of F⁺ by two different subsequently arriving projectiles. Assuming that NaF is not principally different from LiF as already indicated in [1,2,3], we want to apply the time scale given above to our experiment. For LiF, Walkup and Avouris predicted a smaller binding energy of the F⁺ ion in the bridge position than in NaF. To induce F⁺ emission after the rearrangement of the LiF lattice the kinetic energy of the projectiles must exceed the binding energy of the F⁺.

The rearrangement of the lattice due to the interatomic Auger process elongates the Li⁺ and F⁻ ions out of their equilibrium position of the undistorted lattice. This reduces the binding energies of the Li⁺ and F⁻ ions in the surrounding of the F⁺ ion. The reduced binding energies explain the increased sputter yield and reduced sputtering treshold of Li⁺ and F⁻ at kinetic energies of the projectile below 100 eV. At high energies, which are 100 times larger than the change of the binding energy, the influence of the reduced binding energy on the sputtering yield can be neglected.

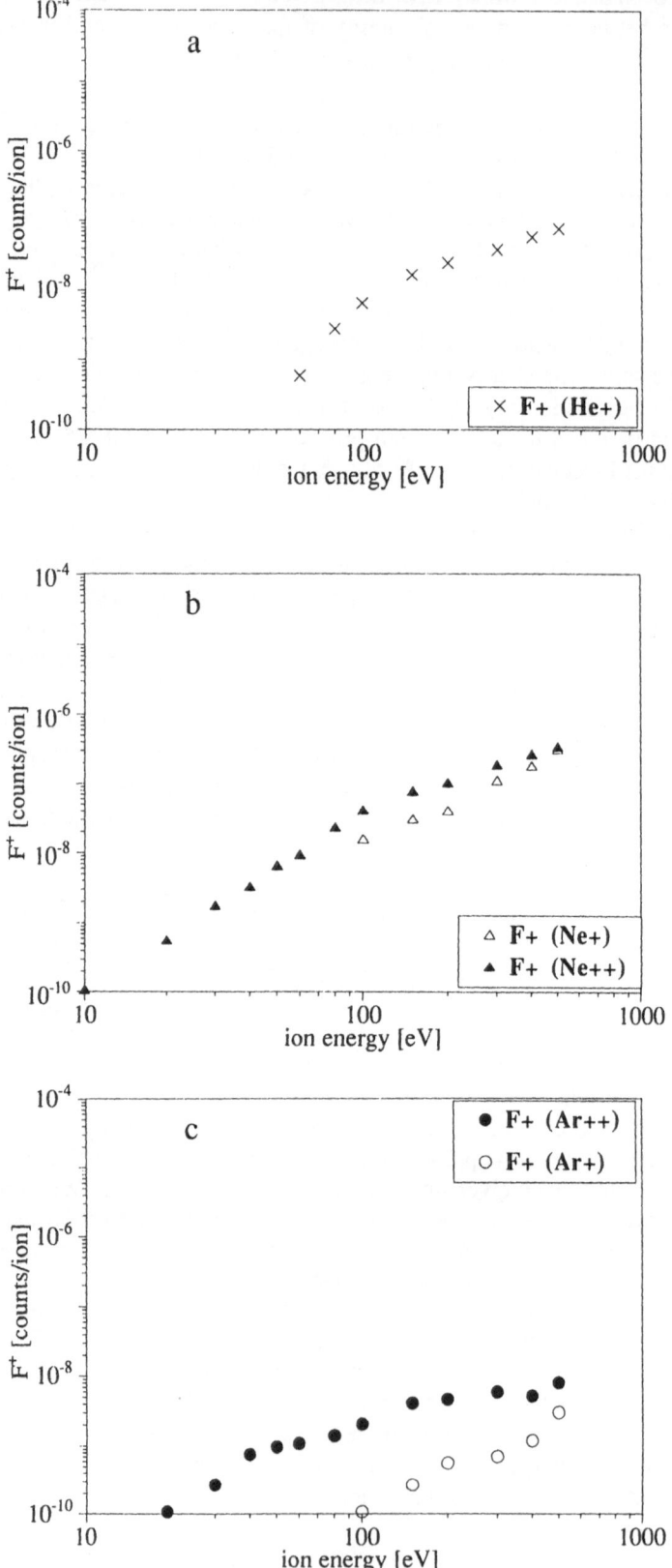

Fig. 3 a-c show the emission of F^+ induced by singly and doubly charged He, Ne and Ar ions in dependence of the kinetic energy of the primary ions.

3. CONCLUSION

Sputtering of the surface of a LiF thin film with hyperthermal singly- (He^+, Ne^+ and Ar^+) and doubly-charged (Ne^{++} and Ar^{++}) ions has been studied. For Li^+ and F^- ejection, only a small influence of the charge state has been observed at impact energies below 100 eV. In contrast to this, it has been shown that sputtering of F^+ strongly depends on the potential energy of the projectile. This can be explained by assuming that in this low energetic impact region, F^+ is formed by an Auger neutralization process between the lattice F^- ion and the He ions, since the potential energy of the projectiles exceeds more than two times the band gap at the surface. From the experimental data it is also evident that for the emission of F^+ ions a minimum value of momentum transfer from the projectile to the ion to be emitted is necessary, and the repulsive potential between the lattice and the F^+ alone is not able to eject the projectile from the surface.

ACKNOWLEDGEMENTS

This work was supported by the Austrian Fonds zur Förderung der wissenschaftlichen Forschung (project P 8969)

REFERENCES

[1]   P.Varga, U.Diebold and D.Wutte, Nucl.Instr.Meth. B 58 (1991) 417

[2]   D.Wutte, U.Diebold, M.Schmid and P.Varga, Nucl.Instr.Meth.B 65 (1992) 167

[3]   T. Neidhart, M. Schmid, P. Varga, DIET 5, Springer Series in Surface Science (1992), to be published

[4]   R.E.Walkup and Ph.Avouris, Phys.Rev.Lett. 56 (1986) 524

[5]   R.Behrisch, ed., Sputtering by Particle Bombardment 1 & 2, Topics Applied Physics Vol . 47 & 52 (1981,1983) Springer

[6]   Sh.S. Radzhabov, R.R. Rakhimov and D.Abdusalomov, Izv. Akad.Nauk SSSR, Ser.Fiz. 40 (1976) [7] S.N.Morozov, D.D.Gruich and T.U.Arifov, Iz.Akad.Nauk SSR, Ser.Fiz., 43 (1979) 612

[8]   I.S.Bitensky, M.N.Murachmedov and E.S.Parilis, Zh.Techn.Fiz. 49 (1979) 1042 I.S.Bitensky and E.S.Parilis, Journ.de Physique C2 (1989) 227

[9]   P.Williams, Phys.Rev.B 23 (1981) 6187

[10] M.L.Knotek and P.J.Feibelmann, Phys.Rev.Lett. 40 (1978) 964

[11] J.A.Schultz, P.T.Murray, R.Kumar, Hsin-Kuei Hu and J.W. Rabalais, Springer Series in Chemical Physics 24 (1983) 191, "Desorption Induced by Electronic Transitions, DIET I", eds.: N.H.Tolk, M.M.Traum, J.C.Tully and T.E.Madey

[12] M. Thompson, W. Taylor, Surface Science 176 (1986) 610-618

[13] U.Diebold, thesis T.U.Vienna 1990, unpublished

[14] U.Diebold and P.Varga, Surface Sci.Lett. 241 (1991) L6

[15] U.Diebold and P.Varga, in Springer Series in Surface Science Vol.19 (1990) 193, "Desorption Induced by Electronic Transitions, DIET IV", eds. G.Betz and P.Varga

[16] P.Wurz, C.H.Becker, Surf.Sci.Lett. 224 (1989) L559

[17] M.P. Tosi, Sol. State. Phys. 16,1 (1964)

[18] M. Piacentini and J.Anderegg, Sol.State Comm.38 (1981) 191

[19] A.B. Kunz, Physical Review B 26,4 (1981) 2056-2069

[20] P.Varga, Appl.Phys. A 44 (1987) 31

[21] P.Appell, Nucl. Instr. Meth. B 23 (1987) 242

PARTICIPANTS

A. Akkerman
SOREQ NRC
Applied Physics and Math Dept.
Yavne 70600, Israel
Ph: 972-8-434583
Fax: 972-8-434315

Fritz Aumayr and Hannspeter Winter
Institut f. Allgemeine Physik
Wiedner Haupstr. 8-10
TU-Wien, A-1040 Wien, Austria
Ph: 43-222-58801-5710
Fax: 43-222-564203
aumayr@eapv38.tuwien.ac.at

Toshiyuki Azuma
College of Arts and Sciences
University of Tokyo
Meguro-ku, Komaba, Tokyo 153, Japan
Ph: 81-3-3467-1171 Ext 268
Fax: 81-3-3467-1281
azuma@tansei.cc.u-tokyo.ac.jp

Raúl A. Baragiola
University of Virginia
Engineering Physics
Charlottesville, VA 22901, USA
Ph: 1-804-982-2907
Fax: 1-804-924-1353
raul@virginia.edu

Asunta Bonanno, Antonino Oliva and
Fang Xu
Dipartimento di Fisica
Universita della Calabria
I-87036 Arcavacata di Rende (CS), Italy
Ph: 39-984-493-132
Fax: 39-984-839-389
oliva%csfisi.infn.it@icineca.cineca.it

Walter L. Brown
AT&T Bell Laboratories
600 Mountain Ave.
Murray Hill, NJ 07974, USA
Ph: 1-908-582-3941
Fax: 1-908-582-4228
wlb@physics.att.com

Mark Cardillo
AT&T Bell Laboratories
600 Mountain Ave.
Murray Hill, NJ 07974, USA
Ph: 1-908-582-2418
Fax: 1-908-582-3958

Jacques Cazaux
Université de Reims
BP 347 Faculté de Sciences
Lab. d'Analyse des Solides, Surfaces et
Interfaces
51062 Reims Cédex, France
Ph: 33-26-05-3223
Fax: 33-26-05-3250

Fernando Flores
Dept. Física de la Materia Condensada,
C-XII
Universidad Autónoma de Madrid
Cantoblanco, 28049 Madrid, Spain
Ph: 34-1-397-5053
Fax: 34-1-397-4950

Alberto Gras-Martí
Dep. Física Aplicada
Universitat d'Alacant
Apt. 99, E-03080 Alicante, Spain
Phone: 34-6-590-3541
Fax: 34-6-590-3464
fapl@ealiun11.bitnet

Edward B. Hale
Dept. Physics
University of Missouri-Rolla
Rolla, MO 65401, USA
Ph: 1-314-341-4353
Fax:: 1-314-341-2071
mrc@umrvmb.umr.edu

Werner Heiland
Universität Osnabrück
Fachbereich Physik
Barbarastrasse 7, Postfach 44 69
W-4500 Osnabrück, Germany
Ph: 49-541-9690
Fax: 49-541-969-2670
wheiland%dosuni1.bitnet

Wolfgang O. Hofer
Project Kernfusion
Forschungszentrum Jülich/KFA
W-5170 Jülich, Germany
Phone: 49-2461-61-6368
Fax: 49-2461-61-5452

Robert E. Johnson
University of Virginia
Engineering Physics
Charlottesville, VA 22901, USA
Ph: 1-804-924-3244
Fax: 1-804-924-1353
rej@virginia.edu

Volker Kempter
Physikalisches Institut der
T. Universität Clausthal
Leibnizstrasse 4
W-3392 Clausthal-Zellerfeld, Germany
Ph: 49-5323-72-2363
Fax: 49-5323-72-3600

Vladas B. Leonas (deceased)
contact: Valery Fine
Academy of Sciences
Institute for Problems in Mechanics
Prospect Vernadskogo 101
117526 Moscow. Russia
Ph: 7-095-434 33 92
Fax: 7-095-938-20 48
fine@main2.jinr.dubna.su

Joseph Macek
Theoretical Physics
University of Tennessee
Knoxville, TN 37996-1501, USA
Ph: 1-615-974-0768
Fax: 1-615-974-2667
macek@utkvx.bitnet

Giovanni Marletta
Dipartimento di Scienze Chimiche
Universita di Catania
Viale A. Doria 6, 95125 Catania, Italy
Ph: 39-95-221-635
Fax: 39-95-580138

Ewa A. Maydell and Derek J. Fabian
University of Strathclyde
48 North Portland Street
Glasgow G1 1XN, Scotland, UK
Ph: 44-41-552-4400
Fax: 44-41-552-0775

Felix Meier
Laboratorium f. Festkörperphysik
ETH
CH-8093 Zürich, Switzerland
Phone: 41-1-377-2360
Fax: 41-1-371-6268
e-mail: meier@czheth5a.bitnet

Arend Niehaus
Buys Ballot Laboratorium
Rijksuniversiteit Utrecht
Princetonplein 5
3584 CC Utrecht, The Netherlands
Ph: 31-30-533302
Fax: 31-30-531601
atphys@hutruu51.bitnet

Rupert Pfandzelter
Sektion Physik
Universität München
Amalienstrasse 54
W-8000 München 40, Germany
Ph: 49-89-2180-3357
Fax: 49-89-2180-3441
pfandzelter@lehrstuhl-sizmann.physik.uni-
muenchen.dbp.de

Carl Rau
Rice University
Dept. Physics
Houston, TX 77251-1892, USA
Ph: 1-713-285-5417
Fax: 1-713-527-9033
rau@ricevm1.bitnet

Curt T. Reimann
Uppsala Universitet
Inst. f. Stralningsvetenskap
Box 535, 751 21 Uppsala, Sweden
Ph: 46-18-183-057
Fax: 46-18-555-736
scooter@tsl.uu.se

Max Rösler
Karl-Pokern-Str. 12
O-1162 Berlin, Germany
Ph: 49-30-6457154
Fax: 49-30-6572700

Hermann Rothard
Institut de Physique Nucléaire de Lyon
Groupe de Collisions Atomiques
Université Claude Bernard - Lyon I
43, Bd. du 11 Novembre 1918
F-69622 Villeurbanne Cédex, France
Ph: 33-72 44 80 00 (Ext. 3115)
Fax: 33-72-44-80-04
rothard@lyolav.in2p3.fr

Gregor Schiwietz
Hahn-Meitner Institut
Department P3
Glienicker Strasse 100
D-1000 Berlin 39, Germany
Ph.: 49-30-8009-2448
Fax: 49-30-8009-2097

Jorgen Schou
Physics Department
Risø National Laboratories
DK-4 Roskilde, Denmark
Ph: 45-42371212 Ext. 4755
Fax:45-42370115

Peter Sigmund
Odense University
Physics Dept.
DK-5230 Odense M, Denmark
Ph: 45-66-158-600
Fax:45-66-158-760
psi@dou.dk

Zdeněk Šroubek
Czechoslovak Academy of Sciences
URE
Chaberska 57, 182 51 Prague 8,
Czechoslovakia
Ph: 42-2-84 37 41
Fax: 42-2-84 06 09

Peter Varga
Institut f. Allgemeine Physik
Wiedner Haupstr. 8-10
TU-Wien, A-1040 Wien, Austria
Phone: 43-222-58801-5710
Fax: 43-222-564203
varga@eapv38.una.ac.at

Helmut Winter
Universität Münster
Institut für Kernphysik
Wilhelm-Klemm-Strasse 9
4400 Münster, Germany
Ph: 49-251-834-991
Fax: 49-251-83-4962
winter@vsikp0.uni-muenster.de

INDEX